The Monocotyledons
A Comparative Study

Botanical Systematics

An Occasional Series of Monographs

Series Editor

V. H. HEYWOOD

Palynotaxonomic Investigation of *Fagus* L. and *Nothofagus* Bl.: Light Microscopy, Scanning Electron Microscopy, and Computer Analyses *by* SHARON L. HANKS and DAVID E. FAIRBROTHERS, *and* Genisteae (Adans.) Benth. and Related Tribes (Leguminosae) *by* R. M. POLHILL, 1976.

The Monocotyledons: A Comparative Study *by* ROLF M. T. DAHLGREN and H. TREVOR CLIFFORD, 1981.

The Monocotyledons:
A Comparative Study

by

ROLF M. T. DAHLGREN

The Botanical Museum, University of Copenhagen, Copenhagen, Denmark

and

H. TREVOR CLIFFORD

Department of Botany, University of Queensland, Brisbane, Australia

in cooperation with U. Hamann, J. B. Harborne, L. Holm, H. Huber, S. R. Jensen, J. A. Nannfeldt, S. Nilsson and F. Rasmussen

1982

ACADEMIC PRESS
A Subsidiary of Harcourt Brace Jovanovich, Publishers
London   New York
Paris   San Diego   San Francisco
São Paulo   Sydney   Tokyo   Toronto

ACADEMIC PRESS INC. (LONDON) LTD.
24/28 Oval Road,
London NW1

United States Edition published by
ACADEMIC PRESS INC.
111 Fifth Avenue
New York, New York 10003

British Library Cataloguing in Publication Data
Dahlgren, R. M. T.
 The monocotyledons.
 1. Monocotyledons
 I. Title   II. Clifford, H. T.
 584   QK495.A14

   ISBN 0–12–200680–1 ✓

   LCCCN 81–67906

Filmset by Northumberland Press Ltd, Gateshead, Tyne and Wear
Printed in Great Britain by Fletcher and Son Ltd, Norwich

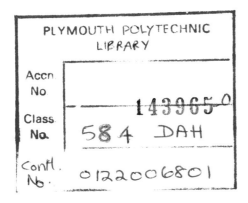

Preface

The present work has developed out of the authors' shared interest in creating a factual basis for a taxonomic treatment of the angiosperms by surveying the distribution of character states and their consistency and homogeneity over major groups. Work on these lines had been done separately by each of us for some time when, at a meeting in Hamburg in September 1976, we planned a closer cooperation. Most of the cooperative work was carried out in Copenhagen during the autumn of 1978 and early spring of 1979. The Danish Natural Science Research Council is here gratefully acknowledged for its economic support to both of us in connection with the project.

It soon proved that the task, even though restricted to the monocotyledons, was too extensive for a team of two persons during a few years, but with the generous help of several colleagues, including substantial contributions from some of them, this book can now be presented to the public.

Various readers will inevitably discover deficiencies in this presentation, and maybe in the long run its chief benefit will be to provoke colleagues and their students to extend the studies of various attributes and reconsider their significance. The deficiencies and conclusions are wholly to be ascribed to us, while the relatively great amount of data and other benefits are largely to be ascribed to our contributors.

November, 1981   *R. M. T. Dahlgren*
                                                              H. T. Clifford

It has been pretended that the characters
of the Natural classes of plants
are not to be ascertained without much
laborious research; and that not a step
can be taken until this preliminary
difficulty is overcome.

But it is hardly necessary to say
that in natural history many facts
which have been originally discovered
by minute and laborious research
are subsequently ascertained to be connected
with other facts of a more obvious nature;
and of this Botany offers perhaps
the most striking proof that can be adduced.

John Lindley (*1853*)

Editor's Preface

The occasional series Botanical Systematics was introduced to provide a vehicle for the publication of extensive monographic studies in the field of systematic botany. The last volume was well received and encouraged the editor and publishers to proceed with the preparation of further issues.

This second volume is devoted entirely to a major review of the systematics of the monocotyledons by Professor Rolf Dahlgren and Dr Trevor Clifford. A unique feature is the presentation of a large body of data, derived from various disciplines, on which their classification and suggested relationships within the monocotyledons are based. This data-base, together with the large number of illustrations, will make this volume a leading reference work on this important group of plants.

For this volume the format has been changed so as to allow a better presentation of the data and to do justice to the excellent illustrations.

November, 1981   *V. H. Heywood*

Acknowledgements

The following contributors have taken the most active part in this work by adding material, text and criticism.

Professor Ulrich Hamann, Bochum, has read the text critically, especially its embryological parts, and has suggested innumerable improvements to the MS. He has also contributed a great many results, largely unpublished, from his own studies and the studies of his associates, Peter Kircher, Traudel Rübsamen, Asta Tiemann and others at his laboratory. His works on "Farinosae" of 1961 and 1962 and specialized studies on Philydraceae, Centrolepidaceae etc. have contributed much information to this study.

Professor Herbert Huber, Kaiserslauten, has contributed material and text on tubers and corms, taken part in the construction of the base diagram used and, besides, has critically examined the character diagrams. His work of 1969 on the Liliiflorae has been of basic importance to the family taxonomy of this group.

Dr Søren Rosendal Jensen, Copenhagen, has contributed the sections on saponins, chelidonic acid and cyanogenic compounds. Part of the information was extracted from Professor Robert Hegnauer's card index in Amsterdam, which he kindly placed at our disposal.

Professor John A. Nannfeldt and Dr Lennart Holm, Uppsala, have contributed the text of the section "Host specificity of parasitizing fungi".

Dr Siwert Nilsson, Stockholm, has contributed to the pollen morphological treatment and with the help of Mr Gamal El-Ghazaly has supplied the scanning electron microphotographs for this book.

Mr Finn Rasmussen, Copenhagen, has critically examined and made substantial contributions to text and diagrams for the order Orchidales throughout this work.

Professor Rolf Berg has contributed data on arils, strophioles and similar structures and to the endosperm formation in some groups. Professor H.-D. Behnke, Heidelberg, has kindly placed unpublished data on sieve tube plastids at our disposal. Professor Per Wendelbo has contributed recent data on genera of the Hyacinthaceae. Professor Vernon H. Heywood in the course of his editorial revision has contributed much valuable advice.

Dr Gertrud Dahlgren, Lund, has read the manuscript critically and Mrs Kirsten Harder, Copenhagen, has typed the manuscript.

Economic support was provided by the Danish Natural Science Research Council, allowing us to work together in Copenhagen for about half a year and covering also some expenses for material and travels in connection with this work.

Contents

CONTENTS

CONTENTS

Introduction

In recent years there has been a re-awakening of interest in the classification of flowering plants and several new but divergent systems have been proposed. Often the taxa within the systems have been only briefly described with stress laid upon those characters most useful for defining and for distinguishing between groups.

The relative reliability of taxonomic characters is usually the subject of considerable debate and often those characters held to be of value in one part of the system may be completely neglected in another. Whether this neglect is by design or otherwise cannot, as a rule, be determined from the text in most published systems.

Accordingly, it was decided to investigate the monocotyledons with respect to a wide range of characters and to determine the distribution of these over the whole group. The characters chosen were selected primarily on the basis of two criteria: (1) it was necessary that they be known for a widely representative sample of monocotyledons; (2) that they were not, or only occasionally, subject to variation within genera. Subject to these criteria the distribution of about a hundred characters was investigated.

It was anticipated that by these procedures characters not previously known to be of taxonomic value might be revealed and that the merit of those commonly accepted to be of taxonomic value could be evaluated.

All data were recorded at generic level and then plotted at ordinal level on a modified form of a diagram of affinity as first proposed by one of us (RD).

From such a diagram it is possible to determine at a glance the distribution of any character and its constancy or otherwise within orders and superorders. Where there is discordancy, reference to the data presented or to the basic diagram will indicate the families involved. It is then a matter of taxonomic judgement as to whether or not particular families would be better placed elsewhere in the system. Indeed, from a consideration of the total data we have drawn up modified schemes.

To avoid difficulties in the interpretation of characters, they have, in most instances, been defined in the text and used in that sense and where appropriate have been illustrated. A fuller understanding of the characters may also be gained from a consideration of the taxa possessing them.

A brief account of some previous classifications has been provided for several reasons. Comparisons of the different systems with the distributions of the characters enable the reader to assess the importance accorded particular attributes in previous classifications. Some of the family groupings proposed by earlier taxonomists are still of considerable interest even if not presently acceptable. Finally, in this way the historical roots of our own classification are revealed.

It is hoped that the diagrams will focus attention on those attributes and taxa that are poorly known and so stimulate research directed at remedying those defects.

Difficulties in the circumscription of families have been avoided by listing the genera in those families whose limits are debatable.

Thus every effort has been made to ensure that the text is self-contained and that the data on which our conclusions have been based are available to the reader.

Economic support for this project has been received from the Danish Natural Science Research Council.

Some Features of Previous Classifications in Monocotyledons

To review all the systems that have been proposed for the classification of the monocotyledons would be of limited value. Mainly those that have contributed to the development of modern systems are discussed below.

PERIOD 1853–1964

Since we are here concerned primarily with higher level categories it was decided to neglect classifications prior to that of Lindley (1853), who may be consulted for a summary of earlier works on the subject.

Lindley (1853) regarded the monocotyledons as comprising two classes—Endogens with "leaves parallel-veined, permanent; wood of the stem always confused" and Dictyogens with "leaves net-veined, deciduous; wood of stem, when perennial, arranged in a circle with central pith" (op. cit., p. 4). Within each class he recognized a series of alliances (orders) and orders (families), whose ranks were indicated by the terminations -ales and -aceae attached in general to generic names.

He regarded the Dictyogens as sharing characters from both the monocotyledons and dicotyledons (op. cit., p. 211) finding the Aristolochiaceae to be their nearest relative amongst the latter.

Class Endogens
  Glumales: Graminaceae, Cyperaceae, Desvauxiaceae (= Centrolepidaceae), Restionaceae, Eriocaulaceae
  Arales: Pistiaceae (incl. Lemnaceae), Typhaceae (incl. Sparganiaceae), Araceae, Pandanaceae (incl. Cyclanthaceae)
  Palmales: Palmaceae
  Hydrales: Hydrocharitaceae, Najadaceae (incl. *Zannichellia, Phyllospadix* etc.), Zosteraceae (incl. *Cymodocea* and *Posidonia*), Triuridaceae
  Narcissales: Bromeliaceae, Taccaceae, Haemodoraceae (incl. Velloziaceae), Hypoxidaceae, Amaryllidaceae (incl. Agavaceae and Alstroemeriaceae), Iridaceae
  Anomales: Musaceae (incl. Heliconiaceae and Strelitziaceae), Zingiberaceae (incl. Costaceae), Marantaceae (incl. Cannaceae)
  Orchidales: Burmanniaceae, Orchidaceae (incl. Cypripediaceae), Apostasiaceae
  Xyridales: Philydraceae, Xyridaceae (incl. Rapateaceae), Commelinaceae, Mayacaceae

  Juncales: Juncaceae (incl. *Narthecium, Astelia, Lomandra, Susum, Calectasia, Kingia* etc.), Orontiaceae (= Araceae p.p.)
  Liliales: Gilliesiaceae (= *Gilliesia, Miersia*), Melanthiaceae (incl. Colchicaceae), Liliaceae (*sensu lato*), Pontederiaceae
  Alismales: Butomaceae, Alismaceae, Juncaginaceae (incl. Scheuchzeriaceae, Aponogetonaceae, Potamogetonaceae)

Class Dictyogens
  Dioscoreales: Dioscoreaceae, Smilacaceae, Philesiaceae, Trilliaceae, Roxburghiaceae
  Many dicotyledons also belong to this class.

Of his orders only two are recognized today in the same circumscription. They are Palmales (Arecales) and Anomales (Zingiberales). The familial composition of the remainder have all changed somewhat, though much of the basic framework shows through in most present day classifications.

Bentham and Hooker (1883). In their scheme it is evident that some rationalization of Lindley's work has been undertaken. The families recognized are much the same but these are grouped into fewer "series" than the orders of Lindley and the series are named after attributes instead of taxa. Furthermore, epigyny has been exaggerated so as to define the groups Microspermae and Epigynae.

The series recognized together with representative families for each are listed below:

Series   I. *Microspermae:* Orchidaceae, Burmanniaceae
      II. *Epigynae:* Bromeliaceae, Iridae, Amaryllidae
     III. *Coronariae:* Liliaceae, Pontederiaceae
     IV. *Calycinae:* Juncaceae, Palmae
      V. *Nudiflorae:* Pandanaceae, Typhaceae, Aroideae
    VI. *Apocarpae:* Alismaceae, Naiadaceae
   VII. *Glumaceae:* Gramineae, Cyperaceae

Of these series the Apocarpae have persisted as Helobiae or Alismatiflorae of later classifications.

Van Tieghem (1891). Unlike his contemporaries van Tieghem divided the monocotyledons into only four orders. Even allowing that these may correspond more closely to superorders than to orders, his groupings are unsatisfactory in that he apparently relied upon too few characters for their definition. For example, in his key to the orders he used only the nature of the perianth, its presence or absence and the position of the ovary as diagnostic attributes.

Order	Family
Gramininées	Graminées, Cypéracées, Centrolépidées, Lemnacées, Naiadacées, Aroïdées, Cyclanthacées, Typhacées, Pandanées
Joncinées	Restionacées, Eriocaulées, Triglochinées, Palmiers, Joncacées
Liliinées	Alismacées, Commelinacées, Xyridacées, (incl. Mayacaceae, Rapateaceae and Philydracae), Pontederiacées, Liliacées (incl. Stemonaceae)
Iridinées	Amaryllidées (incl. Taccaceae, Velloziaceae and Burmanniaceae), Dioscoréacées, Iridées, Hémodoracées, Broméliacées, Scitaminées, Orchidées, Hydrocharidées.

The classification has little merit in itself but it is apparently the first to accord taxonomic significance to the mode of pollen grain formation. In defining the monocotyledons van Tieghem draws attention to the fact that of those then known all followed the Successive type.

Engler (1892). In contrast to the system of Bentham and Hooker (1883), that of Engler ends with instead of beginning with the Microspermae. Instead of seven series the families were organized into ten "*Reihen*" (orders), the additional *Reihen* arising from granting the Palmae (Principes), Aroideae (Spathiflorae) and Cyclanthaceae (Synanthae) ordinal status.

(*Reihen*)
Pandanales: Typhaceae, Pandanaceae, Sparganiaceae
Helobiae: Potamogetonaceae *sensu lato*, Najadaceae, Aponogetonaceae, Juncaginaceae, Alismaceae, Butomaceae, Hydrocharitaceae
Glumiflorae: Gramineae, Cyperaceae
Principes: Palmae
Synanthae: Cylanthaceae
Spathiflorae: Araceae, Lemnaceae
Farinosae: Flagellariaceae, Restionaceae, Centrolepidaceae, Mayacaceae, Xyridaceae, Eriocaulaceae, Rapateaceae, Bromeliaceae, Commelinaceae, Pontederiaceae, Philydraceae, (in his final system of 1930 also Thurniaceae and Cyanastraceae)
Liliiflorae: Juncaceae, Stemonaceae, Liliaceae, Haemodoraceae, Amaryllidaceae, Velloziaceae, Taccaceae, Dioscoreaceae, Iridaceae
Scitamineae (Arillatae): Musaceae, Zingiberaceae, Cannaceae, Marantaceae
Microspermae: Burmanniaceae, Orchidaceae

These ordinal names acquired considerable prestige,

being used in many subsequent editions of his "Syllabus", "Die natürlichen Pflanzenfamilien", "Das Pflanzenreich" and countless taxonomic manuals modelled upon these works.

Wettstein (1901). Although there are marked similarities between the classifications of Engler and Wettstein they differ in a number of important respects. Because he considered the monocotyledons to be derived from Polycarpicae-type ancestors Wettstein placed the apocarpous Helobiae before the other orders. Furthermore, he recognized the considerable differences between the sedges (Cyperaceae) and grasses (Poaceae) by placing them in separate *Reihen*. Finally, in what we would regard as a retrograde step, he combined Engler's Principes and Synanthae with Spathiflorae into a single *Reihe*, thereby making an ordinal grouping quite different from those of earlier systems.

A presentation of Wettstein's somewhat altered system in the fourth edition of the "Handbuch" (1935) is given below.

(*Reihen*)
1. *Helobiae:* Alismataceae, Butomaceae (incl. Limnocharitaceae), Hydrocharitaceae, Scheuchzeriaceae (incl. Juncaginaceae), Aponogetonaceae, Potamogetonaceae (incl. Zosteraceae etc.), Najadaceae, ? Triuridaceae
2. *Liliiflorae:* Liliaceae *sensu lato*, Stemonaceae, Cyanastraceae, Pontederiaceae, Haemodoraceae, Philydraceae, Amaryllidaceae (incl. Agavaceae *pro parte* and Hypoxidaceae), Velloziaceae, Iridaceae, Juncaceae, Thurniaceae, Flagellariaceae, Rapateaceae, Thurniaceae, Bromeliaceae, Dioscoreaceae, Taccaceae, Burmanniaceae
3. *Enantioblastae:* Commelinaceae, Mayacaceae, Xyridaceae, Eriocaulaceae, Centrolepidaceae, Restionaceae
4. *Cyperales:* Cyperaceae
5. *Glumiflorae:* Gramineae
6. *Scitamineae:* Musaceae *sensu lato*, Zingiberaceae, Cannaceae, Marantaceae
7. *Gynandrae:* Orchidaceae
8. *Spadiciflorae:* Palmae, Cyclanthaceae, Araceae, Lemnaceae
9. *Pandanales:* Pandanaceae, Sparganiaceae, Typhaceae.

Lotsy (1911). The principal contribution of Lotsy to the history of the classification of monocotyledons was his suggestion that the subclass is biphyletic. He regarded one order, Spadiciflorae, as having been derived from Piperalean ancestors and presumed the remaining orders to have been derived from pro-Ranalean ancestors. Thus, his system differs from those previously proposed in acknowledging the close agreement between Piperaceae and Araceae on the one hand and between

Nymphaeaceae and other basically apocarpous Polycarpicae (such as Magnoliaceae, Annonaceae, Aristolochiaceae), Alismataceae and other Helobiae on the other. This similarity has been noted repeatedly in many subsequent works. Another interesting feature in Lotsy's system is that the families, especially the Liliiflorae, were split into many small ones, in a way reminiscent of Huber's (1969) treatment.

> *Spadiciflorae* (derived from Piperalean ancestors): Araceae, Lemnaceae, Cyclanthaceae, Palmaceae, Pandanaceae, Sparganiaceae, Typhaceae
> Others (derived from pro-Ranalean ancestors):
> *Helobiae:* Alismataceae, Butomaceae, Hydrocharitaceae, Scheuchzeriaceae, Zosteraceae, Posidoniaceae, Aponogetonaceae, Potamogetonaceae, Najadaceae, Altheniaceae, Cymodoceaceae, Triuridaceae
> *Enantioblastae:* Commelinaceae, Mayacaceae, Rapateaceae, Xyridaceae, Eriocaulaceae, Centrolepidaceae, Restionaceae, Pontederiaceae
> *Liliiflorae:* Melanthiaceae, Asphodelaceae, Aloëaceae, Eriospermaceae, Johnsoniaceae, Agapanthaceae, Alliaceae, Gilliesiaceae, Tulipaceae, Scillaceae, Asparagaceae, Dracaenaceae, Smilacaceae, Luzuriagaceae, Ophiopogonaceae, Lomandraceae, Dasypogonaceae, Calectasiaceae, Juncaceae, Flagellariaceae, Stemonaceae, Cyanastraceae, Iridaceae, Haemodoraceae, Hypoxidaceae, Velloziaceae, Agavaceae, Amaryllidaceae, Bromeliaceae, Dioscoreaceae, Taccaceae, Burmanniaceae
> *Scitamineae:* Musaceae, Cannaceae, Zingiberaceae, Marantaceae
> *Monandrae:* Orchidaceae
> *Glumiflorae:* Graminaceae, Cyperaceae

Warming (1912). In his Danish textbook "Frøplanterne" (The Seed Plants) Warming was greatly influenced by both Wettstein and Engler. His system has many modern features with an almost consistent use of ordinal names ending in *-ales* (cf. Lindley, above). The chief drawback is the scattered treatment of the grass-like plants; Glumales (Juncaceae, Cyperaceae, Gramineae) being placed far apart from Enantioblastae (containing among other families Eriocaulaceae, Restionaceae and Centrolepidaceae). However, Warming acknowledged the distinctness of the Helobiales, the Scitaminales and the Arales, and he placed Orchidales (Orchidaceae) next to Liliales *sensu lato*.

> *Klasse Monocotyledones*
> *Helobiales:* Scheuchzeriaceae (incl. Juncaginaceae), Potamogetonaceae (incl. Ruppiaceae, Zosteraceae, Zannichelliaceae, Cymodoceaceae), Najadaceae, Alismataceae (incl. Butomaceae), Hydrocharitaceae, Triuridaceae
> *Glumales:* Juncaceae, Cyperaceae, Gramineae
> *Pandanales:* Typhaceae (incl. Sparganiaceae), Pandanaceae
> *Palmales:* Palmae, Cyclanthaceae

> *Arales:* Araceae, Lemnaceae
> *Enantioblastae:* Commelinaceae, Mayacaceae, Xyridaceae, Eriocaulaceae, Rapateaceae, Restionaceae, Centrolepidaceae
> *Liliales:* Colchicaceae, Liliaceae, Convallariaceae, Pontederiaceae, Haemodoraceae, Philydraceae, Bromeliaceae, Amaryllidaceae, Taccaceae, Iridaceae, Burmanniaceae, Dioscoreaceae
> *Orchidales:* Orchidaceae
> *Scitaminales:* Musaceae (incl. Heliconiaceae, Strelitziaceae), Zingiberaceae (incl. Costaceae), Cannaceae, Marantaceae

Hallier (1903, 1905a and b, 1912). In the same year as Warming published his treatise, Hallier published the second edition of his system. He considered the monocotyledons to be derived from dicotyledonous ancestors of the lardizabalaceous stock. He considered the diverse orders as derivatives of a liliiflorean basic type, thereby generating a different classification from that of Wettstein and Warming. It may seem that Hallier's treatment of the monocotyledons in 1912 is a "somewhat scrambled assemblage" (Lawrence, 1951). Still it contains some modern features, the Alstroemeriaceae and Agavaceae, for example, being separated from Amaryllidaceae, and the Pontederiaceae and Philydraceae being placed together and noted to share a starchy endosperm. But also here, as in Warming's system, the Cyperales are far separated from the Enantioblastae, though unlike Warming he included the Poales in the latter order. His Spadiciflorae is according to modern views an unnatural mixture of Arales, Typhales and the Areciflorae. The Scitamineae and Orchidaceae form insignificant groups within the Ensatae.

> *Liliiflorae:* Liliaceae, Alstroemeriaceae, Amaryllidaceae
> *Anthorrhizae:* Dioscoreaceae, Taccaceae, Burmanniaceae
> *Ensatae:* Agavaceae, Bromeliaceae, Haemodoraceae, Velloziaceae, Iridaceae, Scitamineae, Orchidaceae
> *Enantioblastae:* Flagellariaceae, Gramineae, Restionaceae, Centrolepidaceae, Mayacaceae, Xyridaceae, Eriocaulaceae, Rapateaceae, Commelinaceae, Pontederiaceae, Philydraceae
> *Spadiciflorae:* Cyclanthaceae, Palmae, Pandanaceae, Typhaceae, Araceae, Lemnaceae
> *Cyperales:* Cyperaceae, Juncaceae (incl. Thurniaceae)
> *Helobieae:* Triuridaceae, Najadaceae, Alismaceae, Hydrocharitaceae

Bessey (1915). Bessey presented several classifications of which the final is discussed here. He laid much stress on whether the ovary was superior or inferior, this feature being the main dividing line in his arrangement of the dicotyledons as well as of the monocotyledons. Thus the monocotyledons were divided into two subclasses, the Strobiloideae (hypogynous) and the Cotyloideae (epigynous). Unfortunately such a division results in an artificial grouping with parallel taxa in the two subclasses, for example Alismatales versus Hydrales and Liliales versus Iridales. Bessey's Liliales has a wide circumscription and besides families with non-starchy endosperm includes, for example, Pontederiaceae, Philydraceae, Commelinaceae, Xyridaceae, Juncaceae and Eriocaulaceae (and also the family Najadaceae) which is a drawback in relation to previous and contemporary systems. This classification therefore appears artificial and intended more as an aid to identification than as an attempt towards a natural system of classification.

Class *Alterniflorae* (*Monocotyledoneae*)
  Subclass *Strobiloideae*
    Alismatales: Alismataceae, Butomaceae, Triuridaceae, Scheuchzeriaceae, Typhaceae, Sparganiaceae, Pandanaceae, Aponogetonaceae, Potamogetonaceae
    Liliales: Liliaceae, Stemonaceae, Pontederiaceae, Cyanastraceae, Philydraceae, Commelinaceae, Xyridaceae, Mayacaceae, Juncaceae, Eriocaulaceae, Thurniaceae, Rapateaceae, Najadaceae
    Arales: Cyclanthaceae, Araceae, Lemnaceae
    Palmales: Palmaceae
    Graminales: Restionaceae, Centrolepidaceae, Flagellariaceae, Cyperaceae, Poaceae
  Subclass *Cotyloideae*
    Hydrales: Vallisneriaceae (= Hydrocharitaceae)
    Iridales: Amaryllidaceae, Haemodoraceae, Iridaceae, Velloziaceae, Taccaceae, Dioscoreaceae, Bromeliaceae, Musaceae, Zingiberaceae, Cannaceae, Marantaceae
    Orchidales: Burmanniaceae, Orchidaceae.

Ankermann (1927). Ankermann used morphological as well as serological methods to trace the "evolutionary history of the monocotyledons". The serological data are of particular interest and may be regarded as belonging to the Mez school of serology, although he claimed to get more reliable results by using artificial sera. His fairly extensive investigation resulted in the phylogenetic tree given in Fig. 1. Among the noticeable features of this tree are the connections between the Orchidales and the Zingiberales, between the Arales and Arecales and between the Pandanales and Typhales, connections that would hardly gain support from modern surveys. The close connections shown between the epigynous Iridaceae, Amaryllidaceae, Velloziaceae and Hypoxidaceae are also striking. However, Ankermann's tree also expresses a number of connections that are in accordance with present findings, such as between Restionaceae and Centrolepidaceae, between Rapateaceae, Xyridaceae and Eriocaulaceae, between Dioscoreaceae and Taccaceae and between many of the alismatiflorean families.

Ankermann's results should be judged against the contemporary concepts and methods and not used as arguments for rejecting serological methods.

Rendle (1930). The scheme of classification proposed by Rendle was based largely on Engler (1892) although unlike that writer Rendle placed Araceae, Lemnaceae and Palmae in the same order. Thus he contributed nothing new to the classification and so will not be discussed further.

Calestani (1933) proposed a classification in which the monocotyledons were grouped into three series. These in no way correspond with the three proposed almost simultaneously by Hutchinson (see below) and with the exception of the Hydranthae are heterogeneous assemblages.

Monocotyleae
  Series I. *Lirianthae*
    1. Principes: Palmae
    2. Farinosae: Juncaceae, Commelinaceae, Bromeliaceae, Scitamineae etc.
    3. Scillinae: Liliaceae, Dioscoreaceae, Amaryllidaceae, Iridaceae, Orchidaceae etc.
  Series II. *Micranthae*
    1. Spadiciflorae: Araceae, Pandanaceae, Cyclanthaceae
    2. Junciformes: Sparganiaceae, Typhaceae, Cyperaceae etc.
    3. Gramineae: Graminaceae
  Series III. *Hydranthae:* Alismaceae, Hydrocharidae, Najadaceae, Lemnaceae

Skottsberg (1940). Writing in Swedish Skottsberg presented a system of classification which recognized 16 orders. His system had the advantage of being more differentiated and having more uniform family assemblages than most contemporary systems. It furthermore has the advantage of treating the orders in four natural groups corresponding best to present superorders. These are

1. Helobieae—Triuridales—Potamogetonales
2. Principes—Synanthae—(Spathiflorae)—Pandanales
3. Enantioblastae—Glumiflorae—Juncales—Cyperales
4. Bromeliales—Liliiflorae—Burmanniales—(Scitamineae)—Gynandrae

This system expresses an essentially modern view of many connections between groups, e.g. that between the Juncales and Cyperales.

1. *Helobiae:* Alismataceae, Scheuchzeriaceae, Butomaceae (incl. Limnocharitaceae), Hydrocharitaceae
2. *Triuridales:* Triuridaceae
3. *Potamogetonales:* Aponogetonaceae, Potamogetonaceae (incl. Juncaginaceae, Zosteraceae, Posidoniaceae), Naiadaceae (incl. Zannichelliaceae and Cymodoceaceae)

4. *Principes:* Arecaceae (= "Palmae")
5. *Synanthae:* Cyclanthaceae
6. *Spathiflorae:* Araceae, Lemnaceae
7. *Pandanales:* Pandanaceae, Sparganiaceae, Typhaceae
8. *Enantioblastae:* Restionaceae, Centrolepidaceae, Mayacaceae, Xyridaceae, Eriocaulaceae, Commelinaceae
9. *Glumiflorae:* Poaceae (= "Gramineae")
10. *Juncales:* Juncaceae, Flagellariaceae, Thurniaceae
11. *Cyperales:* Cyperaceae
12. *Bromeliales:* Bromeliaceae, Rapateaceae
13. *Liliiflorae:* Liliaceae (*sensu lato*), Stemonaceae, Petrosaviaceae, Haemodoraceae, Amaryllidaceae (incl. Agavaceae, Alstroemeriaceae and Hypoxidaceae),

Velloziaceae, Taccaceae, Dioscoreaceae, Iridaceae, Pontederiaceae, Cyanastraceae, Philydraceae
14. *Burmanniales:* Burmanniaceae, Corsiaceae
15. *Scitamineae:* Musaceae (incl. Strelitziaceae and Heliconiaceae), Zingiberaceae (incl. Costaceae), Marantaceae, Cannaceae
16. *Gynandrae:* Apostasiaceae, Orchidaceae (incl. Cypripediaceae).

Hutchinson (1934, 1959). The scheme proposed in 1959 is a modification of that published in 1934. In relation to contemporary botanists Hutchinson is a "splitter" at the family as well as at the ordinal level. Accordingly, there

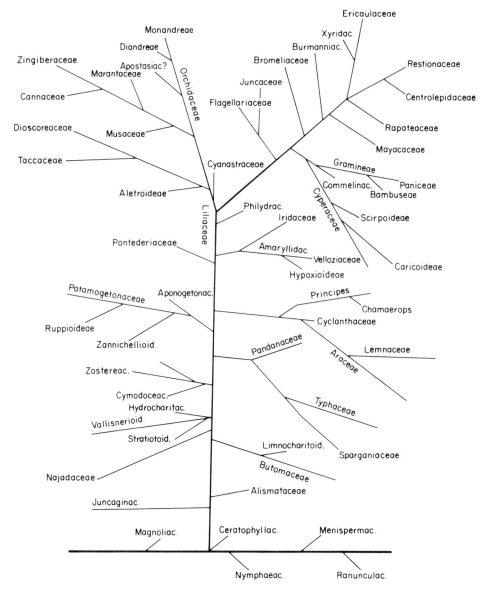

Fig. 1. Phylogenetic tree for the monocotyledons based on serological investigations, according to Ankermann (1927).

are greater chances for these groups to be natural entities than for those proposed by other taxonomists. His outlook was basically evolutionary and he regarded the monocotyledons as derived from Ranalean ancestors. Hence the apocarpous families corresponding largely to Helobiae are placed at the beginning of his system.

Next to these he placed Commelinaceae and related families giving much emphasis to the perianth being differentiated into a sepaloid outer and petaloid inner whorl. Thus Commelinales, Xyridales, Eriocaulales, Bromeliales and Zingiberales follow the first "helo-bialean" orders and all these are treated as the division Calyciferae. In the second division, Corolliferae are placed the "Liliiflorae" orders and also the Arales and Typhales, in accord with Hutchinson's view that these two orders had evolved from the tribe *Aspidistreae* of Liliaceae. Also, but separately derived from the liliiflorous plants, were the Palmales, Pandanales and Cyclanthales, the first of which at least Hutchinson regarded as showing similarities to his Agavales (Agavaceae, Xanthorrhoeaceae). The third division, Glumiflorae, consists of Juncales, Cyperales and Grami-nales, which are far separated from the Xyridales, Eriocaulales and Commelinales. Hutchinson in his division no doubt relied much on habit and certain macroscopic characters. However, his treatment shows some interesting alignments worthy of consideration. His Commelinales comprises Commelinaceae, Cartonemataceae, Mayacaceae and Flagellariaceae, a group likely to show some degree of homogeneity. His Xyridales consists of the undoubtedly related Xyridaceae and Rapateaceae and is placed next to Eriocaulaceae. His Alstroemeriales consists of Petermanniaceae, Philesiaceae and Alstroemeriaceae, the two last-mentioned families having some conspicuous similarities. Alliaceae is included in his Amaryllidaceae rather than in his Liliaceae, which is justifiable if considering the Liliaceae *sensu stricto* but less so if considering the *Scilla* group (our Hyacinthaceae) a part of Liliaceae. In addition, his Dioscoreales seems to be a natural group of families.

The treatment in the second edition (1959) and in the third and unchanged edition (1973) of his monocotyledon classification is given below.

Division 1. Calyciferae
1. *Butomales:* Butomaceae, Hydrocharitaceae
2. *Alismatales:* Alismataceae, Scheuchzeriaceae, Petro-saviaceae
3. *Triuridales:* Triuridaceae
4. *Juncaginales:* Juncaginaceae, Lilaeaceae, Posidonia-ceae
5. *Aponogetonales:* Aponogetonaceae, Zosteraceae

6. *Potamogetonales:* Potamogetonaceae, Ruppiaceae
7. *Najadales:* Zannichelliaceae, Najadaceae
8. *Commelinales:* Commelinaceae, Cartonemataceae, Flagellariaceae, Mayacaceae
9. *Xyridales:* Xyridaceae, Rapateaceae
10. *Eriocaulales:* Eriocaulaceae
11. *Bromeliales:* Bromeliaceae
12. *Zingiberales:* Musaceae, Strelitziaceae, Lowiaceae, Zingiberaceae, Cannaceae, Marantaceae
Division 2. Corolliferae
13. *Liliales:* Liliaceae, Tecophilaeaceae (incl. Cyana-straceae), Trilliaceae, Pontederiaceae, Smilacaceae, Ruscaceae
14. *Alstroemeriales:* Alstroemeriaceae, Petermannia-ceae, Philesiaceae
15. *Arales:* Araceae, Lemnaceae
16. *Typhales:* Sparganiaceae, Typhaceae
17. *Amaryllidales:* Amaryllidaceae (incl. Alliaceae)
18. *Iridales:* Iridaceae
19. *Dioscoreales:* Stenomeridaceae, Trichopodaceae, Roxburghiaceae, Dioscoreaceae
20. *Agavales:* Xanthorrhoeaceae, Agavaceae
21. *Palmales:* Arecaceae
22. *Pandanales:* Pandanaceae
23. *Cyclanthales:* Cyclanthaceae
24. *Haemodorales:* Haemodoraceae, Hypoxidaceae, Velloziaceae, Apostasiaceae, Taccaceae, Phily-draceae
25. *Burmanniales:* Burmanniaceae, Thismiaceae, Cor-siaceae
26. *Orchidales:* Orchidaceae
Division 3. Glumiflorae
27. *Juncales:* Juncaceae, Thurniaceae, Centrolepida-ceae, Restionaceae
28. *Cyperales:* Cyperaceae
29. *Graminales:* Poaceae

Faulks (1964). Although this classification is essentially that of Hutchinson it is worthy of separate consideration because of its method of presentation. All the orders of flowering plants are arranged in a plane in such a manner as to indicate their relative similarities. Each order is regarded as a branch of an evolutionary tree and time is considered to lie as an axis at right angles to and behind the plane. In these respects the presentation is similar to our own (Diagram 1). The boundary between the Mag-noliatae and Liliatae is not sharply defined and the two classes interlock like the pieces of a jigsaw. It is significant that the Magnoliales are very distant from the Liliatae in this scheme which is original in many other respects.

Novák (1954). In this system are found many of the presently accepted features but there are certain group-ings of families that appear unnatural, for example his inclusion of Juncaceae in Liliales. The Nymphaeaceae, Cabombaceae and Ceratophyllaceae as well as the Sauru-raceae and Piperaceae are contained in the giant order

Ranunculales, while Aristolochiales is treated separately and placed nearby.

Alismatales:
  1. Butomineae: Butomaceae, Hydrocharitaceae, Najadaceae, Zannichelliaceae, Cymodoceaceae
  2. Alismineae: Alismaceae, Scheuchzeriaceae
  3. Potamogetonineae: Potamogetonaceae, Ruppiaceae, Zosteraceae
Triuridales: Triuridaceae
Liliales:
  1. Liliineae: Liliaceae, Smilacaceae, Stemonaceae, Haemodoraceae, Amaryllidaceae, Velloziaceae, Taccaceae, Dioscoreaceae
  2. Iridineae: Iridaceae, Geosiridaceae
  3. Juncineae: Juncaceae
Cyperales: Cyperaceae
Commelinales: Commelinaceae, Flagellariaceae, Cyanastraceae, Pontederiaceae, Philydraceae, Bromeliaceae, Restionaceae, Mayacaceae, Xyridaceae, Eriocaulaceae, Centrolepidaceae, Thurniaceae, Rapateaceae
Poales: Poaceae
Zingiberales: Musaceae, Zingiberaceae, Cannaceae, Marantaceae
Orchidales: Apostasiaceae, Burmanniaceae, Thismiaceae, Corsiaceae, Orchidaceae
Arecales: Arecaceae
Cyclanthales: Cyclanthaceae
Arales: Araceae, Lemnaceae
Pandanales: Pandanaceae, Sparganiaceae, Typhaceae.

Deyl (1955). The classification produced by Deyl has little to commend it, for several of the orders are apparently unnatural assemblages. For example Bromeliales include Zingiberales (Scitamineae); Orchidales include Philydraceae, and Xanthorrhoeales combine Eriocaulaceae and related families with Xanthorrhoeaceae. Nonetheless some associations of families within orders are interesting and should not be lightly dismissed. In particular his treatment of Dioscoreales is worthy of consideration.

Hydrocharitales: Lilaeaceae, Najadaceae, Zannichelliaceae, Lemnaceae, Potamogetonaceae, Aponogetonaceae, Scheuchzeriaceae, Butomaceae, Alismaceae, Hydrocharitaceae
Arecales: Araceae, Acoraceae, Pandanaceae, Sparganiaceae, Typhaceae, Cyclanthaceae, Palmae
Juncales: Gramineae, Cyperaceae, Restionaceae, Centrolepidaceae, Thurniaceae, Flagellariaceae, Juncaceae
Xanthorrhoeales: Eriocaulaceae, Xyridaceae, Rapateaceae, Xanthorrhoeaceae
Bromeliales: Commelinaceae, Mayacaceae, Musaceae, Zingiberaceae, Cannaceae, Marantaceae, Bromeliaceae
Dioscoreales: Trilliaceae, Taccaceae, Ruscaceae, Smilacaceae, Alstroemeriaceae, Philesiaceae, Stenomeridaceae, Trichopodaceae, Roxburghiaceae, Petermanniaceae, Dioscoreaceae
Liliales: Pontederiaceae, Hypoxidaceae, Velloziaceae, Haemodoraceae, Tecophilaeaceae (incl. Cyanastraceae),

Agavaceae, Colchicaceae, Iridaceae, Amaryllidaceae, Liliaceae
Orchidales: Triuridaceae, Petrosaviaceae, Philydraceae, Apostasiaceae, Burmanniaceae, Thismiaceae, Corsiaceae, Orchidaceae.

Kimura (1956). In his treatment of the monocotyledons (he neglected the dicotyledons), Kimura placed much emphasis on the degree of fusion of the carpels, and divided the monocotyledons into two main groups: Apocarpae (= Helobiae) and Syncarpae. The latter were further subdivided into Subsyncarpae and Coenocarpae. His classification is very different from the contemporary ones and, like Hutchinson (1934), he has many orders. There is no doubt that the system has many advantages but the wide separation of Liliales (Syncarpae) from Amaryllidales and Iridales (Coenocarpae) is undesirable. As with the system of Bessey (1915) that of Kimura lays too much stress on single attributes, such as epigyny.

  I. *Apocarpae*
    A. *Helobiae*
      1. *Alismatales:* Butomaceae, Limnocharitaceae, Alismataceae
      2. *Hydrocharitales:* Hydrocharitaceae, Thalassiaceae, Halophilaceae, Vallisneriaceae
      3. *Scheuchzeriales:* Triglochinaceae (Juncaginaceae), Scheuchzeriaceae, Aponogetonaceae
      4. *Potamogetonales:* Potamogetonaceae, Zannichelliaceae, Zosteraceae
      5. *Najadales:* Lilaeaceae, Najadaceae
      6. *Triuridales:* Triuridaceae
  II. *Syncarpae*
  II: 1 Subsyncarpae
    B. *Liliiflorae*
      7. *Liliales:* Petrosaviaceae, Liliaceae, Stemonaceae, Xanthorrhoeaceae, Trilliaceae, Pontederiaceae, Smilacaceae (incl. Ruscaceae), Philesiaceae, Agavaceae
    C. *Spadiciflorae*
      8. *Arecales:* Arecaceae
      9. *Pandanales:* Pandanaceae
      10. *Cyclanthales:* Cyclanthaceae
  II: 2 Coenocarpae
    D. *Nudiflorae*
      11. *Arales:* Araceae, Lemnaceae
      12. *Typhales:* Sparganiaceae, Typhaceae
    E. *Sicciflorae*
      13. *Eriocaulales:* Eriocaulaceae
      14. *Restionales:* Restionaceae, Flagellariaceae, Centrolepidaceae
      15. *Juncales:* Juncaceae, Thurniaceae
      16. *Poales:* Poaceae
      17. *Cyperales:* Cyperaceae
    F. *Calyciferae*
      18. *Commelinales:* Commelinaceae
      19. *Xyridales:* Mayacaceae, Xyridaceae, Rapateaceae
      20. *Philydrales:* Philydraceae
      21. *Bromeliales:* Bromeliaceae

22. *Zingiberales:* Strelitziaceae, Lowiaceae, Musaceae, Zingiberaceae, Cannaceae, Marantaceae

G. *Epigynae*

23. *Dioscoreales:* Dioscoreaceae, Stenomeridaceae, Trichopodaceae, Petermanniaceae

24. *Amaryllidales:* Amaryllidaceae (incl. Alstroemeriaceae and Hypoxidaceae), Haemodoraceae, Velloziaceae, Tecophilaeaceae (incl. *Cyanastrum*)

25. *Iridales:* Iridaceae

26. *Burmanniales:* Burmanniaceae

27. *Orchidales:* Orchidaceae

28. *Taccales:* Taccaceae.

Emberger (1960). The similarities between Piperales and

Arales which impressed Lotsy (1911) also impressed Emberger (Chadefaud and Emberger, 1960) who divided the angiosperms into a series of phyla of which the monocotyledons comprise three: the Helobiae, Piperales—Spadiciflorae and Liliiflorae. Each of the three was supposed to have a common origin with dicotyledons amongst the Aristolochiales—Polycarpicae stock (incl. Nymphaeaceae and allied families).

His inclusion of Piperales in the same phylum as his assemblage Spadiciflorae is of interest in that the conventional monocotyledon-dicotyledon boundary is thereby bridged.

Thus Emberger stressed that the distinction between

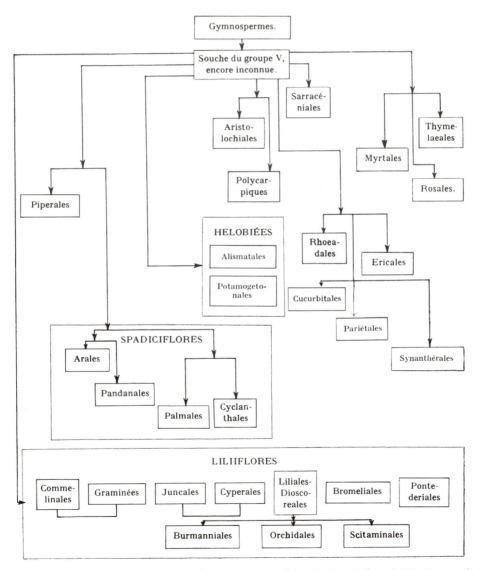

Fig. 2. Diagram illustrating Emberger's derivation of the Piperales as a group parallel to the "Spadiciflorae". It is also seen that he recognized the great resemblance between the Aristolochiales and the monocotyledons. (From Emberger, 1960.)

the monocotyledons and dicotyledons was not as clear-cut as had been traditionally believed, a point that has continued to gain support in recent years.

The family composition of the phyla of monocotyledons plus the dicotyledonous Polycarpicae are given below.

Phylum I. Polycarpiques—Aristolochiales
  Aristolochiales
  Polycarpiques (incl. Nymphaeacaceae, Cabombacaceae, Ceratophyllaceae
Phylum II. Helobiés (Fluviales)
  Alismatales: Butomaceae, Hydrocharitaceae, Alismataceae, Juncaginaceae, Scheuchzeriaceae, Aponogetonaceae
  Potamogetonales: Potamogetonaceae, Ruppiaceae, Lilaeaceae, Zosteraceae, Posidoniaceae, Zannichelliaceae, Najadaceae
  Triuridales: Triuridaceae
Phylum III. Piperales—Spadiciflores
  Piperales: Piperaceae, Saururaceae, Chloranthaceae
  Arales: Araceae, Lemnaceae
  Pandanales: Pandanaceae, Sparganiaceae, Typhaceae
  Palmales: Palmaceae, Phytelephasiaceae, Nipaceae
  Cyclanthales: Cyclanthaceae
Phylum IV. Liliiflores
  Commelinales: Commelinaceae, Mayacaceae, Xyridaceae, Eriocaulaceae, Centrolepidaceae, Restionaceae
  Graminales: Gramineae
  Juncales: Juncaceae, Flagellariaceae, Rapateaceae, Thurniaceae
  Cyperales: Cyperaceae
  Bromeliales: Bromeliaceae
  Pontederiales: Pontederiaceae
  Liliales: Liliaceae, Amaryllidaceae, Haemodoraceae, Velloziaceae, Cyanastraceae (incl. Tecophilaeaceae), Philydraceae, Iridaceae, Geosiridaceae, Roxburghiaceae, Petermanniaceae
  Dioscoreales: Dioscoreaceae, Taccaceae
  Scitaminales: Musaceae, Strelitziaceae, Lowiaceae, Zingiberaceae, Cannaceae, Marantaceae
  Burmanniales: Burmanniaceae, Thismiaceae, Corsiaceae
  Orchidales: Apostasiaceae, Orchidaceae.

Soó (1953, 1961, 1965, 1975). Overall the classification of the monocotyledons proposed by Soó resembles that of Lotsy (see above) in that he regarded them as biphyletic. One group of monocotyledons (Spadiciflorae—Arales sensu lato and Pandanales) and the Piperales he regarded as derived from certain Ranalean or Magnolialean ancestors, the other monocotyledons from Nymphaealean ancestors. He thereby rejected any direct connection between the main part of the monocotyledons and the Spadiciflorae—Arales.

Within the main group of monocotyledons the Liliales retain a central position with the other orders originating separately from their ancestors.

Monocotyledonae
  Series E
    42. Helobiae—Alismatales
    42. a. Triuridales
    43. Liliiflorae—Liliales
    44. Scitamineae—Zingiberales
    45. Gynandrae—Orchidales
    46. Cyperales
    47. Farinosae—Bromeliales
    48. Graminales—Poales
  Series F
    49. Spadiciflorae—Arales
    50. Pandanales.

Melchior et al. (1964). The system proposed by Melchior in the most recent edition of the influential "Syllabus" retains something of the original Englerian characteristics but recognizes several additional Reihen (orders). It begins with the Helobiae instead of Pandanales and like Engler's ends with the Microspermae.

Although the whole approach to the higher level classification is conservative the individual families are treated with considerable expertise.

(Reihen)
  1. Helobiae (Alismatales): Alismataceae, Butomaceae (incl. Limnocharitaceae), Hydrocharitaceae, Scheuchzeriaceae, Aponogetonaceae, Juncaginaceae, Potamogetonaceae (incl. Posidoniaceae and Zosteraceae)
  2. Triuridales: Triuridaceae
  3. Liliiflorae: Liliaceae sensu lato, Xanthorrhoeaceae, Stemonaceae, Agavaceae, Haemodoraceae, Cyanastraceae, Amaryllidaceae, Hypoxidaceae, Velloziaceae, Taccaceae, Dioscoreaceae, Pontederiaceae, Iridaceae, Geosiridaceae, Burmanniaceae, Corsiaceae, Philydraceae
  4. Juncales: Juncaceae, Thurniaceae
  5. Bromeliales: Bromeliaceae
  6. Commelinales: Commelinaceae, Mayacaceae, Xyridaceae, Rapateaceae, Eriocaulaceae, Restionaceae, Centrolepidaceae, Flagellariaceae
  7. Graminales: Gramineae (Poaceae)
  8. Principes: Palmae (Arecaceae)
  9. Synanthae: Cyclanthaceae
  10. Spathiflorae: Araceae, Lemnaceae
  11. Pandanales: Pandanaceae, Sparganiaceae, Typhaceae
  12. Cyperales: Cyperaceae
  13. Scitamineae: Musaceae (incl. Strelitziaceae and Heliconiaceae), Zingiberaceae (incl. Costaceae), Cannaceae, Marantaceae, Lowiaceae
  14. Microspermae: Orchidaceae (incl. Apostasiaceae and Cypripediaceae).

PERIOD 1965–

The year 1968 is of great significance in the history of taxonomic classification. Two new classifications were published, namely by Thorne and by Cronquist, and an English translation became available of Takhtajan's classification in the following year.

Furthermore from this time onwards the majority of classifications recognized superorders as an additional taxonomic rank.

The Takhtajan system had been previously published in Russian and German (1959) and it is apparent that Takhtajan and Cronquist influenced each other in the 1968–1969 versions of their classifications.

Takhtajan (1959, 1969). Takhtajan was convinced that the monocotyledons were derived from dicotyledons having acquired their characteristics while they adapted strongly during a phase of evolution to acquatic or at least marshy habitats. The classification in this respect is not novel in its main features, although the main groups have gained in homogeneity and most orders are placed in a logical sequence. It is noticeable that Zingiberales is inserted between Iridales and Orchidales in Lilianae in spite of its uniformity and considerable differences from the other two orders. The Alismanae and Juncanae are homogeneous groups and in Commelinanae the orders are placed in a series from a more Lilialean to an advanced poalean sequence. Arales and Typhales are placed with Arecanae, which is in agreement with other current systems but hardly with most recent findings.

Class LILIATAE (= Monocotyledones)
  Subclass ALISMIDAE
    Superorder Alismanae
      Alismales: Butomaceae, Limnocharitaceae, Alismaceae
      Hydrocharitales: Hydrocharitaceae
      Najadales: Scheuchzeriaceae, Juncaginaceae (incl. Lilaeaceae), Aponogetonaceae, Zosteraceae, Posidoniaceae, Potamogetonaceae, Ruppiaceae, Zannichelliaceae, Cymodoceaceae, Najadaceae
  Subclass LILIIDAE
    Superorder Lilianae
      Triuridales: Triuridaceae
      Liliales: Liliaceae (incl. Petrosaviaceae and Trilliaceae), Xanthorrhoeaceae, Aphyllanthaceae, Alliaceae, Agavaceae, Amaryllidaceae, Alstroemeriaceae, Haemodoraceae, Hypoxidaceae, Velloziaceae, Philesiaceae, Tecophilaeaceae, Cyanastraceae, Asparagaceae (incl. Ruscaceae), Smilacaceae, Stemonaceae, Dioscoreaceae, Taccaceae, Pontederiaceae, Philydraceae

      Iridales: Iridaceae, Geosiridaceae, Burmanniaceae (incl. Thismiaceae), Corsiaceae
      Zingiberales: Strelitziaceae, Musaceae, Heliconiaceae, Lowiaceae, Costaceae, Zingiberaceae, Cannaceae, Marantaceae
      Orchidales: Orchidaceae
  Subclass COMMELINIDAE
    Superorder Juncanae
      Juncales: Juncaceae, Thurniaceae
      Cyperales: Cyperaceae
    Superorder Commelinanae
      Bromeliales: Bromeliaceae
      Commelinales: Commelinaceae, Mayacaceae, Xyridaceae, Rapateaceae
      Eriocaulales: Eriocaulaceae
      Restionales: Restionaceae (incl. Anarthriaceae and Ecdeiocoleaceae), Centrolepidaceae, Flagellariaceae, Hanguanaceae
      Poales: Poaceae
  Subclass ARECIDAE
    Superorder Arecanae
      Arecales: Arecaceae
      Cyclanthales: Cyclanthaceae
      Arales: Araceae, Lemnaceae
      Pandanales: Pandanaceae
      Typhales: Sparganiaceae, Typhaceae

Cronquist (1968). This system differs in several features from that of Takhtajan. Triuridales is placed in Alismatidae and Commelinidae follows upon this subclass. Then Liliidae is placed at the end of the system, possibly in order to ensure that the highly specialized Orchidales will be at the end. The grasses are placed together with Cyperaceae in Cyperales, making this order somewhat heterogeneous in regard to a number of characters. Typhales in Cronquist's system is placed in sequence with Juncales and Cyperales and not with the Arecanae. In the Liliales there are relatively few families, indicating that Liliaceae is very widely circumscribed, including even the Amaryllidaceae.

Class LILIATAE (Monocotyledoneae)
  Subclass ALISMATIDAE
    Alismatales: Butomaceae, Limnocharitaceae, Alismataceae
    Hydrocharitales: Hydrocharitaceae
    Najadales: Aponogetonaceae, Scheuchzeriaceae, Juncaginaceae, Najadaceae, Potamogetonaceae, Ruppiaceae, Zannichelliaceae, Zosteraceae
    Triuridales: Petrosaviaceae, Triuridaceae
  Subclass COMMELINIDAE
    Commelinales: Rapateaceae, Xyridaceae, Mayacaceae, Commelinaceae
    Eriocaulales: Eriocaulaceae
    Restionales: Flagellariaceae, Anarthriaceae, Ecdeiocoleaceae, Restionaceae, Centrolepidaceae
    Juncales: Juncaceae, Thurniaceae
    Cyperales: Cyperaceae, Gramineae
    Typhales: Sparganiaceae, Typhaceae
    Bromeliales: Bromeliaceae

Zingiberales: Strelitziaceae, Lowiaceae, Heliconiaceae, Musaceae, Zingiberaceae, Costaceae, Cannaceae, Marantaceae

Subclass ARECIDAE

Arecales: Arecaceae

Cyclanthales: Cyclanthaceae

Pandanales: Pandanaceae

Arales: Araceae, Lemnaceae

Subclass LILIIDAE

Liliales: Philydraceae, Pontederiaceae, Liliaceae (incl. Amaryllidaceae), Iridaceae, Agavaceae, Xanthorrhoeaceae, Velloziaceae, Haemodoraceae, Taccaceae, Cyanastraceae, Stemonaceae, Smilacaceae, Dioscoreaceae

Orchidales: Geosiridaceae, Burmanniaceae, Corsiaceae, Orchidaceae.

Cronquist recently revised his classification slightly (given in Jones and Luchsinger, 1979, and also in Cronquist, 1979). The following amendments are made:

Cymodoceaceae is elevated to family rank within the Najadales.

Joinvilleaceae and Hydatellaceae are recognized as separate families in the Restionales, while Anarthriaceae and Ecdeiocoleaceae are withdrawn (probably included in Restionaceae).

Bromeliales and Zingiberales are treated as a separate subclass, the Zingiberidae (see Cronquist, 1978).

Aloëaceae and Hanguanaceae are acknowledged as families in the Liliales.

Thorne (1968, 1976). In the system presented in 1968 Thorne divided the monocotyledons amongst five superorders placed according to a sequence often adopted, viz. with the Alismatiflorae first followed by Triuridiflorae and Liliiflorae. His system is characterized by broad order and family concepts, subordinal and subfamilial ranks being adopted to give the necessary differentiation. Thus Liliaceae is widely circumscribed including numerous subfamilies (e.g. Amaryllidoideae, Alstroemerioideae, Agavoideae and Haemodoroideae), several of which are usually accepted as families or even orders in other systems (see below). Typhaceae and Sparganiaceae are included in Arales, and Commelinales includes such diverse families as Pontederiaceae, Bromeliaceae, Juncaceae, Eriocaulaceae, Commelinaceae and Poaceae, the order being in fact more inclusive than the subclass Commelinidae of Cronquist (1979).

Superorder Alimatiflorae

Alismatales: Butomaceae, Alismataceae, Hydrocharitaceae

Zosterales: Aponogetonaceae, Scheuchzeriaceae (incl. Juncaginaceae), Potamogetonaceae, Posidoniaceae, Zannichelliaceae, Zosteraceae

Najadales: Najadaceae

Superorder Triuridiflorae

Triuridales: Triuridaceae

Superorder Liliiflorae

Liliales: Liliaceae (with the subfamilies Melanthioideae, Herrerioideae, Asphodeloideae, Dracaenoideae, Xanthorrhoeoideae, Wurmbeoideae, Lilioideae, Scilloideae, Allioideae, Alstroemerioideae, Ixiolirioideae, Amaryllidoideae, Agavoideae, Hypoxidoideae, Haemodoroideae, Cyanastroideae, Asparagoideae, Ophiopogonoideae, Aletroideae, Luzuriagoideae, Smilacoideae), Roxburghiaceae, Dioscoreaceae, Taccaceae, Velloziaceae, Iridaceae, Burmanniaceae (incl. Corsiaceae), Orchidaceae (incl. Apostasiaceae)

Superorder Ariflorae

Arales: Araceae, Lemnaceae, Sparganiaceae, Typhaceae

Arecales: Arecaceae

Cyclanthales: Cyclanthaceae

Pandanales: Pandanaceae

Superorder Commeliniflorae

Commelinales: Bromeliaceae, Rapateaceae, Xyridaceae, Pontederiaceae, Philydraceae, Juncaceae (incl. Thurniaceae), Cyperaceae, Commelinaceae, Mayacaceae, Eriocaulaceae, Flagellariaceae, Restionaceae, Centrolepidaceae, Poaceae

Zingiberales: Musaceae (incl. Strelitziaceae and Heliconiaceae), Lowiaceae, Zingiberaceae, Cannaceae, Marantaceae.

In a revised classification, Thorne (1976) has included Velloziaceae as a subfamily of Liliaceae and added Trichopodaceae among the lilialean families. Triuridales is included in Alismatiflorae and Ariflorae divided into Ariflorae *sensu stricto* and Areciflorae. Ecdeiocoleaceae is also added among the families of Commelinales and Strelitziaceae and Heliconiaceae are acknowledged as separate families in Zingiberales.

In the second version of his classification Thorne begins with Liliiflorae followed by Alismatiflorae and Ariflorae. The position of the latter two orders close to each other is important for they share a number of characters though this has not always been appreciated.

Stebbins (1974). A classification based upon those of Takhtajan and Cronquist but differing in minor details from both has been proposed by Stebbins. The monocotyledons are mainly in accordance with Cronquist, although the grasses have been separated from the Cyperales. His superorders, as Cronquist's subclasses, end in *-idae*.

Subclass MONOCOTYLEDONES

Superorder Alismatidae

Alismatales: Butomaceae, Limnocharitaceae, Alismataceae

Hydrocharitales: Hydrocharitaceae

Najadales: Aponogetonaceae, Scheuchzeriaceae, Juncaginaceae, Najadaceae, Potamogetonaceae, Ruppiaceae, Zannichelliaceae, Zosteraceae

Triuridales: Petrosaviaceae, Triuridaceae

Superorder Commelinidae
    Commelinales: Rapateaceae, Xyridaceae, Mayacaceae, Commelinaceae
    Eriocaulales: Eriocaulaceae
    Restionales: Flagellariaceae, Restionaceae, Centrolepidaceae
    Poales: Gramineae
    Juncales: Juncaceae, Thurniaceae
    Cyperales: Cyperaceae
    Typhales: Sparganiaceae, Typhaceae
    Bromeliales: Bromeliaceae
    Zingiberales: Strelitziaceae, Lowiaceae, Heliconiaceae, Musaceae, Zingiberaceae, Costaceae, Cannaceae, Marantaceae
Superorder Arecidae
    Arecales: Palmae
    Cyclanthales: Cyclanthaceae
    Pandanales: Pandanaceae
    Arales: Araceae, Lemnaceae

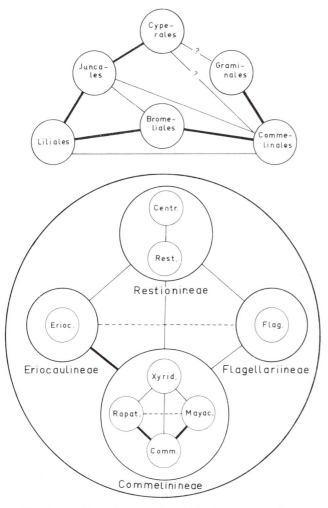

Fig. 3. Diagram illustrating the relationships between the suborders and families of Commelinales (bottom) and the relationships between Commelinales and some other monocotyledonous orders. According to Hamann (1961).

Superorder Liliidae
    Liliales: Philydraceae, Pontederiaceae, Liliaceae (incl. Amaryllidaceae), Iridaceae, Tecophilaeaceae, Agavaceae, Xanthorrheaceae, Velloziaceae, Haemodoraceae, Taccaceae, Cyanastraceae, Stemonaceae, Smilacaceae, Dioscoreaceae
    Orchidales: Geosiridaceae, Burmanniaceae, Corsiaceae, Orchidaceae.

Hamann (1961) made an extensive survey of the monocotyledons with particular consideration to the "Farinosae" *sensu* Engler, comparing the families Bromeliaceae, Centrolepidaceae, Commelinaceae, Cyanastraceae, Eriocaulaceae, Flagellariaceae, Mayacaceae, Philydraceae, Pontederiaceae, Rapateaceae, Restionaceae, Thurniaceae and Xyridaceae. His studies include a great number of attributes, which were thoroughly examined (see also p. 18). Great emphasis was placed on the embryology and in later penetrating studies on Philydraceae and Centrolepidaceae (Hamann, 1966a, 1975) further light was cast especially on embryological details of these groups.

Hamann in the concluding chapter of the 1961 paper suggested the following classification of the families included in his study (Fig. 3).

Order *Commelinales*
    Suborder Commelinineae: Commelinaceae, Mayacaceae, Rapateaceae, Xyridaceae
    Suborder Eriocaulineae: Eriocaulaceae
    Suborder Restionineae: Restionaceae, Centrolepidaceae
    Suborder Flagellariineae: Flagellariaceae (maybe also *Hanguana*)
Order *Bromeliales:* Bromeliaceae
Order *Liliales:* a great number of families including: Stemonaceae, Liliaceae, Haemodoraceae, Amaryllidaceae, Velloziaceae, Taccaceae, Dioscoreaceae, Iridaceae, Cyanastraceae, Philydraceae and Pontederiaceae, which according to Hamann should no doubt be distributed amongst a few suborders
Order *Juncales:* Juncaceae, Thurniaceae

As will be seen, the results in the present study are in good agreement with Hamann's, although his suborders are comparable to our orders. One difference is that Rapateaceae and Xyridaceae were referred to Commelinales rather than Eriocaulales; the evidence seems to be about equal for both alternatives. Flagellariaceae was placed separately, more closely to the Restionaceae than to other monocotyledons. Also in the present study there is evidence that Flagellariaceae is somewhat intermediate between Restionaceae and Poaceae.

Huber (1969, 1977). Another major recent contribution to the classification of monocotyledons has been due to

Huber (1969), who worked principally on the Liliaceae *sensu lato*, a large and heterogeneous assemblage that impinges on many other groups. The study was based primarily upon seeds with special reference to the occurrence of phytomelan, the structure of the testa and endosperm and the size and shape of the embryo, but many other attributes were also taken into consideration. In the analysis an impressive amount of information was utilized and comparisons were made in various directions.

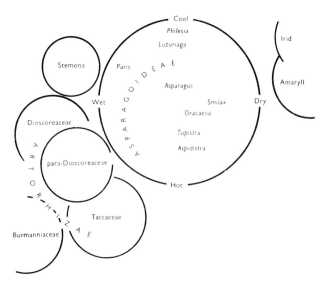

Fig. 4. Burkill's diagram suggesting how the families closest to the Dioscoreaceae evolved out of the "proto-Liliales". (From Burkill, 1960.)

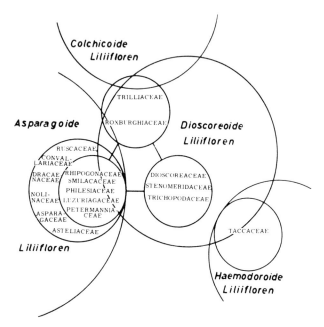

Fig. 5. Huber's diagram showing the relationships between the main evolutionary lines among the Liliiflorae. (From Huber, 1969.)

Huber paid great attention to the close similarity between Dioscoreaceae and some other families (Trichopodaceae, Stenomeridaceae, Roxburghiaceae, Smilacaceae, Philesiaceae etc.) on the one hand and certain dicotyledons on the other, and presented an alternative derivation of monocotyledons from dicotyledons with the Dioscorealean groups as transitional. He also in his liliiflorean synopsis placed Dioscoreales at the beginning followed by the bacciferous (Stemonalean and) Asparagalean families. An interesting division of the central families of the Liliiflorae into Asparagales and Liliales was introduced for the first time. The distribution is supported by a number of conspicuous differences in seed characters, nectaries, endosperm formation, tepal colour pattern etc.

In his classification Huber has narrow family and order concepts aimed at avoiding heterogeneous assemblages, for example among the "woody", trunk-bearing liliaceous groups which most likely have evolved along parallel lines. Beside the main orders Asparagales and Liliales, a series of small orders are recognized most of which seem to be well founded but not as yet acceptable to most taxonomists. These include Roxburghiales (Stemonales), Taccales, Haemodorales, Velloziales, Pontederiales and Philydrales.

An attempt towards a similar classification of a more limited number of families around Dioscoreales (Fig. 4) had earlier been presented by Burkill (1960). The work of Huber's is more discursive than formal and the table presented below is an attempt to formalize his classification. (N.B. Only part of the monocotyledons is covered.)

Several of the small families proposed by Huber have also been acknowledged in the course of the present study which makes it possible to give them an individual status in the comparison with other families and orders.

Dioscoreales: Dioscoreaceae, Stenomeridaceae, Trichopodaceae
Roxburghiales: Roxburghiaceae, Trilliaceae
Asparagales: Ripogonaceae, Smilacaceae, Philesiaceae, Luzuriagaceae, Petermanniaceae, Ruscaceae, Convallariaceae, Asparagaceae, Dracaenaceae, Nolinaceae, Herreriaceae, Asteliaceae, Dianellaceae, Hypoxidaceae, Lanariaceae, Walleriaceae, Eriospermaceae, Cyanastraceae, Tecophilaeaceae, Dasypogonaceae, Xanthorrhoeaceae, Anthericaceae, Aphyllanthaceae, Ixioliriaceae, Asphodelaceae, Agavaceae, (Funkiaceae) Phormiaceae, Doryanthaceae, Alliaceae, Agapanthaceae, Hyacinthaceae, Amaryllidaceae, Hemerocallidaceae
Liliales: Colchicaceae, Iridaceae, (Campynemataceae), Alstroemeriaceae, Tricyrtidaceae, Liliaceae, Calochortaceae, Melanthiaceae
Taccales: Taccaceae
Haemodorales: Haemodoraceae

Velloziales: Velloziaceae
Bromeliales: Bromeliaceae
Pontederiales: Pontederiaceae
Philydrales: Philydraceae.

In elaborating on this scheme at a conference in 1976, Huber (1977) emended his earlier work somewhat and extended the work to all the monocotyledons. (Fig. 6). He dealt in particular with the superorders includ-

ing those dicotyledons showing monocotyledonous tendencies. These results are summarized in Fig. 7.

Dahlgren (1975a). One of us (RD) has proposed a system of classification in connection with the presentation of diagrams constructed to show the distribution of character states in angiosperms. For the classification of the monocotyledons the main features of Huber's Lilii-

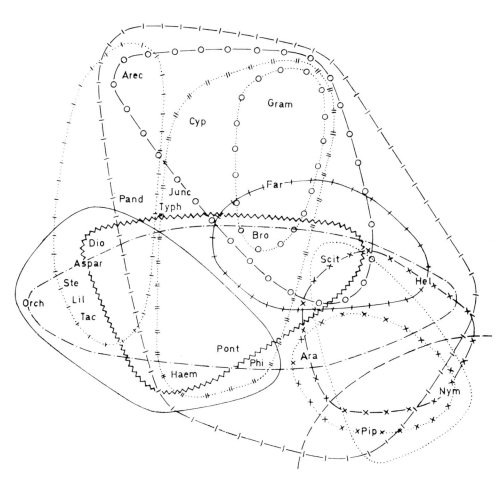

— — — — —	P-type sieve tube plastids
—·—·—·—	vessels only in roots
o—o—o—	silica bodies in particular cells
x—x—x—	laticiferous tubes present
I—I—I—I	stomata with 2 or more subsidiary cells
————————	perianth petaline, not differentiated into calyx and corolla
˧ ˧ ˧ ˧	perianth differentiated into calyx and corolla
wwwwwww	with septal nectaries
··················	seeds with storage nucellus (perisperm)
···x···x···x···x·	endosperm formation cellular
····I····I····I····	endosperm with hemicellulose
····II····II····II·	endosperm with starch
····o····o····o·	embryo lateral

Fig. 6. Diagram showing specific superposition of characters indicative of ordinal and higher level in the monocotyledons. (After Huber, 1977.)

Dicotyledonous features prevailing	Intermediate evolutionary state	Dicotyledonous features rare or absent
a) isolated evolutionary lines: **Piperiflorae** (one order: *Piperales*) **Nymphaeiflorae** (one order: *Nymphaeales*)	**Ariflorae** (one order: *Arales*) **Helobiae** or **Alismatiflorae** (2–4 orders)	**Scitamineae** (one order: *Zingiberales*)
b) reticulate evolutionary lines:	**Liliiflorae** (probably 5 orders: *Dioscoreales, Stemonales, Asparagales, Liliales, Orchidales*) **Tacciflorae** (one family: *Taccaceae*) **Palmiflorae** (with 5 or 6 orders if construed broadly: *Arecales, Cyclanthales, Pandanales, Cyperales, Typhales,* and doubtfully *Juncales*)	**Haemodoriflorae** (one family: *Haemodoraceae*) **Pontederiiflorae** (2 orders: *Pontederiales* and *Philydrales*) **Bromeliiflorae** (possibly 2 orders: apart from *Bromeliales* also *Velloziales*) **Enantioblastae** or **Commeliniflorae** (superorder embracing all families with mealy endosperm and lateral embryo except *Bromeliaceae*; these families have been arranged in one [Thorne, 1968] to 4 [Dahlgren, 1976] orders)

Fig. 7. Conspectus of the monocotyledonous superorders according to Huber (1977).

florae system were adopted although attempts were made to reduce slightly the number of orders and families. This scheme differs somewhat from the classification used in the present work, but the two are similar in essential features. In the 1975 classification, Typhales was placed in a separate superorder, which in the light of the present knowledge does not seem to be apt, and at the onset of this study was regarded as better placed in the superorder Commeliniflorae. Besides, there appeared later to be some evidence for transferring Xyridaceae and Rapateaceae from Commelinales to Eriocaulales and to include Centrolepidaceae, Restionaceae, Flagellariaceae and a few smaller families in a separate order, Restionales, near the grasses. The evidence for this viewpoint was proved to need reconsideration in the course of the present study (see the section on Evaluation). The close connection between the Alismatiflorae and the Ariflorae was not obvious at this stage and thus has been reconsidered.

The system of Dahlgren (1975) is as follows:

Alismatanae
  Alismatales: Alismataceae, Limnocharitaceae
  Hydrocharitales: Butomaceae, Hydrocharitaceae, Aponogetonaceae
  Zosterales: Scheuchzeriaceae, Juncaginaceae, Potamogetonaceae, Zosteraceae, Posidoniaceae, Zannichelliaceae, Cymodoceaceae
  Najadales: Najadaceae
Lilianae
  Dioscoreales: Dioscoreaceae
  Stemonales: Stemonaceae, Trilliaceae
  Asparagales: Smilacaceae, Philesiaceae, Ruscaceae, Convallariaceae, Asparagaceae, Dracaenaceae, Hypoxidaceae, Tecophilaeaceae, Phormiaceae, Xanthorrhoeaceae, Aphyllanthaceae, Asphodelaceae, Anthericaceae, Ixioliriaceae, Agavaceae, Hemerocallidaceae, Hyacinthaceae, Alliaceae, Amaryllidaceae
  Taccales: Taccaceae
  Haemodorales: Haemodoraceae, Pontederiaceae, Philydraceae
  Liliales: Colchicaceae, Iridaceae, Alstroemeriaceae, Liliaceae, Melanthiaceae
  Triuridales: Triuridaceae
  Burmanniales: Burmanniaceae, Thismiaceae, Corsiaceae
  Orchidales: Apostasiaceae, Cypripediaceae, Orchidaceae
  Bromeliales: Bromeliaceae, Velloziaceae
Typhanae
  Typhales: Sparganiaceae, Typhaceae
Zingiberanae
  Zingiberales: Lowiaceae, Heliconiaceae, Musaceae, Strelitziaceae, Zingiberaceae, Costaceae, Cannaceae, Marantaceae
Commelinanae
  Commelinales: Commelinaceae, Cartonemataceae, Mayacaceae, Xyridaceae, Abolbodaceae, Rapateaceae
  Eriocaulales: Eriocaulaceae
  Juncales: Juncaceae, Thurniaceae
  Cyperales: Cyperaceae
  Centrolepidales: Centrolepidaceae
  Poales: Restionaceae, Ecdeiocoleaceae, Flagellariaceae, Joinvilleaceae, Poaceae
Arecanae
  Arecales: Arecaceae
  Pandanales: Pandanaceae
  Cyclanthales: Cyclanthaceae
Aranae
  Arales: Araceae, Lemnaceae.

The above system in an emended form was used in a Danish textbook (Dahlgren, 1976) where the more important families are supplied with descriptions and illustrations. A similar system and diagram (constructed by Dahlgren) was also used by Wagner (1977) to illustrate the distribution of vessels in the xylem of monocotyledons.

Ehrendorfer (1978). In his treatment of the monocotyledons Ehrendorfer regarded these as a class with three subclasses. Within one of these, Liliidae, he recognized four superorders thereby bringing his groups of orders (though not their ranks) into line with other modern classifications. Were the Arales transferred from the Arecidae to the Alismatidae, then the primary division in this classification would coincide with that of our Diagram 108 (p. 326).

An outline of his scheme is given below. (Note, that his textbook does not deal with all families.)

Class MONOCOTYLEDONEAE
  Subclass ALISMATIDAE
    Alismatales: Butomaceae, Alismataceae
    Hydrocharitales: Hydrocharitaceae
    Najadales: Scheuchzeriaceae, Juncaginaceae, Potamogetomaceae, Zosteraceae, Zannichelliaceae, Najadaceae
  Subclass LILIIDAE (incl. Commelinidae)
    Lilianae:
      Liliales: Liliaceae (many subfamilies), Agavaceae, Amaryllidaceae, Iridaceae, Dioscoreaceae, Pontederiaceae
      Orchidales: Orchidaceae
    Bromelianae:
      Bromeliales: Bromeliaceae
      Zingiberales: Musaceae, Zingiberaceae, Cannaceae, Marantaceae
    Juncanae:
      Juncales: Juncaceae
      Cyperales: Cyperaceae
      Typhales: Typhaceae
    Commelinanae:
      Commelinales: Commelinaceae
      Eriocaulales: Eriocaulaceae
      Restionales: Restionaceae
      Poales: Poaceae

  Subclass ARECIDAE
    Arecales: Arecaceae
    Cyclanthales: Cyclanthaceae
    Arales: Araceae, Lemnaceae
    Pandanales: Pandanaceae

All the above classifications are attempts to synthesize the vast amount of knowledge that has accumulated concerning the monocotyledons. Much of the information is scattered in research papers and refers to detailed studies of anatomy, embryology, palynology, cytology, phytogeography. Many new, original observations are encapsulated in "The Monocotyledons" by Arber (1925).

Besides such non-taxonomic literature some papers on critical families have materially assisted in shedding light on relationships in different parts of the system. In particular reference should be made to the works by Hamann (1961, 1962, 1966, 1975) on the "Farinosae" and especially the Philydraceae, Centrolepidaceae and Hydatellaceae, by D. Müller-Doblies (1968) and U. Müller-Doblies (1969) on the Typhales, by Harling (1958) on the Cyclanthaceae and by Moore (1973) on the Arecaceae.

Detailed knowledge of such families provides fixed reference points against which the positions of other families can be measured. These reference points materially assisted us in making the survey presented below.

Numerical Approaches to Classification

PHENETIC

Since the advent of electronic computers, considerable attention has been directed towards establishing numerical methods for generating classifications with the help of lists of taxa and their attributes. As developed, these methods have been applied more or less successfully to many plant and animal groups. The theoretical bases of these methodologies have been summarized by Sneath and Sokal (1973) and Clifford and Stephenson (1975).

With respect to the higher levels of classification of the angiosperms numerical studies have been undertaken on both the dicotyledons (Young and Watson, 1970) and the monocotyledons (Lowe, 1961; Clifford, 1970, 1977). At the higher levels of the hierarchy each analysis produced a number of the widely recognized major groups and otherwise produced assemblages worthy of further consideration. Clifford's results (1977) for the classification of the monocotyledons are summarized in Fig. 8 and Table 1.

The theoretical bases of most current computer-based classificatory strategies are phenetic in outlook but phylogenetic methodologies such as those of Hennig (1965, 1966) and Schlee (1971), are available and might be attempted as complementary.

A detailed survey of the families of the "Farinosae" *sensu* Engler has been published by Hamann (1961), who summarized in tabular form the states of more than 80 characters for the families concerned (Bromeliaceae, Centrolepidaceae, Cyanastraceae, Eriocaulaceae, Flagellariaceae, Mayacaceae, Philydraceae, Pontederiaceae, Rapateaceae, Restionaceae, Thurniaceae and Xyridaceae). By counting and evaluating the similarities he estimated the relative affinities of the families to each other and to selected other families (Cyperaceae, Poaceae, families of the Liliiflorae). This work has much in common with the present survey, and inspired one of us (HTC) to start collecting data for a numerical classification of the monocotyledons (Clifford, 1970).

Whether investigations such as these should be classified as phenetic or phylogenetic is a matter of judgement.

TABLE 1

The Family Compositions of the Ten Groups defined by the Dendrogram in Fig. 8

Mostly aquatic families
1. Alismataceae, Aponogetonaceae, Butomaceae, Corsiaceae, Halophilaceae, Hydrocharitaceae, Limnocharitaceae, Thalassiaceae
2. Cymodoceaceae, Juncaginaceae, Lilaeaceae, Naiadaceae, Posidoniaceae, Potamogetonaceae, Ruppiaceae, Scheuchzeriaceae, Zannichelliaceae, Zosteraceae

Mostly bird-pollinated, tropical, terrestrial families (Zingiberales)
3. Cannaceae, Costaceae, Heliconiaceae, Lowiaceae, Marantaceae, Musaceae, Strelitziaceae, Zingiberaceae

Mostly insect-pollinated, temperate, terrestrial families
4. Agavaceae, Alliaceae, Aloëaceae, Alstroemeriaceae, Amaryllidaceae, Aphyllanthaceae, Apostasiaceae, Gilliesiaceae, Hypoxidaceae, Iridaceae, Orchidaceae, Petrosaviaceae, Philesiaceae, Philydraceae, Thismiaceae, Xanthorrhoeaceae
5. Asparagaceae, Bromeliaceae, Burmanniaceae, Commelinaceae, Dioscoreaceae, Geosiridaceae, Haemodoraceae, Isophysidaceae, Liliaceae, Ruscaceae, Smilacaceae, Tecophilaeaceae, Trichopodaceae, Velloziaceae
6. Araceae, Lemnaceae, Petermanniaceae, Taccaceae, Trilliaceae
7. Croomiaceae, Roxburghiaceae, Stemonaceae

Mostly wind-pollinated, terrestrial families
8. Abolbodaceae, Cartonemataceae, Thurniaceae, Eriocaulaceae, Xyridaceae, Hanguanaceae, Juncaceae, Mayacaceae, Pontederiaceae, Triuridaceae
9. Anarthriaceae, Centrolepidaceae, Ecdeiocoleaceae, Flagellariaceae, Joinvilleaceae, Rapateaceae, Restionaceae
10. Arecaceae, Cyclanthaceae, Cyperaceae, Pandanaceae, Poaceae, Sparganiaceae, Typhaceae

---

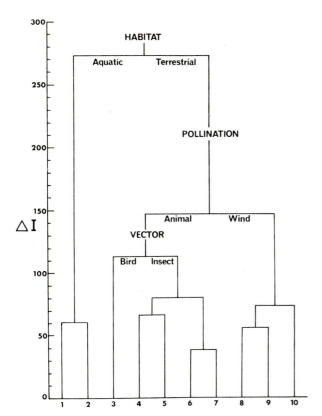

Fig. 8. Dendrogram resulting from clustering of the families of Liliatae, truncated at the 10-group level with the principal habitats and pollination mechanisms of the groups superimposed. The bottom row numbers refer to Table 1. (After Clifford, 1977.)

EVOLUTIONARY

An alternative statistical approach is due to Sporne (1956, 1974) who devised a method of calculating indices of advancement. Before such indices can be calculated it is necessary to determine for each character which of its two to several states is advanced and which is primitive. For example, are superior ovaries to be regarded as primitive or advanced?

In order to solve this question Sporne (op. cit.) proposed that character states that are primitive ought, more often than would occur by chance, to be associated with other primitive character states. Accordingly he calculated the degree of association as measured by χ^2

between numerous pairs of characters to determine which were statistically correlated.

Given these he decided which of the states was primitive or advanced either by reference to the fossil record or by their being associated with attributes known to be primitive from that record. Knowing which character states are primitive enabled him to develop an advancement index for each taxon. This was calculated as the percentage of advanced character states in relation to all characters used in the comparison. It now became possible to construct a putatively phylogenetic framework for any classification combining both the taxa in the groups and their degree of advancement, assuming that the basis of the methodology is acceptable. Such a phylogenetic classification using the basic scheme of Cronquist (1968) was prepared by Sporne and is shown in Fig. 9.

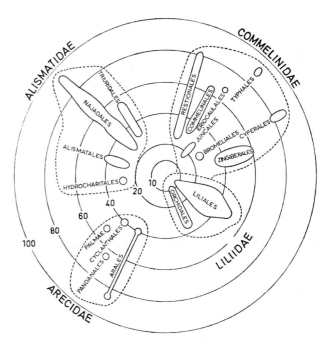

Fig. 9. A phylogenetic classification of the monocotyledons according to Sporne (1974). The relationships are those suggested by Cronquist (1968). The radial extent of each order corresponds to the range of advancement indices of its constituent families.

The Classification Used in the Course of the Present Study

GENERAL

The classification used as the basis for character mapping at the outset of this study was of necessity preliminary in that many of the data presented herein were not then available. However the pattern of variation for 14 characters had been mapped (Dahlgren, 1976) and the distributions of others were known. Hence there is reasonable agreement between the initial and final classifications. The latter, which are summarized in Diagrams 106–108, have arisen out of a consideration of all the data.

The classification adopted for commencing the project was largely influenced by Huber (1969). His classification has the advantage of recognizing small relatively homogenous families. Such a treatment served the purpose of the present study, as the registration of character states has been given mostly on the family level. Differences between small segregate families are then easily demonstrable in the diagram, and moreover differences are more readily discussed for families than for variously described tribes, subtribes, generic groups and other categories of variable rank, no matter how distinct. The classification is summarized in Table 2 and Diagrams 1 and 2.

Though our system has recognized many segregate families it is still restricted in the sense that several families sometimes recognized as distinct have been incorporated in others, e.g.

Thalassiaceae, Halophilaceae and Vallisneriaceae in Hydrocharitaceae
Limnocharitaceae in Alismataceae
Lilaeaceae in Juncaginaceae
Ruppiaceae in Potamogetonaceae
Stenomeridaceae and Trichopodaceae in Dioscoreaceae
Croomiaceae in Stemonaceae (Roxburghiaceae)
Ripogonaceae in Smilacaceae
Luzuriagaceae and Geitonoplesiaceae in Philesiaceae
Ruscaceae in Asparagaceae
Nolinaceae in Dracaenaceae
Asteliaceae, Anthericaceae, Aloëaceae and Ixioliriaceae in Asphodelaceae
Walleriaceae and Lanariaceae in Tecophilaeaceae
Agapanthaceae and Gilliesiaceae in Alliaceae
Petrosaviaceae in Melanthiaceae

Isophysidaceae in Iridaceae
Campynemataceae in Colchicaceae
Conostylidaceae in Haemodoraceae
Abolbodaceae in Xyridaceae.

In the diagrams each family has its own area within the ordinal figure, as shown in Diagrams 1–2. The mutual positions of the families are believed to reflect to a large extent phylogenetic relationships.

A SURVEY OF THE SUPERORDERS

In this chapter the classification at the higher levels that is used throughout the analytic part of this study shall be presented in some detail. The chapter includes descriptions for the orders and superorders provisionally adopted and serves to give their circumscription. For the families where the reader may be uncertain of the genera included, these have been enumerated, but for especially the larger families where there is reasonable agreement (such as Iridaceae, Amaryllidaceae, Orchidaceae, Poaceae and Arecaceae), this has been neglected.

Thus the chapter should comprise a firm reference point for the main part of the book where characters are analysed through the monocotyledon system, orders and families are discussed and taxa at genus or family level are enumerated in base data lists.

Categories

Defining and applying the higher categories, family, order and superorder, provide problems in monocotyledons as in other angiosperms. There is no consensus among taxonomists on definition or level for these categories.

Thus, to take one of the most variably circumscribed orders, the order Liliales in Thorne's system (1968) includes beside the Liliaceae and a number of closely related families also the Taccaceae, Velloziaceae, Haemodoraceae, Burmanniaceae and Orchidaceae, but not the Pontederiaceae or Philydrales, while in Huber's classification (1969) each of the latter families is placed in a

TABLE 2
The Classification Used in the Course of the Present Survey, down to Family Level

Superorder ALISMATIFLORAE

Order Hydrocharitales
Butomaceae
Hydrocharitaceae
Aponogetonaceae

Order Alismatales
Alismataceae

Order Zosterales
Scheuchzeriaceae
Juncaginaceae
Potamogetonaceae
Zosteraceae
Posidoniaceae
Zannichelliaceae
Cymodoceaceae

Order Najadales
Najadaceae

Order Triuridales
Triuridaceae

Superorder ARIFLORAE

Order Arales
Araceae
Lemnaceae

Superorder LILIIFLORAE

Order Dioscoreales
Dioscoreaceae
Stemonaceae
Trilliaceae

Order Taccales
Taccaceae

Order Asparagales
Smilacaceae
Petermanniaceae
Philesiaceae
Convallariaceae
Asparagaceae
Herreriaceae
Dracaenaceae
Doryanthaceae
Dasypogonaceae
Phormiaceae
Xanthorrhoeaceae
Agavaceae

Hypoxidaceae
Asphodelaceae
Aphyllanthaceae
Dianellaceae
Tecophilaeaceae
Cyanastraceae
Eriospermataceae
Hemerocallidaceae
Funkiaceae
Hyacinthaceae
Alliaceae
Amaryllidaceae

Order Liliales
Iridaceae
Geosiridaceae
Colchicaceae
Alstroemeriaceae
Tricyrtidaceae
Calochortaceae
Liliaceae
Melanthiaceae

Order Burmanniales
Burmanniaceae
Thismiaceae
Corsiaceae

Order Orchidales
Apostasiaceae
Cypripediaceae
Orchidaceae

Order Pontederiales
Pontederiaceae

Order Haemodorales
Haemodoraceae

Order Philydrales
Philydraceae

Order Velloziales
Velloziaceae

Order Bromeliales
Bromeliaceae

Superorder ZINGIBERIFLORAE

Order Zingiberales
Lowiaceae
Heliconiaceae
Musaceae

Strelitziaceae
Zingiberaceae
Costaceae
Cannaceae
Marantaceae

Superorder COMMELINIFLORAE

Order Commelinales
Commelinaceae
Cartonemataceae
Mayacaceae

Order Eriocaulales
Rapateaceae
Xyridaceae
Eriocaulaceae

Order Typhales
Sparganiaceae
Typhaceae

Order Juncales
Thurniaceae
Juncaceae

Order Cyperales
Cyperaceae

Order Hydatellales
Hydatellaceae

Order Restionales
Centrolepidaceae
Restionaceae
Anarthriaceae
Ecdeiocoleaceae
Flagellariaceae
Joinvilleaceae
Hanguanaceae

Order Poales
Poaceae

Superorder ARECIFLORAE

Order Arecales
Arecaceae

Order Cyclanthales
Cyclanthaceae

Order Pandanales
Pandanaceae

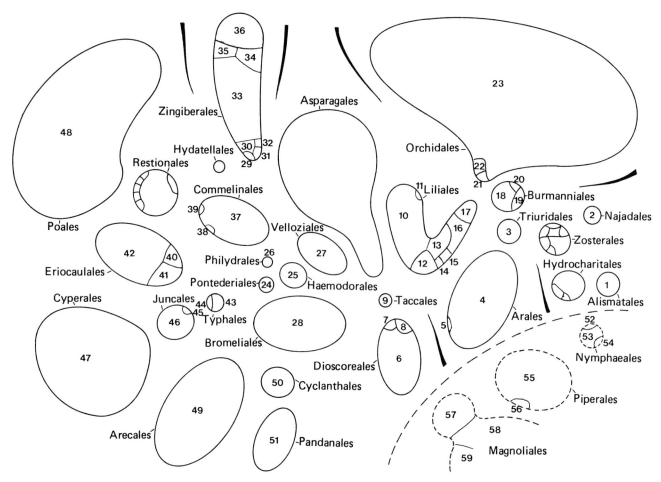

Diagram 1. Positions of families in the orders of monocotyledons followed in the diagrams (3–106) throughout the present study. The numerals refer to the families of the table below. The Asparagales, Zosterales, Hydrocharitales and Restionales are given separately in Diagram 2. Some families of dicotyledons (52–59) are also included for comparison.

1. Alismataceae	21. Apostasiaceae	41. Xyridaceae
2. Najadaceae	22. Cypripediaceae	42. Eriocaulaceae
3. Triuridaceae	23. Orchidaceae	43. Sparganiaceae
4. Araceae	24. Pontederiaceae	44. Typhaceae
5. Lemnaceae	25. Haemodoraceae	45. Thurniaceae
6. Dioscoreaceae	26. Philydraceae	46. Juncaceae
7. Stemonaceae	27. Velloziaceae	47. Cyperaceae
8. Trilliaceae	28. Bromeliaceae	48. Poaceae
9. Taccaceae	29. Lowiaceae	49. Arecaceae
10. Iridaceae	30. Heliconiaceae	50. Cyclanthaceae
11. Geosiridaceae	31. Musaceae	51. Pandanaceae
12. Colchicaceae	32. Strelitziaceae	52. Cabombaceae
13. Alstroemeriaceae	33. Zingiberaceae	53. Nymphaeaceae
14. Tricyrtidaceae	34. Costaceae	54. Ceratophyllaceae
15. Calochortaceae	35. Cannaceae	55. Piperaceae
16. Liliaceae *s. str.*	36. Marantaceae	56. Saururaceae
17. Melanthiaceae	37. Commelinaceae	57. Aristolochiaceae
18. Burmanniaceae	38. Cartonemataceae	58. Annonaceae
19. Thismiaceae	39. Mayacaceae	59. Magnoliaceae
20. Corsiaceae	40. Rapateaceae	

separate order; and in his classification of 1976, the Taccales makes up the Tacciflorae, Haemodorales the Haemodoriflorae, Pontederiales and Philydrales the Pontederiflorae, and Velloziales with the Bromeliales the Bromeliiflorae.

The consensus with regard to the circumscription of families is equally vague. Thorne's concept of the Liliaceae includes about 40 of Huber's families distributed upon at least three orders. These are extreme cases: hardly any family in the angiosperms is so variably interpreted

as Liliaceae. However, a short indication of our concepts for families, orders and superorders may be in its place here.

There is no doubt that quite objective criteria for the categories are impossible to apply. Beside criteria obtained from the number and significance of differences and discontinuity between the groups, general practice among taxonomists and practical reasons, such as number of next subordinate taxa, are more often than here taken into consideration.

At first we must declare that whatever our personal opinions, fairly narrow circumscriptions of the categories facilitate the procedures used in this project, and our approach is therefore basically splitting and more like Huber's than Thorne's.

Families are conceived of here as groups of genera which exhibit obvious similarities indicating relationship in the evolutionary sense or single genera when these are morphologically isolated. Geographical distributions are of importance inasmuch as groups of similar genera frequently but not necessarily are geographically concentrated or connected. On the family level concealed character states often play as great a role or a greater role than easily observed characters connected with environment, or relative size and shape of leaves and floral parts, although in monocotyledons a habitat in many cases seems to be characteristic to the group (for example the aquatic in the Zosterales, rain forest in the Arales and Zingiberales). In this study the criteria of homogeneity have been stressed more than is usual, resulting especially in the Liliiflorae in fairly numerous and often small families with regard to number of genera.

The consistency in circumscription of families is admittedly somewhat poor in that the palms (as in practically all contemporary literature) comprise a single variable

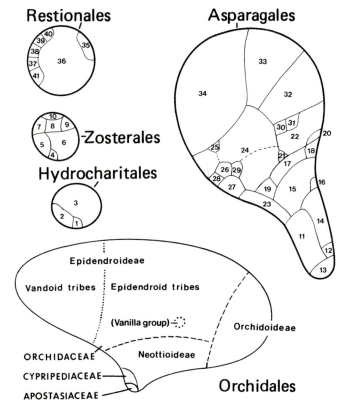

Diagram 2. Positions of families in the orders Hydrocharitales, Zosterales, Asparagales, and Restionales, extracted from Diagram 1. The positions of the families are those followed in all the diagrams used in the present study. The numerals refer to the following familien

1. Butomaceae	15. Asparagaceae	29. Eriospermaceae
2. Aponogetonaceae	16. Herreriaceae	30. Hemerocallidaceae
3. Hydrocharitaceae	17. Dracaenaceae	31. Funkiaceae
4. Scheuchzeriaceae	18. Doryanthaceae	32. Hyacinthaceae
5. Juncaginaceae	19. Dasypogonaceae	33. Alliaceae
6. Potamogetonaceae	20. Phormiaceae	34. Amaryllidaceae
7. Zosteraceae	21. Xanthorrhoeaceae	35. Centrolepidaceae
8. Cymodoceaceae	22. Agavaceae	36. Restionaceae
9. Posidoniaceae	23. Hypoxidaceae	37. Anarthriaceae
10. Zannichelliaceae	24. Asphodelaceae	38. Ecdeiocoleaceae
11. Smilacaceae	25. Aphyllanthaceae	39. Flagellariaceae
12. Petermanniaceae	26. Dianellaceae	40. Joinvilleaceae
13. Philesiaceae	27. Tecophilaeaceae	41. Hanguanaceae
14. Convallariaceae	28. Cyanastraceae	

In the ordinal figure of Orchidales are given the spaces of the families and subfamilies. This is utilized for many diagrams in the book, which is explained in the legends for these diagrams.

family, while in the liliiflorean part of the monocotyledons, grouping scarcely more distinct than the tribes of palms are treated as families.

The rank of **order** is equally subjectively defined than families. Orders can be defined as single, more or less isolated families, or as groups of families standing out as a distinct unit, by having a number of characteristic attributes in common although sometimes showing a stepwise to continuous variation between considerably different extremes. Ordinal characteristics are often such that they are in common for a number of families and indicate more or less strongly an affinity between these. In our study the orders have been dimensioned so as to be practical units but are still believed to represent monopyletic (i.e. homogeneous) groups in an evolutionary sense. As pointed out above, there is no firm consensus among taxonomists about the ordinal level, and at the end of our treatise with our concluding classifications and diagrams we will present alternatives with regard to the circumscription of some orders (such as Liliales *sensu lato* versus Liliales *sensu stricto*, plus Asparagales and Dioscoreales; Haemodorales *sensu lato* versus Haemodorales *sensu stricto* plus Pontederiales and Philydrales; Bromeliales *sensu lato* versus Bromeliales *sensu stricto* plus Velloziales).

The rank of the **superorder**, finally, follows the same criteria as the orders, although the requirements demanded with regard to mutual similarities are lessened. For monocotyledons the coincidence or partial coincidence of a number of attributes with a particular distribution pattern (oxalate raphides, endosperm formation, starchy endosperm, stomatal construction, type of pollination) can be helpful in deciding the delimitation of the superorders, which will be further discussed with the evaluation of the distribution of character states in this book. Not even in this case would one conceive the entities as arbitrary, but a factual basis must be applied, even if the superorder as an intercollated unit has its importance in facilitating the survey in its main features. Therefore such an order as Triuridales is treated here, in the concluding chapter, as a superorder, because we are able to relate it with certainty neither to the Liliiflorae nor to the Alismatiflorae.

The subclass rank *sensu* Cronquist or Takhtajan has been avoided here, mainly because in the dicotyledons this rank is difficult to apply. There, the pattern is more complicated than in the monocotyledons and the relationships not understood well enough to allow for such a coarse division. In case the subclass rank in the Cronquist–Takhtajan sense should be used in monocotyledons, perhaps the following three groups would stand out corresponding to this level: (1) the Alismatiflorae plus the Ariflorae; (2) the Liliiflorae plus the Zingiberiflorae, the Commeliniflorae and, perhaps, the Triuridiflorae, the last superorder being alternatively referred to group 1; (3) the Areciflorae.

However, we find that the rank of subclass is better reserved for the monocotyledons as such, subclass Liliidae being thus equivalent to the Monocotyledonae. This would make the angiosperms comprise a class, the Magnoliophyta ("Magnoliatae") comparable to the gymnosperm classes Pinophyta, Ginkgophyta, Cycadophyta etc.

Superorder Alismatiflorae

Mostly rhizomatous aquatics, marsh herbs or saprophytes. Leaves opposite or alternate; bracteate, band-shaped or differentiated into petiole and lamina. Intravaginal squamules (a type of multicellular hairs) usually present. Stipules or stipule homologues common. Stomata usually lacking or if present the guard-cells enclosed by two subsidiary cells. Laticiferous ducts sometimes present. Silica bodies and oxalate raphides absent. Vessels lacking in stem and leaves.

Flowers actinomorphic or more rarely slightly zygomorphic, outer tepals sometimes sepaloid or perianth often reduced. Stamens from solitary to numerous with basifixed, extrorse or more rarely latrorse anthers. Tapetum amoeboid, with initially uninucleate cells. Microsporogenesis successive. Pollen grains usually single and inaperturate or sulcate to foraminate, generally trinucleate. Gynoecium generally apocarpous or monocarpellate, inferior in Hydrocharitaceae, but otherwise superior, often with bibrachiate stylodia. Stigmas dry. Placentation laminal, lateral or basal; most genera outside Hydrocharitales with one ovule per carpel only. Embryo formation in all cases known according to the Caryophyllad Type. Endosperm formation helobial or (Alismataceae *pro parte*, some families of Zosterales, and Triuridales and Najadales) nuclear, and endosperm consumed in the seeds of all taxa but Triuridales.

Fruits follicles, achenes or rarely drupes with nonstrophiolate, exarillate seeds, containing a straight or curved, linear, often macropodous, frequently chlorophyllous embryo.

The superorder Alismatiflorae roughly corresponds to the "Helobiae" in older literature. It includes marsh plants and freshwater as well as marine aquatics distributed over most of the world. The following five orders may be recognized. Hydrocharitales is often not separated from Alismatales, at least not with the same circumscription.

Hydrocharitales consist of marsh as well as freshwater and marine plants, lacking laticiferous ducts and having hypo- or epigynous flowers, often (in Hydrocharitaceae) with sepaloid outer and petaloid inner tepals.

Reduction in tepal number in Aponogetonaceae. Stamens few to numerous, the pollen grains sulcate or (in genera of Hydrocharitaceae only) inaperturate. Placentation laminal or basal.

Styles sometimes bifurcate, ovules crassinucellate embryo sac monosporic. Fruit variable. Seeds with straight embryo.

Anthocyanin pseudobases often present.

> **Butomaceae** (one genus: *Butomus*), mostly European distribution
> **Aponogetonaceae** (one genus: *Aponogeton*), tropic–subtropic Old World distribution
> **Hydrocharitaceae** (incl. *Thalassiaceae* and *Halophilaceae*; 15 genera). The reduction in rank of Thalassiaceae and Halophilaceae to subfamilies may be questioned but they are included in Hydrocharitaceae largely because of their epigynous perianth. Thalassioideae has globose and Halophiliodeae subfiliform pollen grains. Cosmopolitan distribution.

Alismatales. This order, consisting largely of marsh plants and aquatics is subcosmopolitan. It is related most closely to the Hydrocharitales, especially the Butomaceae, and is characterized by secretory ducts, hypogynous flowers, often green, sepaloid outer and petaloid inner tepals, pollen grains foraminate, with 2–30 circular apertures. Ovules without parietal cell but the epidermis of nucellus dividing periclinally. Carpels numerous, developing into achenes or rarely follicles. Embryo sac bisporic (formed according to the Allium type), and embryo curved, horseshoe-like.

Anthocyanin pseudobases lacking. One family, Alismataceae (incl. Limnocharitaceae; c. 14 genera). Limnocharitaceae has tentatively been included in this family, but may deserve separate status. It has all the ordinal characters and must not be included in Butomaceae, as it has been sometimes.

Zosterales. Like the previous order this contains marsh as well as freshwater and marine plants. Secretory ducts sometimes present. Leaves variable, often stipulate. Hairs (apart from intravaginal squamules) and stomata often lacking.

Flowers often in spike-like inflorescences varying from complete to very reduced. Perianth sometimes considered to be absent, being substituted in some families by structures questionably regarded as "connective appendages"; never differentiated into outer sepaloid and inner petaloid members. Stamens variable in number (6–1),

with inaperturate, globose or filiform pollen grains. Carpels 6–1, free or (in Juncaginaceae) fused in centre. Ovules 1–2 per carpel, anatropous to orthotropous, with or without parietal cell, sometimes with periclinal divisions in the epidermis of the nucellus. Embryo sac monosporic. Fruits variable.

Rhodoxanthin sometimes present.

The order has a world-wide distribution, some of the families being temperate-boreal in distribution; others, especially the marine aquatics, tropical or subtropical. Within the order there is a series of variation from marsh plants with more or less complete flowers to specialized aquatics with most reduced, unisexual flowers.

> **Scheuchzeriaceae** (one genus: *Scheuchzeria*).
> **Juncaginaceae** (incl. Lilaeaceae; 4 genera: *Triglochin, Maundia, Tetroncium, Lilaea*). *Lilaea*, sometimes treated in its own family, has polygamic monomerous flowers and is sometimes regarded as primitive, sometimes as advanced.
> **Potamogetonaceae** (incl. *Ruppiaceae*; 2–3 genera: *Potamogeton* incl. *Groenlandia, Ruppia*). *Ruppia*, sometimes treated in a separate family, differs mostly in having small staminal appendages (? = tepals), 2 rather than 4 stamens and (4 or more) longstipitate fruits.
> **Zosteraceae** (3 genera: *Zostera, Heterozostera, Phyllospadix*).
> **Posidoniaceae** (one genus: *Posidonia*).
> **Zannichelliaceae** (4 genera: *Zannichellia, Althenia, Lepilaena, Vleisia*).
> **Cymodoceaceae** (6 genera: *Cymodocea, Diplanthera, Syringodium, Amphibolis, Halodule, Thalassodendron*).

Najadales consist of fresh or brackish water aquatics with opposite (pseudo-verticillate) leaves which are dentate to entire. Stomata lacking.

Flowers unisexual: the male unistaminate, the stamen enclosed in an apically two-lipped sheath and supported by two scales; the female naked or supported by a spathe, probably monocarpellate. Pollen grains inaperturate, ellipsoidal. Pistil with 2–4 stylar lobes and a single, basal, anatropous, crassinucellate ovule. Endosperm formation nuclear. Fruit an achene. A sub-cosmopolitan genus of water plants; often included in the former order.

> One family: **Najadaceae** (one genus: *Najas*).

Triuridales consist of small, mycorrhizal saprophytes lacking chlorophyll. Leaves small, bract-like or lacking.

Flowers small, actinomorphic, usually unisexual, with 3 or 6 (rarely 4 or 10) tepals sometimes extended as tails. Stamens 2 to 6, with short filaments; anthers bi- or tetrasporangiate, sometimes dehiscing transversely. Pollen grains smooth, inaperturate. Carpels numerous, small, free, each with a single stylodium which may be terminal, lateral or gynobasic. Ovules solitary, basal, anatropous, tenuinucellate, without parietal cell. Endosperm formation nuclear. Fruits small achenes, single seeded with a fatty endosperm.

The taxonomic position of Triuridales is doubtful. It is tentatively placed here though it shows similarities to both Alismatiflorae (e.g. Alismataceae) and Liliiflorae (e.g. Melanthiaceae). A most interesting family of frequently tiny little parasites growing in the tropics of America as well as Africa and Asia.

> One family: **Triuridaceae** (7 genera: *Triuris, Peltoptoyllum* (*Heturis*), *Andruris, Sciaphila, Seychellaria, Hyalisma, Soridium*)

Superorder Ariflorae

Mostly rhizomatous herbs, including many tropical climbers, rarely weakly arborescent. Roots sometimes with velamen. Stem and leaves vesselless; secondary thickening growth lacking. Leaves alternate, often differentiated into petiole and blade, sometimes digitate or palmately lobate or compound (e.g. *Anthurium* spp.). Ligule in *Calla*; intravaginal squamules known in some species of *Philodendron*. Stomata mostly with 4 or more, rarely 2, subsidiary cells. Laticifers common. Silica bodies lacking but oxalate raphides widely distributed. Inflorescence a simple spadix supported by a spathe.

Flowers minute, hypogynous, actinomorphic, bi- or unisexual, with or without perianth. This when present mostly with 3 + 3 or 2 + 2 tepals, which are scale-like to subprismatic. Stamens 0 to 6, with basifixed mostly extrorse, sometimes poricidal anthers. Tapetum amoeboid, initially with 1- or less often 2–4-nucleate cells. Microsporogenesis successive. Pollen grains variable, inaperturate, sulcate or 2–4-sulculate or 3–4-forate, and either 2- or 3-nucleate. Pistil 1–3-carpellate, with as many locules, mostly styleless, with variable placentation. Ovules with or without parietal cell, often with periclinal divisions in the nucellus epidermis. Endosperm formation cellular, generally with chalazal haustorium cell. Embryo formation generally of Caryophyllad (or in Lemnaceae of Onagrad) type.

Fruits generally berries, the seeds with or without endosperm. Endosperm when present at least frequently with starch. Embryo straight or curved, often macropodous.

Arales as above. The order is large and widely distributed, although most concentrated in the tropics. It is more frequently associated with the Areciflorae, but does not seem to have any close affinities with this group.

> **Araceae** (incl. *Acoraceae* and *Pistiaceae* p.p.; c. 110 genera).
> **Lemnaceae** (6 genera: *Spirodela, Lemna, Wolffia, Pseudowolffia, Wolffiopsis, Wolffiella*).

Superorder Liliiflorae

Herbs, rarely shrubs, or woody, sparingly branched "trees", without, or in several families with, secondary thickening growth (of a different type to that of dicotyledons). Stems underground mostly modified as rhizomes, corms or bulbs. Roots usually with vessels, but stems and leaves with vessels only in certain families. Leaves usually alternate or rarely opposite or verticillate, linear to lanceolate, sessile, generally sheathing at base, in some groups petiolate with a distinct lamina, if so this entire or rarely (sometimes in Dioscoreaceae and Taccaceae) compound or digitately lobate. Leaf blade inverted in some genera. Venation mostly parallel, reticulate in few genera. Intravaginal squamules lacking; ligules or analogous structures rare. Stomata generally without subsidiary cells (excepting most orchids and the orders with regularly starchy endosperm). Silica bodies mostly lacking (except in Bromeliaceae and certain orchids), but oxalate raphides commonly present.

Flowers hypo- or epigynous, actinomorphic or zygomorphic, tepals normally all petaline, sometimes conspicuously spotted, only in rare cases the outer perianth whorl sepaline. Nectaries frequently either in septa of pistils or at the bases of tepals or stamens. Stamens mostly 6, 3 or (in Orchidales and Philydrales) 2–1; rarely up to 9 or more (e.g. *Vellozia, Pleea*). Anthers tetrasporangiate or (Smilacaceae) bisporangiate and tapetum generally secretory. Microsporogenesis successive or simultaneous. Pollen grains single but tetrads present in orchids, some Velloziales and a few genera of other orders. Binucleate, sulcate pollen grains present in most families, but aperture conditions and nucleus number variable. Gynoecium syncarpous (approaching apocarpous in, for example, *Petrosavia* and *Protolirion*), stylodia free or fused to a single style; stigmas dry or wet. Placentation variable, the ovules mostly anatropous to campylotropous, with or without a parietal cell, nucellus epidermis with or without periclinal divisions. Endosperm formation helobial or nuclear.

Fruits generally capsules or berries, the seeds few to several, sometimes with elaiosome, those of Asparagales generally with a phytomelan crust. Endosperm absent in Orchidales, otherwise present, non-starchy in most members but starchy in several orders, only rarely ruminate. Embryogenesis mostly of Onagrad or Asterad types. Embryos non-macropodous, mostly straight, chlorophyllous mainly in orchids.

Steroid saponins common.

This superorder can be variously circumscribed. Of the ten orders included here, five have seeds with starchy endosperm. These are the Velloziales, Bromeliales, Pontederiales, Philydrales and Haemodorales, some of which are frequently associated with the orders placed here in the Commeliniflorae.

The Liliiflorae as circumscribed here is subcosmopolitan, but some of its orders have restricted distributions.

Dioscoreales largely consists of terrestrial shade plants as well as climbers and creepers, often having basal or cauline tubers. Bulbs lacking. Vessels often present in the stem, but secondary thickening growth mostly lacking. Leaves alternate, opposite or verticillate, petiolate or sessile, sometimes with "stipules" and with entire or compound, frequently net-veined, lamina. Multicellular hairs of variable types sometimes present, stomata mostly without subsidiary cells, sometimes (in contrast to other monocotyledons) not arranged parallel to the midribs but in different directions. Oxalate raphides common.

Flowers actinomorphic, hypo- or frequently epigynous and bi- or frequently unisexual, in some genera 2-, 4- or up to 7-merous. Tepals often greenish and similar, but sometimes different in the two whorls the outer being sepaline in Trilliaceae; not spotted. Stamens (in bisexual or male flowers) in two whorls or (some being staminodial) in one whorl, often with connate filaments; anthers basifixed, mostly introrse, in some genera with connectives protracted as tails. Microsporogenesis simultaneous or more rarely successive, the pollen grains sulcate or more often 2 (–3)-sulculate. Ovules several per carpel and mostly anatropous, semicrassinucellate, with a parietal cell; endosperm formation except in some Trilliaceae nuclear.

Fruit capsular or baccate, with estrophiolate or strophiolate seeds generally having a copious non-starchy endosperm and a minute embryo.

Steroid saponins common.

This order is dubiously homogenous, Trilliaceae being peripheral. The Dioscoreaceae have a wide, tropical to subtropical distribution, while the Stemonaceae have their centre in Indomalesia (*Croomia* in Florida) and the Trilliaceae in the temperate-boreal regions of the northern Hemisphere.

Dioscoreaceae (incl. Stenomeridaceae and Trichopodaceae; 6–9 genera: *Avetra, Bordera, Dioscorea, Epipetrum. Higinbothamia, Rajania, Stenomeris, Tamus, Trichopus*). The two segregates Stenomeridaceae (*Stenormeris, Avetra*) and Trichopodacea (*Trichopus*) are dubiously distinct; they have bisexual flowers with stamens protracted into connective tails; they also differ from Dioscoreaceae *sensu stricto* in pollen apertures, and some other minor characters.

Stemonaceae (= Roxburghiaceae; incl. Croomiaceae; 3 genera: *Stemona, Srichoneuron, Croomia*).

Trilliaceae (3 genera: *Paris, Trillium* and *Scoliopus;* the last genus according to Berg (1962b) is rather isolated).

Taccales consists of perennial herbs with tubercular starch-rich rhizomes. Stem vesselless and leaves petiolate with simple and entire or palmately lobate to compound lamina. Uniseriate hairs and oxalate raphides present. Inflorescence a pseudo-umbel, supported by an involucre of a few broad leaves and several long filiform bracts.

Flowers epigynous, bisexual, actinomorphic, campanulate, with 6 subequal dull-coloured tepals and 6 epitepalous stamens. Anthers broad, introrse, with rounded connective. Anther wall formation of Dicotyledonous type; tapetum secretory. Microsporogenesis simultaneous. Pollen grains free, sulcate, binucleate. Nectar secretion probably lacking. Style broad, apically trilobate, with dry stigmas. Ovary unilocular, with parietal placentae. Ovules anatropous, with parietal cell. Endosperm formation nuclear.

Fruits baccate or capsular, with prismatic, exarillate seeds having non-starchy endosperm.

Only one, monogeneric family, **Taccaceae** (with the genus *Tacca,* incl. *Schizocapsa*; Drenth, 1972).

The Taccales comprise a pantropic group of peculiar-looking terrestrial herbs, whose affinity is probably closer to the Dioscoreales than to the Amaryllidaceae of the Asparagales, near which they are sometimes placed.

Asparagales, a large order of herbs, shrubs or rosette trees with rhizomes, tubers, corms or often bulbs. Roots sometimes with velamen and containing, as a rule, vessels with scalariform to simple perforation plates. Stems herbaceous or more rarely woody, in certain families with secondary thickening growth; vessels mostly lacking but sometimes present in the stem (and more rarely the leaves). Stomata mostly without (more rarely with 2) subsidiary cells. Trichomes rather rare. Leaves usually linear to lanceolate, sometimes succulent, sessile or more rarely petiolate and then with cordate to lanceolate lamina; stipules generally lacking; ligules uncommon (single taxa of Asphodelaceae and Alliaceae); laticifers very rare (present only in some Alliaceae). Oxalate raphides common. Inflorescence variable, on a leafless scape or a leafy stem.

Flowers mostly with a perianth petaline in both whorls, almost never with a drop-like colour pattern, actinomorphic or zygomorphic, and more often hypogynous than epigynous. Nectaries generally in the septa of the ovary, being absent from tepals or stamens. Stamens generally 6, with introrse or more rarely extrorse tretasporangiate or rarely (Smilacaceae) bisporangiate anthers. Tapetum mostly secretory (perhaps amoeboid in Hypoxidaceae). Microsporogenesis successive except in certain families (especially subfam. Anthericoideae of Asphodelaceae). Pollen grains single, in most cases sulcate (or more rarely inaperturate) and binucleate. Gynoecium generally tricarpellate, mostly with a single style (free stylodia or subsessile stigmas found for example in Smilacaceae), and with wet or dry stigmas. Placentation generally axile and ovules generally anatropous or campylotropous (rarely orthotropous), sometimes with a parietal cell campylotropous (rarely orthotropous), with or sometimes without a parietal cell. Nucellus epidermis rarely with periclinal divisions. Endosperm formation either helobial or nuclear.

Fruits capsular or in certain families baccate or rarely otherwise, with seeds having more often than not a black phytomelan crust. Arils present, for example in some Asphodelaceae, and various elaiosomes not rare. Endosperm normally not starchy; embryo mostly linear, non-macropodous, non-chlorophyllous and only rarely curved.

Steroid saponins frequently present.

The Asparagales is a large order comprising more than half of the species in what is frequently conceived as the order Liliales. The circumscription of the order is somewhat diffuse, and the affinities of families such as Smilacaceae and Petermanniaceae is perhaps more with the Dioscoreaceae, while there are some peculiar similarities between some Philesiaceae and the Alstroemeriaceae of the Liliales. The homogeneity especially of the families Philesiaceae and Asphodelaceae is dubious. The main features of this classification of the order are according to Huber (1969).

Asparagales is a subcosmopolitan order, with typical North Hemisphere as well as extreme South Hemisphere families (e.g. the Convallariaceae and Philesiaceae, respectively).

Smilacaceae (4 genera: *Smilax, Heterosmilax, Pseudosmilax* and *Ripogonum*). *Ripogonum* deviates in lacking tendrils, in having opposite leaves and bisexual flowers in spike-like inflorescences and in having sulcate rather than inaperturate pollen grains and starchy endosperm.

Petermanniaceae (one genus: *Petermannia*).

Philesiaceae (incl. *Luzuriagaceae*; 6 genera: *Lapageria* and *Philesia* forming subfam. Philesioideae; *Behnia, Eustrephus, Geitonoplesium* and *Luzuriaga* forming subfam. Luzuriagoideae.) There are certain conspicuous differences between the two subfamilies. The former has almost sepaline outer tepals with drop-like patterning on the inner tepals, nectaries at the tepal bases, inaperturate pollen grains etc., in which it differs from the subfam. Luzuriagoideae. Thus there is some support for Huber's recognition of both families.

Convallariaceae (22 genera: *Aspidistra, Campylandra,*

Clintonia, Convallaria, Disporopsis, Disporum, Gonioscypha, Liriope, Lourya, Maianthemum, Oligobotrya, Ophiopogon (= *Mondo*), *Peliosanthes, Polygonatum, Reineckea, Rhodea, Smilacina, Speirantha, Streptopus, Theropogon* and *Tupistra*). *Drymophila* often referred here may be better placed in Philesiaceae subfam. Luzuriagoideae. At least *Disporum* and *Clintonia* show embryological features strongly indicating affinity rather with Colchicaceae and Liliaceae of the Liliales (Björnstad, 1970). However, until more genera have been studied a division of the family has not been undertaken.

Asparagaceae (incl. Ruscaceae; 6 genera: *Asparagopsis, Asparagus* and *Myrsiphyllum* forming subfam. Asparagoideae; *Danaë, Ruscus* and *Semele* forming the subfamily Ruscoideae). The subfamily Ruscoideae is distinct in having a staminal column with extrorse anthers, and in having subsessile stigmas and almost orthotropous ovules.

Herreriaceae (one genus: *Herreria*).

Dracaenaceae (incl. Nolinaceae; 6 genera: the Old World *Dracaena* and *Sansevieria* forming subfam. Dracaenoideae; and the New World *Beaucarnea, Calibanus, Dasylirion* and *Nolina* forming subfam. Nolinoideae). The Nolinoideae lack true oxalate raphides and have pseudoraphides, the guard cells of their stomata have oil contents, and their mostly unisexual flowers have free tepals. Moreover the ovaries are unilocular and may form an indehiscent fruit. Possibly the subfamily should be treated as a distinct family as suggested by Huber (1969).

Doryanthaceae (one genus: *Doryanthes*; possibly also *Herpolirion* should be included).

Dasypogonaceae (8 genera: *Acanthocarpus, Baxteria, Calectasia, Chamaexeros, Dasypogon, Kingia, Lomandra* and *Romnalda*). These form a variable assemblage and may prove to be divisible into natural groups, as suggested by Stevens (1978) or by Chanda and Ghosh (1976).

Phormiaceae (1–3 genera: *Phormium*, maybe also *Blandfordia* and *Xeronema*).

Xanthorrhoeaceae (one genus: *Xanthorrhoea*).

Agavaceae (c. 12–14 genera: *Agave, Beschorneria, Bravoa, Furcraea, Littaea, Manfreda, Polianthes* and *Pseudobravoa* forming subfam. Agavoideae; and *Clistoyucca, Hesperaloë, Hesperocallis, Hesperoyucca, Samulea* and *Yucca* forming subfam. Yuccoideae). The generic circumscriptions in the family may need to be reviewed.

Hypoxidaceae (9 genera: *Curculigo, Empodium, Forbesia, Hypoxis, Ianthe, Molineria, Pauridia, Rhodohypoxis* and *Spiloxene*).

Asphodelaceae (incl. Asteliaceae and Anthericaceae; c. 54 genera divisible among the three subfamilies Astelioideae, 4 genera, Asphodeloideae, 18 genera, and Anthericoideae, 32 genera. The genera are as follows: Subfam. Astelioideae: *Astelia, Cohnia, Cordyline* and *Milligania*. Subfam. Asphodeloideae: *Aloë, Alectorurus, Aprica* (*Astroloba*), *Asphodeline, Asphodelus, Bulbine, Bulbinella, Bulbinopsis, Chamaealoë, Chortolirion, Eremurus, Gasteria, Glyphosperma, Haworthia, Kniphofia* (incl. *Nothosceptrum*), *Lomatophyllum, Trachyandra, Verinea*. Subfam. Anthericoideae: *Agrostocrinum, Alania, Ane-*

marrhena, *Anthericum* (excl. *Trachyandra*), *Arnocrinum*, *Arthropodium*, *Borya*, *Bottinaea*, *Caesia*, *Chamaescilla*, *Chlorophytum*, *Corynotheca*, *Dasystachys*, *Debesia*, *Dichopogon*, *Diuranthera*, *Echeandia*, *Eremocrinum*, *Hensmania*, *Hodgsonia*, *Ixiolirion* (?), *Johnsonia*, *Kopakowskia* (?), *Laxmannia*, (*Bartlingia*), *Nanolirion*, *Paradisia*, *Pasithea*, *Simethis*, *Sowerbaea*, *Stawellia*, *Terauchia*, *Thysanotus*, *Tricoryne*, *Verdickia*). The Astelioideae and Asphodeloideae seem to be fairly natural groups (among the latter of these the succulent-leaved Aloë-group is an integral part), while according to Huber (personal communication) subfam. Anthericoideae may be an artificial assemblage. The three groups are treated by Huber (1969) as families and there may be reasons for doing so, although the distinction between them is not altogether sharp. Anthericoideae have articulate flowers (mostly with periclades) and they generally have free, mostly white or rose-coloured tepals, campylotropous ovules, successive microsporogenesis, and exarillate seeds; and steroidine saponins are widely distributed. These characters are in contrast to all or most taxa of subfam. Asphodeloideae. However, the Johnsonia and Caesia groups as well as *Ixiolirion*, *Kopakowskia* and *Simethis* of the Anthericoideae seem to be aberrant or transitional.

Aphyllanthaceae (one genus: *Aphyllanthes*).

Dianellaceae (5 genera: *Dianella*, *Stypandra*, *Styponema*, *Excremis*, *Rhuacophila*). *Dianella* and *Rhuacophila* have baccate fruits and non-hairy filaments, *Styponema*, *Stypandra* and *Excremis* capsules and woolly filaments.

Tecophilaeaceae (6–7 genera: *Conanthera*, *Cyanella*, *Lanaria*, *Odontostomum*, *Tecophilaea*, *Walleria* (?) and *Zephyra*). *Walleria* may be better placed in the Eriospermaceae. The proper position of *Lanaria* in this family is supported by embryological evidence (de Vos, 1963).

Cyanastraceae (one genus: *Cyanastrum*). The peculiar feature of the chalazosperm and other details seem to justify a separate position in relation to the Tecophilaeaceae, which the genus may approach most closely.

Eriospermaceae (one genus: *Eriospermum*).

Hemerocallidaceae (one genus: *Hemerocallis*).

Funkiaceae (= Hostaceae) (two or three genera: *Hosta*, *Hesperocallis* and probably *Leucocrinum*).

Hyacinthaceae (= "Scilloideae") (c. 44 genera: *Albuca*, *Alrawia*, *Amphisiphon*, *Androsiphon*, *Bellevalia* (incl. *Strangweya*), *Bowiea*, *Brimeura*, *Camassia*, *Chionodoxa*, *Chlorogalum*, *Daubenya*, *Dipcadi*, *Drimia*, *Drimiopsis*, *Eucomis*, *Fortunatia*, *Galtonia*, *Hemiphylacus*, *Hyacinthella*, *Hyacinthoides* (= *Endymion*), *Hyacinthus*, *Lachenalia*, *Ledebouria*, *Liriothamnus*, *Litanthus*, *Massonia*, *Muscari* (incl. *Botryanthus*, *Leopoldia* and *Pseudomuscari*), *Neobakeria*, *Neopatersonia*, *Ornithogalum*, *Periboea*, *Polyxena*, *Prospero*, *Pseudogaltonia*, *Puschkinia*, *Rhadamanthus*, *Rhodocodon*, *Scilla* (should probably be divided), *Schoenolirion*, *Schizobasis*, *Thuranthus*, *Urginea*, *Veltheimia*, *Whiteheadia*). Of these genera, *Bowiea* forms one deviating group and *Schoenolirion* and *Chlorogalum* another; the latter may prove so distinct that it may deserve separate family rank, but seems to be better placed here than in any of the other families here recognized.

Alliaceae (incl. *Agapanthaceae* and *Gilliesiaceae*; c. 32 genera divisible into 3 or more subfamilies: Subfam. Agapanthoideae: *Agapanthus*, *Tulbaghia*. Subfam. Allioideae: *Allium*, *Androstephium*, *Behria*, *Bezzera*, *Bloomeria*, *Brevoortia*, *Brodiaea*, *Dandya*, *Dichellostemma*, *Diphalangium*, *Latace*, *Leucocoryne*, *Milula*, *Muilla*, *Nectaroscordum*, *Nothoscordum*, *Petronymphe*, *Steinmannia*, *Stropholirion*, *Tristagma*, *Triteleia*, *Triteleiopsis*. Subfam. Gilliesioideae: *Ancrumia*, *Eriuna*, *Gethyum*, *Gilliesia*, *Ipheion* (= *Milla*), *Miersia*, *Solaria*, *Speea*, *Trichlora*).

Amaryllidaceae (excl. the Ixiolirieae; c. 50 genera). See for example Traub (1963).

Liliales consist of herbs or rarely shrublets; a few genera (of Iridaceae subfam. Aristeoideae) with secondary thickening growth. Underground stems modified to corms, bulbs or rhizomes. Vessels confined mainly to the roots, lacking in stems and leaves. Stomata usually without subsidiary cells; multicellular trichomes rare. Laticifers rare or absent. Leaves usually linear to lanceolate, sessile, parallel-veined, often ensiform, without stipules or ligules. Oxalate raphides only in certain groups; silica bodies absent. Inflorescences variable, only rarely on a leafless scape.

Flowers hypogynous or very often epigynous, with both whorls of the perianth petaline and rather often with a drop-like colour pattern, actinomorphic to zygomorphic. Nectaries often present at the base of tepals and stamens, septal nectaries present in Iridaceae subfam. Ixioideae, in Alstroemeriaceae and in some Melanthiaceae. Stamens 6 or 3, with introrse or more often extrorse, tetrasporangiate anthers. Tapetum secretory; microsporogenesis successive or (in, for example, Iridaceae) simultaneous. Pollen grains never in tetrads, generally sulcate (but other types occur) and binucleate. Carpels 3, generally fused in the ovary region but often apically with free stylodial branches, these sometimes bifurcate or further branched. Stigmas wet or (especially in Iridaceae) dry. Placentation generally axile, with as a rule anatropous ovules with (Iridaceae and many Melanthiaceae) or without parietal cell (tissue). Endosperm formation nuclear or more rarely (Melanthiaceae) helobial.

Fruits nearly always capsular, with few to numerous seeds. These never with phytomelan and only rarely with starchy endosperm; embryo linear.

Steroid saponins present in some families (rarer than in Asparagales).

This order is relatively homogeneous although showing trends in the directions of the Dioscoreales, Asparagales, Burmanniales and Orchidales. The Melanthiaceae deviate in some features, and the Alstromeriaceae in others, from the other families. *Medeola* and *Scoliopus* seem to connect Liliaceae of this order with Trilliaceae of Dioscoreales.

Liliales with the families Iridaceae and Alstroemeriaceae show South Hemispheric concentration (mainly South Africa and exclusively South America, respectively), while the other families are mostly North Hemispheric.

Iridaceae (incl. Isophysidaceae; c. 70 genera). We have excluded *Campynema* and *Campynemanthe*, which have 6 stamens but epigynous flowers, from Iridaceae and placed them tentatively in Colchicaceae. *Isophysis* (*Hewardia*) has hypogynous flowers but is, with some hesitation, included in Iridaceae.

Geosiridaceae (one genus: *Geosiris*). A saprophyte, dubiously distinct from Iridaceae.

Colchicaceae (incl. Campynemataceae, see Iridaceae above; c. 20 genera: *Androcymbium, Anguillaria, Baeometra, Burchardia, Campynema, Campynemanthe, Colchicum* (incl. *Bulbocodium*), *Dipidax, Gloriosa, Hexacyrtis, Iphigenia, Kreysigia, Littonia, Merendera, Neodregea, Ornithoglossum, Reya, Sandersonia, Schelhammera, Uvularia, Wurmbea*). It is very likely that this family (or a separate one) should also include a few genera now placed in Convallariaceae, such as *Clintonia* and *Disporum*, which is strongly indicated by their embryological details (Björnstad, 1970; see further on p. 310).

Alstroemeriaceae (4 genera: *Alstroemeria, Bomarea, Leontochir, Schickendantzia*).

Calochortaceae (one genus: *Calochortus*).

Tricyrtidaceae (one genus: *Tricyrtis*, syn. *Brachycyrtis*).

Liliaceae s.str. (14 genera: *Cardiocrinum, Eduardoregelia, Erythronium, Fritillaria, Gagea, Giraldiella, Korolkovia, Lilium, Lloydia, Medeola, Nomocharis, Notholirion, Rhinopetalum, Tulipa*). *Medeola* is referred here with some misgivings on the grounds presented by Berg (1962a and b).

Melanthiaceae (incl. *Petrosaviaceae*; c. 24 genera: *Aletris, Amianthium, Chamaelirium, Chionographis, Helonias, Heloniopsis, Melanthium, Metanarthecium, Narthecium, Nietneria, Oceanoros, Petrosavia, Plea, Protolirion, Schoenocaulon* (= *Sabadilla*), *Stenanthella, Stenanthium, Tofieldia, Toxicoscordion, Tracyanthus, Veratrum, Ypsilandra, Xerophyllum, Zigadenus*). The recognition of the family Petrosaviaceae for the two saprophytic genera *Petrosavia* and *Protolirion* does not seem to be justified.

Burmanniales. Small, saprophytic and colourless or more rarely green and autotrophic herbs, the perennial taxa with rhizomes or tubers. Vessels present in roots and in some Burmanniaceae also in stem and leaves. Leaves linear and parallel-veined or bracteate. Ligules and stipules lacking. Hairs lacking or unicellular. Stomata lacking or without subsidiary cells. Oxalate raphides sometimes present. Inflorescence cymose or racemose.

Flowers actinomorphic or in some Burmanniaceae and in Corsiaceae zygomorphic, half epigynous to epigynous, generally with two whorls of tepals; these similar or dissimilar and sometimes fused to a tube. Longitudinal wings sometimes present on the ovary. Tepals coloured or white to hyaline, sometimes bizarre in shape. Septal or placental nectaries sometimes present. Stamens 6 or 3, with basifixed tetrasporangiate, extrorse or introrse anthers. Tapetum secretory; microsporogenesis successive. Pollen grains free, normally sulcate or ulcerate, bi- or trinucleate. Stylodia free or fused into a style, placentation most often parietal, rarely axile. Ovules numerous, small, anatropous, lacking parietal cell. Endosperm formation helobial.

Fruit capsular, with numerous, minute, exarillate seeds lacking phytomelan and having an at least sometimes starchy endosperm and a minute embryo.

The members of this interesting order, being in the chlorophyll-less state a parallel of the Triuridales, are pantropic and grow mainly in rain forests. The three families are sometimes treated as tribes of one family, the Burmanniaceae, but at least the Corsiaceae seem to comprise a good family.

Burmanniaceae (up to c. 12 genera, some being probably synonymous): *Apteria, Burmannia, Campylosiphon, Cymbocarpa, Desmogymnosiphon, Dictyostegia, Dipterosiphon, Gymnosiphon, Hexapterella, Marthella, Miersiella, Ptychomeria*).

Thismiaceae (up to c. 10 genera, some being probably synonymous: *Afrothismia, Glaziocharis, Mamorea, Myostoma, Ophiomeris, Oxygyne, Sarcosiphon* (incl. *Bagnisia, Geomitra, Rodwaya*), *Scaphiophora, Thismia, Triscyphus, Triurocodon*).

Corsiaceae (2 genera: *Arachnites, Corsia*).

Orchidales, a large order of terrestrial or epiphytic herbs, in some cases climbers (*Vanilla* etc.) rarely growing to a length of many metres; saprophytes in various genera. Roots often forming carnose storage structures. Velamen common. Stems often fleshy and swollen, but bulbs lacking. Vessels present in roots and rarely (?) in stems (e.g. *Vanilla*), but not known in leaves. Leaves alternate to subopposite, distichous or not, generally linear to broadly ovate or circular, sometimes falsely petiolate, never compound, rarely ensiform. Stipules and ligules lacking. Stomata variable, with or without subsidiary cells. Spherical silica bodies sometimes present and oxalate raphides widely distributed. Laticifers lacking. Multicellular or unicellular hairs frequent, but many taxa glabrous.

Flowers zygomorphic, epigynous, usually resupinated by twisting of pedicel or ovary. Tepals petaline and frequently with spotted colour pattern, upper median one usually enlarged to form a labellum, this frequently but other tepals more rarely spurred. Nectar sometimes secreted at the tepal bases, but osmophores with volatile substances etc. probably more widespread in attracting insects. Stamens 3–1, usually only the median upper one fertile, fused to the style into a gynostemium (column). Anther(s) tetrasporangiate, pollen grains single or more often in tetrads, these frequently coherent in massulae or pollinia. Tapetum secretory; microsporogenesis simultaneous. Pollen grains binucleate, generally inaperturate but sometimes with one aperture. Style simple, two stigma lobes usually functional, with wet surface, one transformed apically into a "rostellum". Placentation mostly parietal with innumerable, small, mostly anatropous, tenuinucellate ovules. Parietal cell lacking. Endosperm not formed (or reaching only to a few-nucleate stage). Embryo minute.

Fruit a many-seeded capsule, with diminutive seeds lacking endosperm and having a small and at least often chlorophyllous embryo.

Alkaloids common, of various kinds.

The members are usually treated in a single family, Orchidaceae, from which Apostasiaceae may be distinguished as separate. Here also Cypripediaceae has been distinguished as a separate family (Vermeulen, 1966). The two anthers of this are of the outer staminal whorl, neither thus corresponding to the single anther developed in Orchidaceae *s. str.* (classification: see, for example, Dressler and Dodson, 1960; Dressler, 1974.)

The Orchidaceae, including or excluding the Cypripediaceae, together with the Asteraceae or Compositae comprise the largest family of flowering plants as regards number of species. They are chiefly tropical but numerous are subtropical, temperate and boreal, and some species are even arctic (occurring for example in Greenland).

Apostasiaceae (2 genera: *Apostasia* and *Neuwiedia*).
Cypripediaceae (4 genera: *Cypripedium, Paphiopedilum, Phragmipedium, Selenipedium*).
Orchidaceae (c. 750 genera).

Pontederiales. Rhizomatous and sometimes stoloniferous free floating or usually rooted herbs in swamps and waters, without secondary thickening growth. Leaves alternate, distichous, generally differentiated into a sheath, a petiole and an entire lamina. Ligules and stipule-like lobes sometimes present. Glandular hairs widespread. Stomata with two subsidiary cells. Vessels with scalariform perforation plates often present. Laticiferous vessels and silica bodies lacking; oxalate raphides sometimes present but styloids (pseudoraphides) more dominant. Inflorescences variable.

Flowers hypogynous, actinomorphic to zygomorphic, with petaline, usually white or blue tepals fused basally. Stamens 6 or by reduction 3 or 1, anthers tetra-sporangiate, basi- or dorsifixed, introrse, opening longitudinally or by pores. Tapetum secretory or amoeboid, microsporogenesis successive. Pollen grains free, binucleate, sulcate or bi- or tri-sulculate. Pistil syncarpous, often with septal nectaries (these may be absent or rudimentary), a single style and a dry stigma. Ovary trilocular (two locules sometimes empty) with axile placentation. Ovules one or several per locule, anatropous, with parietal cell; endosperm formation helobial.

Fruit a loculicidal capsule or (in *Pontederia*) a nut, the seeds with starchy endosperm and a linear straight embryo.

A pantropical order of aquatics, frequently with attractive, mostly bluish flowers. It probably has its closest relatives among the Philydrales and Haemodorales.

Pontederiaceae (9 genera: *Eichhornia, Eurystemon, Heteranthera, Hydrothrix, Monochoria, Pontederia, Reussia, Scholleropsis, Zosterella*).

Haemodorales. Erect, terrestrial, often hairy herbs with subterranean rhizomes or tubers. Roots and sometimes also stems with vessels. Leaves linear, distichous, often ensiform, parallel-veined. Hairs unicellular or uniseriate, sometimes branched and multicellular. Stomata paracytic. Silica bodies lacking but oxalate raphides common.

Flowers hypogynous or epigynous, actinomorphic to zygomorphic, with 3 + 3 more or less petaline tepals; these sometimes fused into a tube. Stamens 6 or 3, with tetrasporangiate, basifixed or dorsifixed, introrse anthers. Tapetum amoeboid; microsporogenesis successive. Pollen grains single, binucleate and usually sulcate or 2 (–8)-foraminate. Ovary tricarpellate, trilocular, with septal nectaries. Style single, bearing a small stigma with dry or wet surface. Ovules orthotropous to hemianatropous, with parietal cell (tissue). Endosperm formation helobial.

Fruit a loculicidal capsule, with variably shaped, sometimes hairy seeds with starchy endosperm and a small, globose-ovoid, non-chlorophyllous embryo.

No steroid saponins reported, but chelidonic acid present.

The order consists of a single family, which is frequently placed right among asparagalean families like Cyanastraceae, Tecophilaeaceae and Hypoxidaceae, and some incompletely known genera are maybe still incorrectly placed among the families. The distribution is largely South Hemispheric-disjunct, although in the New World it is scattered also over parts of North America.

Haemodoraceae (incl. Conostylidaceae: 16 genera divisible into two subfamilies. Subfam. Haemodoroideae *Barberetta, Dilatris, Haemodorum, Hagenbachia, Lachnanthes,*

Phlebocarya, Pyrrorhiza, Schiekia, Wachendorfia, Xiphidium. Subfam. Conostylidoideae: *Anigozanthos, Blancoa, Conostylis, Lophiola (?), Macropidia, Tribonanthes.*) *Lophiola* (with 3 stamens), which is here referred to subfam. Conostylidoideae, deviates from the 6-staminate genera also in having stomata without subsidiary cells and in having hairs and vascular conditions more similar to *Aletris* (Asphodelaceae); it also has sulcate pollen grains and erect ovules. Its seeds deserve further study. Perhaps it is best placed with *Lanaria* in or near Tecophilaeaceae (de Vos, 1963).

Philydrales. Erect, often large, perennial marsh or water herbs with rhizomes or "tubers". Vessels absent in stem and leaves. Leaves alternate, distichous, linear, usually ensiform, flat, parallel-veined, eligulate and exstipulate. Glandular hairs and uniseriate hairs with a long end cell common. Stomata paracytic or sometimes tetracytic. Oxalate raphides present in tapetal cells; styloids common in vegetative parts, silica bodies lacking. Inflorescences racemose.

Flowers bracteate, hypogynous, zygomorphic, with petaline tepals, the lateral of the outer whorl and median of inner whorl fused and forming a large upper lip and the median of the outer whorl forming a large lower lip, the lateral inner tepals being smaller. Septal nectaries lacking. Androecium with a single functional stamen having a dorsifixed, basically introrse (in *Philydrum* helically twisted) anther with 4 microsporangia. Tapetum secretory; microsporogenesis successive. Pollen grains free or in tetrads, binucleate, sulcate. Pistil tricarpellate, tri- or unilocular, with simple style and dry stigmatic surface. Placentation axile or intrusive-parietal; ovules anatropous, with parietal cell (tissue) and with helobial endosperm formation.

Fruit capsular (or in *Helmholtzia* a dry berry); seeds with endosperm starchy but containing also oil (and thus not "mealy"); embryo small, straight.

The order is restricted to south-eastern Asia, New Guinea and parts of Australia and comprises a single small family.

> **Philydraceae** (4 genera: *Helmholtzia, Orthothylax, Philydrella* and *Philydrum*).

Velloziales. Perennial, more or less xerophytic, sometimes arborescent with up to 6 m tall stem covered with persistent leaf sheaths and adventitious roots. Roots and also normally the leaves but more rarely the stem with vessels. Leaves generally spirally set, linear, sclerenchymatous, parallel-veined, often slightly dentate along the margins. Trichomes uni- or multicellular, often in tufts. Stomata para- or more rarely tetracytic. Oxalate raphides and silica bodies lacking, but mucilage cells and rhomboidal crystals present in some species.

Flowers epigynous, actinomorphic, with 3 + 3 petaloid and basally connate tepals. "Corona" structures opposite stamens common. Stames 3 + 3 or more (up to > 60 by "dedoublement"), in groups of 3 or more, with basi- or medifixed, introrse anthers. Tapetum secretory; microsporogenesis successive. Pollen grains solitary or rarely in tetrads, sulcate, binucleate. Septal nectaries present. Ovary globose, inferior, trilocular; style simple; stigma trilobate. Placentation axile, with numerous anatropous ovules. Parietal cell not formed. Endosperm formation probably helobial.

Fruit capsular, many-seeded; seeds with copious starchy endosperm and a small, ovoid embryo.

Velloziales comprise a single family distributed in tropical South America (mainly Brazil) and Africa. They are frequently associated with the asparagalean Hypoxidaceae, but this is not supported by recent studies.

> **Velloziaceae** (5 genera: *Barbacenia, Barbaceniopsis, Talbotia, Vellozia, Xerophyta*).

Bromeliales. Perennial herbs or rarely (*Puya* spp.) arborescent, terrestrial or epiphytic plants, with spirally set rosette leaves. Stems sometimes with vessels. Leaves generally stiff, succulent, sheathing at base, linear to ovate, often with dentate or spinulose margins. Leaves and stem glabrous or more often clothed with peltate or stellate, multicellular, water-absorbing hairs. Stomata often in furrows, with 2 subsidiary cells. Schizogenous ducts of mucilage occasionally present. Calcium oxalate raphides abundant and rounded silica bodies widespread. Inflorescence spicate, often with large, coloured bracts.

Flowers generally actinomorphic, hypo- or epigynous, with 3 + 3 petaline tepals, the outer generally much smaller than the inner, free or fused, often with fringed appendages between the stamens. Stamens 3 + 3, with dorsifixed, introrse, tetrasporangiate, longitudinally dehiscent anthers. Tapetum secretory; microsporogenesis successive. Pollen grains free (tetrads in *Cryptanthus*), sulcate or (in part of the subfamily Bromelioideae) bi- or triforaminate, in most cases binucleate. Septal nectaries present. Pistil tricarpellate, trilocular with superior or inferior ovary and with a single style and three often contorted stigmatic branches. Stigmatic surface dry or more often wet. Ovules anatropous, parietal tissue present; and endosperm formation helobial.

Fruits mostly septicidal capsules or berries with seeds having copious starchy endosperm and a small, straight embryo.

The order is a very natural and homogenous one. It is restricted to Southern and Central America with one (probably recent) outpost in western Africa (*Pitcairnia feliciana*). The single family is, however, rich in genera and species.

> **Bromeliaceae** (c. 50 genera divisible into three subfamilies). No problems seem to occur in the circumscription of this family.

Superorder *Zingiberiflorae*

Small to very large, generally perennial herbs, rarely shrubs or trees often with starch-rich rhizomes. Velamen absent. Vessels present in roots, more rarely in stems (some Strelitziaceae and Zingiberaceae and most Marantaceae) and probably rarely in leaves. No secondary thickening growth. Vertical aerial stem often short, when inflorescence-bearing often covered with bracteate leaves. Leaves alternate, frequently distichous, sheathing at base, usually petiolate, with a large and simple (or secondarily split) often broad and pinnately veined lamina. Ligules present in many Zingiberaceae. Intravaginal squamules lacking. Hairs lacking or unicellular or very rarely bicellular or uniseriate. Stomata generally with 2–6 subsidiary cells. Oxalate raphides present in families with 5–6 stamens, otherwise lacking. Silica bodies widely scattered in epidermal or other tissue, of variable shape and size. Inflorescence variable, often a spike or (as in Marantaceae) complex and possibly determinate.

Flowers epigynous or rarely (Lowiaceae) half-epigynous, zygomorphic or asymmetric, mostly with 3 + 3 petaline, usually basally connate tepals. These generally more inconspicuous than the petaloid staminodia in the unistaminal families (see below); in Musaceae 5 tepals fused to a sheath and one free; in Strelitziaceae the tepals sometimes strongly differentiated. Androecial members 3 + 3 or fewer, the functional stamens 5–6 in the first 4 families, one only in the remaining families, the others lacking or transformed into conspicuous petaloid staminodia. Anthers basifixed or in some Zingiberaceae dorsifixed, introrse, tetrasporangiate or in Cannaceae and Marantaceae bisporangiate (monothecic). Tapetum secretory or more rarely amoeboid; microsporogenesis successive. Pollen grains simple, usually (or always?) inaperturate (dubiously ulcerate or spiraperturate in few cases), binucleate or rarely trinucleate. Style always simple, with wet stigma. Ovary in all families except Zingiberaceae with septal nectaries, trilocular or more rarely unilocular, with axile or parietal, seldom basal placentas. Ovules anatropous (campylotropous in some Marantaceae), with parietal cell, and with helobial or nuclear endosperm formation. Embryo formation of Onagrad or Asterad types.

Fruit usually a loculicidal capsule, rarely a berry, nut or schizocarp; seeds generally arillate, with sometimes starchy endosperm in the 5–6-anthered families, without or with little endosperm but with copious starchy perisperm in one-anthered families. Embryo linear or capitate, sometimes curved.

This superorder contains a single order:

Zingiberales. This order generally has had the same circumscription in most taxonomic treatments during the last century. Heliconiaceae and Strelitziaceae are often included in Musaceae, and Costaceae in Zingiberaceae. There is some indication that the first four of the families, enumerated below, which have 5–6 functional stamens, form one group, and the last four families, with but one or a half functional stamen, form another.

The distribution for all families is mainly tropical: Costaceae, Zingiberaceae and Marantaceae being pantropical, Heliconiaceae and Cannaceae neotropical and Lowiaceae, Strelitziaceae (except *Phenacospermum*) and Musaceae paleotropical.

> **Lowiaceae** (one genus: *Orchidantha*, syn. *Lowia*).
> **Heliconiaceae** (one genus: *Heliconia*).
> **Musaceae** (2 genera: *Musa, Ensete*).
> **Strelitziaceae** (3 genera: *Ravenala, Phenakospermum, Strelitzia*).
> **Zingiberaceae** (45 genera).
> **Costaceae** (4 genera: *Costus, Dimerocostus, Monocostus, Tapeinocheilos*).
> **Cannaceae** (one genus: *Canna*).
> **Marantaceae** (c. 30 genera).

Superorder *Commeliniflorae*

Herbaceous, generally rhizomatous plants, including many "graminoids", none with secondary thickening growth, no saprophytes and but few specialized aquatics. Shrub habit occasional and low "trees" in a few genera. Roots never with velamen. Lateral roots in certain cases opposite phloem strands in pericycle (never in other monocots). Vessels usually present in roots, stems and leaves (exception Hydatellaceae). Leaves never opposite or verticillate, always scattered, alternate, di- or tristichous or with other phyllotaxy, in most cases differentiated into sheath and "lamina", only rarely petiolate, never compound, and always parallel-veined. Stipule-like lobes rare; ligules present in several groups. Intravaginal squamules lacking. Hairs lacking or unicellular or uniseriate. Stomata usually with two subsidiary cells (4 or more in most Commelinales). Oxalate raphides present only in Commelinales and Typhales, but silica bodies of characteristic shape present in several (most of the other) orders. Laticifers lacking. Inflorescences variable, often spike-like.

Flowers hypogynous, the perianth often greenish, scarious, hyaline or transformed into bristles or hairs or lacking altogether, but inner tepals in some groups (Commelinales, Eriocaulales) petaline, actinomorphic or zygomorphic; outer sometimes sepaline (Commelinales). Septal nectaries lacking. Stamens usually 6, 3 or solitary, rarely numerous or in dimerous whorls. Anthers dorsi- or basifixed, latrorse or introrse (extrorse in some Xyridaceae), tetrasporangiate or mostly bisporangiate in Restionales and some Eriocaulaceae. Tapetum secretory in most groups, periplasmodial or amoeboid in Commelinales and Typhales; microsporogenesis generally successive but simultaneous in Rapateaceae and in the Juncales (incl. perhaps Thurniaceae) and Cyperales. Pollen grains free or in tetrads (some Typhales, all Juncales and Cyperales), sulcate or much more often ulcerate, sometimes inaperturate or spiraperturate, bi- or trinucleate. Pistil tri- or by reduction bi- or monocarpellate, tri- to unilocular, with free stylodia or branched or simple style. Stigma dry in all families studied except in some Commelinaceae. Placentation variable; ovules often few or solitary, anatropous or orthotropous; parietal cell (tissue) present or not; nucellus epidermis sometimes with periclinal divisions. Endosperm formation nuclear, more rarely (Juncales, Typhales) helobial, or very rarely (Hydatellales) cellular. Embryo formation according to the Asterad or Onagrad types.

Fruits mostly few-seeded capsules or nutlets, the seeds always with copious starchy endosperm (except in Hydatellales which has starchy perisperm). Phytomelan crust lacking. Embryo small, straight (curved in some grasses and sedges), usually broad or capitate, in the grasses lateral to the endosperm, non-chlorophyllous.

Steroid saponins rare or lacking.

This superorder seems to form a fairly uniform assemblage, although Hydatellales and Typhales are different in various respects from the other orders. Also Commelinales, the nominal order, deviates in several respects from the others. The superorder approaches especially Bromeliales and other orders with starchy endosperm in the Liliiflorae, and it may be argued that such orders belong better in Commeliniflorae, as suggested by Thorne (1968, 1976).

This superorder with the exception of the Commelinales and probably of many (most?) Eriocaulales are characterized by the wind-pollination syndrome of characters (see p. 43). But also other features which have nothing to do with pollination (p. 299) connect the Commeliniflorean orders. In spite of this the superorder approaches certain orders with starchy endosperm of the Liliiflorae, in particular Bromeliales.

The Commeliniflorae form a supposedly uniform assemblage although Hydatellales and Typhales are so different in various respects from the other orders that

their position here may be doubted.

Climax groups in the superorder are especially the grasses (Poales) and sedges (Cyperales) which are probably not as closely related with each other as often believed. The Commeliniflorae are cosmopolitan and include taxa forming much of the grassland vegetation and the vegetation in moist habitats of the world.

Commelinales. Herbs with jointed, succulent stem and alternate leaves, which are flat, sheathing, non-petiolate, linear or lanceolate to ovate and parallel-veined. Vessels in stems and leaves with simple or (in Mayacaceae) scalariform perforation plates (lacking? in Cartonemataceae). Trichomes variable, unicellular or uniseriate; stomata usually with 4–6 subsidiary cells. Silica bodies sometimes present; spinulose, spherical or minute. Oxalate raphides of common occurrence. Inflorescences mostly cymose. Flowers actinomorphic or zygomorphic with outer tepals sepaline, inner coloured and petaline, often with oblique symmetry plane, one petal often reduced in size or suppressed. Stamens free, 3 + 3 or 3 (rarely 2 or 1), 3 (the upper) being often staminodial or one whorl being missing. Filaments often hairy. Anthers basifixed, introrse, tetrasporangiate, rarely poricidal. Tapetum periplasmodial. Microsporogenesis successive Pollen grains simple, usually sulcate, bi- or rarely trinucleate. Pistil 3-carpellate, 3- (rarely 2-) locular (in Mayacaceae unilocular) with a single, unbranched style and a punctuate or 3-lobate, dry or wet stigma. Placentas axile or (in Mayacaceae) parietal, with one to numerous anatropous, hemianatropous or orthotropous ovules with or rarely without parietal cell; nucellus epidermis sometimes with periclinal divisions. Endosperm formation nuclear. Fruit a capsule, rarely a berry. Seeds sometimes arillate. Embryo often conical, under a disc-like structure (embryostegia) beneath the seed coat. Embryological data known in detail only for Commelinaceae.

The Commelinales consist of two or three families. The order is somewhat unusual in the superorder Commeliniflorae by the showy, insect-pollinated flowers, which may even be zygomorphic. The distribution is wide and mainly tropical to subtropical, the two small families being restricted to Australia (Cartonemataceae) and Central and Southern America and part of Southern Africa (Mayacaceae) respectively. Cartonemataceae is mostly included in Commelinaceae.

Commelinaceae (c. 38 genera).
Cartonemataceae (one genus: *Cartonema*).
Mayacaceae (one genus: *Mayaca*).

Eriocaulales. Mostly perennial rhizomatous, scapose

herbs, sometimes with a short vertical subterranean stem, and generally with a distinct basal leaf rosette in distichous or other phyllotaxy. Vessels in stems with simple or (most Rapateaceae) scalariform perforation the leaves in most Rapateaceae without vessels. Leaves linear, sheathing, sometimes with a short pseudopetiole, lamina linear to lanceolate, parallel-veined, the sheath in Rapateaceae oblique. Ligule lacking. Hairs variable, unicellular or uniseriate. Stomata paracytic. Oxalate raphides lacking, but rounded druses of silica bodies present in at least Rapateaceae. Inflorescence capitate or spicate.

Flowers trimerous or in Eriocaulaceae often bimerous, actinomorphic or bisymmetic or because of the unequal outer tepals zygomorphic, outer tepals often bracteate, inner petaloid, thin and brightly coloured; especially the inner often fused basally. Stamens in two whorls or only the inner retained; anthers basi- or dorsifixed, introrse or extrorse, sometimes poricidal (1, 2 or 4 pores). Tapetum secretory (or probably periplasmodial in *Abolboda;* Tiemann, unpublished); microsporogenesis successive in Xyridaceae and Eriocaulaceae, simultaneous in Rapateaceae. Pollen grains simple, sulcate, bisulculate, zonisulculate or spiraperturate, trinucleate at least in Rapateaceae, otherwise binucleate. Pistil bi- or trilocular, either with simple or with basally simple and apically branched style; stigma dry (one record only). Ovules orthotropous or more rarely campylotropous or anatropous, with or without parietal cell (tissue); endosperm formation nuclear or probably helobial in *Abolboda* (Tiemann, unpublished).

Fruit capsular; seeds with copious endosperm and a small, lenticular or conical embryo located at the micropylar end.

The order consists of three families, all with tropical distributions. The position of the three families in one order can be doubted (cf. the extensive study of Hamann, 1961), but the families possess several features in common and no doubt form more or less allied, parallel lines near the Commelinales.

Xyridaceae and Eriocaulaceae are pantropic with the greatest concentration in South America, and Rapateaceae with the exception of one West African genus is restricted to this region with the greatest concentration in the Guayana Highland.

Rapateacea (16 genera).
Xyridaceae (incl. Abolbodaceae) (4 genera: *Abolboda, Achlyphila, Orectanthe, Xyris*). (See Maguire and Wurdeck, 1958). The sometimes distinguished Abolbodaceae (*Abolboda* and *Orectanthe*) has not been acknowledged here, but may prove distinct enough when embryological data are considered.
Eriocaulaceae (c. 14 genera).

Typhales. Perennial, rhizomatous, monoecious marsh or water plants with erect or floating stem. Vessels with scalariform perforation plates present in stems and leaves. Leaves cauline, distichous, linear, flat or triangular in transection, sheathing at base. Hairs generally lacking. Stomata paracytic. Oxalate raphides present in the vegetative parts but silica bodies lacking. Inflorescences compound, globose or cylindrical with secondary (and sometimes tertiary) minute axes.

Flowers unisexual; male in upper and female in lower inflorescences, actinormorphic, with (1–) 3–4 (–6) -bracteate tepals (in Sparganiaceae) or these substituted by hairs situated at some distance below the stamens (in Typhaceae). Stamens (in male flowers) 1–6 (–8) per flower, sometimes with fused filaments. Anthers basifixed, broadening apically, tetrasporangiate. Tapetum amoeboid; microsporogenesis successive. Pollen grains single or more rarely dispersed in tetrads, ulcerate, binucleate (or sometimes trinucleate). Pistil normally monomerous, with one locule and one stylodial branch, stigmatic surface elongate, dry. Locule with one, pendulous anatropous ovule; parietal cell (tissue) formed. Endosperm formation helobial.

Fruit drupaceous or nearly nut-like (in Sparganiaceae) or nut-like but ultimately dehiscent (in Typhaceae), the latter condition approaching a follicle. Seed with copious endosperm, partly with starch; embryo linear or cuneate-fusiform.

Typhales consists of one or two families, the Sparganiaceae being often included in Typhaceae. The position of the order is a frequently disputed subject, and will have attention paid to it later in this study. Its distribution is sub-cosmopolitan with the greatest concentration on the Northern Hemisphere.

Sparganiaceae (one genus: *Sparganium*).
Typhaceae (one genus: *Typha*).

Juncales. Perennial or annual, often rhizomatous herbs with "graminoid" habit, rarely (*Prionium*) with a distinct aerial trunk. Vessels in stem and leaves with scalariform, or simple plus scalariform perforation plates. Leaves alternate, usually tristichous, but sometimes distichous, linear, flat, canaliculate, terete or laterally compressed, sheathing, often with stipule-like ears at the base of the lamina, parallel-veined, glabrous or with ciliate margins. Stomata paracytic. Oxalate raphides lacking; silica bodies mostly lacking in Juncaceae, but present in Thurniaceae. Culm erect, leafless or leafy, with a mostly cymose inflorescence.

Flowers actinomorphic, with 3 + 3 green, brown or hyaline, free tepals, and 3 + 3 or rarely 3 stamens and a tricarpellate pistil. Anthers basifixed, latrorse, tetrasporangiate. Tapetum secretory; microsporogenesis simultaneous. Pollen grains in tetrahedral tetrads, all functional, ulcerate, 3-nucleate. Pistil trilocular or (in *Luzula*, for example) unilocular, with axile or basal placentation. Style usually basally simple, tribrachiate, with dry stigma surface. Ovules 3 to numerous, anatropous, with parietal cell (tissue). Endosperm formation helobial.

Fruit a loculicidal capsule, rarely indehiscent (*Oxychloë*). Seeds sometimes provided with elaiosome; endosperm copious; embryo small, basal, ovoid, with a large cotyledon.

The order has occasionally been placed in the Liliiflorae by virtue of the complete lilialean flower construction and the capsular fruit. However there are more numerous features connecting the Juncales with the Cyperales.

The order consists of two families. It is widely distributed especially in the Northern Hemisphere, but the centre of variation of Juncaceae, and the total distribution area of Thurniaceae, are in Southern America. Ecologically the Juncaceae, the rushes, are important in moist habitats.

> Juncaceae (8–10 genera; *Distichia, Juncus, Luzula* (incl. *Ebingeria*), *Marsippospermum, Oxychloë* (incl. *Andesia*), *Patosia, Prionium, Rostkovia* and ? *Voladeria*, the last mentioned genus dubiously a member of Juncaceae).
> Thurniaceae (one genus: *Thurnia*). Thurniaceae was included in Juncaceae for example by Thorne (1968), but is probably distinct, approaching perhaps Rapateaceae as well as Juncaceae.

Cyperales. Annual or generally perennial, frequently rhizomatous, "graminoid" herbs, often tufted, with leafy or leafless, terete, biconvex or triangular stems and mostly linear, sheathing leaves. Stem and leaves with vessels having simple and/or scalariform perforation plates. Ligules present in some genera. Leaves flat or canaliculate, rarely terete or filiform. Uniseriate hairs usually missing; stomata paracytic. Oxalate raphides lacking. Silica bodies present in epidermal cells, generally conical, simple or compound. Inflorescences mostly consisting of spikes or spikelets in larger, mostly cymose clusters.

Flowers bi- or unisexual, with perianth lacking or reduced to bristles or hairs, rarely present as 3 + 3 bracteate scales (e.g. *Oreobolus*). Stamens normally 3 or less, with thin filament and tetrasporangiate, basifixed, introrse, longitudinally dehiscent anthers. Tapetum secretory; microsporogenesis specialized, apparently simultaneous, three nuclei in each tetrad degenerating. Pollen grains (in tetrads with one functional pollen grain) smooth, ulcerate or ulcerate-forate, tri- or less often bi-

nucleate. Gynoecium bi- or tricarpellate, unilocular, generally with a basally single style and 2–3 long stylodial branches, sometimes with 3 free stylodia. Ovule single, basal, anatropous, with parietal cell. Endosperm formation nuclear.

Fruit a nutlet sometimes enclosed in a flask-like utricle. Seed with copious endosperm and a basal (not lateral), capitate or broad embryo.

The *Mapania* group has flowers (or inflorescences) deviating from this description in that there are several anthers, each subtended by a scale-like bract. Their interpretation is not yet settled.

Cyperales consist of the family Cyperaceae, the sedges, which make up a large and variable complex and has a subcosmopolitan distribution.

> Cyperaceae (c. 90 genera).

Hydatellales. Minute, entirely or mainly annual herbs submerged in shallow fresh water. Leaves tufted, thin, filiform, without distinct sheath. Stomata without subsidiary cells or lacking. Hairs, silica bodies and oxalate raphides apparently lacking. Roots and culms with vessels having scalariform perforation plates, leaves vesselless, or with vessels having scalariform perforation plates. Inflorescence terminal, few-flowered, subtended by 2–4 (–6) hyaline bracts, bisexual or unisexual.

Flowers (according to current interpretation) unisexual, minute, naked, with either a single stamen or a stipitate pistil. Stamen with a stout filament and a tetrasporangiate, basifixed anther. Pollen grains sulcate, apparently binucleate. Pistil unilocular, mono- or pseudomonomerous (3-carpellate in *Trithuria*?), utricle-like, uniovulate, with some sessile uniseriate stigmatic hairs. Ovule apical, pendulous, anatropous. Nucellus epidermis with (some) periclinal divisions. Endosperm formation cellular.

Fruit indehiscent or (*Trithuria*) opening by 2–3 slits. Seed with rudimentary endosperm but with copious, starchy perisperm. Embryo minute, lens-shaped.

This order has only recently been recognized, Hydatellaceae being erected by Hamann as late as 1976. Its position in the Commeliniflorae is dubious, though quite probable. Several features serve to distinguish the family from Centrolepidaceae, in which its genera were previously included.

> Hydatellaceae (2 genera: *Hydatella, Trithuria*).

Restionales. Mainly perennial, "graminoid" herbs, growing tufted or otherwise, sometimes as scramblers. Vessels in stems with single and scalariform perforation plates, vessels in leaves lacking or present and with various kinds of perforations. Leaves alternate, generally not distichous, sheathing, linear or lanceolate, non-petiolate

(*Hanguana* petiolate), parallel-veined, sometimes ligulate (*Joinvillea*, some genera of Restionaceae) but in Restionaceae generally reduced to scarious bracts only. Uniseriate hairs occasionally present, stomata generally paracytic. Oxalate raphides lacking; amorphous silica usually but silica bodies more rarely present (*Joinvillea, Flagellaria, Hanguana,* some Restionaceae). Inflorescence spike-like or otherwise, sometimes very complex.

Flowers bi- or unisexual, actinomorphic (or reduced), in most cases with a perianth of 3 or 3 + 3 free, bracteate, similar or dissimilar tepals, or tepals lacking altogether. Stamens 3 + 3, 3 (most usual) or 1; anthers usually dorsifixed, introrse (or latrorse), bisporangiate or rarely tetrasporangiate. Tapetum secretory; microsporogenesis successive. Pollen grains single, smooth, usually ulcerate, tri- or rarely binucleate. Pistil tri- to monocarpellate and uni- to trilocular with one pendulous (rarely axile), orthotropous (*Hanguana* hemianatropous) ovule per locule. Stylodial branches generally separate from the base, but Joinvilleaceae and Hanguanaceae with three sessile stigma crests on the ovary. Ovules without parietal cell in at least Restionaceae and Centrolepidaceae. Epidermal cells of nucellus generally elongating radially. Endosperm formation in the cases known nuclear.

Fruit capsular, indehiscent and dry, or baccate to drupaceous, in Centrolepidaceae membraneous, dry and dehiscent (follicular). Seed(s) exarillate, with copious endosperm and a small embryo.

This order may prove to be heterogeneous although no doubt Centrolepidaceae, Restionaceae, Ecdeiocoleaceae and Anarthriaceae almost certainly form a natural group.

The distribution of Restionaceae and Centrolepidaceae are both South Hemispheric with great concentration in Australia and some representatives in South America, Restionaceae having besides a rich occurrence in Southern Africa. Anarthriaceae and Ecdeiocoleaceae are Australian segregates from Restionaceae. Flagellariaceae is found in the tropics of the Old World, and Joinvilleaceae and Hanguanaceae are concentrated in the tropics of Asia, and Malaysia.

Centrolepidaceae (4 genera: *Aphelia, Brizula, Centrolepis, Gaimardia*).
Restionaceae (c. 30 genera).
Anarthriaceae (one genus: *Anarthria*). This is very doubtfully distinguished from Restionaceae, but has more well-developed leaves, tetrasporangiate anthers etc.
Ecdeiocoleaceae (one genus: *Ecdeiocolea*), likewise doubtfully distinct from Restionaceae, with bilocular ovule (*Anarthria* has trilocular, Restionaceae uni-, bi- or trilocular).
Hanguanaceae (one genus: *Hanguana*; syn. *Susum*).
Flagellariaceae (one genus: *Flagellaria*).
Joinvilleaceae (one genus: *Joinvillea*).

Poales. Perennial or annual herbs or (in Bambusoideae) shrubs or trees. Rhizomes and stolons common. Stem hollow or compact, usually with prominent nodes. Vessels in stems and leaves with simple or both simple and scalariform perforation plates. Leaves sheathing, usually linear (-lanceolate), rarely petiolate (in some Bambusoideae), ligulate, with paracytic stomata and often with unicellular or small bicellular ("micro-") hairs. Oxalate raphides lacking, but silica bodies of variable shapes present and mostly typical to the subfamily, deposited in epidermal short cells. Inflorescences consisting of distichous spikelets of determinate or indeterminate construction, these mostly in larger aggregates, panicles; branches of panicles without supporting leaves (exception *Anomochloa*).

Flowers of each spikelet in the axil of a bract, the lemma, and with a palea representing the prophyll or two fused members of an outer whorl of tepals. Inner whorl of perianth consisting of 3 or more often 2 members (lodicules), which are generally small. Stamens 3 + 3 or usually 3 (more rarely 1–2 or numerous; up to 120 in *Ochlandra*). Filaments thin; anthers tetrasporangiate, basifixed, latrorse; tapetum secretory; microsporogenesis successive. Pollen grains simple, smooth, ulcerate, trinucleate. Pistil unilocular, with 2 or less frequently (some Bambusoideae) 3 stylodial branches, these usually free to the base or more rarely fused basally to a style. Stigma lobes plumose, dry. Ovule solitary, lateral or apical, orthotropous or rarely hemianatropous or campylotropous, without parietal cell but often with periclinal divisions in the nucellus epidermis. Endosperm formation nuclear.

Fruit generally a caryopsis (nutlet with pericarp fused to the seed coat), but in some Bambusoideae a nutlet or berry (*Melocanna*). Seed consisting in the main of a starch-rich endosperm; embryo lateral to this, its cotyledon forming a haustorial tissue and a tubular structure (scutellum and coleoptile).

Cyanogenic compounds and alkaloids common.

Poaceae is the economically most important of all plant families and contributes considerably to the plant cover of the earth. Its resemblance with the Restionales is so manyfold that the justification of the former as a separate order can be questioned. The grasses have a cosmopolitan distribution and show a variety of life forms but never grow as submerse marine aquatics as do some of the Alismatiflorae.

Poaceae (= Gramineae; c. 650 genera); the third family, in terms of species numbers of angiosperms if "Leguminosae" are divided into three families.

Superorder Areciflorae

Giant herbs, lianas or trees, the latter frequently high, not or little branched and with a terminal crown of leaves. Secondary growth lacking even in the large woody palms. Vessels always present in roots, often (except in Cyclanthales) in stems, and (in all groups) in leaves, with simple and/or scalariform perforation plates. Aerial roots common especially in Pandanales. Leaves spirally set, rarely distichous, petiolate or sessile; lamina simple or plicate, often splitting up along veins, bifurcate or pinnately or digitately compound. Venation not reticulate. Hastulae (ligule-like structures) sometimes present in palms; intravaginal squamules lacking. Stomata usually with 4(–6) subsidiary cells, one at each end of the stoma often being smaller than the lateral; paracytic stomata rare. Unicellular or multicellular hairs with multicellular base sometimes present. Silica bodies present in Arecales, variable in shape. Oxalate raphides widely distributed.

Flowers small or medium-sized, hypogynous to epigynous often unisexual, with 3 + 3, 3 or no tepals, these when present sometimes petaline, those of the outer and the inner whorl similar or dissimilar. Septal nectaries common in Arecales and Pandanales. Stamens usually 6, more rarely numerous or 4 or less, sometimes with connate filaments, e.g. in Cyclanthaceae. Anthers basifixed or often dorsifixed in palms, mainly latrorse and tetrasporangiate. Tapetum glandular (as far as known); microsporogenesis simultaneous (most palms) or successive. Pollen grains single, sulcate or bisulcate, trichotomosulcate or ulcerate, binucleate. Carpels usually 3–4, in Pandanales apparently variable in number, free (certain palms) or more or less fused, with separate stylodia, a single style or sometimes with subsessile stigma lobes. Stigma surface dry or wet. Placentation basal, parietal or axile. Ovules solitary or few per locule, anatropous or hemianatropous (rarely orthotropous), with or (in Cyclanthales) without parietal cell; often with periclinal divisions in the nucellus epidermis. Endosperm formation nuclear or (in Cyclanthales) helobial. Embryo formation according to the Onagrad or Asterad types.

Fruit a berry or drupe. Seeds exarillate, without perisperm, but with copious, often ruminate endosperm containing fat, protein and hemicellulose, rarely also starch in Cyclanthales. Embryo mostly linear, straight, non-chlorophyllous.

The superorder in the present circumscription seems to form a natural group in the evolutionary sense, and is recognized in this circumscription in many current classifications, although the Arales and Typhales are some-times associated with it without justification. Its three families are also usually treated as separate orders, which seems acceptable in the light of their great distinctness.

Their geographical distribution is mainly tropical.

Arecales. Mostly trees with single trunk, also short-stemmed or slender trees or lianas. Leaves petiolate, principally simple, plicate, but usually split up into feather- or fan-shaped, compound laminae, rarely doubly compound (*Caryota*). Silica bodies and oxalate raphides widely distributed. Inflorescences diverse, making up large panicles, simple spikes or thick spadices.

Flowers bisexual or unisexual (plants di- or monoecious or polygamous). Perianth usually of 3 + 3 rarely 3 tepals, these imbricate or valvate in each whorl. Stamens usually 6, 9 or numerous, rarely 3, with filaments free or fused basally. Anthers basifixed or dorsifixed, generally latrorse. Microsporogenesis simultaneous or rarely successive. Pollen grains sulcate, bisulculate or trichotomosulculate. Carpels usually 3, free or fused with tri- or rarely unilocular ovary, one style or three stylodia. Placentation mostly basal.

Endosperm large, often hard, non-starchy.

The Arecales comprise a single family, the Arecaceae (or Palmae), which is very variable, containing a number of recognizable tribes or subfamilies (some of which are possibly worthy of family rank).

The palms have a tropical distribution (Fig. 92 C).

Arecaceae (= Palmae; c. 210 genera).

Cyclanthales. Large, perennial herbs or climbers, rhizomatous, with mucilage cells in all parts. Vessels in roots and leaves, with scalariform perforation plates. Leaves alternate, distichous or spirally set, petiolate and with a bifurcate or divided lamina. Inflorescence a spadix or spike. Male and female flowers arranged as the squares of a chessboard on the thick spadices or rarely (*Cyclanthus*) in alternating, unisexual superposed whorls.

Male flowers naked or with cup-shaped perianth with short to obsolete lobes. Stamens numerous, with basally fused filaments. Anthers basifixed, latrorse. Microsporogenesis successive. Pollen grains sulcate or ulcerate. Female flowers tetramerous, without or with a 4-lobate carnose perianth, with 4 staminodes and with a 4-carpellate, unilocular, mostly inferior ovary with 4 sessile stigma crests. Ovules numerous, on one or 4 apical or 4 parietal placentas. Ovules anatropous, without parietal cell, but the nucellus epidermis dividing periclinally. Endosperm formation helobial.

Fruits baccate, often laterally coherent. Seeds with a succulent, carnose seed coat and with copious, generally non-starchy endosperm and a minute, linear embryo.

This order, as the palms, consists of but one, tropical family. This is however restricted in its distribution to South and Central America, where it is most concentrated in the Amazon basin.

Cyclanthaceae (11 genera: *Asplundia, Carludovica, Cyclanthus, Dicranopygium, Evodianthus, Ludovia, Pseudoludovia, Schultesiophytum, Sphaeradenia, Stelestylis, Thoracocarpus*).

Pandanales. Dioecious trees, often with considerable trunk and generally sparingly branched stems, but also shrubs, climbers and large herbs, the climbers supported by aerial roots. Vessels (with scalariform perforation plates) present in roots, stems and leaves. Leaves more or less distinctly tristichous, long, linear, sessile, with sheathing base, stiff or generally tough or grasslike, occasionally with lateral spines. Oxalate raphides widespread but silica bodies apparently absent. Inflorescences racemose spadices (or in *Saranga* panicles) subtended by white or coloured spathes.

Flowers unisexual, naked or with a reduced perianth. Male flowers with numerous, variously arranged, free or basally fused stamens. Anthers basifixed. Pollen grains ulcerate. Female flowers naked; ovary with (one–) few to numerous carpels situated in rows or rings to form "phalanges". Pistil with one or more locules, these with one to numerous anatropous ovules on basal or parietal placentas; stylodia either short and free or fused or obsolete. Nucellus with parietal cell (tissue). Endosperm formation nuclear.

Fruits consisting of multilocular drupes with one-seeded locules or of berries containing small seeds with fleshy, presumably non-starchy endosperm and a minute, linear embryo.

Pandanales consists of the palaeotropic family Pandanaceae.

Pandanaceae (3 genera: *Freycinetia, Pandanus, Sararanga*). *Pandanus* and *Sararanga* have drupes, *Freycinetia* berries. For this and other reasons there might be justification in according *Freycinetia* family status (Fagerlind, personal communication).

Character Syndromes and Taxonomy

Plants are fully integrated organisms and so there is mutual interdependence between their character states. Hence a change in any one of these, be it due to mutation or recombination, will influence either favourably or unfavourably the survival potential of the individual. Accordingly, it is not by chance that certain features tend to occur in combination and so it is that similar syndromes of character states are typical of different species which for example grow in the same environment or are dependent upon the same pollinators.

In order to illustrate this principle of co-evolution two families will be considered in some detail. The first of these are the animal pollinated orchids and the second the wind pollinated grasses.

ORCHIDS (Fig. 10)

The orchids are an enormously richly differentiated group of plants many of which are epiphytes. Their greatest concentration is in the tropics. In vegetative characters the variation is extraordinarily great, and the ecological range is also considerable. Vessel characters, stomata, presence of silica bodies, chemical constituents etc. are all variable.

In the reproductive region the specialization is very diverse. The inflorescence is racemose and the flowers thus lateral which condition is often connected with zygomorphy. The tepals are petaline and the labellum may or may not be supplied with a spur. The flowers are always epigynous. Epigyny may have evolved as a means of protection of the gynoecium and possibly also generated greater stability of the perigone. Nectar production

or probably more frequently the production of various volatile substances lead to the attraction of particular groups of insects which carry out the pollination. Specific relationships orchid–insect have yet been demonstrated in relatively few cases, some of which are, however, famous. Specialization in the shape, size, colour and pubescence of the perigone indicates a long period of co-evolution with the pollen vector. Further adaptations of style, stigma and androecial parts have secured a position of the functioning anther and the stigmatic areas to facilitate pollination. The pollen grains are numerous and usually cohere in massulae or pollinia, which are further adapted for dispersal by the occurrence of stipes, caudicule and a retinaculum which serves to keep the pollen aggregated. As the pollen is often sticky there is no need for a thick, sculptured exine or any distinct apertures.

Successful pollination events are few but this is compensated for by the numerous pollen grains transferred. As these are all transferred to one flower, this must be adapted accordingly. The ovary has thousands of small ovules massed on parietal placentas. The ovules are small, fewer-celled than normal and tenuinucellate, and furthermore endosperm formation is suppressed. The great number of seeds would be compatible neither with large seed size nor with a well-developed endosperm. The diminutive seeds therefore have practically no nutrient supply. They are dispersed by wind from a capsule with slits or from a fruit which merely disintegrates. The seedlings are associated with mycorrhiza.

Practically all of the above character states are operating in interaction and are logical consequences of the specialized mode of pollination.

Fig. 10. Advanced insect pollination syndrome as expressed in orchids: complicated, zygomorphous flower with chemical as well as visual attraction, pollen grains in tetrads and cohering in pollinia with specialized structures attached, comprising pollinaria; ovary inferior, with thousands of very small, tenuinucellate ovules, endosperm formation inhibited, embryos minute, fruits capsular with minute wind-dispersed seeds having extended wing-like testa. A–B: *Ophrys sphegodes*, upper part of plant and flower. C: gynostemium with pollinia partly concealed. D: two pollinaria. E: pollinarium *Orchis militaris*. F: flower of *Herminium monorchis*, transverse section, showing pollinia, here two for each theca. G: ovary of *Orchis militaris*, sectioned at middle. H: ovary of *Cephalanthera damasonium*, transverse section. I: ovule of *Orchis maculata*. J: embryo sac of *Corallorhiza maculata*, showing fertilization of egg cell and degenerating second sperm cell. K: same of *Bletia shepherdii*, showing triple fusion nucleus in early stage of degeneration (no endosperm formation). L: embryogenesis in *Cymbidium bicolor*, with large sac-like suspensor cells and a small embryo. M–N: seeds of *Stanhopea* and *Epipactis* respectively. (A–D from Ross-Craig, 1972; E, G and M from Weberling and Schwantes, 1972; F, H and N from Hagerup and Petersson, 1956; I from Hagerup, 1944; J–K from Sharp, 1912; L from Swamy, 1949b.)

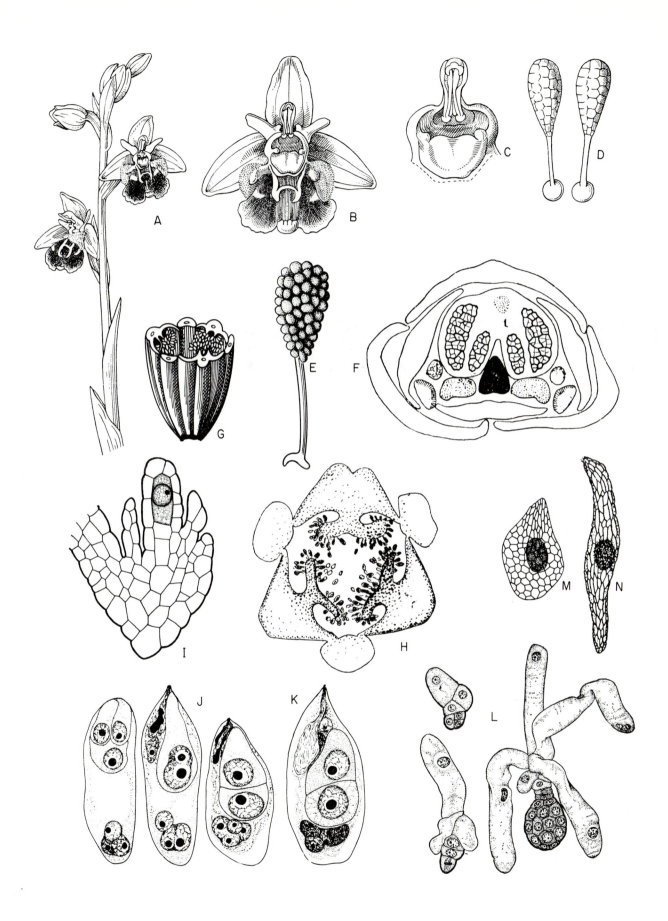

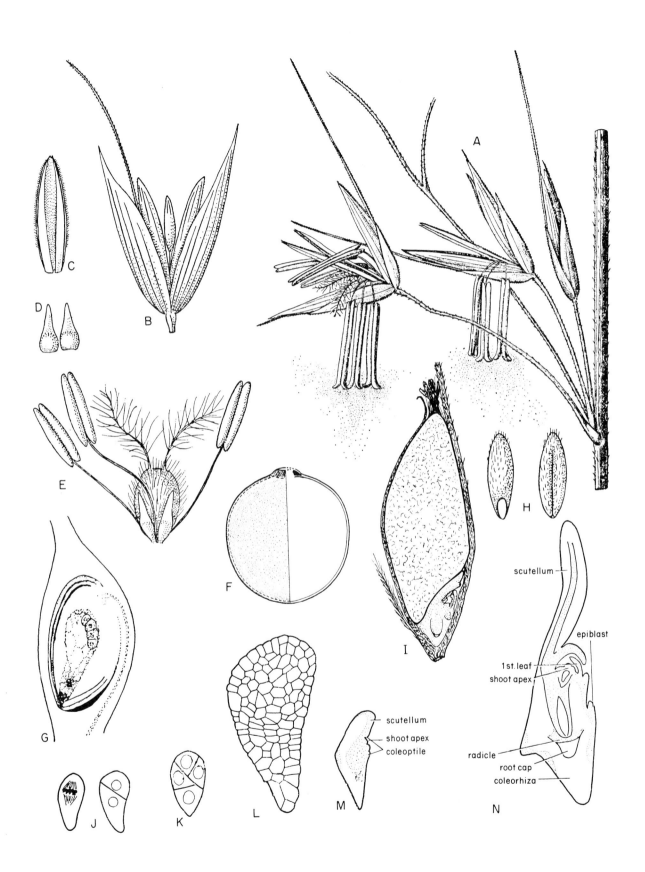

scutellum

epiblast

1st.leaf
shoot apex

radicle
root cap
coleorhiza

scutellum
shoot apex
coleoptile

GRASSES (Fig. 11)

A totally different syndrome of character states is found in the grasses. These are terrestrial plants often with effective vegetative propagation. While in the majority of orchids the flowers are large and relatively few per inflorescence, they are small and inconspicuous in the grasses, where the small size (and single seed) is compensated for in part by the great number of flowers. This as a consequence influences the inflorescence structure which is generally complex and much branched.

The perianth in grasses is strongly reduced, and may be represented only by the two, or rarely three, lodicules, perhaps also by the palea. The stamens generally have more or less weak, often long and slender filaments and the pollen grains are free from each other, smooth and provided with a circular aperture. Pollination (in allogamous taxa) is mainly by wind and thus the chance that several or many pollen grains shall land on the same stigma is not great, even though it is relatively well developed. A great number of ovules in each ovary would therefore be a waste of reproductive energy. The grasses have a single ovule in the unilocular ovary. A corresponding reduction in size or number of stigmatic branches would be of negative selective value! The single seed has a copious starchy endosperm which gives it a good nutrient supply during its early life, enabling the seedling to survive and compete successfully in semi-closed plant associations. Few seeds would germinate successfully in a closed grass association. As there is but one seed there is no need for a mechanism to enable the fruit to open and it forms a caryopsis or rarely nutlet or berry.

DISCUSSION

The above examples are representative of two groups which are each highly specialized to a particular pollination mechanism and in connection with this also to a particular mode of dispersal and seedling life. The more one investigates the various attributes of each plant group studied the more evident it is that each attribute is greatly influenced by and interacts with the remainder and the general biology of the plant.

It may be questioned from this point of view whether the character states of a group are at all indicative of an ancestry and if its similarity with other groups indicate phylogenetic relationships rather than a common (parallel) adaptation to, for example, a habitat in combination with the pollination vectors available.

It is, however, our experience that in monocotyledons it is possible in most cases to survey the main lines of specializations that have happened, but to separate the adaptations described from any phylogenetically basic character states which do *not* reflect the respective adaptations is difficult because these adaptations are probably old and have occurred successively in evolution.

In the absence of fossils it is by comparison with other extant groups, which are less specialized, that one will be able to estimate the relationships and the conditions under which deviating (specialized) attributes have evolved.

Thus there is no doubt that a strongly reduced perianth, smooth pollen grains, single-ovuled ovary, and dry indehiscent fruits have developed in several groups independently. Some of these similarities between Poaceae and Cyperaceae have certainly developed along parallel lines from ancestors which most likely had a double perianth, two whorls of stamens and a tricarpellate, trilocular ovary with several or numerous ovules and a capsular fruit such as in extant Juncales. However, many shared features in Poaceae and Cyperaceae (seeds with a starchy endosperm, stem with vessels etc.) indicate that the groups are yet relatively closely allied and have not achieved all their similarities as a result of convergent evolution.

It is accordingly by comparison with less specialized groups exhibiting a great number of fundamental similarities that one has the opportunity of estimating probable relationships. The orchids fundamentally belong to the liliiflorous assembly of monocotyledons with two whorls of petaline perianth. A comparison of numerous of its features gives no precise clue as to which extant family it is particularly closely allied (see, however, p. 307) and we can not, perhaps, be totally convinced that some of the more conspicuous similarities to the

Fig. 11. The wind pollination syndrome as met with in grasses: flowers with reduced perianth, slender filaments, pendulous anthers, smooth pollen grains, large plumose stigmas, unilocular uniovulate ovary, caryopsis-fruit, seed with copious endosperm. A: part of panicle of *Arrhenatherum elatius*. B: spikelet of *Avena sativa*. C: palea (homologous to bracteole or two fused outer tepals) of *Avena sativa*. D: lodicles (homologous to inner tepals) of same. E: flower of same, showing 2 lodicles, 3 stamens and pistil. F: pollen grain of *Dendrocalamus strictus*. G: ovary, longitudinal section of *Glyceria declinata*. H: caryopsis of *Avena sativa*. I: caryopsis of *Hordeum distichum*, longitudinal section, showing embryo at the base. J–N: embryogenesis and (N) embryo of mature seed in *Triticum*. (A–E and H from Hubbard, 1954; F from Erdtman, 1952; G and I from Hagerup and Petersson, 1956; J–N from Batygina, 1969.)

families of Burmanniales or Liliales may not have evolved along parallel lines. However, a rather close relationship between Orchidales, Liliales, Burmanniales and also Asparagales seems to be more likely. Obvious convergence in floral appearance is demonstrable between the flower of the Orchidales and the Zingiberaceae. However, while in the Zingiberaceae the labellum consists of petaloid staminodia, it represents the inner median tepal in the Orchidales. Pollen, vessel characters, arillus, perisperm, and numerous other details of the zingiberaceous plant indicate that the Zingiberaceae are only distantly related to the Orchidales.

Therefore it can be concluded that there is a significant connection and interdependence between the attributes and the biology of a plant group, a connection that turns out to be more profound than believed even in so-called less specialized plants. However, such interdependence and the resulting syndrome tendency does not prevent us from drawing important conclusions from comparative studies of the different taxa.

A discussion of the phylogenetic diversity in relation to factors which determine this diversity is given by Sachs (1978). Some (? most) families are shown to have specific adaptive specializations which give them competitive advantages for part of their environmental resources only, which ensures their co-existence and the maintenance of phyletic diversity. This has great relevance to the above discussion.

A Survey of the Distribution of Selected Characters and their States

CHARACTERS AND CHARACTER STATES

The reader may be unfamiliar with the terms characters and character states so they need to be explained here. More commonly encountered are terms such as *attributes*, *characteristics* or *features* which express a certain property of a plant. The attributes may refer to the absence or presence of a structure or chemical compound or it may express different shapes, positions or other conditions of certain structures. Two or more different but corresponding attributes referring to the same structure or group of properties of a plant are regarded as states of one and the same character.

We may conversely state that a character expresses itself in different states.

Thus absence, versus presence, of a chemical compound represents different states of one character. With increasing exactness we may distinguish more states, such as absence, presence in low quantity and presence in rich quantity of the compound, low and rich being preferably defined in terms of relative quantity, and proper statements being made as to possible insecurity with regard to exactness.

With regard to shapes the character "ovule type" can express itself as anatropous, hemi(ana)tropous, campylotropous and orthotropous states, but the ovule may also be classified in terms of some other character such as number of integuments, in which case bitegmic, unitegmic and ategmic states are commonly recognized.

In this chapter we will deal with numerous characters and their states, the distribution of each character state being mostly illustrated by shading figures representing groups (orders, families and parts of these) in a diagram. In such a case one state can be shaded and if only one alternative is recognized, this is left unshaded. The other alternative may also be differently shaded, in which case the unshaded groups may represent those yet unknown.

Generally, each character with its states will be treated under a common title in this chapter, but in some single cases, where the variation pattern needs much attention, separate character states are treated under separate titles.

This is the case with the pollen grains, where the sulcate, bi- to oligosulculate, trichotomosulcate, ulcerate, foraminate, spiraperturate and inaperturate states are treated separately.

THE SURVEY

In the following pages a considerable number of characters are analysed with respect to the distribution of their different states.

It needs to be pointed out that there has been a great number of limitations as regards the completeness of the data presented. The following guide-lines have been followed.

1. In most cases the survey has been restricted to characters, the states of which show a pattern of distribution from which conclusions can be drawn as to phylogeny (i.e. also for taxonomy) and evolution. Occasional biological adaptations are therefore in most cases omitted.

2. Characters have been omitted which are not known for representatives of most major groups of monocotyledons. The reliability of data, however, is very variable from one character to another. It has not been possible to apply general standards, as each character poses its own problems. Often a character is of great interest—and may be well known—in some major groups, but may be little known or virtually unknown in others. In cases where the data have the character of spot tests in a number of orders or families we have generally used circles and other symbols to represent these. Where the cases are considered for one reason or another to be more or less representative of a group or part of a group, shading (dotting, hatching, checking) has been chosen.

3. The fact that some characters have been omitted does not necessarily mean that they are not useful in monocotyledon classification. It may have been beyond our ability to collect sufficient data for them or they may not have been demonstrated yet to be sufficiently useful. We are thus convinced that various embryological, anatomical and even gross morphological characters

shall prove very useful in future phylogenetic work.

4. The work has been generalized and simplified considerably so as to suit mainly the discussions on the ordinal and superordinal levels. The very interesting details of variation in many characters within, for example, the Asparagales (i.e. between its families) have largely been neglected in our discussions.

5. Most characters presented contain a degree of incorrectness. In certain cases, such as for embryo sac formation, this has been very high in the literature, but critical revisions may have reduced this to a minimum. It is presumed that specialists in various fields will discover faults or misinterpretations. For each character a revised future presentation based on extensive and thorough basic studies is therefore desirable, and one of the main outcomes of this book may be to provoke such studies. However, as the characters included here are numerous and the records for each of them frequently numerous as well, it is expected that individual faults will not jeopardize the conclusions drawn. Often in-

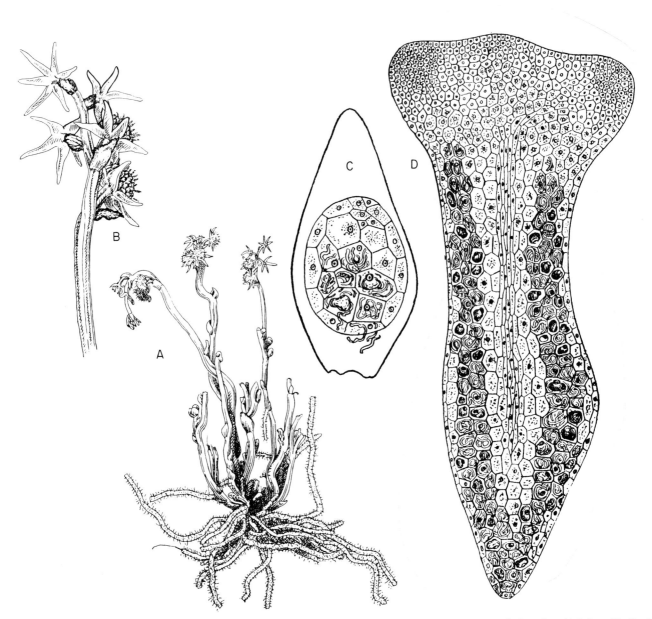

Fig. 12. Mycotrophy in combination with lack of chlorophyll. A–B: *Sciaphila thaidanica* (Triuridaceae), a saprophyte c. 5 cm high from Thailand (Larsen 1961). C–D: Embryos of *Neottia nidus-avis* (Orchidaceae) with mycorrhiza-bearing cortical layer. (From Bernard, 1909.)

dividual data which may be erroneous stand out as exceptions in such diagrams where the character shows a definite pattern of variation.

The above reservations are important, because only by concessions with regard to completeness of data has it been possible to carry through this survey. Criticism may be raised with regard to lack of data as regards inflorescence types, chromosome numbers, contents of non-protein amino acids, alkaloids etc., but each of these was neglected for one or another of the foregoing reasons.

In the analyses of the distribution of character states in monocotyledons some dicotyledonous groups are also included. These in the discussions and base data lists are called "Dicotyledons associated with the monocotyledons". They comprise: the order Nymphaeales, here provisionally placed in a separate superorder, the Nymphaeiflorae; the order Piperales consisting in the present

circumscription of the families Saururaceae and Piperaceae (incl. Peperomiaceae); and the order Magnoliales, which is here very widely circumscribed including the Aristolochiaceae beside Annonaceae, Eupomatiaceae, Cannellaceae, Myristicaceae, Magnoliaceae, Degeneriaceae, Winteraceae etc., the order not being treated here in detail at all, except for the Aristolochiaceae and, partly, the Annonaceae, which show considerable similarities with the monocotyledons.

Piperales and Magnoliales, besides Laurales, Illiciales and Rafflesiales (the latter three orders neglected in this study), are no doubt closely allied and make up the superorder Magnoliiflorae. Further differentiation of Magnoliales is made elsewhere (Dahlgren, 1980a) but is of little interest in connection with the monocotyledons. At the end of this study, the dicotyledonous orders mentioned here and included in the study will be discussed further.

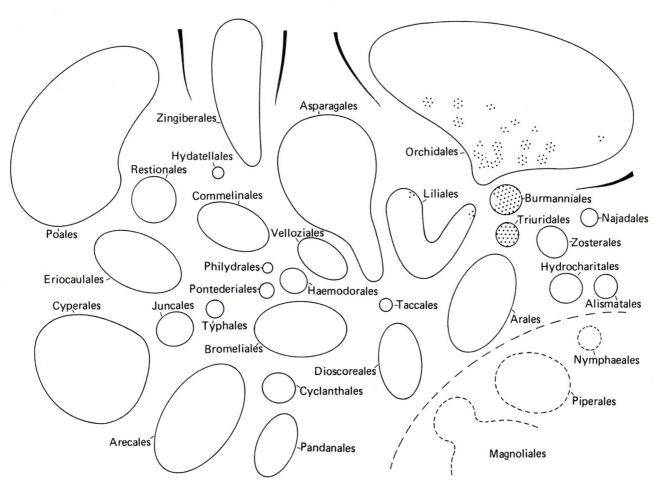

Diagram 3. Mycotrophy in combination with loss of chlorophyll. The chlorophyll-less Orchidales are according to Diagram 2.

MYCOTROPHY IN COMBINATION WITH LACK OF CHLOROPHYLL
(Fig. 12, Diagram 3)

All chlorophyll-less, monocotyledonous so-called "saprophytes" are probably mycotrophic. The distribution of chlorophyll-less mycotrophic plants is restricted and limited to scattered families of the superorders Alismatiflorae and Liliiflorae.

In the Triuridales (Alismatiflorae) all species possess this character and also in the Burmanniales (Liliiflorae) most taxa are chlorophyll-less (although several species of *Burmannia* are autotrophic and chlorophyllous), whereas in the Liliales and Orchidales chlorophyll-less species are limited to certain genera. Within the Liliales such species occur in the monotypic Geosiridaceae and in the genera *Petrosavia* and *Protolirion* of the Melanthiaceae. Amongst the orchids there are a number of chlorophyll-less, mycotrophic genera, such as *Galeola*, *Neottia*, *Corallorhiza*, *Limodorum* and *Epipogium*, while *Dipodium* includes both chlorophyllous and non-chlorophyllous species.

Another, but probably dubious, record of a saprophyte ("parasite") is *Brachychilus* (Zingiberaceae) reported by Loesner (1930). This requires further investigation for it may turn out to be the chlorophyll-less inflorescence of a plant the green leaves of which are developed at a different season.

The taxonomic value of lack of chlorophyll has to be considered for each of the orders wherein it occurs. In orchids it is no doubt of very little value, the chlorophyll-less species being rather scattered in the family and sometimes occurring in the same genus. Geosiridaceae has sometimes been placed in Burmanniales, a position supported also by its numerous small seeds. If this is so, mycotrophy and lack of chlorophyll here is of considerable significance. Besides, *Petrosavia* and *Protolirion*, often placed in a separate family (Petrosaviaceae), have sometimes been associated with the Triuridales. Even though there is undoubtedly anatomic similarities between the groups (Stant, 1970), this is likely to be by convergence.

Base Data

ALISMATIFLORAE: TRIUR.: all.
LILIIFLORAE: LILIA.: Geosi.; Melan.: *Petrosavia*, *Protolirion*.—BURMA.: Burma.: most, except some spp. of *Burmannia*; Thism.; Corsi.—ORCHI.: Orchi.: several genera, such as *Corallorhiza*, *Dipodium*, *Epipogium*, *Limodorum*, *Neottia*.
(Uncertain: ZINGIBERIFLORAE: ZINGI.: Zingi.: *Brachychilus*; not in the diagram.)

TREE AND SHRUB HABIT
(Fig. 13, Diagram 4)

Plants with a woody, simple or branched axis giving a tree or shrub habit are widely distributed as may be seen from Diagram 4. Sometimes the woody habit is accompanied by secondary growth (Diagram 19) but usually the habit is acquired in other ways. For example in palms the columnar habit of the stem results from a cap meristem involving the leaf bases, whereas that of bamboos (Poaceae) results from the incorporation of extra sclerenchyma around the periphery of the hollow culms. In the Velloziaceae the stems are supported by adventitious roots which pass down the outside of the stem amongst the dead leaf bases.

Although the tree or shrub habit occurs in all superorders other than the Alismatiflorae the condition is widespread only in the Areciflorae and especially in the Arecales. Here there are many species with massive columnar trunks. In the Pandanales in contrast branching of the trunk is common, giving rise to a habit similar to that of *Hyphaene* (Arecaceae). Many species of *Freycinetia* are not trees or shrubs, however, but comprise mostly epiphytic, rhizomatous herbs. Amongst the largely herbaceous Cyclanthales shrub or liana habit is developed in some genera.

The Ariflorae, i.e. the Arales, are predominantly herbaceous and largely restricted to rainforest habitats and like that order have a few members with a shrubby habit. These include species of *Philodendron* and *Alocasia* which have columnar stems up to 2 m tall surmounted by a crown of leaves. Also *Xanthosoma* may be enumerated under the "woody" Arales.

The Asparagales have nine of the families with woody taxa. Here belong the tree lilies (*Dracaena*, Dracaenaceae), the grass trees (*Xanthorrhoea*), the rosette trees (*Yucca*, *Agave*, *Furcraea* etc. in Agavaceae). Small shrubby members include *Calectasia* (Dasypogonaceae), *Ruscus* (Asparagaceae) and *Tricoryne* (Asphodelaceae). Included here are several genera of scandent shrubs such as *Smilax* and *Ripogonum* (Smilacaceae), *Petermannia* (Petermanniaceae) and *Philesia* and other genera of Philesiaceae. Of the Liliales only Iridaceae has any woody members which actually have secondary growth; they are species of *Nivenia*, *Witsenia* and *Klattia*, from southern

Fig. 13. Tree-like habit in different and distantly related groups of monocotyledons. A: *Puya berteroniana* (Bromeliaceae). B: *Cocos nucifera* (Arecaceae). C: *Yucca* sp. (Agavaceae). (A from Pizarro, 1959; B from Lawrence, 1951; C from Lotsy, 1911.)

African genera and a few Australian species. Among the orchids (Orchidales) the terrestrial genus *Sobralia* is almost shrubby (J. T. Waterhouse, personal communication). In the Velloziaceae (Velloziales) there are many shrubby forms including *Vellozia* which may be branched and up to 6 m tall. The Bromeliaceae are mostly rosette plants but may produce massive tall columnar forms, as in *Puya*.

Although the Zingiberiflorae produce such tall plants as the herbaceous *Musa* (Musaceae) the woody habit has only rarely been produced, viz. in the three genera of Strelitziaceae, incl. the well known *Ravenala*.

Among the Commeliniflorae the woody habit has only been developed to any extent in the Poaceae. Here the Bambusoideae (Poaceae) are largely tall and woody but a branched shrubby habit is widespread especially in grasses of drier regions.

The Restionales include only one shrubby genus, *Flagellaria* (Flagellariaceae), which is a scandent vine; the Cyperales and Juncales each include single shrubby genera, e.g. *Microdracoides* and *Prionium* respectively, and the Eriocaulales possess a few shrubby species in *Paepalanthus*.

From the scattered distribution of the woody habit and the diversity of mechanisms by which it has been achieved it can be concluded to have evolved independently on several occasions.

Base Data

ARIFLORAE: ARAL.: Arac.: occasionally, as in spp. of *Alocasia*, *Philodendron* and *Xanthosoma*.

LILIIFLORAE: ASPAR.: most or all Phile.; Peter.; Smila.; Aspar.; Dracae.: most genera except *Sansevieria*; most Dasypo.; Xanth.; most Agava. (*Yucca*, *Agave*, *Furcraea* etc.); Hypox.: *Hypoxis* spp.; Aspho.: spp. of *Aloë*, *Laxmannia*, *Tricoryne* etc.—LILIA.: Irida.: *Klattia*, *Nivenia*, *Witsenia*.— ORCHI.: Orchi.: maybe *Sobralia* spp.—VELLO.: Vello.: spp. of *Barbacenia*, *Vellozia*, *Xerophyta*.—BROME.: Brome.: few spp. of *Puya*.

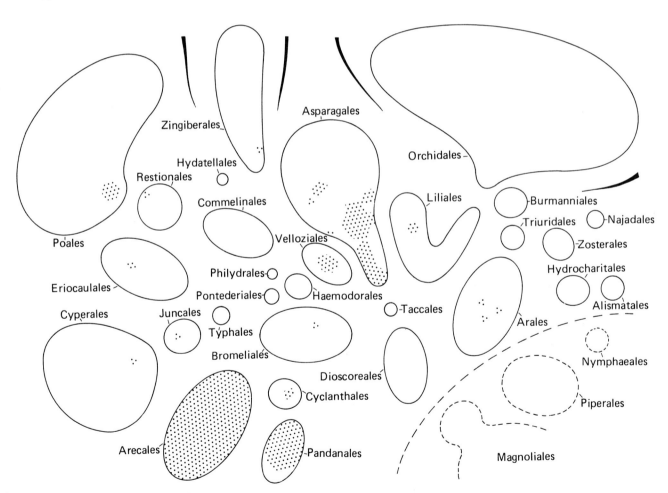

Diagram 4. Tree and shrub habit (shaded).

COMMELINIFLORAE: ERIOC.: Erioc.: *Paepalanthus* (rarely).—JUNCA.: Junca.: *Prionium.*—CYPER.: *Microdracoides.*—RESTI.:Flage.—POAL.: Poac.: many members of subfam. Bambusoideae.

ZINGIBERIFLORAE: ZINGI.: Zingi.: Strel.: *Phenacospermum, Ravenala, Strelitzia* p.p.

ARECIFLORAE: ARECA.: all.—CYCLA.: Cycla.: some genera.—PANDA.: Panda.: most spp. but not all in *Freycinetia.*

ORIGIN OF LATERAL ROOTS
(Fig. 14, Diagram 5)

In the majority of angiosperms the lateral roots arise in the pericycle opposite the xylem poles. However, it was observed long ago by van Tieghem (1887) and van Tieghem and Duliot (1888), that in those species where there is no pericycle between the xylem and the endo-

dermis the lateral roots arise opposite the phloem. The data plotted in our diagram are largely from the two sources cited.

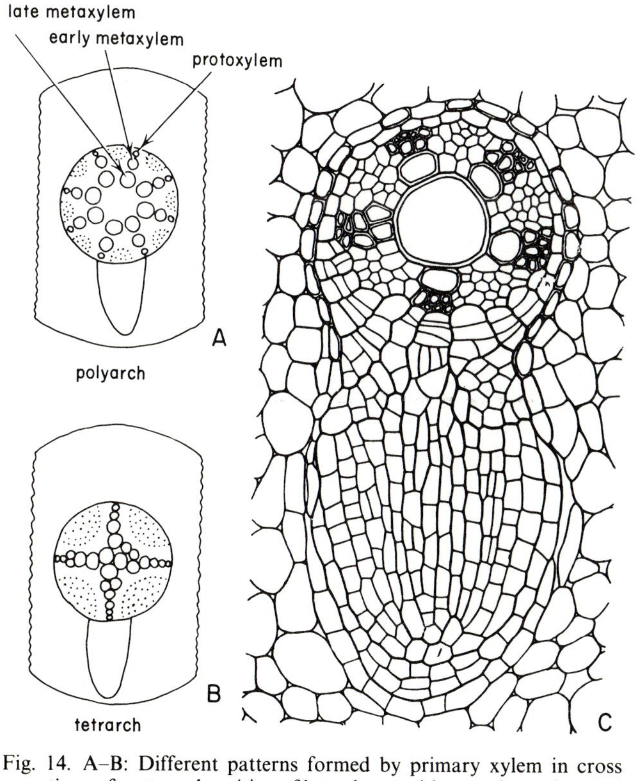

Because the taxa investigated in each order are few it is not possible to draw conclusions as to the constancy of phloem originating lateral roots in the Mayacaceae, Xyridaceae and Centrolepidaceae; these are the only three families reported to possess exclusively the attribute. However, it is certain that both types of lateral root origin are present in the Eriocaulaceae, Juncaceae, Poaceae and Cyperaceae, all of which are likewise members of the Commeliniflorae and plants generally growing in unshaded habitats.

Base Data

Opposite xylem

ALISMATIFLORAE: HYDRO.: Hydro.: *Hydrocharis*; Apono.: *Aponogeton.*—ZOSTE.: Potam.: *Potamogeton.*

ARIFLORAE: ARALE.: Arac.: *Alocasia, Monstera, Richardia.*

LILIIFLORAE: ASPAR.: Dasyp.: *Lomandra*; Xanth.: *Xanthorrhoea*; Diane.: *Dianella.*—LILIA.: Irida.: *Iris*; Lilia.: *Lilium.*—ORCHI.: Orchi.: *Cymbidium, Oncidium.*—HAEMO.: Haemo.: *Anigozanthos.*

ZINGIBERIFLORAE: ZINGI.: Musac.: *Musa*; Zingi.: *Amomum, Hedychium*; Costa.: *Costus*; Canna.: *Canna.*

COMMELINIFLORAE: COMME.: Comme.: *Commelina, Tradescantia.*—TYPHA.: *Typha.*—RESTI.: Resti.: *Chaetanthus, Elegia, Hypodiscus, Leptocarpus, Lepyrodia, Lyginia, Restio, Thamnochortus, Willdenowia*; Anart.: *Anarthria*; Flage.: *Flagellaria.*

ARECIFLORAE: ARECA.: *Areca, Chamaedorea, Hypophorbe.*—PANDA.: *Pandanus.*

Opposite xylem as well as phloem

COMMELINIFLORAE: ERIOC.: Erioc.: *Eriocaulon, Paepalanthus.*—JUNCA.: Junca.: *Juncus.*—CYPER.: Cyper.: *Cyperus*, other genera are variable.—POAL.: variable.

Opposite phloem

COMMELINIFLORAE: COMME.: Mayac.: *Mayaca.*—ERIOC.: Xyrid.: *Xyris*; Erioc.: *Lachnocaulon, Philodice.*—RESTI.: Centr.: *Aphelia, Centrolepis (Alepyrum), Gaimardia.*—POAL.: Poac.: *Elymus, Saccharum, Zea*, and maybe more genera (variable).

Fig. 14. A–B: Different patterns formed by primary xylem in cross sections of roots and position of lateral root with regard to xylem and phloem of the main root. The type seen in A is found in certain monocotyledons, primarily in the Commeliniflorae; while that of B is found in dicotyledons and (but with polyarch stele) in most monocotyledons, also together with the former in the Commeliniflorae. C: Illustration of the condition seen in *Allium giganteum.* (A–B from Esau, 1953; C from von Guttenberg, 1960.)

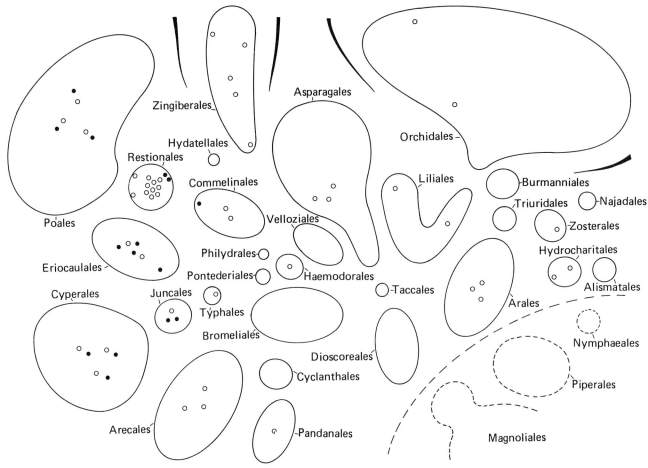

Diagram 5. Origin of lateral roots. Solid circles: lateral roots arising opposite phloem. Hollow circles: lateral roots arising opposite xylem.

DEVELOPMENT OF ROOT HAIRS (Fig. 15, Diagram 6)

The trichomes (root hairs) which arise from the outer walls of the superficial layer (piliferous layer) of vascular plant roots are of two kinds. With the first (Type I) any cell of the piliferous layer may develop a root hair whereas in the second type (Type II) only specialized cells bear hairs. These specialized cells are distinguished from other cells of the piliferous layer by their smaller size and denser cytoplasmic contents (Leavitt, 1904).

The distributions of these two hair types are shown in Diagram 6 from which it is clear the character is of considerable taxonomic importance. Type I root hairs predominate in the Liliiflorae, Ariflorae and Areciflorae, whereas Type II predominate in the Commeliniflorae and are found constantly in the Alismatiflorae and Zingiberiflorae. Root hairs in Anarthriaceae and Ecdeiocoleaceae arise from cells similar in size to the remaining epidermal cells (Cutler and Airy Shaw, 1965).

In commenting upon the two orders possessing both types of root hair, Leavitt (op. cit.) says of the Orchidales that Type II root hairs are very restricted and that in the Poales Type I hairs are admitted to occur only with a certain amount of misgiving. Hence the variation in these two orders does not seriously challenge the taxonomic value of the character.

Amongst the dicotyledon orders with monocotyledonous tendencies the Nymphaeales, in common with the Alismatales, has Type II root hairs and the Piperales, in common with most Arales, has Type I root hairs.

Base Data

Type I

ARIFLORAE: ARAL.: Arac.: *Aglaonema, Anthericum, Arisaema, Caladium, Dieffenbachia, Zantedeschia.*

LILIIFLORAE: ASPAR.: Conva.: *Aspidistra, Polygonatum*; Aspar.: *Asparagus*; Drac.: *Dracaena*; Dasyp.: *Lomandra*; Xanth.: *Xanthorrhoea*; Agava.: *Yucca*; Aspho.: *Cordyline*; Hypox.: *Hypoxis*; Diane.: *Dianella*; Funki.: *Hosta;* Hyaci.: *Hyacinthus, Muscari*; Allia.: *Allium, Brodiaea*; Amary.: *Amaryllis* (? = *Hippeastrum*), *Eucharis, Leucojum, Pancratium.*—LILIA.: Irida.: several genera; Lilia.: *Lilium.*—ORCHI.: Cypri.: *Selenipedium*; Orchi.: probably most genera, at least *Goodyera, Liparis, Odontoglossum, Vanda, Vanilla.*—BROME.: *Billbergia, Nidularium, Tillandsia.*

COMMELINIFLORAE: TYPHA.: Sparg.: *Sparganium*; Typha.: *Typha.*—RESTI.: Anart.; Ecdei.—POAL.: Poac. (with reservations): maybe 7 out of 50 genera.

ARECIFLORAE: ARECA.: Areca.: *Oreodoxa.*—

CYCLA.: Cycla.: *Carludovica.*

Dicotyledons: MAGNOLIIFLORAE: PIPER.: Piper.: *Peperomia.*

Type II

ALISMATIFLORAE: ALISM.: *Alisma, Limnocharis, Sagittaria.*—HYDRO.: Apono.: *Aponogeton*; Hydro.: *Hydrocharis, Hydromystria, Limnobium, Stratiotes.*—ZOSTE.: Junca.: *Lilaea, Tetroncium, Triglochin* (Solereder and Meyer, 1933); Potam.: *Ruppia*; Zoste.: *Zostera*; Zanni.: *Zannicellia*; Cymod.: *"Cymodocea".*—NAJAD.: Najad.: *Najas.*

ARIFLORAE: ARAL.: Arac.: *Anthurium, Monstera.*

LILIIFLORAE: ORCHI.: Orchi.: rel. few: *Arethusa, Calopogon, Pogania.*—HAEMO.: *Anigozanthos.*

ZINGIBERIFLORAE: ZINGI.: Musac.: *Musa*; Zingi.: *Ellettaria*; Costa.: *Costus*; Canna.: *Canna*; Maran.: *Maranta.*

COMMELINIFLORAE: COMME.: Comme.: *Commelina, Tradescantia.*—ERIOC.: Xyrid.: *Xyris*; Erioc.: *Eriocaulon, Lachnocaulon, Paepalanthus.*—JUNCA.: Junca.: *Juncus, Luzula.*—CYPER.: Cyper.: 11 genera.—POAL.: Poac.: 43 out of 50 genera.

Associated dicotyledons: NYMPHAEIFLORAE: NYMPH.: Cabom.: *Brasenia.*

VELAMEN IN ROOTS
(Fig. 16, Diagram 7)

The aerial roots of epiphytic orchids and certain aroids are characterized by the presence of a parchment-like outer layer, the velamen. This tissue has thickened cell walls, is non-living and capable of absorbing water (Barthlott, 1976). It is epidermal in origin and is not restricted to epiphytic plants as is shown in Diagram 7, where the data are mostly from Guttenberg (1968). Data for orchids are also from Lavarack (1971) and Withner *et al.* (1974). The data on absence are taken largely from the series "Anatomy of the Monocotyledons" (Metcalfe, 1961), where the use of the term hypodermis is not regarded as equivalent to velamen, in that his use of the term refers either to parenchymatous cells or suberized cells (that are waterproof). Clearly further study is required of this interesting tissue.

In Orchidales, unlike in other monocotyledonous orders, the negative records for velamen are given for individual genera in our diagram. As is seen in this, velamen is much commoner in subfam. Epidendroideae, which has many epiphytes, than in the Neottioideae and Orchidoideae (cf. Diagram 2).

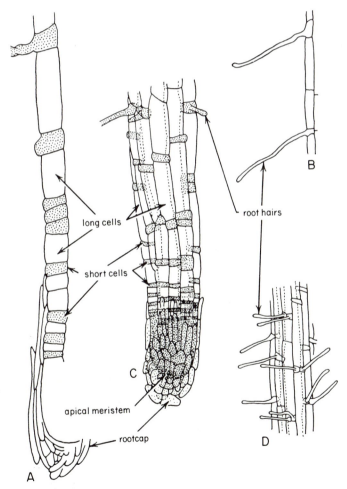

Fig. 15. Development of root hairs in cases where these are formed from specialized protodermal cells (short cells or trichoblasts). A and B from *Cyperus* (Cyperaceae) and C and D from *Anigozanthos* (Haemodoraceae). (According to Leavitt, 1904; from Esau, 1953.)

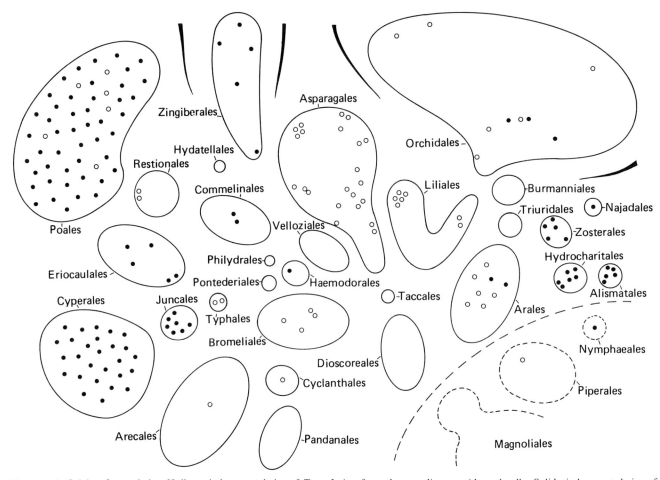

Diagram 6. Origin of root hairs. Hollow circles: root hairs of Type I, i.e. formed on ordinary epidermal cells. Solid circles: root hairs of Type II, formed on predetermined short epidermal cells. The nine orchid genera mentioned by Leavitt are placed according to Diagram 2.

From Diagram 7 it is obvious that a velamen is not restricted to the roots of orchids (indeed the tissue is absent from the roots of many terrestrial orchids); rather it is a common feature of the roots of several members of Asparagales, whereas it has been reported from only one member (*Lilium*) of Liliales. The tissue has also been recorded from *Anthurium* in Araceae (Arales), a family rich in epiphytic members. The velamen may be single-layered to multi-layered; both types being included in our diagram.

It is not unlikely that the velamen in aerial roots represents a somewhat different phenomenon than in subterranean roots. Therefore a taxonomic or phylogenetic evaluation of this character should be made in connection with studies on their function.

Base Data

Multiple velamen present
    ARIFLORAE: ARAL.: Arac.: *Anthurium*.

Fig. 16. Several-layered velamen in root of *Clivia nobilis* (Amaryllidaceae), seen in longitudinal section. (From von Guttenberg, 1940.)

LILIIFLORAE: ASPAR.: Conva.: *Aspidistra, Ophiopogon, Polygonatum, Tupistra*; Aspar.: *Asparagus, Semele*; Agava.: *Agave*; Aspho.: *Aloë, Anthericum, Arthropodium, Chlorophytum*; Hemer.: *Hemerocallis*; Allia.: *Agapanthus*; Amary.: *Amaryllis, Ammocharis, Brunsvigia, Buphane, Clivia, Crinum, Cyrtanthus (Vallota), Haemanthus, Hippeastrum, Nerine.*—LILIA.: Lilia.: *Lilium.*—ORCHI.: Cypri.: *Paphiopedilum*; Orchi.: many, e.g. *Aeranthis, Arachnis, Bulbophyllum, Calanthe, Campylocentrum, Cattleya, Cryptopheranthus, Dendrocolla, Epidendrum, Eria, Eulophia, Haemaria, Oberonia, Phajus, Phalaenopsis, Pholidota, Polyrrhiza, Sobralia, Taeniophyllum.*

Single-layered velamen present
LILIIFLORAE: DIOSC.: Diosc.: *Dioscorea (Testudinaria).*—TACCA.: *Tacca.*—ASPAR.: Dorya.: *Doryanthes*; Aspho.: *Bulbine, Gasteria*; Cyana.: *Cyanastrum*;

Hyaci.: *Bowiea, Eucomis, Ornithogalum, Veltheimia*; Amary.: *Hymenocallis.*—ORCHI.: Orchi.: many, e.g. *Ophrys, Spiranthes.*
COMMELINIFLORAE: COMME.: Comme.: *Rhoeo.*

Velamen absent
LILIIFLORAE: BROME.
COMMELINIFLORAE: COMME.: Carto.; Mayac. —ERIOC.: Rapat.; Xyrid.; Erioc.—JUNCA.: Junca.; Thurn.—CYPER.: Cyper.—RESTI.: Resti.; Centr.; Anart.; Ecdei.—POAL.: Poac.
ZINGIBERIFLORAE: ZINGI.: all families.

BULBS (Fig. 17, Diagram 8)

As considered here, bulbs are perennating organs com-

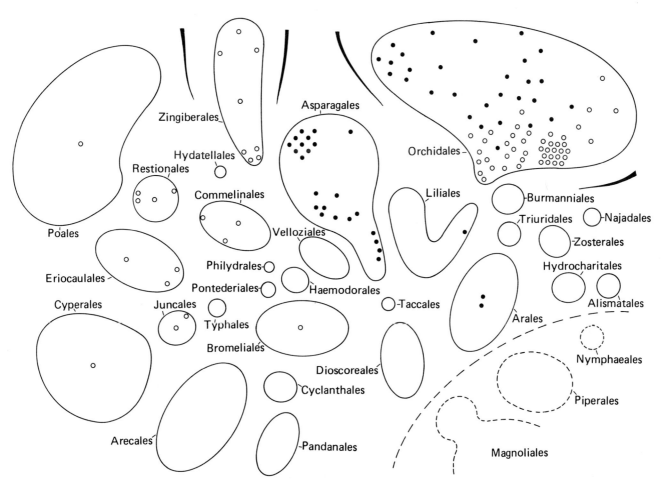

Diagram 7. Velamen in roots. Solid circles: multi- or single-layered velamen. Hollow circles: velamen documented as absent. The solid circles, and in orchids also the hollow circles, represent individual genera; while the hollow circles except in the orchids represent families proved to lack velamen (a variable number of genera investigated). The genera of orchids distributed according to Diagram 2.

prising a short vertical axis bearing a series of fleshy modified leaves (Fig. 17). Each season a new bulb is formed from an axillary bud enabling the plant to survive the unfavourable season undergound. Hence, bulbs primarily enable a single plant to persist in a given locality and so are only secondarily involved in dispersal. The replacement of flowers by bulbs or bulbils is a special phenomenon as is the production of scaly turions in *Potamogeton*. Both of these are neglected here.

The distribution of bulbs is shown in Diagram 8, from which it is clear they are almost exclusively restricted to the Liliiflorae. Furthermore, within this superorder they are confined to the Liliales and Asparagales.

In the Liliales the Liliaceae and Calochortaceae have many bulbous species in contrast to the Iridaceae and Melanthiaceae which have few. In the Asparagales most **Alliaceae, Amaryllidaceae and Hyacinthaceae** are

bulbous. In the Tecophilaeaceae bulbs are rare but occur for example in *Odontostomum*.

Bulbs are also found in *Triglochin bulbosum* (Zosteraceae) where they may be associated with the unusual habit, saline swamps subject to seasonal drying-out.

There are no bulbous species amongst the dicotyledons showing monocotyledonous tendencies though bulb-like structures are common in the Oxalidaceae.

Base Data

(ALISMATIFLORAE: ZOSTE.: Junca.: *Triglochin bulbosum*.)

LILIIFLORAE: ASPAR.: Tecoph.: *Odontostomum*; Hyaci.: most genera incl. *Bowiea*, *Schizobasis* and *Chlorogalum* (but not *Schoenolirion*); Allia.: most genera except *Agapanthus*, *Tulbaghia*, although some may be considered

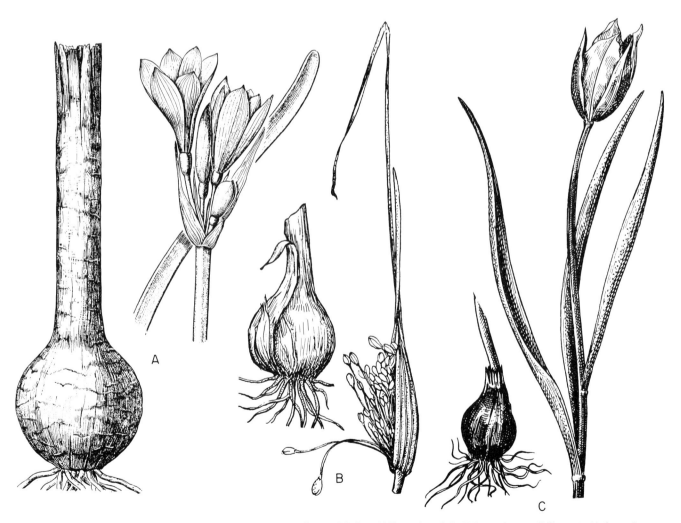

Fig. 17. Bulbs in A: *Rhodophiala elwesii* (Amaryllidaceae), B: *Allium pulchellum* (Alliaceae) and C: *Tulipa sylvestris* (Liliaceae). (A from Correa, 1969; B from Moore, 1955; C from Baillon, 1894.)

to have tunicated corms (p. 58); Amary.: practically all, except for example some spp. of *Scadoxus* and *Clivia* having transitional structures between bulb and rhizome.—LILIA.: Irida.: at least spp. of *Iris* and *Tigridia*; Caloc.: *Calochortus*; Lilia.: all taxa; Melan.: *Amianthium*, *Stenanthium*, *Zygadenus* spp. (some with tunicated corms?).

CORMS AND TUBERS
(with contributions from H. Huber)
(Fig. 18, Diagram 9)

It is often difficult to distinguish strictly between corms, bulbs and ordinary tubers in dried specimens without destroying them. Literature is also generally unreliable in regard to this information. Apart from thickened, tuber-like roots, which are common, for example, in

Asparagaceae and Asphodelaceae but apparently absent from the bulbous Asparagales (Hyacinthaceae, Alliaceae, Amaryllidaceae) and from most or all Liliales, there are besides the rhizomes the following types of underground storage organs.

Hypocotyledonary tubers. These form a general feature in the Dioscoreaceae *s.str.*, and seem to occur in many Araceae. At least in *Tamus* also the epicotyl takes part in the formation of the tuber, but remains subordinate. There is no information available as to what extent this may be the case in other Dioscoreaceae.

Hypocotyledonary tubers are also characteristic of *Eriospermum* (Eriospermaceae, Fig. 18B, C) and most probably for *Walleria* (Tecophilaeaceae, Fig. 18A), but absent from other Tecophilaeaceae. A tuber ("Sprossknolle") is also found in the monotypic genus *Philydrella* (Philydraceae) according to Hamann (1966).

Among the dicotyledons related to monocotyledons hypocotyledonary tubers are found in herbaceous species

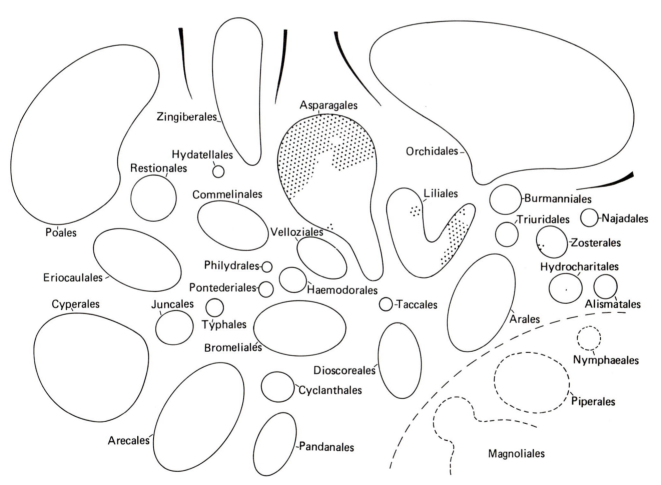

Diagram 8. Bulbs (shaded). The structure in the Zosterales (*Triglochin*) is dubiously classified as a bulb.

of *Aristolochia*, but are absent from the less derived members of Aristolochiaceae. Furthermore, they are not rare in Piperaceae, a typical example being *Peperomia peruviana*.

Corms are here regarded synonymous with stem-tubers or rhizomatous tubers with basal innovation. It is the basal innovation that distinguishes corms from ordinary tubers, in which growth is continued by a bud from near its apex. This distinction is difficult to establish convincingly, except on living specimens.

Corms as defined here seem to be common in the Tecophilaeaceae but are probably absent in the Hypoxidaceae and Haemodoraceae, members of which have at times been associated with the Tecophilaeaceae in literature. Corms are also known from most Colchicaceae (Fig. 18D) (except *Uvularia*) and are of more or less general occurrence in the Iridaceae (Fig. 18E–I; Goldblatt, personal communication), although tuberous

rhizomes also occur in this family; bulbs have been ascribed to species of Iridaceae repeatedly but perhaps mostly wrongly (see p. 57).

Strictly speaking, the "pseudo-bulbs" (albeit green, aerial and water-storing) of many orchids come close in their morphology to the concept of corm as defined above. According to Dressler and Dodson (1960), the "pseudo-bulbs" in the majority of the Bletiinae and Cyrtopodiinae and related orchids "seem to have been derived phyletically from more or less corm-like structures, as in *Bletia* and *Phajus*".

In the Alliaceae several genera are regarded by Moore (1953) possessing a membranous- or fibrous-coated corm, rather than a bulb, although their distinction from bulbs may be vague; *Dichellostemma*, for example, although thought by Moore (1953) to have a corm, possesses a typical bulb. The underground structure in such genera of the Alliaceae as *Androstephium*, *Bessera*, *Bloomeria*, *Brodiaea*, *Dandya*, *Milla*, *Muilla*, *Petro-*

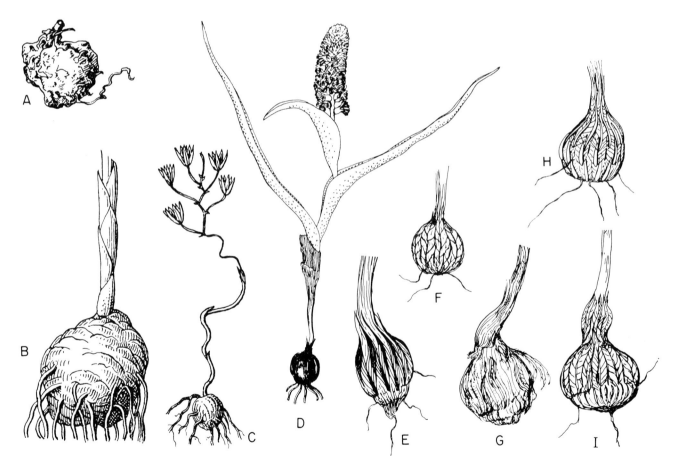

Fig. 18. Tubers and corms in the Liliiflorae. A: tuber of *Walleria mackenzii* (Tecophilaeaceae). B–C: tuber and whole plant of *Eriospermum majanthemifolium* and *E. spirale* (Eriospermaceae) respectively. D: plant with corm of *Wurmbea compacta* (Colchicaceae). E–I: tunicated corms in *Moraea margaretae* (E), *M. barkerae* (F), *M. saxicola* (G), *M. lurida* (H) and *M. fugax* (I) (Iridaceae). The tunics provide specific characters. (A from Carter, 1966; B–C from Krause, 1930; D from Nordenstam, 1964a; E–I from Goldblatt, 1977.)

nymphe, Triteleia and *Triteleiopsis* classed either as corms or bulbs may be transitional structures and deserve further morphological studies.

The taxonomic implications of the distribution of hypocotyledonary tubers as well as corms may be limited, although the corms in the strict sense may indicate affinity between Iridaceae, Colchicaceae and maybe Orchidaceae.

One probably can not attach any considerable importance to the tubers for proving relationships. The character has arisen along different lines of evolution, e.g. in the Dioscoreales, but only in the most derived (dioecious) groups, and apparently does not occur in the groups with bisexual flowers. In the Piperales tubers occur in the most derived genus (*Peperomia*) only, and here especially in semi-xerophytic or xerophytic sections. Similarly in the Aristolochiaceae they occur only in the most highly derived group.

The same can be seen in the Ranunculaceae (*Eranthis*), Fumariaceae (*Corydalis*), Berberidaceae (*Leontice*) and Apiaceae (*Bunium*). In the Eriospermaceae and Teco-

philaeaceae the tubers seem to be a helpful taxonomic feature (*Walleria* being perhaps best placed in the former of these families or in or near the Colchicaceae).

Base Data

The available data are restricted to those given in the text above.

"OPPOSITE" AND "VERTICILLATE" LEAVES
(Fig. 19, Diagram 10)

The distributions of "opposite", i.e. mostly pseudo-opposite, and "verticillate", i.e. mostly pseudo-verticillate, leaves are similar (Diagram 10), both being restricted to Alismatiflorae and Liliiflorae.

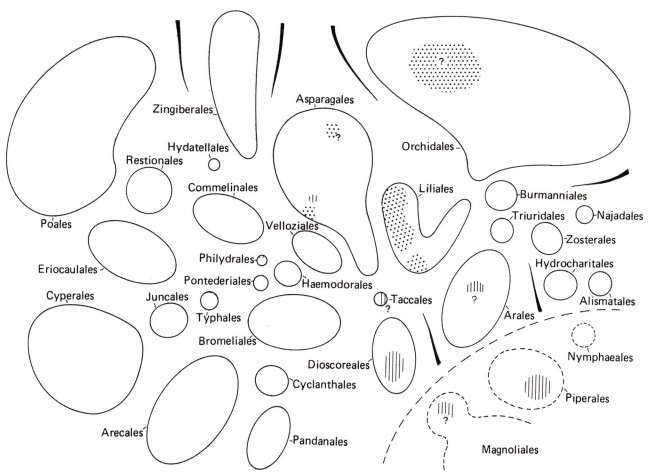

Diagram 9. Corms (dots) and hypocotyledonary tubers (vertical hatching). The corms mapped here may be taken as corms in a very strict sense, other similar structures being classified as variably short, oblique or almost vertical rhizomes.

Of the two types of leaf-insertion the possession of opposite leaves has little taxonomic significance except at lower levels. For example, it may be constant within genera as with *Halophila* (Hydrocharitaceae), subgenera as with *Potamogeton* subgenus *Groenlandia* (Zosterales), and individual species as in the genus *Dioscorea* (Dioscoreales).

In contrast the possession of verticillate leaves may be a characteristic of a family, as in Trilliaceae (Dioscoreales, Liliiflorae). In Orchidaceae verticillate leaves occur in at least one species of *Pogonia*, in both species of *Isotria* and in the monotypic *Codonorchis*. Elsewhere, verticillate leaves occur only in a few species of genera that otherwise have alternate leaves. For example *Lilium* and *Fritillaria* (Liliales, Liliiflorae), *Polygonatum* (Asparagales, Liliiflorae) and *Stemona* (Dioscoreales, Liliiflorae) all have some species with verticillate leaves. *Hydrothrix* in Pontederiales has dissected, falsely verticillate leaves (see Cook, 1978).

That opposite and whorled leaves often occur in species whose close relatives have alternate leaves suggests that the character has little or no phylogenetic significance.

Base Data

Opposite (or pseudo-opposite) leaves
ALISMATIFLORAE: HYDRO.: Hydro.: *Halophila*.

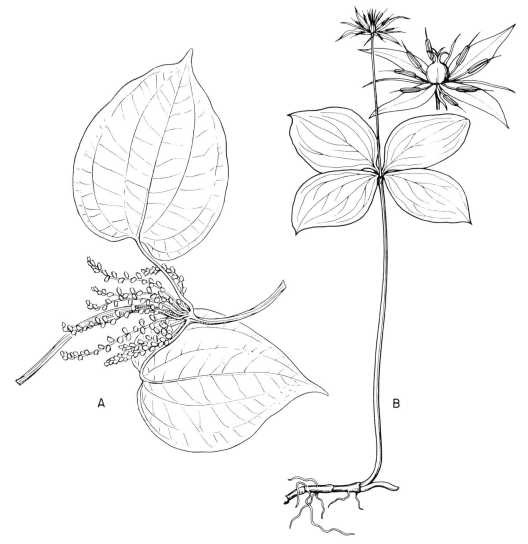

Fig. 19. A: opposite leaves in *Dioscorea longicuspis* (Dioscoreaceae) (from Milne-Read, 1975). B: verticillate leaves in *Paris quadrifloria* (Trilliaceae) (from Larsen, 1977).

—ZOSTE.: Potam.: *Potamogeton* (incl. *Groenlandia*) spp.; Zanni.: *Zannichellia* spp.—NAJAD.: Najad.: *Najas* (pseudo-verticillate).

LILIIFLORAE: DIOSC.: Diosc.: *Dioscorea* (several spp.); Stemo.: *Stemona* spp.—ASPAR.: Smila.: *Smilax*, few spp., *Ripogonum.*—LILIA.: Colch.: *Gloriosa* (sometimes).—ORCHI.: Orchi.: *Listera* and other genera; mostly subopposite only.

Verticillate (or pseudo-verticillate) leaves

ALISMATIFLORAE: ZOSTE.: Cymod.: *Cymodocea* spp.—NAJAD.: Najad.: *Najas* spp. (pseudo-verticillate).

LILIIFLORAE: DIOSC.: Stemo.: *Stemona* spp.; Trill.: *Paris, Trillium.*—ASPAR.: Conva.: *Polygonatum* spp.—LILIA.: Lilia.: *Lilium* spp., *Medeola, Fritillaria* spp.—ORCHI.: Orchi.: *Codonorchis, Isotria, Pogonia.*—PONTE.: *Hydrothrix* (falsely verticillate).

COMPOUND AND PSEUDO-COMPOUND LEAVES
(Fig. 20, Diagram 11)

Truly compound leaves in the sense of the lamina being replaced by leaflets are unknown for monocotyledons. However, an approach to such a leaf form is achieved either by irregular growth producing a deeply dissected lamina or by dissolution of rows of specialized cells to form a more or less regular series of "leaflets". Tearing of the lamina also occurs but the term leaflet is scarcely appropriate to the resulting segments.

The distributions of each of these three types of leaves is summarized in Diagram 11, from which it is clear that the range of each is restricted.

Dissection by irregular growth is limited to the Arales, where about 25% of the species are involved, Taccales,

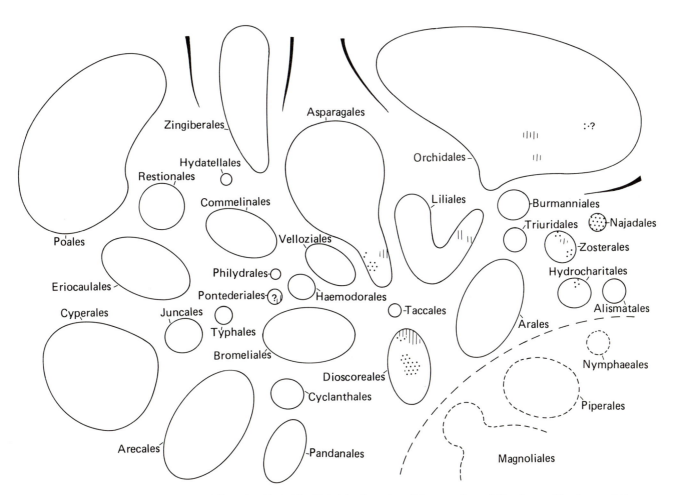

Diagram 10. Leaves opposite (dots) or verticillate (hatching). The Najadales (here dotted) has pseudo-verticillate leaves. The three orchid genera are placed according to Diagram 2.

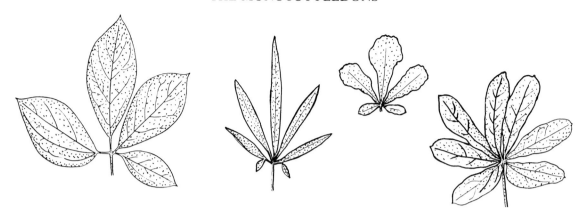

Fig. 20. Compound leaves in *Dioscorea quartiniona*. (From Burkill, 1960.)

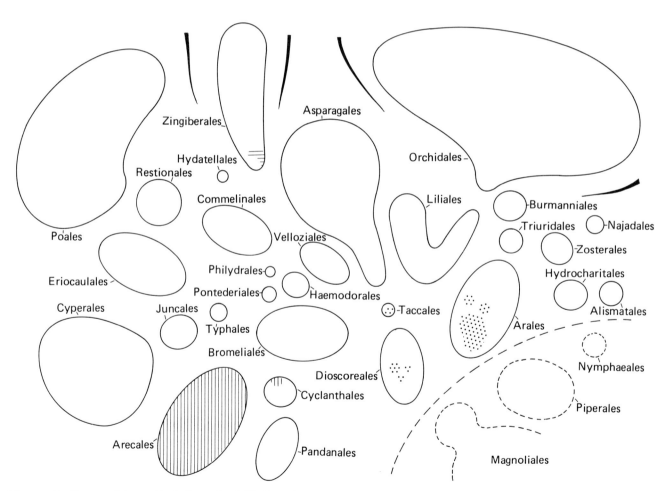

Diagram 11. Compound leaves (shaded) may be divided into three types: deeply dissected leaves (dots) occur in the Arales, Dioscoreales and Taccales; the arecalean and cyclanthalean leaves (vertically hatched) are initially entire but frequently split up into a compound state in the course of the development; other pseudo-compound leaves (horizontally hatched) occur in certain taxa of the Musaceae, Heliconiaceae and Strelitziaceae, where they are torn (dissoluted) by wind actional forces.

a few of which have dissected leaves, and Dioscoreales, where some species such as *D. quartiniona* have digitately compound leaves (Fig. 20). Decompound leaves occur in a few Araceous genera, such as *Amorphophallus*.

Dissolution of cells leading to a series of regular "leaflets" is widespread in the Arecales and Cyclanthales, two orders of Areciflorae.

Tearing rather than dissolution leads to tattered leaves in the Musaceae, Heliconiaceae and Strelitziaceae, all of the Zingiberiflorae.

The latter categories are here conceived as "pseudo-compound" leaves. Because of their different origins the three forms of "compound" and "pseudo-compound" leaves are totally different and so conclusions must not be drawn from a comparison between their distributions.

Nonetheless each of the leaf types has a restricted distribution. Those resulting from the dissolution of cells are confined to the Areciflorae and those from tearing to the Zingiberiflorae, whilst those due to irregular growth occur only in the Ariflorae and Liliiflorae. The occurrence of compound leaves resulting from irregular growth in the Taccales supports a possible close relationship between this order and both the Dioscoreales and Arales.

Amongst the dicotyledons showing affinities with the monocotyledons, none has compound leaves.

Base Data

ARIFLORAE: ARAL.: Arac.: several genera, representing between 15% and 25% of the species. Also pinnatisect and trifid leaves occur in the family (*Amorphophallus*, *Dracontium*).

LILIIFLORAE: DIOSC.: Diosc.: *Dioscorea* (some spp.).—TACCA.: *Tacca* (some spp.).

ZINGIBERIFLORAE (torn leaves): ZINGI.: occasionally in Musa., Helic. and Strel.

ARECIFLORAE (by dissolution of cells): ARECA. (generally).—CYCLA. (dissolution of cells common, but leaves mostly bifid only).

DISTICHOUS LEAVES
(Diagram 12)

Leaves inserted on the stem at 180 degrees from one another, and thus lying in one plane are described as distichous. Their distribution is shown in Diagram 12. Distichy is far more common in monocotyledons than in dicotyledons, but its ecological significance is somewhat uncertain. When the leaves are both equitant and

ensiform, the plants become flat and, as in many Iridaceae, may be compressed as a whole. Dorsiventral leaves when distichous are often arranged loosely on a stem, but they may also be strongly tufted as in many grasses. In the latter case the distichous phyllotaxis is not always obvious unless culms or lateral shoots are studied separately.

Leaves with a broad lamina, when concentrated in a basal rosette would have the disadvantage of shading each other if distichously arranged, consequently distichous leaves are rare or absent in broad-leaved members of such families as the Alismataceae, Hydrocharitaceae and Araceae as well as in most terrestrial rosette-leaved orchids (e.g. in Orchidaceae subfam. Orchidoideae). Nonetheless distichous broad-leaved rosette plants do occur in some Amaryllidaceae, e.g. *Haemanthus*. A stem with loosely scattered distichous leaves is typical of many grasses, many Zingiberales, many tropical Orchidales and also, for example, most aquatic taxa of Zosterales.

Although the distichous leaves apparently represent an ecological adaptation, their distribution shows some taxonomically interesting features.

In the Alismatiflorae distichous leaves are concentrated in the Zosterales (*Najas* and *Potamogeton* having sometimes opposite leaves which may be derived from these), while most taxa of Hydrocharitales and Alismatales have spirally set leaves. In the Ariflorae leaf insertion is variable.

The Liliiflorae are also highly variable in this character. Distichous leaves are for example found quite frequently in the Amaryllidaceae and in some Asphodelaceae of the Asparagales and in particular in the Iridaceae of the Liliales. In the Burmanniales many Burmanniaceae have distichous leaves, a condition also common in most tropical orchids, many of which belong to the Orchidaceae subfam. Epidendroideae. Spirally set leaves are concentrated mainly in the more temperate subfamilies Orchidoideae and (part of) Neottioideae.

The Haemodorales and Philydrales mostly have linear leaves which are distichously arranged, while most of the more broad-leaved taxa of the Pontederiales tend to have spirally set leaves . In the Bromeliales the particular functions of the leaf sheaths as water-storing structures favour a spiral arrangement.

Regarding the Commeliniflorae one will find a conspicuous difference between the Poales and Typhales, with distichous leaves, and most other orders, which generally have other phyllotaxy. Juncales and Cyperales agree in having almost constantly tristichous leaves, although there are certain exceptions in both orders (e.g. *Distichia* in Juncaceae). Of the Eriocaulales about half the species may have distichous leaves.

The Zingiberiflorae usually have distichous leaves, the most spectacular example being perhaps *Ravenala* (Strelitziaceae). In contrast the rather closely related *Musa* has spirally set leaves. The Costaceae differ from all Zingiberaceae in having non-distichous, spirally inserted leaves.

Finally, the Areciflorae have but few taxa with distichous leaves, mostly in the Cyclanthales.

It does not seem likely that the supposedly broad-leaved ancestors of the monocotyledons possessed distichous leaves, hence the condition may have evolved early is a secondary feature. It is probable that at least in some evolutionary lines distichy is to be regarded as having later given way to other forms of leaf arrangement in that with many genera, such as *Beaucarnea*, the seedlings have distichous leaves and the older plants leaves with other phyllotaxy. Hence it would appear that the condition of spirally set (non-distichous) leaves may be either primitive or derived.

Base Data

ALISMATIFLORAE: ZOSTE.: most taxa.
ARIFLORAE: ARAL.: Arac. maybe c. 10%.
LILIIFLORAE: ASPAR.: Smila.: *Smilax* spp.: Phile.: *Eustrephus, Geitonoplesium;* Dasyp.: *Lomandra;* Phorm.; Aspho.: e.g., in *Caesia, Kniphofia, Sowerbaea;* Diane.; Hemer.; Allia.: espec. subfam. Agaphanthoideae; Amary.: perhaps c. 50%.—LILIA.: Irida.: most taxa; Geosi.; Colch.: several genera, e.g. *Ornithoglossum;* Melan.: some genera, e.g. *Narthecium, Tofieldia.*— BURMA.: Burma. p.p.—ORCHI.: Orchi.: most spp., nearly all of subfam. Epidendroideae.—PONTE.: most taxa.—HAEMO.: all.—PHILY.: all.

ZINGIBERIFLORAE: ZINGI.: Lowia.; Strel.; Zingi.; Maran.

COMMELINIFLORAE: COMME.: Comme. p.p.— ERIOC.: Rapat.: ? most spp.; Xyrid.: most spp.; Erioc.: rarely.—TYPHA.: all.—JUNCA.: rarely, e.g. *Distichia.*

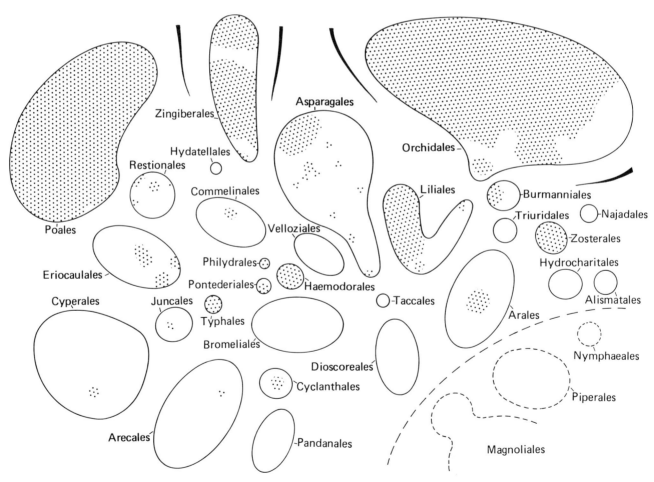

Diagram 12. Distichous leaves (shaded). The diagram is very approximate. In the Orchidaceae, non-distichous leaves are concentrated mainly in the subfamily Orchidoideae.

—CYPER.: rarely.—RESTI.: Centr.: rarely, but in *Centrolepis aristata*; Resti. p.p.; Flage.; Joinv.—POAL.: nearly all (except *Micraira*).

ARECIFLORAE: ARECA.: *Wallichia disticha.*—CYCLA.: some genera.

PETIOLES AND LEAF BASES
(Fig. 21, Diagram 13)

The presence of a petiole intercalated between the leaf base and the lamina is of widespread occurrence as is seen in Diagram 13. However amongst plants with petiolate leaves three distinct morphological groups may be recognized with regard to the leaf base. The encircling leaf base may be evident only on young shoots so that on mature shoots the petiole appears to be inserted directly on the axis, as in *Dioscorea*; the petiole may be inflated and channelled above, loosely investing the axis as in *Alisma*, or there may be a definite sheath investing the stem for some distance with a distinctly petiolate lamina as in many bamboo species. The distributions of these three types are quite distinct.

Dioscorea type

Distinctly petiolate leaves apparently lacking a sheathing base are restricted to the Liliiflorae. They are common in the Dioscoreales and the bacciferous Asparagales (Philesiaceae, Petermanniaceae and Smilacaceae) and occur also in Liliales (Alstroemeriaceae). This is also the petiole type of most dicotyledons associated with monocotyledons.

Alisma type

Petioles channelled above and expanding into a distinct

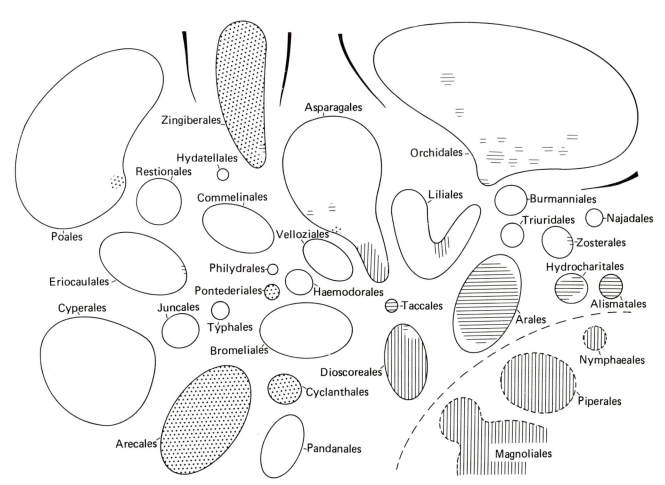

Diagram 13. Different types of petioles and leaf bases. The Dioscorea type: vertical hatching; the Alisma type: horizontal hatching; the Bambusa type: dots.

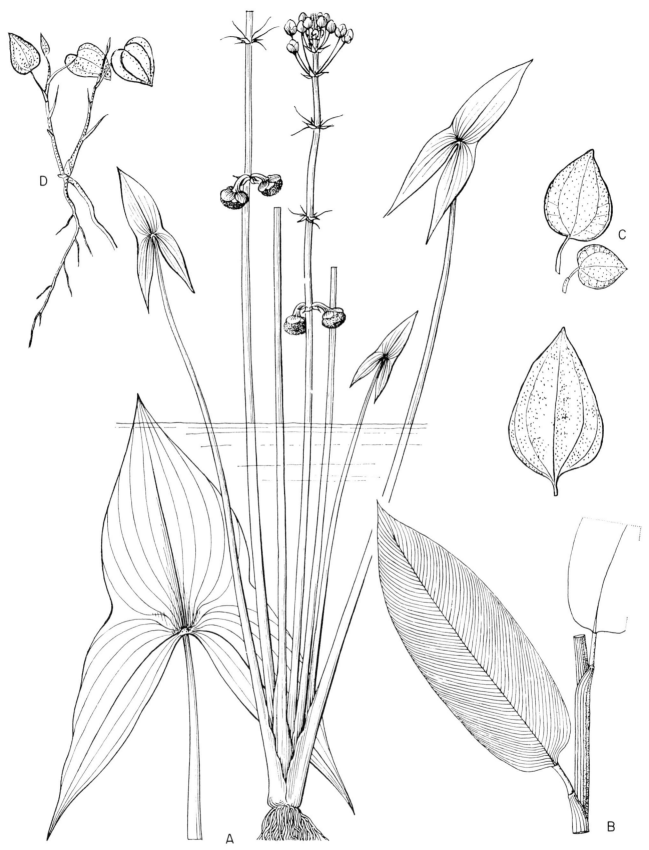

Fig. 21. Types of leaf petioles and leaf bases. A: Alisma type in *Sagittaria montevidensis* (Alismataceae). B: Bambusa type in *Hypselodelphys poggeana* (Marantaceae). C–D: Dioscorea type in *Smilax* sp. (A from Carter, 1960b; B from Koechlin, 1964; C–D from Holm, 1891.)

lamina are common in Alismatiflorae, occurring in three of the orders, Alismatales (Alismataceae), Hydrocharitales (Hydrocharitaceae and Aponogetonaceae) and Zosterales (Potamogetonaceae), but also in the Ariflorae (Arales), viz. in about 90% of the species. In the Liliiflorae, the Alisma-type leaves occur in the Taccales and some Orchidales, in the Commeliniflorae in a few species of *Rapatea*, and in Zingiberiflorae in the family Strelitziaceae. In Orchidales a well developed petiole is found in for example Apostasiaceae (at least three species), several Goodyerinae, some Diurineae, some genera of Bletinae, two of Sobraliinae, some species of *Malaxis* etc.

Bambusa type

Leaves with a well developed petiole between the sheath and lamina are widespread in Areciflorae (Arecales and Cyclanthales) and Zingiberiflorae (except for Strelitziaceae) and they also occur amongst the grasses, and especially the bamboos, in Commeliniflorae. The inclusion here of the majority of Zingiberiflorae may be questioned because the leaf margins overlap considerably (e.g. *Musa*) but if not placed here they become a separate group.

Though the origin of the petiole in each of the three groups may be similar the overall leaf morphology is different, suggesting three or more separate lines of evolution. The similarity of the Arales to Alismatales reinforces the possible phylogenetic affinities between the two and also hints at a link between the Ariflorae and the Dioscoreales and Taccales of the Liliiflorae. Likewise leaf form draws attention to the possible connection between the Liliales and Asparagales via Alstroemeriaceae and Philesiaceae.

Base Data

The petiole types below may be somewhat generalized, especially the border between the first two types which is weak.

Dioscorea type

LILIIFLORAE: DIOSC.: Diosc.: most genera, e.g. *Borderia, Dioscorea, Stenomeris, Tamus, Trichopus*; Stemo.: all genera.—ASPAR.: Smila.: most taxa; Peter.; Phile.: *Lapageria, Luzuriaga*; Conva.: rarely, e.g. *Majanthemum.*—LILIA.: Alstr.: spp. of *Alstroemeria* and *Bomarea.*

Alisma type

ALISMATIFLORAE: HYDRO.: Hydro.: most genera; Apono.—ALISM.: Alism.: most taxa.—ZOSTE.: Potam.: *Potamogeton* (several spp.).

ARIFLORAE: ARAL.: Arac.: c. 90% of the genera.
LILIIFLORAE: DIOSC.: Trill.: some spp. of *Paris* and *Trillium.*—TACCA.: *Tacca* spp. (perhaps Dioscorea type).—ASPAR.: Cyana.: *Cyanastrum* spp.; also few taxa in other families, e.g. in Aspho. subfam. Anthericoideae. —ORCHI.: Apost.: some spp.; Orchi.: scattered genera, mostly considered falsely petiolate (see the text).
COMMELINIFLORAE: ERIOC.: Rapat.: few spp.
ZINGIBERIFLORAE: ZINGI.: Strel.: all.

Bambusa type

LILIIFLORAE: ASPAR.: Hypox.: *Molineria, Curculigo.*—PONTE.: all; long petioles in spp. of *Eichhornia, Heteranthera, Monochoria* and *Pontederia*, inflated in certain *Eichhornia.*
COMMELINIFLORAE: POAL.: Poac.: common in subfam. Bambusoideae, otherwise scattered.
ZINGIBERIFLORAE: ZINGI.: all or most taxa except Strel. (see *Alisma* type).
ARECIFLORAE: ARECA.: all.—CYCLA.: all.

ENSIFORM LEAVES
(Fig. 22, Diagram 14)

Leaves with an equitant base and an isobilateral lamina are known as ensiform. For an account of their anatomy and references to several examples, see Arber (1925). Genera with species possessing ensiform leaves are widely distributed amongst monocotyledons and occur in four of the six superorders.

However only in the Liliiflorae are ensiform leaves at all common. This is due to their abundance in the large family Iridaceae. Ensiform leaves also occur in several genera of the Orchidaceae subfam. Epidendroideae *sensu lato*, e.g. in *Liparis, Malaxis, Oberonia, Dendrobium, Appendicula, Podochilus, Lochhartia, Dichaea, Oncidium, Angraecum* and *Bolusiella*, but in only a few of these genera is it consistent. Elsewhere in the superorder they occur in the Melanthiaceae (*Tofieldia, Narthecium*), Haemodoraceae (e.g. *Wachendorfia, Anigozanthos*) and Philydraceae (all genera).

In the Commeliniflorae, ensiform leaves are common amongst members of Xyridaceae and occur also in the Anarthriaceae (Cutler and Airy Shaw, 1965) and rarely in Centrolepidaceae (*Centrolepis aristata*; Hamann, personal communication). Leaves of this type also occur in *Acorus* (Araceae; Ariflorae) and *Tetroncium* (Juncaginaceae; Alismatiflorae).

Such a scattered distribution of ensiform leaves in

Fig. 22. Ensiform leaves. A: a plant of *Sisyrinchium arenarium* (Iridaceae), showing the typically ensiform leaves of many taxa in this family. B: *Iris pseudacorus* (Iridaceae), transverse section of leaf in the sheath region (to the left) and in the middle part (to the right). C: *Iris spuria*, transverse section of leaf. D: *Tetroncium* sp. (Juncaginaceae) plant and transverse sections from sheath, junction of sheath and lamina, and lamina of a leaf. (A from Correa, 1969; B–C and E–F from Arber, 1925; D from Pizarro, 1959.)

families as well as in orders suggests the attribute has evolved independently in a number of evolutionary lines.

Base Data

ALISMATIFLORAE: ZOSTE.: Junca.: *Tetroncium*, rarely and only partly in *Triglochin*.

ARIFLORAE: ARAL.: Arac.: at least in *Acorus*.

LILIIFLORAE: ASPAR.: Diane.: *Dianella* (only partly).—LILIA.: Irida.: practically all Isophysioideae, Aristeoideae and Ixioideae and most Iridoideae; Melan.: at least *Narthecium* and *Tofieldia*.—ORCHI.: Orchi.: numerous genera (see the text above).—HAEMO.: some genera, e.g. *Anigozanthos*, *Conostylis*, *Phlebocarya* and *Wachendorfia*.—PHILY.: all.

COMMELINIFLORAE: ERIOC.: Xyrid.: most taxa, e.g. *Achlyphila* and *Xyris*.—JUNCA.: rarely and atypically in some *Juncus* spp.—RESTI.: Centr.: *Centrolepis aristata*; Anart.: *Anarthria*.

INVERTED LEAF BLADES
(Fig. 23, Diagram 14)

The inversion of the leaf blade by a twisting through 180 degrees of the petiole occurs in three families of Liliiflorae. These are Alstroemeriaceae, Philesiaceae and Convallariaceae (*Drymophila*).

The restriction of such an unusual attribute to *Drymophila* alone in Convallariaceae along with other attributes suggests that the genus may be taxonomically misplaced in that family. Furthermore the joint occurrence of the attribute in Liliales and Asparagales supports a close relationship between the Philesiaceae and Alstroemeriaceae notwithstanding their being placed in different orders.

In this connection it should be added that leaf blades twisting through only 90 degrees are typical of the Rapateaceae (Eriocaulales).

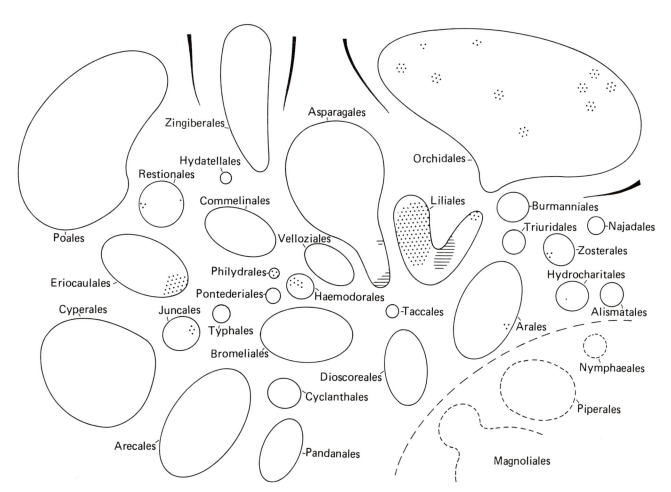

Diagram 14. Ensiform leaves (dots) and inverted leaf blades (hatched).

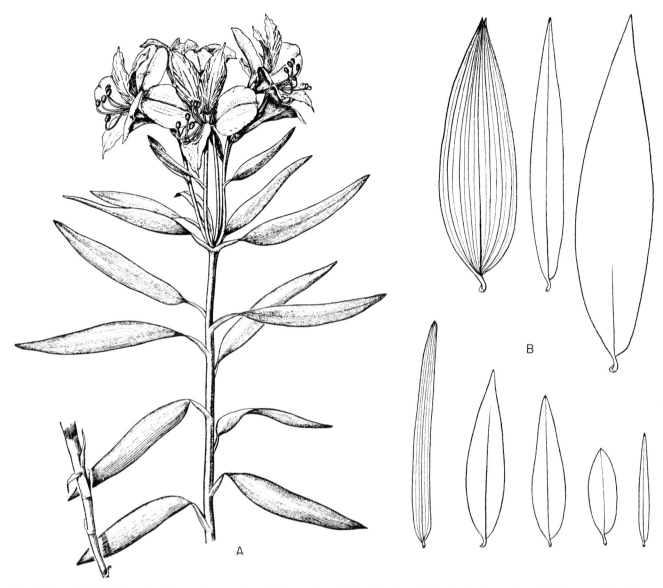

Fig. 23. Inverted leaf blades. A: *Alstroemeria aurantiaca* (Alstroemeriaceae) flowering branch showing the leaves turned 180 degrees at their bases. B: *Geitonoplesium cymosum* (Philesiaceae), variation of leaves within a single species. (Note the leaf base.) (A from Correa, 1969; B from Schlittler, 1951.)

Base Data

LILIIFLORAE: ASPAR.: Philes.: common feature though possibly not general; Conva.: *Drymophila* possibly misplaced here; see above).—LILIA.: Alstra.: general feature.

PTYXIS (Fig. 24, Diagram 15)

The term ptyxis is used to define the manner in which separate parts are folded in bud, vernation being the equivalent of aestivation for vegetative leaves.

The present data are based on a recent study by Cullen (1978), who we have followed in definitions as well as data given. No additions have been given to this account except for the orchids where we have followed Pfitzer (1887, 1888–1889), Mansfeld (1937) and Rosso (1966).

The types of ptyxis are illustrated in Fig. 24, the flat, curved, conduplicate, conduplicate-plicate, plicate, involute and (rarely) supervolute-curved and explicative types being applicable to monocotyledonous leaves.

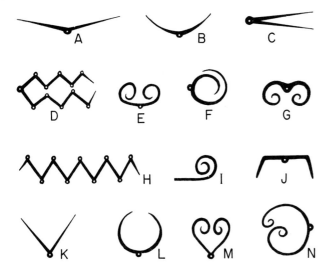

Fig. 24. Types of ptyxis. All the leaves are shown in transverse section except one (I) shown in longitudinal section A: flat*; B: curved*; C: conduplicate*; D: conduplicate-plicate*; E: involute*; F: supervolute*; G: revolute; H: plicate*; I: circinate; J: explicative*; K: conduplicate-flat; L: supervolute-curved; M: conduplicate-involute; N: supervolute-involute. Those represented in the monocotyledons have an asterisk. In the diagram the flat and curved are treated together, as are also the plicate and conduplicate-plicate, which gives somewhat more homogeneous groups. (From Cullen, 1978.)

From the outset it should be stressed that the homology of the leaves in different groups of monocotyledons and related dicotyledons are of crucial importance for any conclusions to be drawn from ptyxis similarities. For example, conduplicate or supervolute leaves in grasses must not be equated with the same types in the Arales or Piperales as the leaf blades are differentiated in different ways and, therefore, are non-homologous structures.

Despite such limitations the pattern of variation in monocotyledons shows some interesting features (Diagram 15). First, a similarity is apparent between those members of the Arales and Alismatales which have well differentiated petiolate leaves with a broad lamina. In these the ptyxis is supervolute, as is the condition in many taxa of the Asparagales and Liliales with non-petiolate leaves. The leaves of certain species of *Galanthus* (sect. *Plicatae*) and of *Phaedranassa* have explicative leaves (asterisked in the diagram), a feature of diagnostic value (Cullen, 1978; Stern, 1956).

Furthermore, Cullen noted that the ptyxis in some Nymphaeaceae (Nymphaeales) is not supervolute as in the Alismatales but rather involute as in most species of *Piper* and some of *Peperomia* (Piperaceae) and those in *Saururus* (Saururaceae), all members of the Piperales. For other indications of relationship between Piperales and Nymphaeales, see seed structure, starchy perisperm etc.

Most taxa of Iridaceae (Liliales) but fewer of Asparagales tend to have conduplicative (more rarely conduplicative-plicate) ptyxis as do the few taxa recorded for the Velloziaceae (Velloziales) and Philydraceae (Philydrales).

The orchids (Orchidales) according to Cullen were found to have conduplicative or, more rarely, conduplicative-plicate ptyxis. However Pfitzer (1887; 1888–1889) and Mansfeld (1937) distinguished between duplicate (= conduplicate) and convolute (= possibly rather supervolute *sensu* Cullen) leaves in the Orchidaceae, and it seems that most taxa of Orchidaceae subfam. Neottioideae and Orchidoideae have convolute ptyxis, and Apostasiaceae and Cypripediaceae (Rosso, 1966) duplicate or plicate ptyxis. Probably the discrepancy here is mainly a matter of concept, but may deserve a separate study. (In the diagram the orchids have been mapped according to Pfitzer and Mansfeld, op. cit.)

The Zingiberales stand out as being quite uniform in having supervolute leaves, a type which is also well represented in the Commelinaceae (Commelinales), which otherwise have leaves that are involute.

The Arecales (and, no doubt, the Cyclanthales though not recorded by Cullen) have plicate leaves, a noticeable difference from the Arales, which serves, together with many other attributes, to distinguish the palms and aroids from each other. The sessile leaves of Pandanales have conduplicative leaves.

The orders of the Commeliniflorae are somewhat variable in ptyxis although the supervolute type seems dominant at least in the Poales (and maybe in Restionales), Commelinales and Juncales while the leaves at least in some grasses (*Poa*, *Glyceria*, *Paspalum*) and in a number of sedges (Cyperales) have conduplicative ptyxis.

Some of the dicotyledons commonly associated with the monocotyledons have been mentioned above, viz. the Nymphaeales and some Piperales agreeing in having involute ptyxis. The ptyxis type, dominant in the Magnoliales and common in the Piperales (*Peperomia*), is however the conduplicative.

Thus the question, what type of ptyxis is the basic (i.e. more ancestral), is highly problematic. Is it the Magnolialean conduplicative, the Piperalean–Nymphaealean involute or the Alismatalean–Aralean–Lilialean supervolute type? Each of these types of ptyxis occurs in leaves with a well differentiated petiole and lamina. No doubt each type should be seen against the habit of the plants in combination with their leaf shapes and the habitats they occupy. Thus the Magnoliales are trees or tree-derived vines or herbs, the Nymphaeales are water-plants,

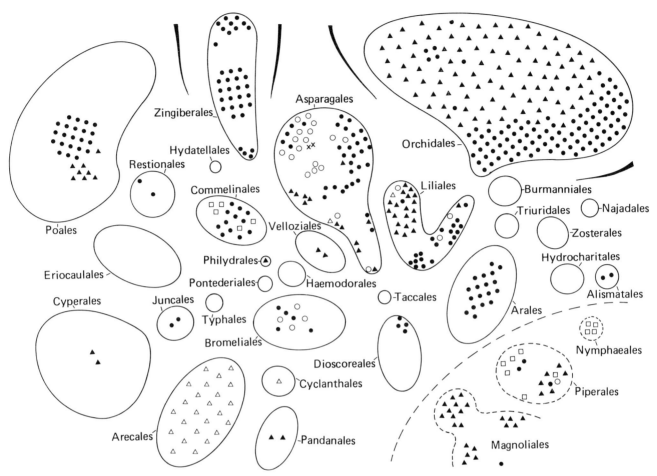

Diagram 15. Ptyxis. Flat or curved: hollow circles; supervolute: solid circles; conduplicate: solid triangles; plicate and conduplicate-plicate: hollow triangles; involute: hollow squares; explicative: crosses.

the Piperales vines or herbs, and the Alismatales–Arales, hydro- or hygrophytic herbs or climbers. Among these the habit of the Magnoliales was probably ancestral.

Base Data

Data given in the diagram are derived mainly from Cullen (1978).

RETICULATE VENATION
(Fig. 25, Diagram 16)

Leaves with reticulate venation are very restricted amongst monocotyledons (Diagram 16) there being only two superorders, the Ariflorae and the Liliiflorae with species possessing such leaves. In neither superorder is the condition common.

Of the Ariflorae only about 10% of the species have reticulate venation as exemplified by *Arum maculatum*. The remaining species have convergent or parallel venation even where the leaves are deeply dissected.

Amongst the Liliiflorae reticulate venation occurs in five of the orders. These are the Taccales and Dioscoreales where the condition is found in nearly all members, the Liliales where in *Tricyrtis* (Tricyrtidaceae) the veins may be reticulate, and the Asparagales and Orchidales. In the Asparagales five families are involved. They are the Smilacaceae (Koyama, 1963), Petermanniaceae, Funkiaceae, Convallariaceae and Philesiaceae (Schlittler, 1949). Lateral connections between parallel veins are not rare; they occur in broad leaves of genera such as *Veratrum* (Melanthiaceae) etc. In the Orchidales a few genera, probably as a secondary attribute, have broad leaves, and reticulate venation.

Such a dicotyledonous character as this has previously attracted the attention of phylogenists, and Hallier (1912) stressed it as one of the characters held in common by the Araceae and Piperaceae. Huber (1969) pointed out that the character is also held in common by the Dioscoreaceae and woody "Polycarpiceae".

Base Data

ARIFLORAE: ARAL.: Arac.: several or numerous genera, e.g. *Arum* and *Pothos* (Fig. 25).

LILIIFLORAE: DIOSC.: Diosc. and Stemo. (all genera) also in Trill.—TACCA.: *Tacca* (unpublished observations).—ASPAR.: Smila.: *Smilax* etc., Peter.: *Petermannia*; Phile.: *Behnia, Lapageria, Philesia*; Conva.: *Diosporum*; Funki.: *Hosta*.—LILIA.: Tricy.: *Tricyrtis*.—ORCHI.: Orchi.: rarely.

Dicotyledons associated with the monocotyledons: most taxa.

STIPULES AND SIMILAR STRUCTURES (Fig. 26, Diagram 17)

The definition and homologies of stipules are perhaps more difficult in monocotyledons and associated dicoty-ledons than in any other part of the system. Thus it is highly important to know whether the stipular or stipule-like structures in some taxa of Piperales and of Nymphae-ales are homologous, or whether the stipules in Nym-phaeales (e.g. in *Nymphaea*) correspond to those in some Alismatiflorae, where they occur in some Hydro-charitales, the Zosterales and the Najadales. In the Dioscoreales, "stipular" structures are found at the base of the pedicel in *Tamus* and several species of *Dioscorea*, and in *Smilax* (Smilacaceae, Asparagales) the petioles bear a pair of lateral tendrils (Yates and Duncan, 1970), having an important function for this climbing plant.

Finally, there are structures which are alternately described as ligules and as stipules although they are ambiguous in both categories, e.g. the axillary hyaline structures in the Pontederiaceae and the lateral hyaline lobes found on each side of the sheath in species of, for example, *Juncus* (Juncales) or in *Joinvillea* (Restionales). These are not accepted as stipules in Diagram 17.

In monocotyledons (Domin, 1911) the stipules have their least disputable occurrence among the Alismati-florae. In the Hydrocharitales (Glück, 1919), stipules occur in Hydrocharitaceae, at least in *Hydrocharis*, but their distribution is neglected in literature. In the

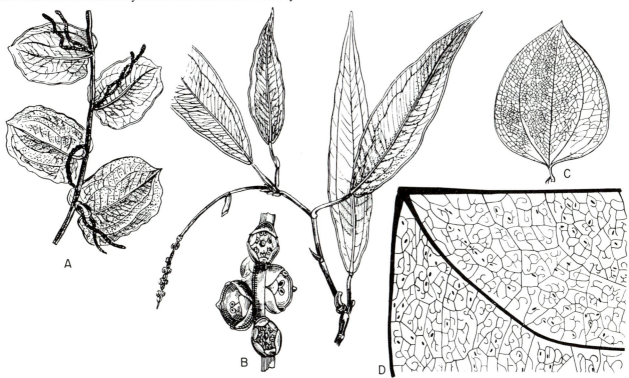

Fig. 25. Reticulate venation in monocotyledons. A–B: *Pothos beccarianus* (Araceae) showing leaf dimorphism with a climbing shoot (A) and a flowering shoot (B). C: a leaf of *Smilax rotundifolia* (Smilacaceae). D: details of venation in *Tamus communis* (Dioscoreaceae). (A–B from Engler, 1919; C from Holm, 1891; D from Burkill, 1960.)

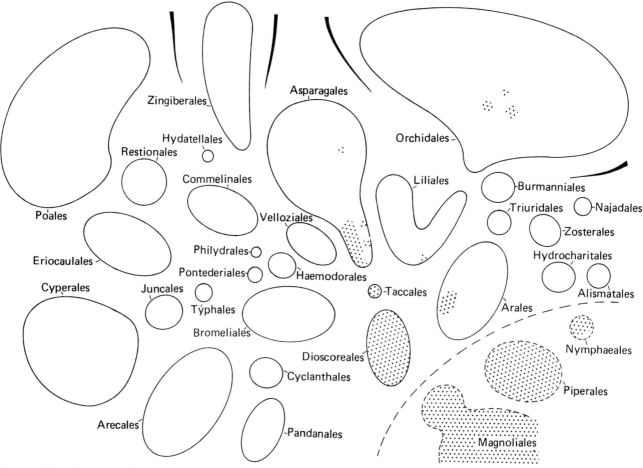

Diagram 16. Reticulate venation (shaded).

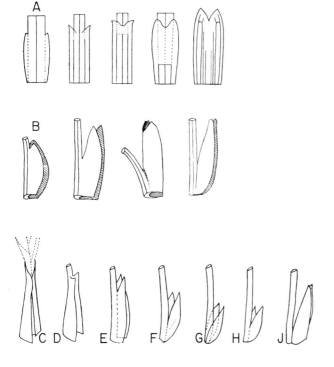

Zosterales, stipule-like or stipular lobes are found in *Scheuchzeria* (Scheuchzeriaceae) and in at least *Triglochin* and *Lilaea* of the Juncaginaceae, but their richest development is met with in *Potamogeton* and *Ruppia* of the Potamogetonaceae, where they form intravaginal "stipular sheaths". These may be conspicuous although they are more or less hyaline structures. In *Potamogeton natans*, they may exceed 50 mm in length. Also in *Najas* stipular or stipule-like structures occur on the leaves.

The occurrence of stipules in the Alismatiflorae as well as Nymphaeiflorae has been used, among other things, to support an affinity between these groups. Glück (1901) drew attention to this in an illustration where also the gradual transition between lateral and intravaginal

Fig. 26. Stipules and stipule-like structures in monocotyledons and certain dicotyledons (Nymphaeales). A: stipulate base of some successive leaves in *Potamogeton rufescens* (Potamogetonaceae). B: the same in *Nymphaea alba* (Nymphaeaceae). C–J: leaf bases with sheath and/or stipules or stipule-like lobes. C: *Aglaonema* (Araceae); D: *Anomochloa* (Poaceae); E–F: *Nymphaea* (Nymphaeaceae); G: *Najas* (Najadaceae); H–J *Nymphaea* (Nymphaeaceae). (All from Emberger, 1960.)

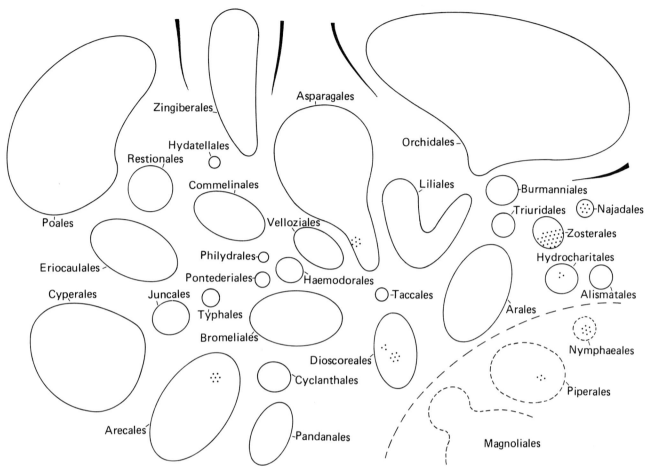

Diagram 17. Stipules and stipule-like structures (shaded). Basal leaf auricles as a rule are not included.

stipules was shown in *Potamogeton* and *Nymphaea* (Fig. 26).

Some doubts might be raised as to the importance of stipules in supporting a close affinity between the Nymphaeiflorae and the Alismatiflorae, but as other characters indicate a close agreement this also should be pointed out.

Base Data

ALISMATIFLORAE: HYDRO.: Hydro.: *Hydrocharis* and maybe other genera.—ZOSTE.: Scheuch.: *Scheuchzeria*; Junca.: at least *Lilaea* and *Triglochin*; Potam.: *Potamogeton, Ruppia.*—NAJAD.: *Najas.*

LILIIFLORAE: DIOSC.: Diosc.: *Dioscorea* spp., *Tamus.*—ASPAR.: Smila.: *Smilax.*

(Other structures of stipule-like nature in monocotyledons, viz. in Juncaceae, Pontederiaceae, Zingiberaceae etc. are mentioned under "ligules" below.)

Dicotyledons associated with monocotyledons.
NYMPHAEIFLORAE: NYMPH.: Nymph. p.p.
MAGNOLIIFLORAE: MAGNO.: Magno.—
PIPER.: Piper.: rarely (not really stipules according to Weberling, 1970).—LAUR.: Chlor.; Austr., Lacto.

LIGULES (Fig. 27, Diagram 18)

A ligule is an upgrowth of tissue from the adaxial surface of the leaf near to the junction of the sheath and lamina. Its function is uncertain and in origin it varies from being an outgrowth of the lamina as in grasses (Poaceae) or an extension of the sheath as in aroids (Araceae). The "ligule" in seagrasses (Zosterales) is here considered a derivative of the stipules and is not included here. The

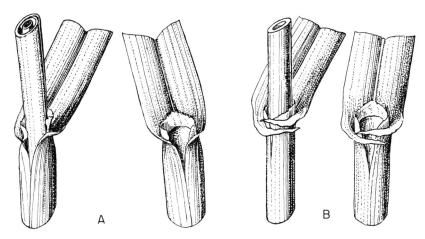

Fig. 27. Ligules. A: *Avena sativa*; B: *Hordeum sativum*. (From Hegi, 1935.)

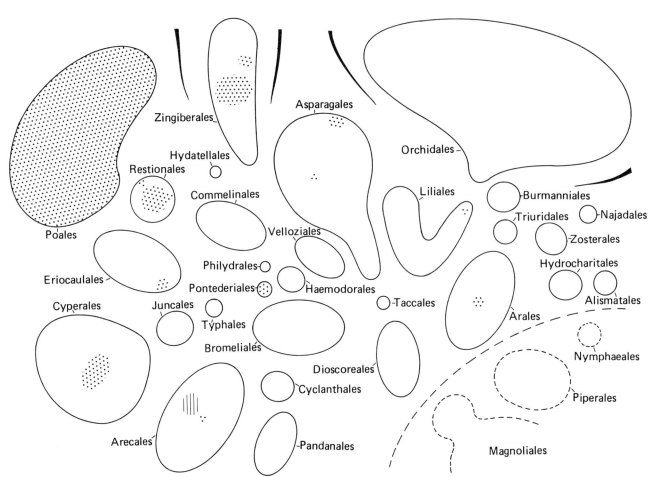

Diagram 18. Ligules (shaded) and hastulas (hatched).

hastula of palms has not been accepted as a form of ligule either, because it may occur on both leaf surfaces and is not associated with the leaf sheath (see Arber, 1925, p. 70). However, an inflated ligule or ochrea-like structure is sometimes present in Lepidocaryoid palms (Moore, 1973).

The distribution of ligules is shown in Diagram 18 from which it is clear that although ligulate species occur in most superorders they are widespread in none. Furthermore in only two orders, the Poales and Pontederiales, is the ligulate condition universal.

Nonetheless within individual orders where the ligules are almost certainly homologous their presence is useful for suggesting relationships. Thus the occurrence of a

ligule in *Joinvillea* supports the viewpoint that this genus is closely allied to the Poaceae (Poales). Likewise the occurrence of ligules in numerous species of the Zingiberaceae and Costaceae suggests they are more closely related to each other than to other families of the Zingiberales.

Structures resembling ligules are unknown amongst those dicotyledons considered as having affinities with the monocotyledons.

Base Data

("Ligules" in our sense probably represent partly non-homologous structures.)

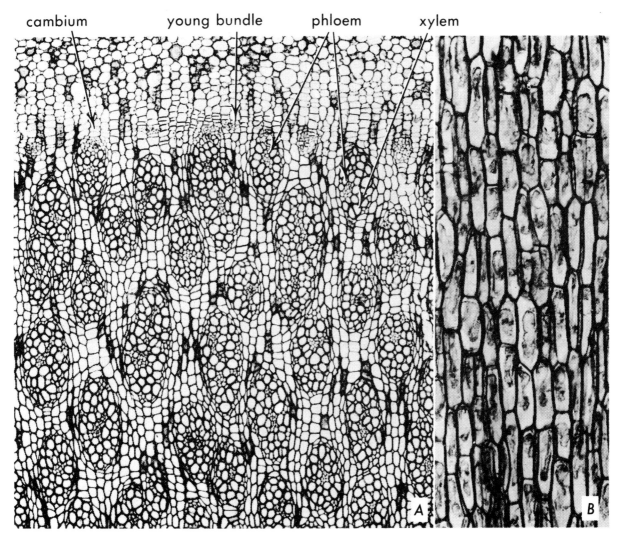

Fig. 28. Secondary growth in a stem of *Cordyline* (Asphodelaceae subfam. Astelioideae). A: cross section of the secondary vascular tissues. B: tangential section of the cambial tissue. (From Cheadle, 1943a and 1937.)

ARIFLORAE: ARAL.: Arac.: at least *Calla*.

LILIIFLORAE: ASPAR.: Aspho.: *Sowerbaea*; Allia.: *Allium* spp.—LILIA.: Melan.: *Pleea*.—PONTE.: *Eichhornia, Pontederia* etc.

ZINGIBERIFLORAE: ZINGI.: Zingi. and Costa.: at least 30% of the genera, e.g. in *Achasma, Aframomum, Alpinia, Elettaria, Hedychium, Phaemeria, Renealmia, Riedelia, Thylacophora, Zingiber*.

COMMELINIFLORAE: ERIOC.: Xyrid.: *Xyris* (most spp., Hansen, 1979) (not in *Achlyphila* or *Abolboda*).—JUNCA.: Junca.: rarely and dubiously typical.—CYPER.: few genera (c. 10%), e.g. in *Fuirena, Scirpus* p.p., *Scleria*.—RESTI.: Resti.: many; Centr.: rarely, in *Gaimardia*; Joinv.: *Joinvillea*.—POAL.: practically all.

ARECIFLORAE: ARECA.: inflated stipular "ochrea" sometimes in Lepidocaryoid palms (hastulas in many palm genera).

(ABERRANT) SECONDARY THICKENING GROWTH IN STEM
(Fig. 28, Diagram 19)

Secondary thickening growth is usually lacking in the stems of the monocotyledons even where the stem is thick as in palms. However, secondary thickening growth occurs in woody and herbaceous stems of certain monocotyledons of the superorder Liliiflorae, viz. mainly in the Dioscoreales and Asparagales, but also, although rarely, in the Liliales (Diagram 19). It is found in single to several genera in the families Dioscoreaceae, Agavaceae, Asphodelaceae (Fig. 28), Dasypogonaceae, Xanthorrhoeaceae and Iridaceae.

The vascular cambium in these plants is continuous with the primary thickening meristem if the latter is discernible, but functions in the part of the stem that has completed elongation. The cambium arises in the paren-

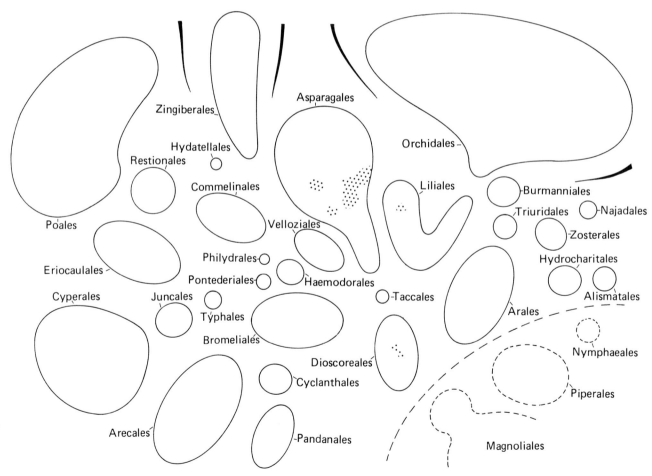

Diagram 19. Secondary growth in stems (shaded).

chyma outside the vascular bundles and produces vascular bundles and ground parenchyma towards the inside and a small amount of parenchyma towards the outside.

Plants with this type of secondary thickening growth usually have an arborescent or shrubby habit. Their taxonomic distribution has been reviewed by Tomlinson and Zimmermann (1969), Tomlinson (1970) and Philipson *et al.* (1971), from which the present data are taken.

As this type of cambium is different from the one in the dicotyledons, including those associated with the monocotyledons, i.e. mainly the Magnoliales, it is usually considered to be a separate evolutionary achievement and not, therefore, a relict character. It is then of interest to estimate whether it appeared in one or several evolutionary lines in the Liliiflorae.

The families of Asparagales having secondary growth are all more or less closely related, but most of them include taxa without this type of cambium.

This makes the problem ambiguous. Possibly the ability to form this cambium evolved along one line in the Dioscoreales, in one line in Liliales (the Iridaceae), and in one or a few lines in the Asparagales, but was lost in certain derivatives as these retained or returned to a herbaceous habit.

Base Data

(The data are probably not complete, but may be regarded as representative.)

LILIIFLORAE: DIOSC.: Diosc.: *Dioscorea* (incl. *Testudinaria*), *Tamus.*—ASPAR.: Dracae.: *Beaucarnea, Dasylirion, Dracaena, Nolina, Sansevieria*; Dasyp.: *Kingia, Lomandra*; Xantho.: *Xanthorrhoea*; Agav.: *Agave, Furcraea, Yucca* and probably more genera; Aspho.: *Aloë, Cordyline, Kniphofia.*—LILIA.: Irid.: *Aristea, Klattia, Nivenia, Witsenia, Patersonia* (J. T. Waterhouse, personal communication).

VESSELS IN ROOTS, STEMS AND LEAVES
(Fig. 29, Diagrams 20–22)

The presence and types of vessels in monocotyledons have been commented upon and surveyed by Cheadle (1943a and b) and Wagner (1977), whose data are summarized in Diagrams 20, 21 and 22.

Vessels may be absent from various vascularized parts of the plant, in which case the vascular elements consist of tracheids only. When the vascular elements have perforated end walls and the cell contents are non-living they are defined as vessels. These vary much in length, in the obliquity of the end walls, in the number and shape of perforations etc. (Fig. 29). The most primitive vessel types are considered to be those which are long and narrow with oblique walls perforated by numerous narrow slits (scalariform perforation plates); the most

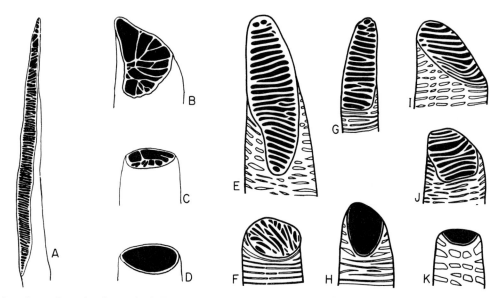

Fig. 29. Perforation plates of vessels of roots in A–D: *Musa sapientum* (Musaceae); E and I: *Agapanthus orientalis* (Alliaceae); F: *Gethyllis afra* (Alliaceae); G: *Cyrtanthus mackenzii* (Amaryllidaceae); H: *Allium christophii* (Alliaceae); J: *Agapanthus patens* (Alliaceae); K: *Tulbaghia violacea* (Alliaceae). (A–D from Tomlinson, 1969; E–K from Cheadle, 1969.)

advanced are short and broad and have transverse end walls with a single large circular perforation (simple perforation plates). Between these extremes there is a continuous transition and the perforation plates of the end walls may also be reticulate. Scalariformly perforated and simply perforated vessels may occur in the same part of the plant in variable proportions, but in most cases one type is exclusive or dominant.

For the monocotyledons it should be remembered that the stem has a so-called atactostele with scattered vascular bundles thus precluding the formation of secondary xylem along the path followed by dicotyledons. Therefore comparisons with the secondary vascular tissue of dicotyledons are irrelevant, especially for drawing phylogenetic conclusions based on homologies.

As is seen from the diagrams, vessels are lacking in the stems and leaves of many families of monocotyledons, while they are generally present in the roots (although even here lacking in the roots of many Alismatiflorae). When vessels are present in stems and leaves as well as in the roots they tend to be more "advanced" in the roots than in the stems and leaves. They may, for example, have simple or scalariform perforation plates with few bars in the roots, but scalariform perforations with more numerous bars in stems and leaves.

The following trends are observable. In the Alismatiflorae and Ariflorae stems and leaves are constantly devoid of vessels, whereas the roots often have vessels. These have simply perforated end walls in the Alismataceae (Alismatales) and Butomaceae (Hydrocharitales), but scalariform perforations in several genera of Hydrocharitaceae (Hydrocharitales) and in the Scheuchzeriaceae, Juncaginaceae and part of Potamogetonaceae (Zosterales); vessels with scalariform perforation plates also occur in the roots of most (?) Arales.

Many taxa of the Hydrocharitaceae and all of the

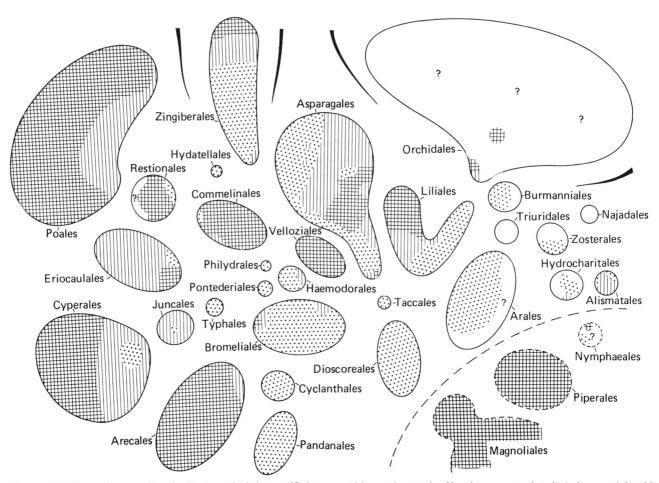

Diagram 20. Vessels in roots. Vessels absent: unshaded; unverified groups with question marks. Vessels present and exclusively or mainly with scalariform perforation plates: dots. Vessels present, types with scalariform and simple perforation plates both present: hatching. Vessels present, with mainly or exclusively simple perforation plates: cross-hatching. Vessel conditions in orchids are incompletely known and the blank areas in Orchidales do not represent absence of vessels.

Aponogetonaceae (Hydrocharitales), and of the marine families as well as the Zannichelliaceae (Zosterales), Najadaceae (Najadales) and Triuridaceae (Triuridales) lack vessels, as also do all Lemnaceae and possibly also some Araceae (Arales).

In the Liliiflorae the pattern is more variable. The roots probably always contain vessels, except in some saprophytic taxa such as *Petrosavia* (Melanthiaceae). It is interesting to note that the root vessels have simple perforation plates in the Velloziaceae (Velloziales) and in many of the tree-like or shrubby taxa of Asparagales (e.g. in the Asparagaceae, Dracaenaceae, Xanthorrhoeaceae, Dasypogonaceae and Agavaceae), which thereby agree with many Commeliniflorae in having advanced root vessels. Several large families, including the Iridaceae, Asphodelaceae, Alliaceae, Hyacinthaceae and Convallariaceae show a variation from scalariform to simple perforation plates in root vessels, whereas scalariform

perforation plates only (and hence primitive vessels) occur in the Dioscoreales, Taccales, most Liliales except Iridaceae, and in most or all taxa of Philesiaceae, Smilacaceae, Petermanniaceae, Hypoxidaceae, Cyanastraceae, Tecophilaeaceae and Amaryllidaceae in the Asparagales. Predominantly or exclusively scalariform perforation plates are also found in root vessels of the Haemodoreales, Bromeliales, Pontederiales and Philydrales. The orchids are poorly known, but simply perforated vessels occur in several of the taxa studied.

In the Liliiflorae vessels are more restricted in the stems and leaves and when present almost invariably have scalariform perforations. They occur in the Dioscoreaceae and Stemonaceae (Ayensu, 1968) (both of Dioscoreales), many of the bacciferous Asparagales (though only rarely in Convallariaceae) and in many Asphodelaceae. An interesting connection between the Alstroemeriaceae and bacciferous Asparagales is indicated by

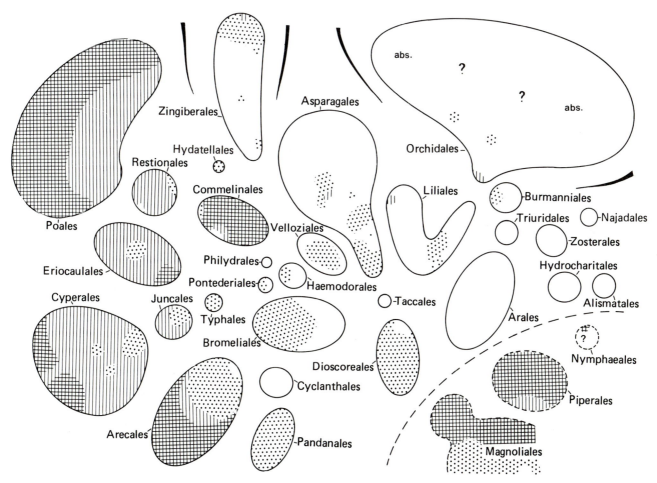

Diagram 21. Vessels in stems. Vessels absent: unshaded; unverified groups with question marks. Vessels present and exclusively or mainly with scalariform perforation plates: dots. Vessels present, types with scalariform and simple perforation plates both present: hatching. Vessels present, with mainly or exclusively simple perforation plates: cross-hatching.

both having scalariformly perforated vessels in the stem, but such vessels are also found in *Tricyrtis* (Tricyrtidaceae), *Sandersonia* (Colchicaceae; Cheadle and Kosakai, 1971) and in a variable proportion of the taxa studied in the Bromeliales, Haemodorales, Velloziales and Pontederiales, and also in certain orchids although the frequency in this group is little known. Unexpectedly, simply perforated vessels are known in *Sisyrinchium* of the Iridaceae and *Apostasia* (*Adactylus*) of the Apostasiaceae, this family otherwise being considered a primitive member of Orchidales.

Vessels are sometimes present in leaves, but missing in the stem, for example in some Velloziaceae, in species of *Xanthorrhoea* (Xanthorrhoeaceae), Dasypogonaceae and *Nolina* of Dracaenaceae. Vessels here may be associated with the formation of secondary vascular tissue in the form of tracheids in the stem. Also of interest is the presence of vessels in the stems and leaves of at least one species of *Burmannia*.

The Zingiberiflorae, as far as known, have vesselless leaves, while vessels with scalariform (rarely single) perforation plates occur in the stems of most members of the Marantaceae and isolated genera in the Costaceae, Zingiberaceae and (probably more often) in the Strelitziaceae. The same tendency is seen in the roots, where most Zingiberales except the Marantaceae have vessels with scalariform perforation plates, some genera of Zingiberaceae and certain of the Marantaceae have scalariform to simple perforation plates, and most genera of Marantaceae only simple perforation plates. This gives a variation series along the "vertical axis" of the order as illustrated in Diagram 21.

In the Commeliniflorae the roots are always provided with vessels. These have simple perforations in the Commelinaceae (Commelinales), Restionaceae (Restionales), Xyridaceae (Eriocaulales) and many grasses (Poales) and sedges (Cyperales). Vessels with scalariform perforation plates only are known in the Centrolepida-

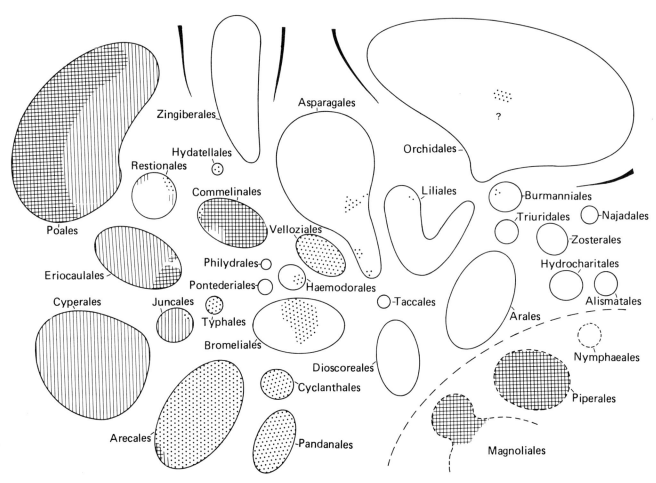

Diagram 22. Vessels in leaves. Vessels absent: unshaded; unverified groups with question marks. Vessels present and exclusively or mainly with scalariform perforation plates: dots. Vessels present, types with scalariform and simple perforation plates both present: hatching. Vessels present, with mainly or exclusively simple perforation plates: cross-hatching.

ceae (Restionales), Hydatellaceae (Hydatellales), most genera of Rapateaceae and both genera of Typhales. In the remaining families of Commeliniflorae vessel perforation varies from scalariform to simple.

Equally informative are the vessel conditions in the stems. Here again the Commelinaceae and Xyridaceae show an advanced stage in having vessels with simple perforation plates only, as do many genera of grasses and some of sedges. Most of the remaining families show a variation in vessel perforation between simple and scalariform plates, but scalariform perforation plates only are known in the Mayacaceae (Commelinales), Centrolepidaceae (Restionales), Hydatellaceae (Hydatellales), Thurniaceae (Juncales), Typhaceae and Sparganiaceae (Typhales), these families being accordingly less advanced in vessel structure. In addition, *Hanguana* (Hanguanaceae, here in Restionales) and *Cartonema* (Cartonemataceae, Commelinales) have been noted to lack vessels in the stem. Smithson (1957), used this observation and some others when arguing that *Hanguana* is more closely related to the Xanthorrhoeaceae and other Asparagales than to the Flagellariaceae and other grasslike families here placed in the Restionales. The lack of

vessels in stems and leaves of *Cartonema* is remarkable in the light of their constant occurrence and advanced state in the closely related Commelinaceae.

The Areciflorae are somewhat heterogeneous in regard to vessel characters. Vessels are always present in the roots having simple perforation in the Arecales and scalariform in the Cyclanthales and Pandanales. The stem likewise has vessels in most of the Arecales and all the Pandanales studied, the former with scalariform perforation plates only, but in the Cyclanthales (as in *Phytelephas* of the Arecales) the stems are vesselless. Nonetheless, the leaves in the Cyclanthales, as in the other two orders, have vessels with scalariform perforation plates. In this superorder, the Cyclanthales stand out as having more primitive vessels than taxa of the other two orders.

The dicotyledonouss groups connected with the monocotyledons show great diversity in vessel conditions. Nymphaeiflorae (Nymphaeales), which in the present treatise do not include *Nelumbo*, appear to lack vessels altogether, even in the roots, thus showing a close agreement with the aquatic Alismatiflorae, e.g. taxa of Hydrocharitaceae. However, Inamdar and Aleykutty (1979)

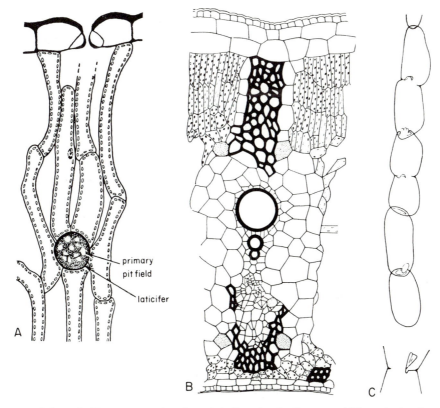

Fig. 30. Laticifers in monocotyledons. A: Transverse section through bulb scale of *Allium cepa* (Alliaceae), showing mesophyll cells and an articulated laticifer with the end wall in surface view. B: articulated laticifers (stippled) in *Musa* (Musaceae) accompanying a vascular bundle. C: diagrammatic illustration of a portion of a laticifer in *Musa*, showing the articulation; below is seen the opening and flap between two adjacent laticifer cells. (A from Esau, 1953; B–C from Fahn, 1967.)

report vessels with simple perforation plates in roots, rhizomes and young aerial stems of *Cabomba aquatica* (Cabombaceae). The Piperales have an atactostele but this contains vessels with either simple or scalariform perforation plates with few bars.

In the Magnoliales, which are broadly circumscribed here, the Winteraceae, for example, lack vessels even in the secondary wood, other families have vessels with scalariform perforations (such as Degeneriaceae and many Magnoliaceae) whereas vessels with simple perforations occur in Annonaceae and Aristolochiaceae. As stated above the vessels of the monocotyledons should not be compared to those in the secondary wood of dicotyledons.

There has been some argument as to whether the vessels originated in the monocotyledons along a separate line from that (or those) in dicotyledons and whether in this case the vesselless forms (as in many Alismatiflorae) are primitively vesselless or have become so by "reduction".

Another view would be that primitive vessels (with scalariform perforations) were already present in roots, stems and leaves of the early monocotyledons and became either further differentiated (into types with simple perforations) as in Commelinales, reduced in stem and leaves (as in most Liliiflorae and Ariflorae), or altogether lost as in the specialized aquatic Alismatiflorae and Ariflorae (and, perhaps, the Nymphaeiflorae). This view appears to us to be very likely although it seems uncertain as to whether the early monocotyledons had vessels in the roots only or also in the shoots and leaves.

Base Data

See Wagner (1977) for detailed documentation. To this may be added the account on Hydrocharitaceae by Ancibor (1979), reporting vessels with scalariform perforation plates in the roots of *Blyxa*, *Enhalus*, *Hydrocharis*, *Limnobium*, *Ottelia*, *Stratiotes* and *Vallisneria*, whereas only tracheids were found in *Egeria*, *Elodea*, *Hydrilla*, *Lagarosiphon*, *Nechamandra* and *Thalassia*.

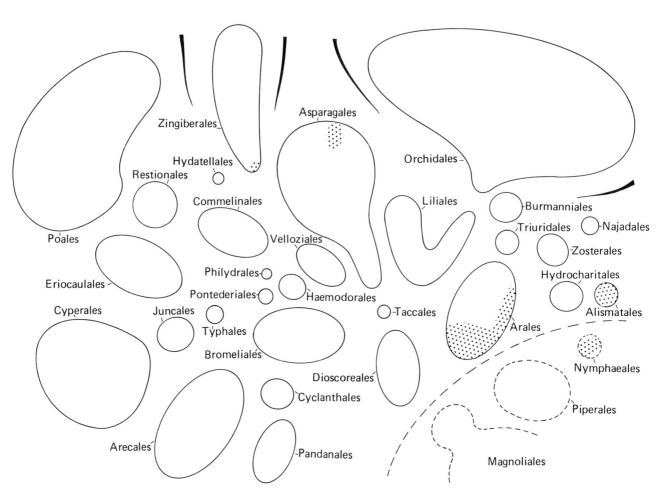

Diagram 23. Laticifers (shaded).

LATICIFERS (Fig. 30, Diagram 23)

The term latex applies to fluids with a somewhat milky appearance due to the suspension of many small particles in a liquid medium with a very different refractive index. Although in most instances the latex is white there are species in which it is more or less translucent. The latex is produced in tubes or cells collectively known as laticifers. These may be either segmented or articulated.

Amongst the monocotyledons latex-producing species are known from five families, distributed among four of the superorders. The distribution of these families is shown in Diagram 23.

The occurrence of laticifers in the Araceae (Ariflorae) and Alismataceae (Alismatiflorae) may represent parallel evolution of the character but may equally well be a further example of the many associations which suggest a phylogenetic link between these two superorders. However, there seems to be no reason for regarding the occurrence of laticifers in the Musaceae (Zingiberiflorae) and the Alliaceae (Liliiflorae) as other than due to parallel evolution. Ancibor (1979) noted that there seems to be an inverse correlation between tannin cells and laticifers in the Alismatiflorae, the former being found in the Hydrocharitaceae and Butomaceae, the latter in the Alismataceae.

The reports of a milky juice in all tissues of the Cyclanthales (Chant, 1978) requires confirmation, because Harling (1958), noting earlier reports of latex for the order, doubted the occurrence of laticifers. He did, however, record an abundant presence of mucilage canals.

In turning to the dicotyledons it is interesting to note the presence of laticifers in the Nymphaeales, an order with several characters that are common in the monocotyledons but rare in the dicotyledons.

The independent evolution of this character in several monocotyledon and dicotyledon families with such diverse ecology suggests it may have evolved as a deterrent to grazing.

Base Data
(Mainly from Metcalfe, 1967)

ALISMATIFLORAE: ALISM.: all (laticifers may be segmented; laticifers in all other groups stated to be articulated).

ARIFLORAE: ARAL.: Arac. subfam. Calloideae, Lasioideae, Philodendroideae, Colocasioideae and Aroideae.

LILIIFLORAE: ASPAR.: Allia.: *Allium* (? all).

ZINGIBERIFLORAE: ZINGI.: Musac.
Dicotyledons associated with monocotyledons:
NYMPHAEIFLORAE: NYMPH.: Nymph. p.p.

SIEVE TUBE PLASTIDS (Fig. 31, Diagram 24)

The sieve elements contain plastids which accumulate ergastic substances consisting either of starch or protein. The same plastids may accumulate either or both protein and starch. Behnke (1969 etc.) has given extensive data on this character, and has classified the sieve elements into the *P*-type (protein type) where protein (and often also starch) is present and the *S*-type (starch type) where there is starch only (Behnke, 1968, 1969, 1971, 1972, 1975; Behnke and Dahlgren, 1976).

Protein is deposited as one or more crystalloid bodies, the shape and size of which are characteristic of various groups. The condition as to whether or not they are solitary is also taxonomically important.

Interestingly enough all monocotyledons investigated have the same type of sieve tube plastids: a *P*-type with cuneate (triangular) crystalloid bodies, generally in a considerable number per plastid. This type is absent from all dicotyledons studied except for the genus *Asarum* (Aristolochiaceae, Magnoliales) (Fig. 31 B). The constancy, distinctness and restricted distribution of the sieve element plastids in the monocotyledons is conspicuous in the light of the great number of monocotyledonous features found in the dicotyledonous groups shown in our diagram. However, all taxa studied in Nymphaeales and Piperales have *S*-type plastids, while in Magnoliales (incl. Aristolochiaceae) the variation is greater than in any other order of angiosperms. In this order *S*- as well as *P*-types are present. When present the protein bodies vary in shape, size and number. Even thin filaments of protein may be present, such as in Cannelaceae. The occurrence of triangular protein bodies, several in number, in Aristolochiaceae is thus not surprising; rather is it surprising that this kind is not found in other members of Magnoliales *sensu lato*, e.g. in Annonaceae, or in any other dicotyledons at all.

This attribute is one of the few which can be used as an argument for retaining the "classical" circumscription of the monocotyledons.

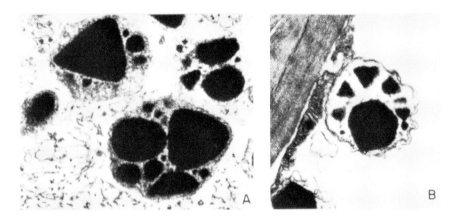

Fig. 31. Cuneate protein crystalloids in the sieve tube plastids of A: *Tradescantia albiflora* (Commelinaceae); B: *Asarum arifolium* (Aristolochiaceae). *Asarum* is the only genus of dicotyledons known to have this monocotyledonous *P*-type sieve tube plastid. (From Behnke, 1975.)

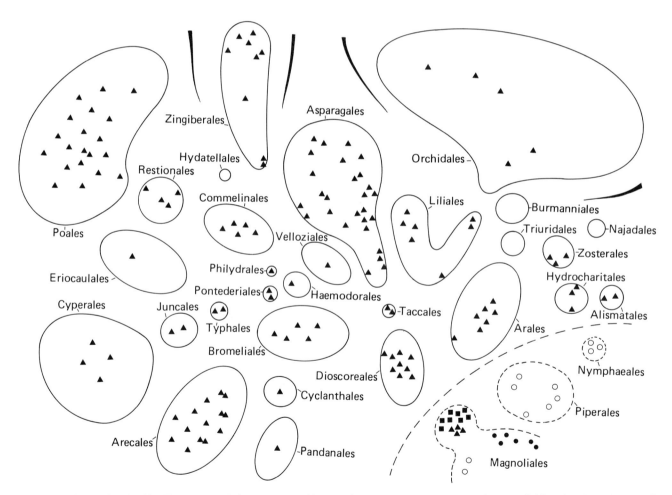

Diagram 24. Sieve tube plastids. The monocotyledonous type, with several to numerous cuneate protein crystalloids; triangles. Absence of protein crystalloids, starch grains only: hollow circles. Square to polygonal protein crystalloids, in *Aristolochia*: squares. Other shapes of protein crystalloids: solid circles.

Base Data

(From Behnke, 1969, and personal communication)
Asterisks indicate observation from other sources than Behnke and his associates.

Monocotyledonous type (see above)

ALISMATIFLORAE: HYDRO.: Butom.: *Butomus*; Hydro.: *Elodea*,* *Thalassia*.*—ALISM.: *Echinodorus, Hydrocleis*.—ZOSTE.: Scheu.: *Scheuchzeria*; Junca.: *Triglochin*; Potam.: *Potamogeton*.

ARIFLORAE: ARAL.: Arac.: *Anubias, Arum, Colocasia, Cryptobergia, Dieffenbachia, Typhonodorum*; Lemna.: *Lemna*.*

LILIIFLORAE: DIOSC.: Diosc.: *Dioscorea* (7 spp.), *Tamus, Trichopus*.—TACCA.: *Tacca* (incl. *Schizocapsa*; 2 spp.).—ASPAR.: Smila.: *Smilax*; Conva.: *Liriope, Ophiopogon, Smilacina*; Aspar.: *Asparagus, Danaë, Ruscus*; Dracae.: *Dracaena, Sansevieria*; Dasyp.: *Lomandra*; Phorm.: *Phormium*;* Agava.: *Agave, Yucca*;* Aspho.: subfam. Astelioideae: *Cordyline*; subfam. Asphodelioideae: *Eremurus, Kniphofia*; subfam. Anthericoideae: *Anthericum*; Cyana.: *Cyanastrum*; Hemer.: *Hemerocallis*; Funki.: *Hosta*; Hyaci.: *Muscari*;* Allia.: *Agapanthus, Allium*;* Amary.: *Brunsvigia, Crinum, Sprekelia*.—LILIA.: Irida.: *Aristea, Gladiolus, Iris, Orthosanthus*; Tricy.: *Tricyrtis*; Melan.: *Narthecium, Veratrum*.—ORCHI.: *Anoectochilus, Epidendrum, Platylepis, Vanda, Vanilla*.—PONTE.: *Eichhornia, Pontederia*.—HAEMO.: *Xiphidium*.—PHILY.: *Philydrum*.—VELLO.: *Vellozia*.—BROME.: *Billbergia*,* *Cryptobergia, Orthophytum, Tillandsia, Vriesia*.*

ZINGIBERIFLORAE: ZINGI.: Musa.: *Musa* (2 spp.): Zingi.: *Hedychium*;* Costa.: *Costus, Tapeinocheilos*; Canna.: *Canna*; Maran.: *Calathea, Ctenanthe, Thalia*.

COMMELINIFLORAE: COMME.: Comme.: *Dichorisandra, Floscopa, Rhoeo, Spironema, Tinantia*.—ERIOC.: Erioc.: *Eriocaulon*.—TYPHA.: Sparg.: *Sparganium*; Typha.: *Typha*.—JUNCA.: Junca.: *Juncus, Luzula*.—CYPER.: *Carex, Eriophorum, Rhynchospora, Scirpus*.—RESTI.: Resti.: *Leptocarpus, Restio*; Centr.: *Centrolepis*; Flage.: *Flagellaria*.—POAL.: *Aegilops*,* *Arrhenatherum, Avena*,* *Bambusa*,* *Bromus, Cortaderia*,* *Hordeum* (2 spp.),* *Oryza*,* *Phleum*,* *Poa, Saccharum, Schizostachyum*,* *Secale*,* *Setaria*,* *Sorghum*,* *Stenotaphrum, Stipa, Triticum*,* *Zea*.*

ARECIFLORAE: ARECA.: *Arenga*,* *Caryota*,* *Chamaedorea* (7 spp.),* *Chamaerops*,* *Cocos*,* *Elaeis*,* *Livistonia* (2 spp.),* *Nypa*,* *Phoenix* (3 spp.),* *Prestoea*,* *Rhaphis*,* *Roystonea*.*—CYCLA.: *Carludovica*.—PANDA.: *Pandanus*.

The only dicotyledonous genus with the same type of protein inclusions is:

MAGNOLIIFLORAE: MAGNO.: Arist.: *Asarum* (4 spp.).

Other types of protein accumulations in dicotyledons associated with the monocotyledons

MAGNOLIIFLORAE: MAGNO.: Arist.: *Aristolochia* (several spp.); Annon.: *Annona, Asimina, Cananga, Monodora*; Canel.: *Canella*; Myris.: *Myristica* etc.

S-type of plastids

NYMPHAEIFLOEAE: NYMPH.: Cabom.: *Cabomba*; Nymph.: *Nuphar, Nymphaea*.

MAGNOLIIFLORAE: PIPER.: Sauru.: *Houttuynia*; Piper.: *Peperomia, Piper*.—MAGNO.: Magno.: *Magnolia*; Winte.: *Drimys* etc.

For other dicotyledons see Behnke and Dahlgren, 1976.

SILICA BODIES
(Fig. 32, Diagram 25)

Hydrated silica is deposited in plant cells either as distinct bodies, which may be relatively large and few or solitary or may be small and in great number ("silica sand"). The cells containing silica bodies are often epidermal or belong to the mesophyll adjacent to vascular strands. Epidermal cells containing silica bodies are often different from other epidermis cells and may be termed stegmata ("short cells" in, for example, the grasses).

The silica bodies may have a shape typical of the group (order, family) where they occur as will be described further below. Their distribution in the monocotyledons is concentrated in the following orders, where silica bodies are common and typical, viz. the Poales, Restionales (partly), Cyperales, Arecales, Zingiberales and Bromeliales. In addition, silica bodies are known in Thurniaceae but not in Juncaceae (Juncales), Rapateaceae and single member(s) of Eriocaulaceae (Eriocaulales), a few genera of the Commelinaceae (Commelinales) and many genera of mainly tropical orchids (Orchidales). Amongst grasses the shape of the silica bodies as seen in surface view is of taxonomic importance, especially at subfamily level. For example whereas members of the Panicoideae have cross and dumb-bell shaped or nodular silica bodies those of the Eragrostoideae are usually saddle-shaped and only rarely similar to those of Panicoideae. In contrast the silica bodies of Pooideae are elongate parallel to the long axis of the leaf and have sinuate or crenate edges (Clifford and Watson, 1978).

In the *Restionales* the presence and appearance of silica bodies is very variable (see Cutler, 1969; Tomlinson, 1969). The Centrolepidaceae seem to lack them completely, although rectangular bodies in *Gaimardia australis* may be siliceous. In leaves of Restionaceae spheroidal silica bodies occur in occasional epidermal cells of *Lepyrodia*, while in other genera they are found as granular amorphous bodies in other tissue, or as spheroidal-nodular bodies in cells of the outer or inner bundle sheaths. Large silica bodies have not been observed in the Anarthriaceae or Ecdeiocoleaceae (Cutler and Airy Shaw, 1965), but silica sand is found in some chlorenchyma cells. In *Flagellaria* silica is present as small irregular bodies occurring singly in small cells above and below the fibrous sheaths of veins of the lamina. They are usually in longitudinal files, but are lacking in the epidermis. *Hanguana* also may have granular silica bodies in endodermal cells around the vascular bundles and larger silica bodies in cells in the abaxial hypodermis and mesophyll. In *Joinvillea* they occur in epidermal short cells of the leaf as more or less cubical small bodies filling the cell lumen or as smooth irregular bodies in cells adjacent to fibres, which indicates poaceous affinity.

In the Eriocaulales, silica occurs mainly in cells of the leaf epidermis of the Rapateaceae, where it forms several to many spheroidal bodies with rough or spiny surfaces. The silica bodies vary in size from small to half the width of the cell. Within the Eriocaulaceae silica bodies seem to be lacking as a rule and are not mentioned by Tomlinson (1969), but Hegnauer (1963) reported silica bodies in the subepidermal tissue of the leaves of *Paepalanthus xeranthemoides*.

The *Commelinales* also are mostly devoid of silica bodies, but they occur in specialized epidermal silica cells in *Callisia*, *Coleotrype*, *Forrestia*, *Gibasis*, *Hadrodemas* and *Tripogandra* in the Commelinaceae. The silica cells in the epidermis are either shallow and contain spinulose spherical bodies or unmodified and contain minute bodies. Both types occur in *Callisia*, *Hadrodemas* and *Tripogandra*, whereas the other genera mentioned only have the former type.

In the *Zingiberales* silica bodies occur in all of the

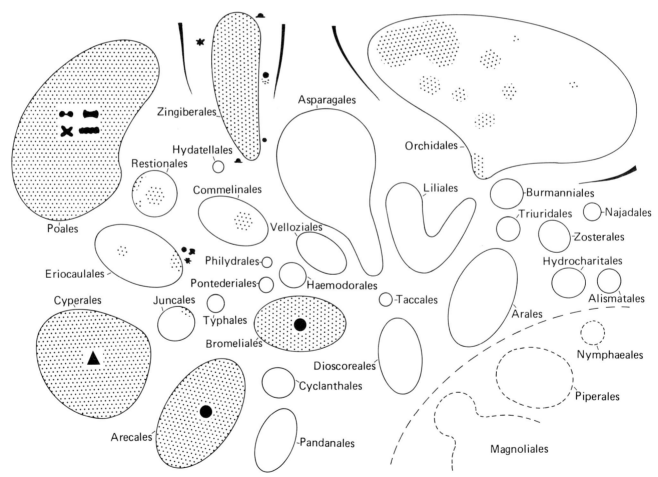

Diagram 25. Silica bodies (shaded). Common shapes in some orders are indicated.

families, but frequently only in internal cells, in addition they may also occur in epidermal cells in the Strelitziaceae and Zingiberaceae. The silica bodies vary considerably in shape, being trough-like in the Musaceae, rectangular in Heliconiaceae, hat-shaped in the Lowiaceae and stellate or druse-like in the Strelitziaceae, Zingiberaceae, Cannaceae and Costaceae, irregular-spherical (in epidermal cells) in the Zingiberaceae, and of various shapes in the Marantaceae, where the variation is perhaps greatest.

In Juncales, silica bodies are lacking in Juncaceae, and thus are restricted to the Thurniaceae, where they are small and nodular, occurring in epidermal cells above fibre strands (*Thurnia jemanii*) or in most epidermal cells (*T. sphaerocephala*).

In the Cyperales (Metcalfe, 1971) the silica bodies are of almost universal occurrence (but there are some exceptions: *Chorizandra, Evandra, Hypolytrum, Lepironia, Mapaniopsis*). They are usually restricted, as in grasses, to the epidermal cells overlying the sclerenchyma accompanying the vascular strands. Their shape is also characteristic: they form cones with their bases resting on the inner walls of the epidermal cells with their apices directed outwards. Their number in each cell varies from

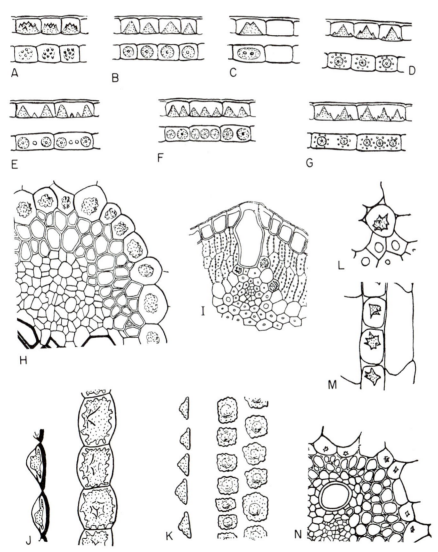

Fig. 32. Shapes of silica bodies. A–G: conical silica bodies in various taxa of Cyperaceae. H: transection of bundle of rhizome in *Sarcophrynium brachystachyum* (Marantaceae) with silica bodies. I: transverse section of culm in *Dielsia cygnorum* (Restionaceae), with silica druses. J: stegmata in *Orchidantha longiflora* (Lowiaceae). K: silica bodies from tissues below veins in *Marantochloa mannii* and *Donax grandi* (Marantaceae). L–M: stegmata from aerial stem of *Costus lucanusianus* (Costaceae). N: midrib of lamina with silica bodies in *Costus malortieanus* (Costaceae). (A–G from Metcalfe, 1971; H and J–N from Tomlinson, 1969; I from Culter, 1969.)

one to numerous and they may be separate or "fused", and equal or unequal in size. The shape may also be warty to nodular in some groups. Wedge-shaped (*Scirpodendron*, *Thoracostachyum*) or bridge-shaped (*Mapania*), silica bodies of a totally different type, embedded in the outer periclinal walls, also occur in the family.

In the Arecales the Arecaceae (palms) (Tomlinson, 1961) frequently have small, solitary, either spherical, conical or hat-shaped silica bodies which are contained in hypodermal cells along the vascular strands.

Finally, the silica bodies in the *Bromeliales* (Bromeliaceae) are solitary and spherical and occur in epidermal cells, as they do in numerous tropical orchids. The orchids with silica bodies were listed by Solereder and Meyer (1930). The occurrence shows a striking pattern: the subfamilies Neottioideae and Orchidoideae and the subtribes Liparidinae and Bulbophyllinae of subfam. Epidendroideae seem to lack silica bodies, while they occur in most other Orchidaceae and in the Cypripediaceae and Apostasiaceae.

The distribution of the silica bodies seems to be largely vicarious in relation to the oxalate raphides. Both accumulations occur, however, in the Arecaceae, Bromeliaceae and also in many orchids. Both silica bodies and oxalate raphides are lacking in all Juncaceae, in some of the Liliiflorae and in all members of the Alismatiflorae.

Moreover, the distribution of the silica bodies in the plants (epidermal, hypodermal, in connection with vascular bundles) is often characteristic of the main orders, the palms agreeing to some extent with members of the Zingiberales in having the silica less concentrated in their epidermis than is the case in the Poales, Restionales and Cyperales. The shapes of the silica bodies in these three orders differ markedly and a consideration of the shape and occurrence of silica bodies is that the Restionales (and possibly the Cyperaceae) may be heterogeneous groups. Besides, it seems very probable that the silica bodies have evolved independently at least in the orders Cyperales and Poales, because Juncales, which is closely allied to Cyperales, mostly lack silica bodies and their shape in the Cyperales (largely conical) is generally very divergent from that in other groups. Silica bodies are also known to occur occasionally in the leaves of various Magnoliales.

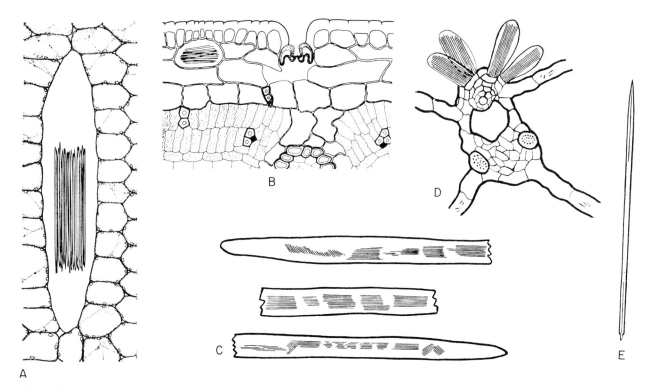

Fig. 33. Raphide bundles in monocotyledons. A: raphide bundle in an elongate cell of *Agave americana* (Agavaceae). B: subepidermal cell with raphide bundle in *Pandanus* (Pandanaceae). C: parts of raphide "vessel" in *Pinellia tuberifera* (Araceae). D: raphide sacs in the inner tissue of a leaf petiole of *Typhonodorum madagascariensis* (Araceae). E: raphide needle of *Gonatopus Boivinii* (Araceae). (A from Frohne and Jensen, 1973; B from Huynh, 1974; C–E from Solereder and Meyer, 1928.)

Base Data

See publications cited above.

CALCIUM OXALATE RAPHIDES
(Fig. 33, Diagram 26)

Calcium oxalate may be accumulated in several different crystal types: as simple short crystals, as simple elongate prismatic crystals which may attain an almost needle-like form (pseudo-raphides), as crystal druses, and as bundles of thin, needle-like crystals, the so-called *raphide bundles* (see Fig. 33).

The raphide bundles are normally contained in mucilage-filled unsuberized cells or "sacs" which are from short to conspicuously elongate. They have an interesting distribution in the angiosperms, being more common in monocotyledons than in dicotyledons and having long been accepted as of taxonomic significance (Gulliver, 1864). Styloides or pseudoraphides (see above) occur in some groups either where oxalate raphides may occur (Pontederiaceae, Philydraceae) or where they are missing (as in the Dracaenaceae subfam. Nolinoideae, in the Phormiaceae and in most genera of the Agavaceae). The pseudo-raphides are often contained in cells with suberized walls.

The distribution of oxalate raphides is summarized in Diagram 26 from which the following facts emerge.

In the superorder Commeliniflorae, oxalate raphides are not reported from the Poales, Restionales, Hydatellales, Eriocaulales, Juncales and Cyperales, and thus are present only in the Commelinaceae (Commelinales) and Typhales.

They are absent from the Alismatiflorae.

In the Liliiflorae they are widespread in the Dioscoreales, Taccales, Asparagales (Stenar, 1949), Orchi-

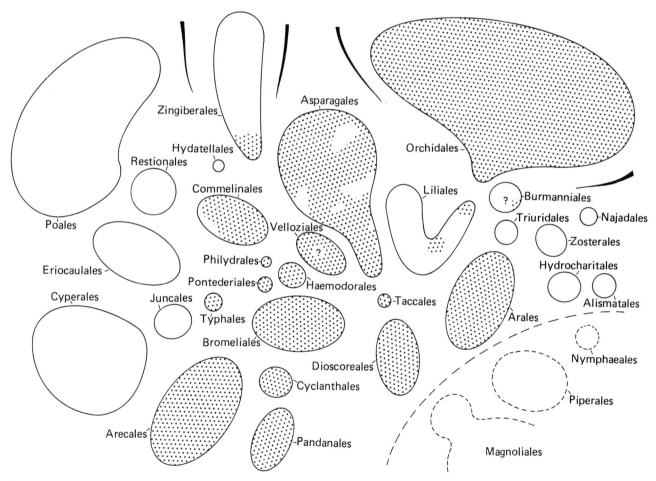

Diagram 26. Oxalate raphides (shaded). The occurrence in orchids especially is incompletely known, but probably almost universal. The raphides may occur in various parts of the plants, such as exclusively in tapetum cells, as in Philydraceae (Fig. 512A).

dales, Haemodorales, Pontederiales, Philydrales and Bromeliales, while they are rare in the order Liliales (in which they seem to be restricted to the Alstroemeriaceae and, perhaps, to a few genera in Melanthiaceae). The genera of Convallariaceae show inconsistency in the occurrence of oxalate raphides. According to Björnstad (1970) they are lacking in *Disporum* and *Clintonia*, which also deviate in embryological and other characters from other Convallariaceae, and perhaps have their proper place among the Colchicaceae (Liliales). Within the Alliaceae, oxalate raphides seem to be largely missing in the two allyl sulphide containing genera *Tulbaghia* and *Allium*. In the Burmanniales they are probably lacking. Data for the Velloziales are somewhat contradictory: Hegnauer (1963) reported absence of oxalate raphides, Smith and Ayensu (1976) mention "crystals present mostly as raphide bundles".

In the Zingiberales they are present in the taxa with 6 or 5 anthers while they are not reported from the families with a single or half anther (Tomlinson, 1969).

Within the Areciflorae oxalate raphides occur amongst most members of Arecales and Pandanales and in all members of Cyclanthales.

The species of Areciflorae are often rich in oxalate giving the tissues a highly pungent taste (see below).

It is interesting to note that in the monocotyledons oxalate raphides are largely vicarious in relation to silica bodies (see Diagram 25), although both substances occur in many palms, in the bromeliads, in the Zingiberalean families with 5–6 stamens and probably in many orchids.

The presence of oxalates is probably of selective value in that, if ingested in sufficient quantity, they are toxic. In the form of raphides they are particularly effective for protecting plants against herbivores, as the sharp needle-like crystals when released from their sacs penetrate the mucous membranes of the mouth of the herbivores causing them to swell and also penetrate the taste buds. The resulting pungent effect deters further grazing.

Amongst the dicotyledons, oxalate raphides are found in such different families as Dilleniaceae, Aizoaceae, Nyctaginaceae, Onagraceae and Balsaminaceae, but they are absent from all the families of Magnoliales, Piperales and Nymphaeales, which otherwise possess many of the attributes of the monocotyledons.

Base Data

ARIFLORAE: ARAL.: Arac. (general, often in great amount); Lemna.: present or absent.

LILIIFLORAE: DIOSC.: Diosc.: most or all; Stemo.: all except *Croomia* (Ayensu, 1968); Trill.: all.—ASPAR.: Smila.; Peter.; Phile.: at least *Eustrephus*; Conva.: com-

mon but irregularly (see discussion); Aspar.: all incl. Rusca.; Dracae. subfam. Dracaenoideae (only pseudoraphides in subfam. Nolinoideae); Dorya.: at least in petals of *Doryanthes*; Dasyp.: absent or (at least in *Lomandra*) present; (Phorm.: only pseudoraphides;) Xanth.; (Agava.: information controversial, maybe only pseudoraphides;) Hypox.: probably all, at least in *Curculigo*; Aspho.; Aphyl.; (Diane.: absent at least in *Dianella*; Tecoph.: not known; Cyana.: probably absent; Erios.: not known;) Hemer.; Funki: present at least in fruit wall; Hyaci.; Allia.: in some genera (but not in *Allium* or *Tulbaghia*); Amary.—LILIA.: (Irida.: absent; Geosi.: ?; Colch.: absent;) Melan.: ? some genera (lacking in a number of genera studied by Sterling, 1978, 1979).—BURMA.: Burma.: ?; Thism.: present in some spp.; Corsi.: ?.—ORCHI.: present in most or all taxa.—PONTE.: present at least in some genera (but styloids more common).—HAEMO.: common.—PHILY.: in tapetal cells (otherwise styloids only).—VELLO.: probably present.—BROME.: abundant.

ZINGIBERIFLORAE: ZINGI.: Lowia.; Helic.; Musac.; Strel. (absent in the other four families).

COMMELINIFLORAE: COMME.: Comme.: common (Carto. and Mayac. absent).—TYPHA.: all.—Probably absent in all other orders.

ARECIFLORAE: present in all three orders.

HAIR TYPES (Fig. 34, Diagram 27)

Hairs are much less frequent in monocotyledons than in dicotyledons, but still present a great and varied category of structures. Some groups are prevailingly or entirely glabrous, as are most Arales, Typhales and Taccales. Others have a variety of hair types, like Commelinaceae, Eriocaulaceae and Dioscoreaceae and in some groups they are conspicuous and important in water absorbtion, e.g. the peltate hairs of Bromeliaceae.

The following account is based on, for example, Staudermann (1924), Stant (1964), Ayensu (1972), Cutler (1969), Tomlinson (1961, 1969), Metcalfe (1971) and other sources. The hairs may be divided into (1) unicellular, (2) multicellular with a unicellular base, and (3) multicellular with a multicellular, sometimes cushion-like, base. There is no doubt that taxonomic and phylogenetic conclusions on various levels *can* be drawn from the occurrence on various types, although, for example, stellate hairs, as seen in species of Dioscoreaceae, Hypoxidaceae, Bromeliaceae, Eriospermaceae and Alismataceae (*Echinodorus, Limnophytum*), have undoubtedly evolved independently in different lines.

The taxonomic value of the "squamulae intravaginales" (the axillary, often paired glands) in the Alismatiflorae and some Ariflorae is treated separately below (Diagram 28). Furthermore, there is some regularity in the distribution of other hair types and some remarks on these follow.

Unicellular hairs have a wide distribution and are probably of little taxonomic significance. The hairs with a multicellular base are also likewise varied and range from marginal "ciliate" hairs to complex glands, or hispid hairs with a cushion-base.

However, the multicellular (or bicellular) hairs with a base of a single cell (Diagram 27) show an interesting distribution which will be commented on here. These hairs vary from uniseriate and non-glandular to uniseriate and gland-tipped and further include multicellular glands, peltate hairs (culminating in Bromeliaceae) and

T-shaped or otherwise branched hairs (as in *Hanguana*, *Gaimardia* etc.).

These types, i.e. the multicellular hairs with an unicellular base, show the following distribution: They are absent (or almost absent) in the Alismatiflorae, and are almost absent in the Ariflorae, occurring however in *Pistia* which has uniseriate "jointed" hairs.

In the Liliiflorae the pattern is variable and hairs are often totally lacking, e.g. in most Amaryllidaceae and Hyacinthaceae. In the Dioscoreales, such hairs are lacking in the Stemonaceae, Trilliaceae and many Dioscoreaceae incl. *Trichopus* and *Stenomeria*, but occur sometimes in various types in *Dioscorea* species, as T-shaped hairs, stellate hairs, normal uniseriate hairs etc. (Fig. 34 F–M). In Liliales hairs occur in *Medeola* and *Lilium* of Liliaceae, in *Sisyrinchium*, *Crocus* and several other genera of Iridaceae and also in Alstroemeriaceae;

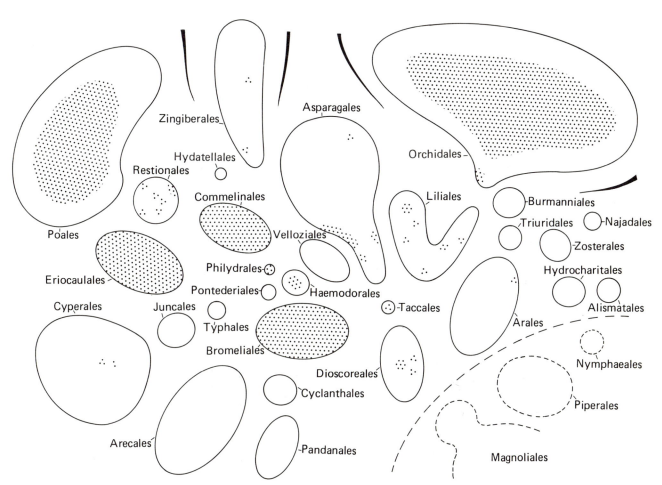

Diagram 27. Multicellular or bicellular hairs (with a single basal cell). Such hairs are of various types and do not form a uniform group, nevertheless they are abundant in some orders of the Liliiflorae and Commeliniflorae. The Bromeliales have stellate to peltate types, while the Poales, for example, has bicellular "microhairs".

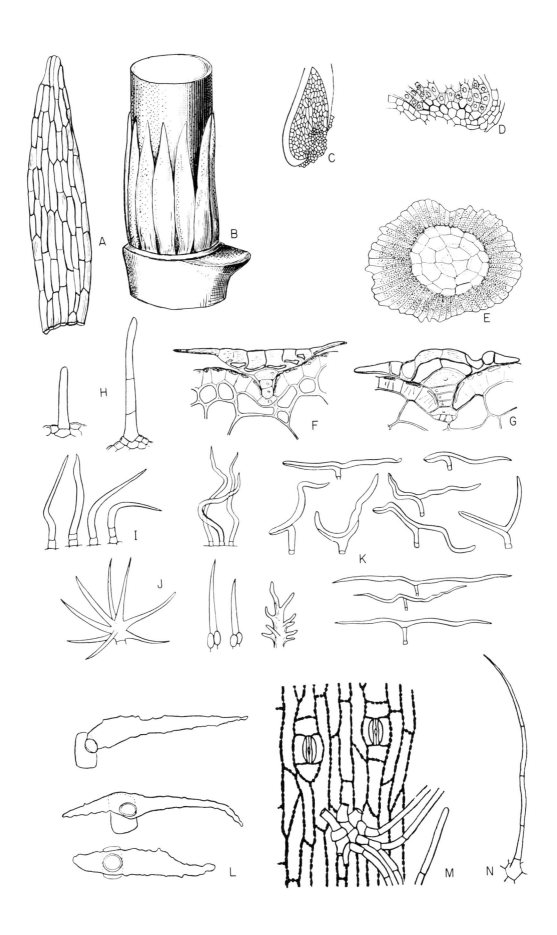

and within Asparagales especially in Hypoxidaceae, where they are uniseriate (also branched hairs with multicellular base are abundant). Multicellular uniseriate hairs with or without an apical gland-cell occur in numerous Orchidales (incl. Cypripediaceae).

The Liliiflorae also include the Bromeliales (Bromeliaceae) where the hairs are mostly peltate, resting on a short uniseriate stalk. The Haemodoraceae, like the Hypoxidaceae, have branched hairs but mostly with multicellular base, while the Philydraceae have uniseriate hairs with a long end cell and sunken glandular hairs in combination.

The plants of the Zingiberiflorae are largely glabrous, though sometimes unicellular hairs may occur, especially in the Marantaceae and Zingiberaceae. Branched uniseriate, multicellular hairs of "candelabra-type" occur rarely in *Heliconia* (*H. distans*, *H. illustris*) and bicellular hairs are found on the rhizomes of *Curcuma* (Zingiberaceae).

In the Commeliniflorae the variation is great. Many Commelinaceae and Cartonemataceae possess uniseriate hairs, some cells of which may be thin-walled and even branched (*Palisota*). In the Eriocaulales the hairs are also very variable. Uniseriate hairs of various types, including T-shaped hairs, and hairs with single enlarged cells occur in the Eriocaulaceae; members of the Xyridaceae often have few-celled uniseriate hairs, the terminal cell of which is mucilage-filled ("glandular"); while taxa of Rapateaceae have 2–5-celled, uniseriate and non-glandular hairs. The Restionales are variable, with some small families being devoid of hairs, viz. Flagellariaceae, Ecdeiocoleaceae and Anarthriaceae, while *Hanguana* (Hanguanaceae) has branched hairs with a uniseriate sunken base and *Joinvillea* likewise branched uniseriate hairs. Rarely there are uniseriate hairs in Restionaceae, viz. T-shaped in, for example, *Lepidobolus* and unbranched in *Loxocarya*. Also the Centrolepidaceae have uniseriate hairs in at least four genera, of which *Gaimardia* has branched hairs. In the grasses unicellular hairs are relatively common, and widespread also are the small, thin-walled 2(–several)-celled hairs, "microhairs", which may correspond to the 2–5-celled hairs sometimes present in the Restionaceae and Commelinaceae. The Hydatellales and Juncales are hairless, the multicellular fringe hairs of *Luzula* being excepted. Likewise the Typhales and most genera of Cyperales are glabrous or at least lack multi-

cellular hairs, some exceptions in the Cyperaceae being species of *Fuiraena* and *Everardia*.

Palms and other Areciflorae may have unicellular or multicellular hairs but the latter do not seem to have a unicellular base.

It is difficult to evaluate the taxonomic value of the above data, but some features are of interest, such as the occurrence of branched hairs in *Hanguana*, *Joinvillea* and *Gaimardia* in the Restionales and the almost total lack of uniseriate hairs in the Typhales, Juncales and Cyperales.

Base Data
See publications cited above.

INTRAVAGINAL (AXILLARY) SQUAMULES
(Fig. 34 A–B, Diagram 28)

The intravaginal or axillary, non-vascularized glands (scales), often called "squamulae intravaginales" are small scale-, gland- or finger-like trichomes occurring in pairs or larger numbers in the axils of vegetative leaves. Their wide occurrence in the Alismatiflorae was acknowledged by Arber (1923) and Staudermann (1924), but they were already reported by Irmish (1858) and Sanio (1865).

They occur in only two superorders, the Alismatiflorae and the Ariflorae (Diagram 28). In the former, they probably occur in all families other than the Triuridales and in the latter are explicitly recorded for only a single species of *Philodendron* (Engler, 1919).

Intravaginal squamules are generally two-layered and multicellular varying from almost circular or broadly ovate, as in *Elodea densa*, to finger-like, as in *Potamogeton* (Fig. 34 A, B). Ancibor (1979) reported multi-layered, intravaginal squamules in *Enhalus* and *Stratiotes* (Hydrocharitaceae). The squamules apparently secrete a protective mucilage, this being perhaps their function, although they have been regarded as protective in their own right. Tannin cells may occur at their base (*Halophila*), in the middle parts (*Enhalus*), at the distal end (*Vallisneria*) or at their tooth-like border (*Hydrilla*) as reported by Ancibor (1979).

Fig. 34. Squamulae intravaginales (A–D) and other hair types in monocotyledons. A–B: squamulae intravaginales in *Potamogeton* (Potamogetonaceae). C–D: same structures in *Triglochin* (Juncaginaceae). E–G: peltate trichomes in species of *Tillandsia* (Bromeliaceae). H–K: variation in hair structures within *Dioscorea* (Dioscoreaceae). L: uni- to bibrachiate hairs in *Paepalanthus* (Eriocaulaceae). M: uniseriate hairs (damaged) in *Hanguana* (Hanguanaceae). N: uniseriate hair in *Costus* (Costaceae). (A–B Ascherson, 1889; C–D from Emberger, 1960; E–G and L–N from Tomlinson, 1969; H–K from Burkill, 1960.)

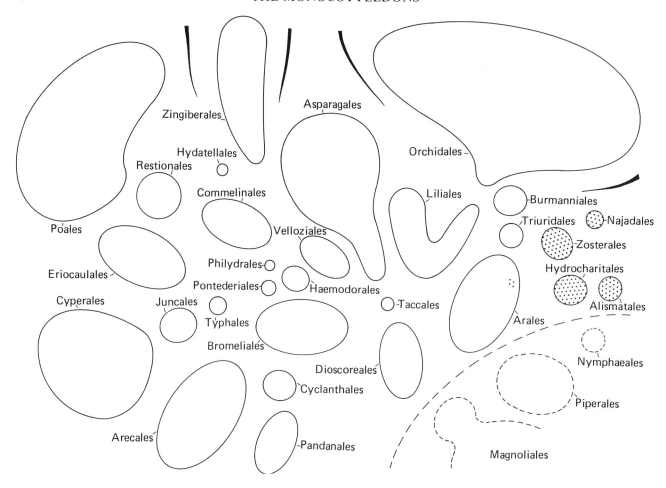

Diagram 28. Intravaginal squamules (shaded). These represent multicellular hair structures (with multicellular bases) and are situated in the leaf axils. They must not be confused with stipules (see Diagram 17).

It is noticeable that they occur in all non-saprophytic orders of the Alismatiflorae and outside these families only in the Araceae (Arales, Ariflorae) indicating a possible connection between these two superorders. (Their further occurrence in the Araceae and their possible occurrence in the Triuridales and Nymphaeales should be investigated).

Caspary (1858), who first described the axillary squamules, considered them to be stipular structures, but they are no doubt trichomes. In, for example, *Potamogeton* they occur associated with stipular sheaths from which they are morphologically different.

Base Data

ALISMATIFLORAE: Probably all taxa except (?) Triuridales. Explicit documentation exists for at least the following genera: HYDRO.: Butom.: *Butomus*; Hydro.: *Egeria, Elodea, Enhalus, Hydrilla, Hydrocharis, Lagaro-*
siphon, Limnobium, Ottelia, Stratiotes, Triamia, Vallisneria.—ALISM.: *Alisma, Caldesia, Hydrocleys, Sagittaria.*—ZOSTE.: Scheu.: *Scheuchzeria*; Junca.: *Triglochin*; Potam.: *Potamogeton, Ruppia*; Zoste.: *Zostera*; Cymod.: *Cymodocea.*—NAJAD.: Najad.: *Najas*.

ARIFLORAE: ARAL.: Arac.: *Philodendron* sp.

NUMBER OF SUBSIDIARY CELLS IN STOMATAL COMPLEXES (Fig. 35, Diagram 29)

Stomata are classified morphologically according to whether the guard-cells are surrounded by epidermal cells of the same shape as the other epidermal cells, or whether

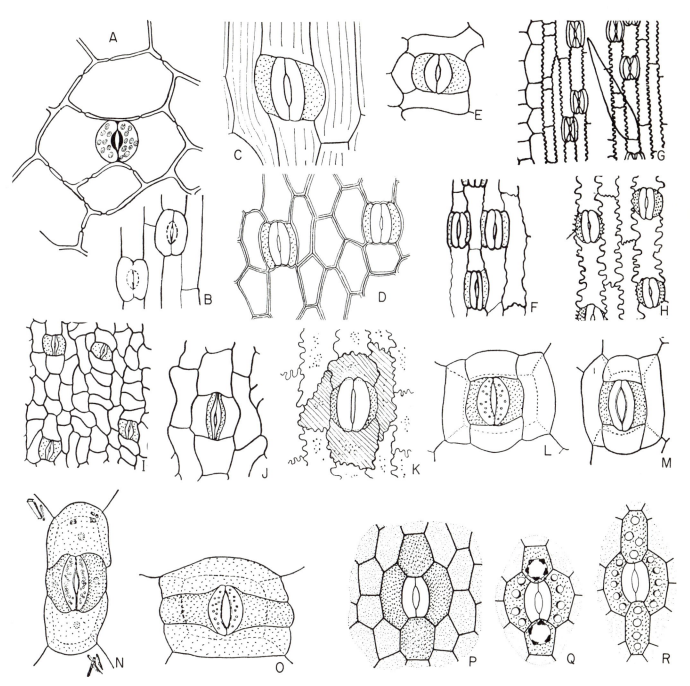

Fig. 35. Different types of stomata in monocotyledons. A–B: typical anomocytic type (lacking well differentiated subsidiary cells). C–H: paracytic stomata in various groups, C–E representing the Alisma subtype, F–H the grass subtype. I–M: transitional types between paracytic and tetra- to hexacytic, where there are two clearly differentiated subsidiary cells complemented with variable distinctive other epidermal cells which may or may not be regarded as subsidiary cells. N–R: Tetracytic types. See further details in the text. The stomata are from the following taxa: A: *Luzuriaga* (Philesiaceae). B: *Trithuria submersa* (Hydatellaceae). C: *Cuthbertia ornata* (Commelinaceae). D: *Elegia obtusifolia* (Restionaceae). E: *Costus lucanusianus* (Costaceae). F: *Juncus gerardi* (Juncaceae). G: *Joinvillea borneensis* (Joinvilleaceae); grass type of stomata. H: *Carex poly-phylla* (Cyperaceae). I–J: *Orchidantha longiflora* (Lowiaceae). K: *Fosterella penduliflora* (Bromeliaceae). L–O: the commelinaceous species *Stanfieldia impertorata* (L), *Murdannia simplex* (M), *Cyanotis arachnoidea* (N) and *Tradescantia blossfeldiana* (O). P–R: different species of *Pandanus* (Pandanaceae). (A from Schlittler, 1951; C–E, G, I–O from Tomlinson, 1969; B, F, H from Cutler, 1969; P–R from Huynh, 1974.)

there are two or more cells deviating from other epidermal cells which surround each pair of guard-cells. These deviating cells are called subsidiary cells.

In the former case, where there are no subsidiary cells, the stomata are generally called anomocytic (or the Ranunculaceous type in dicotyledons).

When the subsidiary cells are present the stomata are classified according to their number and orientation. "Subsidiary cells" are so called irrespective of whether they are derived from the guard-cell initials (very rare in monocotyledons: certain orchids) or from epidermal cells adjacent to them. Accordingly, this superficial classification of stomata is therefore somewhat inadequate, as has been pointed out by various authors, and an alternative classification of monocotyledonous stomata has been proposed by Tomlinson (1974). It is presented here on p. 101.

Stomata with two subsidiary cells lying parallel to and on each side of the guard-cells, paracytic stomata, are widely distributed in monocotyledons but may be divisible into two subtypes: the grass subtype, where the subsidiary cells are widely different from and normally much smaller than the ordinary epidermal cells, and the Alisma subtype where the subsidiary cells are rather similar to the epidermal cells. Between these transitional variants may occur.

Stomata which tend to have four (tetracytic) or six (hexacytic) subsidiary cells are likewise divisible into two types, viz. those in which the four cells are of about equal size, and those where the lateral are larger than those situated at each end of the stomatal complex. The former have a somewhat scattered distribution while the latter type is the dominant in the Areciflorae, but there seems to be a gradual transition between the two types. Nonetheless the pattern of stomatal variation in monocotyledons is not without taxonomic significance.

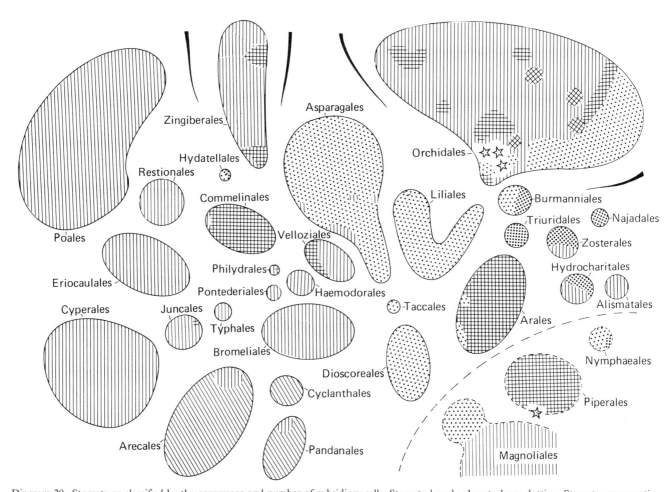

Diagram 29. Stomata as classified by the occurrence and number of subsidiary cells. Stomata largely absent: dense dotting. Stomata anomocytic, i.e. lacking subsidiary cells: sparse dotting. Stomata largely paracytic: vertical hatching. Stomata tetracytic to hexacytic, with mostly more or less equal-sized subsidiary cells: cross-hatching. Tetracytic stomata, mostly with the subsidiary cells at the ends of the stoma of a smaller size, or with a shorter wall, bordering on the stoma, than the lateral subsidiary cells: oblique hatching. The orchids are classified according to Diagram 2.

The distribution of all these stomatal types is shown in Diagram 29.

In the Alismatiflorae stomata are often absent in aquatic and saprophytic plants, for example all Triuridales (?) and Najadales, most Zosterales and about half of the Hydrocharitales. Paracytic stomata (Alisma subtype) occur in the Alismataceae (Alismatales) (Stant, 1964), Butomaceae, Aponogetonaceae, and members of the Hydrocharitaceae (Hydrocharitales), and in the Juncaginaceae, Scheuchzeriaceae, and members of the Potamogetonaceae (Zosterales).

In the Ariflorae the stomata are mostly of the tetracytic type with subequally large, sometimes weakly defined subsidiary cells, while well defined subsidiary cells are missing in the Lemnaceae and at least some species of *Arisaema* of the Araceae.

The Liliiflorae are a very variable group with respect to stomatal characters. It seems that stomata are missing in many saprophytic taxa of the Burmanniales, while in other taxa of Burmanniales and most members of the Orchidales, Liliales, Asparagales, Taccales and Dioscoreales anomocytic stomata are present. Rarely they may be oriented in a non-parallel, scattered manner similar to that of the anomocytic stomata in many dicotyledons. In the Asparagales there are a few families, however, with paracytic stomata. These include the Hypoxidaceae, *Doryanthes* (Doryanthaceae), *Xanthorrhoea* (Xanthorrhoeaceae) and *Astelia* (but not *Milligania*, in Asphodelaceae subfam. Astelioideae). In this feature these groups agree with most members of those orders of the Liliiflorae with a starchy endosperm, viz. most Velloziales and all Haemodorales, Pontederiales, Philydrales and Bromeliales. In some members of the Velloziales (Velloziaceae) the stomata are hexa- or tetracytic (Ayensu, 1974), and in the Bromeliales (Fig. 35 K) they are also often described as tetracytic, but the two guard-cells are accompanied by one subsidiary cell on each side which is very narrow in superficial view. These two subsidiary cells often have their lumina more lowly situated than those of the guard-cells. The guard-cells and two well differentiated subsidiary cells are surrounded as a rule by four epidermal cells somewhat different from the other epidermal cells. If these are conceived as subsidiary cells the stomata are hexacytic, but it seems to us more reasonable to classify them as paracytic.

For the Orchidales, which exhibit an unusually great variability in stomatal complexes, the classification according to number of subsidiary cells is particularly difficult due to the vague definition of "subsidiary cells". Thus, for example, the configuration of cells surrounding the stomata of *Paphiopedilum* (Cypripediaceae) have been described as anomocytic, as surrounded by two subsidiary cells and two ordinary cells, or, finally, as supplied with four subsidiary cells! In the tribe Cranichideae (the "succulent Neottioideae") the stomata seem to be largely mesoperigenous, i.e. the "subsidiary cells" are derived from the guard-cell initials as well as from the surrounding epidermal cells. In spite of their apparent uniformity they can be variously described as having 2–6 subsidiary cells in the mature stomatal complex. These mesoperigenous stomata are given a special symbol in the diagram for the orchids(*). In addition, it seems as if most of the subfam. Orchidoideae and tribe Diurideae of subfam. Neottioideae have a characteristic type of stomata of the anomocytic type, and thus lack subsidiary cells, as do practically all members of Liliales. Most other orchids have subsidiary cells. In the tribes Vandeae and Oncidieae these tend to be two in number, but 4, 6 or more subsidiary cells of unknown ontogeny have been observed in a number of genera (Williams, 1979). (Information mainly from F. N. Rasmussen.)

In the Zingiberiflorae there is a basic similarity in stomata to those of the Commeliniflorae, the guard cells being accompanied by two thin-walled subsidiary cells more or less clearly differentiated from the "normal" epidermal cells. In addition, in the Musaceae, Strelitziaceae, Heliconiaceae and Costaceae, there are generally two (terminal) or more subsidiary cells, sometimes vaguely differentiated from the other epidermal cells so that the definition of stomatal type is difficult. In the Marantaceae the condition is comparable to that in the Bromeliaceae (see above) although the slightly modified epidermal cells around the stomata are fewer; the stomata being here classified as paracytic. The Zingiberaceae, Lowiaceae and Cannaceae according to Tomlinson (1969) have paracytic stomata.

In the Commeliniflorae the basic stomatal type is paracytic (grass subtype or otherwise) except in the Commelinales, where the Commelinaceae and Cartonemataceae have tetracytic or hexacytic stomata of the type seen in Fig. 35 L–O, while the Mayacaceae have stomata rather transitional between para- and tetracytic types. In most of the other groups there are two lateral thin-walled subsidiary cells clearly differentiated from other epidermal cells, and no additional subsidiary cells (e.g. Solereder and Meyer, 1933). Stomata in the Thurniaceae have a tendency to be tetracytic, that is with two lateral and two terminal subsidiary cells (Cutler, 1969).

The Restionales have paracytic stomata throughout. They generally resemble those of the grasses closely (having thick-walled guard-cells) except in *Hanguana* (Hanguanaceae), which has thin-walled guard-cells (Alisma subtype). Anomocytic stomata are present in *Trithuria* of the Hydatellales (*Hydatella* seems to lack stomata altogether) and are very aberrant in the Commeliniflorae. Whether they have evolved in response to the submerged habitat or have retained their original appearance is not yet certain.

Finally, in the Areciflorae, the stomata are generally tetracytic, the terminal cells being often smaller than the lateral (Tomlinson, 1961; Huynh, 1974; Harling, 1958). However, there is great variation in the relative size of lateral and terminal subsidiary cells. Paracytic stomata occur in *Sararanga* (Pandanaceae) and in *Caryota* and *Calamus* (Arecaceae).

In the orders of dicotyledons bordering on the monocotyledons anomocytic stomata are found in Nymphaeales (except for the Ceratophyllaceae which lack stomata). In the Piperales, the Saururaceae has a "rosette" of subsidiary cells and the Piperaceae, at least in many cases, have tetracytic stomata (cf. the Arales), while in the Magnoliales the Aristolochiaceae have anomocytic stomata whereas most other families in the order have paracytic stomata.

Base Data

Stomata absent

ALISMATIFLORAE: HYDRO.: Hydro. (most genera).—ZOSTE.: Potam.: p.p.; Zoste.: Posid.; Zanni.; Cymod.—NAJAD.: Najad.—TRIUR.: Triur.

LILIIFLORAE: BURMA.: many taxa.

COMMELINIFLORAE: HYDAT.: at least *Hydatella inconspicua*.

Associated dicotyledons:

NYMPHAEIFLORAE: NYMPH: Cerat.

Stomata anomocytic (lacking subsidiary cells)

ARIFLORAE: ARAL.: Arac.: at least *Arisaema*; Lemna.

LILIIFLORAE: DIOSC.: all.—TACCA.: all.—ASPAR.: all except Dorya., Xantho., Aspho.: *Astelia*, and Hypox. (see below paracytic stomata).—LILIA.: all or nearly all taxa.—BURMA.: Burma.: many.—ORCHI.: Cypri. p.p.; Orchi.: esp. subfam. Orchidoideae; stomatal condition otherwise variable, see the text).

COMMELINIFLORAE: HYDAT.: *Trithuria*, (*Hydatella* ?).

Associated dicotyledons:

NYMPHAEIFLORAE: NYMPH.: all except Cerat.

MAGNOLIIFLORAE: MAGNO.: Arist. and other families, most of them being however paracytic.

Stomata paracytic (two parallel subsidiary cells)

ALISMATIFLORAE: HYDRO.: Butom.; Hydro.: p.p.; Apono.—*Alism.:* all.—ZOSTE.: Junca.; Potam. p.p.

LILIIFLORAE: ASPAR.: Dorya.: *Doryanthes*; Xanth.: *Xanthorrhoea*; Hypox.: most or all; Aspho.: at least *Astelia.*—Orchi.: Orchi.: of widespread occurrence, especially in subfam. Epidendroideae.—PONTE.: all.—HAEMO.: most or all.—PHILY.: *Philydrella*, *Philydrum.*—VELLOZ.: common, also tetracytic types occur.—BROME.: widespread, may also be interpreted as tetra- or hexacytic (see the text).

ZINGIBERIFLORAE: ZINGI.: Lowia.; Maran.; Zingi. and Canna. may be considered paracytic, but also 4- or 6-cytic if the more or less deviating neighbouring cells are conceived as subsidiary cells.

COMMELINIFLORAE: COMME.: Mayac. (or tetracytic of type 2).—ERIOC.: all.—TYPHA.: all.—JUNCA.: Thurn.: some; Junc. all.—CYPER.; all.—RESTI.: all (incl. *Hanguana*).—POAL.: all.

ARECIFLORAE: ARECA.: present in some taxa, e.g. spp. of *Caryota* and *Calamus.*—PANDA.: rare, e.g. in *Sararanga*.

Associated dicotyledons:

MAGNOLIIFLORAE: MAGNO.: many families, e.g. in Annon. and Magno.

Stomata tetracytic (or hexacytic) with more or less similar subsidiary cells

ARIFLORAE: ARAL.: Arac.: general, but with a few exceptions, e.g. in *Arisaema*.

LILIIFLORAE: ORCHI.: Orchi. (scattered in subfam. Epidendroideae).—VELLO.: common, along with paracytic.—PHILY.: *Helmholtzia*, *Orthothylax.*—BROME.: the stomata may be interpreted as either para- or hexacytic, the outer 4 neighbouring cells ("subsidiary cells") being variously interpreted.

ZINGIBERIFLORAE: ZINGI.: Musac.; Helic.; Strel.; Costa; the stomata in Zingi. and Canna. may also be interpreted as 4-6-cytic (see above).

COMMELINIFLORAE: COMME.: Comme.; Carto.—JUNCA.: Thurn. sometimes.

Associated dicotyledons:

MAGNOLIIFLORAE: PIPER.: Piper.: most taxa.

Stomata tetracytic (or hexacytic) with a tendency for the polar subsidiary cells to be narrower and elongated in the stoma direction.

Areciflorae: most taxa but with exceptions, see under paracytic stomata; the distinction between this and the previous type is diffuse in respect to size and shape of subsidiary cells.

STOMATA CLASSIFIED ACCORDING TO EVEN VERSUS OBLIQUE DIVISIONS IN SURROUNDING CELLS
(Figs 36–37, Diagram 30)

Tomlinson (1974) has criticized the classification of stomatal types based solely on number of subsidiary cells, proposing instead another based on the patterns of divisions of the cells surrounding the meristemoid giving rise to the guard cells. He found a certain degree of regularity of these divisions, and according to his studies the monocotyledons could be classified into the four groups seen in Diagram 30. Further data, especially on orchids are given by Williams (1979).

Group 1. In this the stomata are constantly or at least often lacking. Included here are many specialized water plants in the Alismatiflorae, such as the families Zosteraceae, Posidoniaceae, Cymodoceaceae, Zannichelliaceae and Najadaceae and also in Hydatellaceae (in part). They are also lacking in the saprophytic members of the orders Triuridales and Burmanniales. Stomata are also usually lacking in Hydrocharitaceae, but they are known

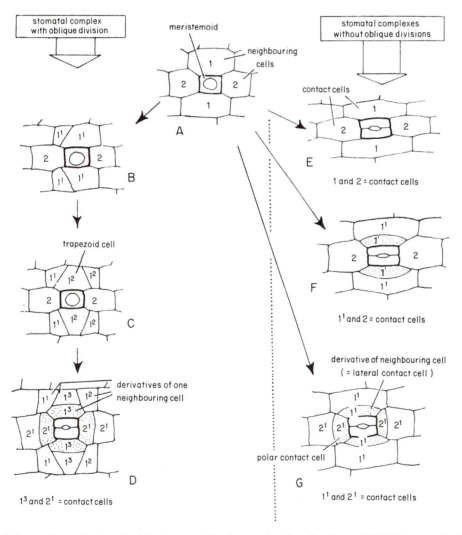

Fig. 36. Stomata classified according to their mode of development. The figure, taken from Tomlinson (1974), illustrates the formation of stomata surrounded by cells with oblique divisions (left row) and without oblique divisions (right row). A: guard cell mother cell. Its appearance and the fact that it is delimited by four (not six) cells is typical of monocotyledons. B–D: stomatal complex with oblique divisions resulting in trapezoid cells (palm type). E: stomatal complex without derivatives of neighbouring cells. F: derivatives of one pair of neighbouring cells (paracytic "grass" type). G: derivatives of both pairs of neighbouring cells. (From Tomlinson, 1974.)

to occur on floating leaves of *Hydrocharis, Ottelia* and *Limnobium* and on some leaves in *Stratiotes* (Ancibor, 1979).

Group 2. Here only the meristemoid divides once to produce the stomata (guard cells). This group corresponds mainly to the "anomocytic type" as defined by Metcalfe and Chalk (1950) and is distributed widely amongst the Ariflorae (Arales) and Liliiflorae, in the latter of which it is the most common in the Liliales, Dioscoreales and Asparagales. In contrast, amongst the Asparagales the neighbouring cells in many woody members (*Agave, Beaucarnea, Cordyline, Doryanthes, Dracaena, Lomandra, Xanthorrhoea* etc.) divide and these divisions are often oblique in relation to the stoma (see Group 4).

Group 3. In this the stomata are surrounded by a series of cells which divide by even (non-oblique) walls in

relation to the stomata. The cells bordering on the guard cells are usually two (as in grasses) or four. This type is the only or primary one registered in the Alismatiflorae and in Poales, Eriocaulales, Cyperales and Juncales; it is also the common type in the Restionales, Zingiberales and Commelinales. Besides, this type occurs in certain Liliiflorae.

In this connections it may be of interest to note (Cutler, 1969; Hamann, 1975) that the Hydatellaceae have either "anomocytic stomata" or lack stomata at all. In any event they do not have Group 3 type stomata, as are found in the Centrolepidaceae.

Group 4. Here the cells surrounding the stoma meristemoid divide by oblique walls, so that "trapezoid" and triangular cells are formed. The simplest situation is where by two oblique divisions a trapezoid (encircling) cell is formed on each side of the stoma (Fig. 36). Such stomatal complexes are found in the Pandanaceae, some

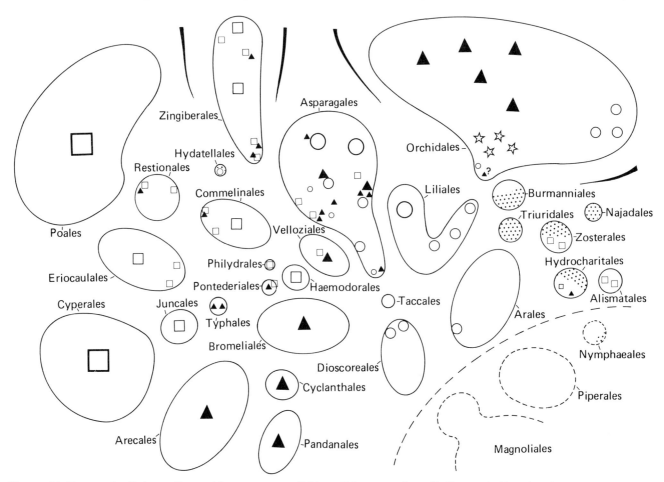

Diagram 30. Stomata classified according to oblique versus even divisions of the surrounding cells. Stomata lacking: dots. Stomata surrounded by non-dividing cells: hollow circles. Stomata surrounded by cells with even divisions: squares. Stomata surrounded by cells with oblique divisions: triangles. The orchids placed according to Diagram 2 are after Williams (1973, 1979) and Rasmussen (personal communication); in these, meso-perigenous stomata are symbolized by asterisks. (Size of symbols refer to size of groups only.)

"tree lilies" (Agavaceae, Xanthorrhoeaceae etc.). Subsequent divisions of the cells of the stomatal complex may lead to increasing complexity. Divisions in the neighbouring polar cells occur in the palms (Arecaceae), in *Typha* (Typhaceae) and *Heliconia* (Heliconiaceae), where from the original four neighbouring cells of the guard cell meristemoid as many as 12 derivatives may be formed. In other groups the second oblique division is placed so that it intersects the first ("intersecting oblique divisions"), but this has not been distinguished here as a separate group. Such divisions occur, for example, in the Cyclanthaceae, in *Sparganium*, *Yucca* and in some orchids. The peculiar condition in *Butomus* (Hydrocharitales) is described by Gupta *et al.* (1975).

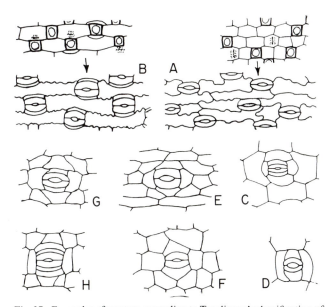

Fig. 37. Examples of stomata according to Tomlinson's classification of 1974. A: *Lilium* (Liliales), no derivatives of neighbouring cells (anomocytic stoma). B: *Joinvillea* (Restionales), derivatives of one pair of neighbouring cells, "grass type", no oblique divisions. C: *Commelina* (Commelinales), derivatives of both pairs of neighbouring cells, no oblique divisions. D: *Rhoeo* (Commelinales), derivatives of one pair of neighbouring cells, no oblique divisions. E: *Pandanus* (Pandanales), divisions of each lateral neighbouring cell has produced a trapezoid contact cell and also polar neighbouring cells have divided. F: *Sparganium* (Typhales), with intersecting oblique divisions, after a first oblique division and a second intersecting this. G: *Cocos* (Arecales), after oblique divisions of each lateral cell trapezoid cells are formed, which have divided to produce lateral contact cells. H: *Strelitzia* (Zingiberales), as in *Cocos*, but with additional division in the lateral cells. (All from Tomlinson, 1974.)

The distribution of these four types of stomatal complexes are summarized in Diagram 30, from which it can be seen that, if the tendency is confirmed after further studies, they distribute themselves according to a definite pattern. The orders of Commeliniflorae (the Hydatellales and Typhales excepted) tend to have the even divisions of neighbouring cells and this is also the commonest condition in the Zingiberales.

The Typhales deviate from most other Commeliniflorae in having oblique divisions in the cells surrounding the meristemoids and agree with the main complex of the Areciflorae and many Liliiflorae. To the latter group also belong at least many members of the Bromeliales, the Pontederiales and the "tree-like members" of the Asparagales, but also many orchids and some genera of Amaryllidaceae and Hypoxidaceae, some succulent-leaved genera of the Asphodelaceae, certain Velloziaceae, and some of the five-staminate taxa of the Zingiberales.

The orchids (Williams, 1979), indeed seem to be divisible to some extent by the stomata, the subfamily Orchidoideae lacking divisions, certain Neottioideae having mesoperigenous stomata, and most taxa of the Epidendroideae (incl. *Vanilla*) having oblique divisions in the neighbouring cells.

Most Liliiflorae with non-starchy endosperm seem to lack divisions in the cells surrounding the stoma initial.

Base Data
(Mainly from Tomlinson, 1974)

The data below are to be taken as representatives; after ordinal abbreviations are given either genera only, or abbreviations of families (especially where these seem to be homogeneous).

Neighbouring cells not dividing
ARIFLORAE: ARAL.: Lemna.
LILIIFLORAE: DIOSC.: Stemo.; Trill.—ASPAR.: Smila.: *Smilax*, irreg.; Phile.: *Geitonoplesium*; Aspar.: *Asparagus*; Dasyp.: *Lomandra* spp.; Aspho.: *Bulbine*; Aphyl.: *Aphyllanthes*; Hyaci.: *Lachenalia*; Amary.: *Narcissus*.—LILIA.: Irid.: several; Alstr.: *Alstroemeria*; Lilia.: *Lilium*, *Medeola*; Melan.: *Veratrum*.—ORCHI.: Cypri.; Orchi.: subfam. Orchidoideae, few taxa of subfam. Epidendroideae.

Neighbouring cells with non-oblique divisions
ALISMATIFLORAE: HYDRO.: (? Butom.: *Butomus*); Hydro.: a few genera.—ALISM.: *Alisma, Limnocharis*.—ZOSTE.: Junca.: *Triglochin*; Potam.: *Potamogen*.
LILIIFLORAE: ASPAR.: Hypox. (heterogeneous); Cyana.: *Cyanastrum*; Funki.: *Hosta*.—PONTE.: Ponte.—HAEMO.: Haemo.—PHILY.: at least *Philydrella, Philydrum*.
ZINGIBERIFLORAE: ZINGI.: Strel.: *Phenacospermum* (heterogeneous), *Ravenala* (heterogeneous),

Strelitzia (heterogeneous); Zingi.; Costa. (heterogeneous); Canna; Maran.

COMMELINIFLORAE: COMME.: Comme.: several genera; Carto.: *Cartonema*; Mayac.: *Mayaca* (heterogeneous).—ERIOC.: Rapat.: *Rapatea*; Xyrid.: *Xyris*; Erioc.: several genera.—JUNCA.: Junca.—CYPER.: Cyper.—RESTIO.: Centr.; Joinv.: *Joinvillea*.—POAL.: Poac.: all.

Neighbouring cells with oblique divisions (non-intersecting where not otherwise stated)

ALISMATIFLORAE: HYDRO.: Butom.: *Butomus* (Gupta *et al.*, 1975).

LILIIFLORAE: ASPAR.: Peter.: *Petermannia*; Phile.: *Eustrephus* (occasionally); Dracae.: *Dracaena, Beaucarnea*; Dorya.: *Doryanthes*; Dasyp.: *Lomandra* p.p.; Xanth.: *Xanthorrhoea*; Agava.: *Agave, Yucca* (with intersecting oblique divisions); Hypox. (heterogeneous); Aspho.: *Aloë, Gasteria, Haworthia* (these heterogeneous), *Cordyline*; Amary.: *Crinum* (heterogeneous?).—ORCHI.: Orchi. (most Epidendroid tribes).—PONTE.: all.—VELLO.: p.p.—BROME.: maybe all.

ZINGIBERIFLORAE: ZINGI.: Streli.: *Phenacospermum, Ravenala* and *Strelitzia*, all these heterogeneous; Helic.: *Heliconia* (with intersecting divisions); Costa. (heterogeneous).

COMMELINIFLORAE: COMME.: Mayac.: *Mayaca* (sometimes with intersecting divisions).—TYPHA.: Sparg. (sometimes with intersecting oblique divisions); Typha.—JUNCA.: Thurn.— RESTI.: Flage.: *Flagellaria*.

ARECIFLORAE: ARECA.—CYCLA. (sometimes with intersecting oblique divisions).—PANDA.

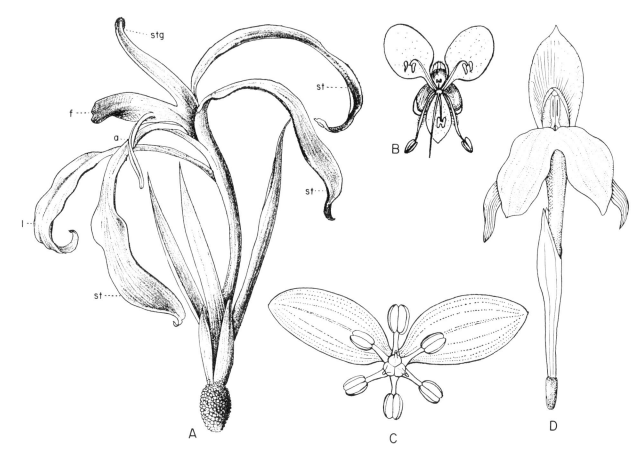

Fig. 38. Zygomorphy and asymmetry in perianth and/or androecium. A: the classical illustration of *Canna violacea*, an asymmetric flower, where the petaloid staminodia are diverse and asymmetric, while the tepals are trisymmetric. Abbreviations: st = petaloid staminodia of outer whorl; l = "labellum" (a petaloid staminodium of the inner whorl); f and a = sterile lobe and functional anther, respectively, of the fertile stamen of the inner whorl: stg = stigma. B: *Aneilema lanceolatum* (Commelinaceae), zygomorphic flower with only two functional stamens, the other being staminodial. C: *Aponogeton ranunculiflorus* (Aponogetonaceae), zygomorphic flower, only two tepals developed. D: *Roscoea purpurea* (Zingiberaceae) a strongly zygomorphic flower. Taking the unsymmetrical outer tepals ("sepals") into consideration this flower may even be taken as asymmetric. (A from Kränzlin, 1912; B from Morton, 1966; C from Guillarmod and Marais, 1972; D from Burtt and Smith, 1972.)

ZYGOMORPHY OR ASYMMETRY IN PERIANTH AND/OR ANDROECIUM
(Fig. 38, Diagram 31)

The distributions of plants with a petaline perianth and of plants with an outer sepaline and an inner petaline whorl in the perianth are presented in Diagram 32. The transition between the petaline and the bracteate, hyaline, scarious, greenish, and reduced types of perianth is gradual. In Diagram 31 the petaline types have been interpreted in a rather wide sense. In these cases zygomorphy, and more rarely asymmetry, have evolved to a variable extent in a variety of groups, mostly in the Liliiflorae and Zingiberiflorae.

Zygomorphy may be very weak and only influence the degree of curvature of the stamens and the relative size (and sometimes the colour pattern) of the tepals, as in the Funkiaceae, Hemerocallidaceae, many members of the Amaryllidaceae, many taxa of the *Aloë* group in Asphodelaceae, *Doryanthes* (Doryanthaceae) and *Phormium* (Phormiaceae), which all belong to the Asparagales, and besides in many members of the Iridaceae, Liliaceae and Alstroemeriaceae of the Liliales. Even strong zygomorphy not connected with reductions influencing the floral diagram (in relation to actinomorphic taxa) occurs within several of the families mentioned, as in *Sprekelia* (Amaryllidaceae), *Gasteria* (Asphodelaceae), *Antholyza*, *Anapalina* and *Gladiolus* (all in Iridaceae; Goldblatt, 1971), *Chionographis* (Melanthiaceae) etc.

Marked zygomorphy occurs in nearly all Orchidales especially with regard to the androecium which has only one, rarely two or three functional stamen(s). In the last-

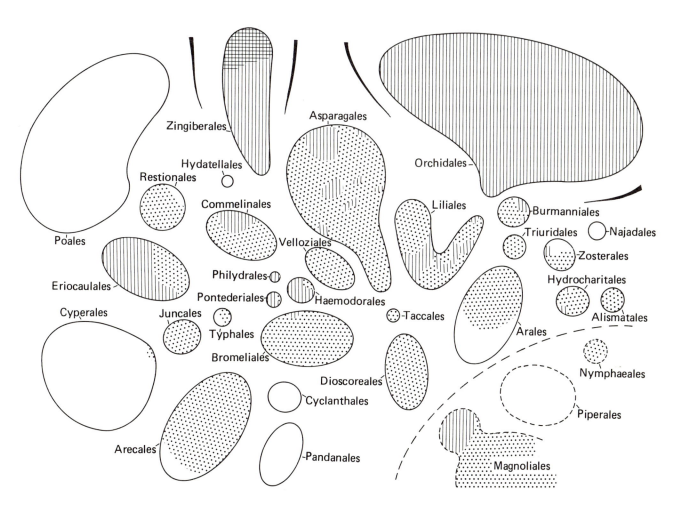

Diagram 31. Zygomorphy and asymmetry in perianth and/or androecium. Groups with reduced perianth not included and left unshaded in the diagram. Radial symmetry: dots. Zygomorphy: vertical hatching. Asymmetry: cross-hatching.

mentioned case however, the stamens belong to two different whorls. The perianth is also generally strongly zygomorphic in this order (especially the labellum being more or less different from the remaining five tepals, which are often subequal).

In the Philydraceae a parallel and equally strong reduction has occurred in the androecium in combination with strong zygomorphy in the perianth, and in the Tecophilaeaceae and Haemodoraceae there are various stages in such a reductionary trend. A similar trend is seen in the subfamily Gilliesioideae of the Alliaceae (Asparagales).

The order Zingiberales is divisible into two groups, four families having 5–6 functional stamens, the remaining four families having only one, or a half functional stamen. In the latter group the stamen sometimes and the staminodia always are petaloid thereby providing mutually dissimilar structures making the flowers strongly zygomorphic or asymmetric. In the Cannaceae and Marantaceae there is only one theca on one stamen, and the staminodia being dissimilar make the flower asymmetric (Fig. 38). Also in the families with 5–6 functional stamens the perianth and androecium are zygomorphic.

Within the Commeliniflorae petaline inner tepals are found mainly in Commelinales and Eriocaulales. In the Commelinales the outer whorl is usually distinctly sepaline. In the subfamily Commelinioideae of Commelinaceae, one "petal" is often reduced in size and different in colour, and the three upper stamens are often reduced to staminodes (Fig. 38B). Zygomorphy in the flowers of many genera of Eriocaulaceae and Xyridaceae is caused mainly by the outer tepals ("sepals"), the lateral of which may be fused medially.

The Alismatiflorae are interesting in respect of floral symmetry. Though they are mostly either actinomorphic or reduced, there is rarely weak to strong zygomorphy. Weak zygomorphy occurs in *Vallisneria* of the Hydrocharitaceae, and strong zygomorphy in certain species of Aponogetonaceae where one or more of the tepals may be large and the others lacking or reduced in size (Fig. 38C). In the Zosterales there is dubiously a well developed petaline perianth; some families are included here by virtue of the "connective appendages" according to some taxonomists (tepals of others, e.g. Hutchinson, 1934). These are several and they are sub-equal in most members of the Juncaginaceae and Potamogetonaceae, but there is only one per flower developed in the Zosteraceae (large in *Phyllospadix*).

The orders of dicotyledons here associated with the monocotyledons may have reduced flowers (Piperales), actinomorphic or "spiromorphic" flowers (Nymphae-ales) or "spiromorphic" as well as actinomorphic and zygomorphic flowers (Magnoliales incl. Aristolochiaceae). The zygomorphic flowers mapped in the diagram are restricted to *Aristolochia* (Aristolochiaceae).

It is apparent from the above that the zygomorphy is not a character to which any great taxonomic importance may be attached as many families contain both actinomorphic and zygomorphic taxa. Furthermore, the variation in symmetry may be considerable even within a genus, as in *Gladiolus*, *Lilium* and *Cyanella*.

Base Data

ALISMATIFLORAE: HYDRO.: Apono. (mostly); Hydro.: rarely and very weak as in *Vallisneria*.—ZOSTE.: Junca. (rarely); Zoste.

LILIIFLORAE: ASPAR.: Dorya. sometimes weak; Phorm.: weak; Aspho.: weak, in some genera of subfam. Asphodeloideae, e.g. *Aloë* and *Haworthia*, strong in *Gasteria*, rarely in subfam. Anthericoideae, e.g. spp. of *Chlorophytum*; Tecop.: often, especially in androecium; Hemer. and Funki. weak; Allia.: subfam. Gilliesioideae sometimes strong; Amary.: often weak, strong in *Sprekelia*.—LILIA.: Irid.: weak to strong in many genera of subfam. Ixioideae, e.g. *Anapalina*, *Antholyza*, *Crocosmia*, *Freesia*, *Gladiolus*, *Tritonia*; Alstr.: at least *Alstroemeria*; Lilia.: many spp. of esp. *Lilium*; Melan.: rarely, as *Chionographis*.—BURMA.: Thism.: some genera; Corsi.—ORCHI.: all.—PONTE.: most.—HAEMO.: most but not all taxa (*Macropidia* being, for example, actinomorphic).—PHILY.: all.—BROME.: few, with weak zygomorphy, e.g. *Pitcairnia*.

ZINGIBERIFLORAE: ZINGI.: all; Canna. and Maran. being asymmetric.

COMMELINIFLORAE: COMME.: Comme.: *Commelina* and a number of allied genera (frequently in androecium).—ERIOC.: Erioc.: frequently (also bisymmetric); Xyrid.: partly.—Reduced flowers in other orders may be zygomorphic by reduction (e.g. grasses) but are not included here.

Dicotyledons associated with the monocotyledons:

MAGNOLIIFLORAE: MAGNO.: Arist.: *Aristolochia* spp.

TEPALS PETALINE IN BOTH WHORLS (Diagram 32)

The distribution of flowers with two petaline whorls is summarized in Diagram 32, from which it is evident that the character is widespread in four of the superorders.

It is absent from the Ariflorae and relatively rare in the Alismatiflorae, where the flowers of *Butomus* (Butomaceae, Hydrocharitales) and the Triuridales approach this condition.

Petaline tepals occur in all Zingiberiflorae (in some families of which they are often smaller than the petaloid staminodia) and the Liliiflorae (except several Trilliaceae). However the condition is less common in the Areciflorae and Commeliniflorae in both of which groups wind as well as insect pollination occurs.

In some Commeliniflorae such as the Juncales and Restionales it is largely a matter of definition as to whether or not the perianth is regarded as petaline. As the tepals are usually brown or white but not green they may be accepted as such, but for the most part have not been so here. Amongst the Eriocaulales the inner perianth members in particular are quite petaline, while the outer vary from mostly hyaline (Xyridaceae, Eriocaulaceae) to mostly sepaline (Rapateaceae). In the Commelinaceae both perianth whorls may be petaline, as in *Rhoeo*, but this condition is uncommon; the outer tepals are normally typically green and sepaline.

Amongst the Areciflorae, flowers with a double tepaline structure are in the majority, and in most species the inner and outer whorls are dissimilar. Though the palms are sometimes regarded as primarily wind pollinated this may be a misconception for many palms are certainly animal pollinated (Moore, 1973).

Within all the petaline groups two subgroups may be recognized according to whether the inner and outer whorls are similar or dissimilar. When present the difference between the whorls is a reflection of specialization for attracting pollen vectors. Sometimes the difference is combined with zygomorphy of the flower as in most orchids (Orchidaceae) or a marked contrast in the colour of the two whorls as in many Bromeliales (Bromeliaceae), where the outer tepals are often distinctly sepaline.

Where wind pollination occurs as in most Restionales and Juncales the perianth whorls are similar and rather smaller than is usual for animal pollinated flowers suggesting they may have been derived therefrom by reduction.

Base Data

Both whorls of tepals petaline (similar or dissimilar in size and shape).

In this survey petaline has been conceived in a "generous" sense including white, brownish and hyaline, but not green or bracteate, reddish or brown tepals when these are not particularly large and conspicuous.

ALISMATIFLORAE: HYDRO.: Butom. (outer somewhat sepaline).—TRIUR.: all.

LILIIFLORAE: DIOSC.: Diosc. p.p. (tepals often green); Stem.; Trill. p.p.—TACCA.: all.—ASPAR.: all except *Philesia* (exceptional cases with greenish tepals occur).—LILIA.: all.—BURMA.: all.—ORCHI.: all.—PONTE.: all.—HAEMO.: all.—PHILY.: all.—VELLO.: all.—BROME.: partly (the outer usually sepaline).

ZINGIBERIFLORAE: all, although the outer tepals may be semihyaline and inconspicuous.

COMMELINIFLORAE: ERIOC.: most or all, although the outer tepals usually hyaline.—JUNCA.: perhaps a few. (RESTI.: in a generous sense some except Centr. could be included, although the tepals mostly have a bracteose character.)

ARECIFLORAE: ARECA.: if widely interpreted, petaline tepals have wide occurrence.

Associated dicotyledons:

NYMPHAEIFLORAE: NYMPH.: Cabom.; Nymph. p.p.

MAGNOLIIFLORAE: MAGNO.: conditions highly variable, tepals often in one, two or three whorls or spirally set.

Both whorls of tepals petaline; the two whorls conspicuously different in size, shape and/or colour.

ALISMATIFLORAE: HYDRO.: Butom.—TRIUR.: p.p.

LILIIFLORAE: DIOSC.: Trill. p.p.—ASPAR.: rarely, the few cases not registered here.—LILIA.: Irida.: frequently in, especially, subfam. Iridoideae; Alstr.: nearly all taxa; Tricy.; Caloch.; Lilia (rarely); Melan.: some, e.g. *Petrosavia*.—BURMA.: all except Corsi.—ORCHI.: many, the labellum neglected.—PHILY.: all.—BROME.: partly (when the outer tepals are petaline).

ZINGIBERIFLORAE: all.

COMMELINIFLORAE: ERIOC.: Xyrid.; Erioc.—JUNCA.: differences in size and shape of tepals common. (—RESTI.: Resti. p.p.)

ARECIFLORAE: ARECA.: many genera.

Associated dicotyledons:

NYMPHAEIFLORAE: NYMPH.: Nymph.: some genera, e.g. *Barclaya*.

MAGNOLIIFLORAE: MAGNO.: Annon.: most taxa having different size and shape between the outermost perianth whorl and the two inner; Arist.: at least *Saruma*.

DIFFERENTIATION OF TEPALS INTO OUTER SEPALINE AND INNER PETALINE
(Fig. 39, Diagram 32)

Flowers with two perianth whorls, the outer of which is green and sepaline and the inner of which is petaline are uncommon in monocotyledons, as may be seen from Diagram 32.

In the Alismatiflorae such flowers occur in species with aerial inflorescences and are restricted to the Alismatales and some Hydrocharitales, viz. members of the subfamily Hydrocharitoideae. In Butomaceae the two whorls are slightly different but the outer hardly sepaline. The flowers of the Aponogetonaceae are reduced to only 1–3 perianth members or none at all and do not exhibit differentiation between whorls.

The records for the Liliiflorae outside the Bromeliales are restricted to Trilliaceae, Philesiaceae (*Philesia*) and some single orchid genus, but in none of these is the outer tepal whorl as sepaline as in the Alismatiflorae or Commeliniflorae. In Bromeliales many genera have sepaline outer tepals, these being often "smaller, more rigid (sometimes slightly leathery to horny, stiff) in texture and green or whitish green in colour" (Harms, 1930).

In the Commeliniflorae flowers with markedly different perianth whorls are restricted to the Commelinales, all three families of which have the character.

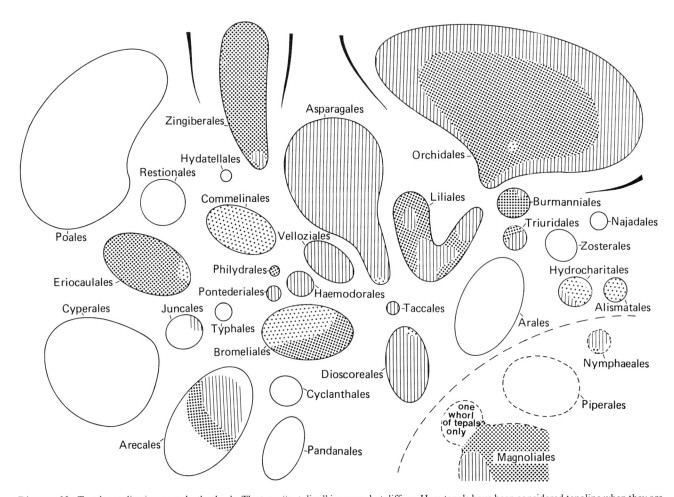

Diagram 32. Tepals petaline in one or both whorls. The term "petaline" is somewhat diffuse. Here tepals have been considered tepaline when they are white or otherwise coloured except when greenish, brownish or scarious. Thus many palms are considered as having petaline tepals although in this group they are generally less conspicuous than in, for example, Asparagales or Liliales. The tepals are not considered to be petaline in any Restionales and only in few Juncales, where they are more often green, or brown scarious, though sometimes reddish or even white (and thus might technically fall under the concept petaline). Tepals petaline and largely similar in both whorls: vertical hatching. Tepals petaline in both whorls but the whorls distinctly dissimilar in size or shape, colour or a combination of these: dense dots. Outer tepals sepaline, inner petaline: sparse dots.

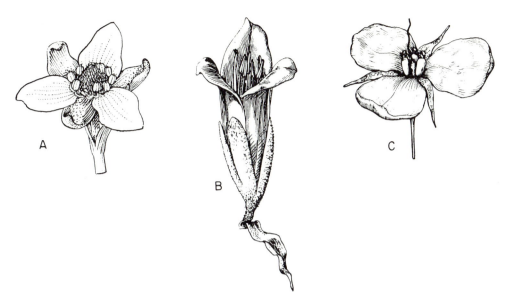

Fig. 39. Sepaloid outer tepals. A: *Ranalisma humile* (Alismataceae). B: *Puya berteroniana* (Bromeliaceae). C: *Mayaca fluviatilis* (Mayacaceae). (A from Carter, 1960b; B from Pizarro, 1959; C from Lourteig, 1965.)

The occurrence of flowers with well marked green sepals in three such widely separated regions of the classification suggests that the trait has evolved independently on separate occasions; cf. however Hutchinson's classification. Furthermore the restriction of the character in the Alismatales–Hydrocharitales and the Commelinales supports the close relationships of the taxa involved in each complex.

Base Data

ALISMATIFLORAE: HYDRO.: Hydro.: most genera of subfam. Hydrocharitoideae.—ALISM.: Alism.
LILIIFLORAE: DIOSC.: Trill. (*Trillium*).—ASPAR.: Phile.: *Philesia*.—ORCHI.: Orchi.: *Isotria*, maybe more.—BROME.: many.
COMMELINIFLORAE: COMME.: all.—ERIOC.: Rapat.

TEPALS SPOTTED
(Fig. 40, Diagram 33)

Among taxa with petaline tepals, these may be distinguished according to whether they are uniformly coloured or patterned with drop-like areas of a different colour.

Species with such flowers have a restricted distribution. They are almost limited to the Liliiflorae (except in

Tetroncium and *Rapatea*) where they are concentrated in the Liliales occurring with varying frequencies in all families (no data are available for Geosiridaceae). Within the Orchidales perhaps only 10–15% of the species have spotted tepals and in many, maybe most, of these

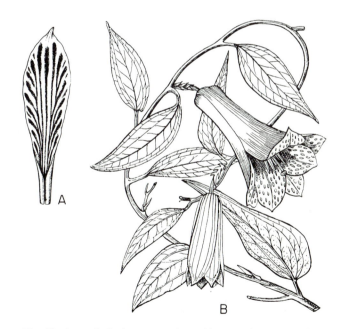

Fig. 40. A: tepal of *Alstroemeria diazii* (Alstroemeriaceae) showing a striated to "drop-like" colour pattern. (From Correa, 1969.) B: *Lapageria rosea* (Philesiaceae) likewise showing a distinct colour pattern on the inner side of the inner tepals. (From Good, 1947.)

the variegation is restricted to the labellum. In the tribus Vandeae and in such large genera as *Habenaria* as well as in many small-flowered tropical orchids the variegation is absent. In the Asparagales the tepal insides are variegated at least in *Lapageria* and *Philesia* (both in Philesiaceae), which in several respects show affinity to the Alstroemeriaceae. Also a few taxa of Pontederiaceae have spots or streaks (serving as nectar guides) on the inside of the inner tepals.

The occurrence of spotted tepals in some Philesiaceae but not elsewhere in the Asparagales raises the problem as to whether this family would be better placed close to Alstroemeriaceae (Liliales). Support for this viewpoint comes from the occurrence of, for example, inverted leaf blades in species of both Philesiaceae and Alstroemeriaceae.

The presence of patterning on the tepals is presumably associated with the signalling of pollen vectors and so is often associated with nectaries or osmophores. Likewise spotting may occur on organs that functionally replace tepals, for example the petaloid staminodes of *Canna* and the spathes of certain Araceae (e.g. *Symplocarpus foetidus*). The former of these has nectaries and the latter osmophores.

Base Data

ALISMATIFLORAE: ZOSTE.: Junca.: *Tetroncium.*
LILIIFLORAE: ASPAR.: Phile.: *Lapageria, Philesia.*
—LILIA.: Irida. (many genera); Colch.: (some genera; e.g. in species of *Colchicum*); Alstr.: most taxa; Tricy.; Caloch. p.p.; Lilia.; frequently, e.g. in *Fritillaria*; Melan.: rarely.—ORCHI.: maybe in c. 10–15% of the genera.— PONTE.: sometimes.
COMMELINIFLORAE: ERIOC.: Rapat.: very rarely.

Dicotyledons associated with the monocotyledons:

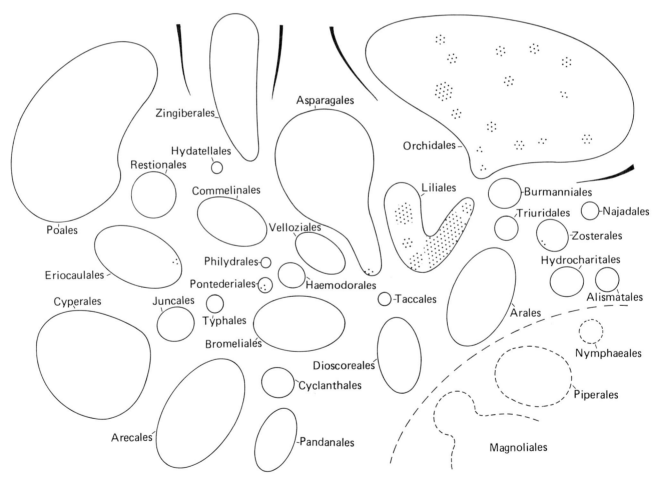

Diagram 33. Tepals spotted (shaded). The variegation in typical cases may consist of drop-like spots merging often into streaks, but also other types of dotted variegation are included.

MAGNOLIIFLORAE: MAGNO.: Arist. (esp. many spp. of *Aristolochia*).

HYPOGYNY VERSUS EPIGYNY
(Fig. 41, Diagram 34)

Flowers in which the perianth and androecium arise below the ovary are known as hypogynous whereas those in which the ovary is beneath the perianth and androecium are known as epigynous.

The epigynous condition is regarded as derived, arising in response to a need for affording the ovules additional protection especially from predation or damage by pollinators.

In monocotyledons inferior ovaries are totally absent in the Commeliniflorae, Areciflorae (except Cyclanthales) and Ariflorae and are confined to a single family, Hydrocharitaceae, in the Alismatiflorae, while they are found in practically all families (Lowiaceae hemiepigynous) in the Zingiberiflorae and are widespread in the Liliiflorae.

This distribution is no doubt explained by the evolution of the perianth. In most of the Commeliniflorae the perianth is not petaline in character but frequently bracteate, scarious, or at least inconspicuous, not infrequently developed as hairs or bristles or altogether absent by reduction. Where the inner tepals are petaline as in many taxa of the Commelinales and Eriocaulales specialization in the direction of zygomorphy and animal pollination has sometimes occurred: here it is likely that the petaline nature of the inner perianth in these two groups developed late in their evolution and so the perianth has not yet become attached to the ovary in either group.

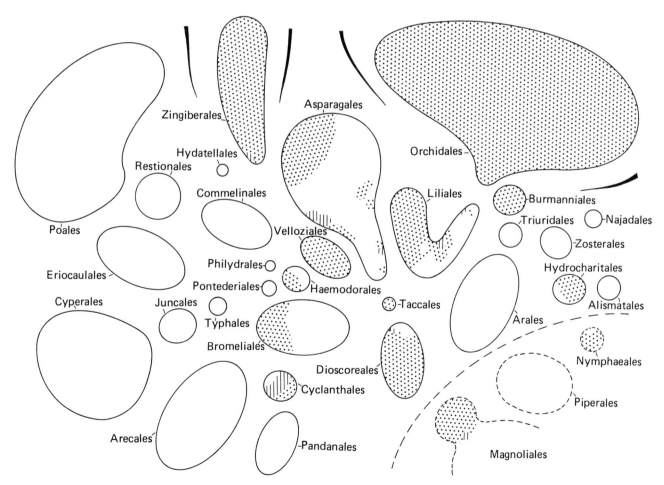

Diagram 34. Epigynous flowers (dots), half-epigynous flowers (vertical hatching) and hypogynous flowers (unshaded).

In the Arales, a largely animal pollinated group, the spathe and not the perianth has assumed the role of attracting vectors. Here the ovules are further protected in a number of ways other than by fusion of the floral whorls to the ovary. The distal portion of the ovary may be massively thickened (*Monstera*); the stamen filaments may be thickened and overarch the pistil (*Pothos*), or the ovary may be partially embedded in spadix (*Symplocarpus*); the female flowers may be crowded together affording one another protection (*Montrichardia*). Furthermore, in all cases the presence of oxalate raphides in the tissues is likely to deter animals from extensive eating (and nectar secretion when present occurs from the stigmatic papillae and not from the ovary or near its base).

Amongst the Areciflorae the palms are a partly wind and partly animal pollinated group. It has been suggested (Moore, 1973) that the increased stamen number in some genera is a response to pollen loss from insect feeding and the presence of nectaries and scent indicates a long history of entomophily. Like the aroids, the palms have protected the ovules by a combination of thick ovary walls, close perianth, or embedding of the flowers in the inflorescence axes and by developing an armoury of chemical defences including tannins as well as oxalates. In the Cyclanthales (Harling, 1958) the ovary is generally inferior or subinferior and more or less sunken in the spadix. In some species, however, of *Sphaeradenia* and *Stelestylis* especially, it is superior.

The Alismatiflorae are predominantly epigynous and often apocarpous. Here in spite of its mainly apocarpous nature the gynoecium has been enveloped by the outer floral parts in the Hydrocharitaceae to develop an epigynous condition. Still the carpels in this family retain a great degree of individuality inside the envelope. An early adaptation to aquatic life (with wind or water pollination!) has led to reduction of the perianth in other groups.

It is mainly in the pronouncedly entomogamous or ornithogamous superorders Zingiberiflorae and Liliiflorae that the perianth (and/or androecium), probably after having been well developed and petaline during a long period, has become attached to the ovary.

In the Zingiberiflorae all families have wholly epigynous flowers except Lowiaceae, where they are only incompletely epigynous. While the perianth may be rather poorly developed (in the Zingiberaceae, Costaceae, Cannaceae and Marantaceae) the androecium to a large extent has been transformed into petaloid structures. Pollination is varied but occurs by insects (butterflies, wasps etc.), birds or bats.

In the Liliiflorae the variation in ovary position is great. The flowers in the Orchidales, Burmanniales, Velloziales and Taccales are exclusively of the epigynous type. In the Dioscoreales epigyny occurs in the Dioscoreaceae, and the flowers are also epigynous in *Stemona* (Ayensu, 1968) and half-epigynous in *Stichoneuron* while they are hypogynous in the third genus, *Croomia*, of the Stemonaceae. In the Asparagales epigyny is found in all members of the Petermanniaceae, Amaryllidaceae, Hypoxidaceae and Doryanthaceae and in most Agavaceae (i.e. in the subfamily Agavoideae) and in *Ixiolirion* and *Kopakowskia* of the Asphodelaceae. Moreover, the probably closely allied families Cyanastraceae and Tecophilaeaceae have half-epigynous flowers, which also occur in, for example, *Peliosanthes* and *Ophiopogon* of the Convallariaceae (Rao and Kaur, 1979). There are few reasons to believe that all these families are closely related to each other, and we do not support the idea of including in the Amaryllidaceae such discordant elements as the Hypoxidaceae, Alstroemeriaceae, Agavaceae subfam. Agavoideae or the Ixiolirieae. Inflorescence, floral construction, seeds and other structures, as well as chemical contents (e.g. alkaloids) are against such a treatment. It is thus more likely that epigyny developed along several separate lines in the Asparagales, that in the Hypoxidaceae being sometimes regarded as allied to the Velloziaceae (Velloziales). In the Liliales inferior ovaries are found in all members of the Iridaceae except *Isophysis* (a monotypic Tasmanian genus), in the monotypic Geosiridaceae, in the Alstroemeriaceae, where this attribute has probably arisen independently from that in the Iridaceae, and in *Aletris* and *Nietneria* of Melanthiaceae. The epigyny although possibly giving some advantages may be thought to prevent the emergence of nectar from septal nectaries, but this is apparently not so, septal nectaries being present, for example, in the Amaryllidaceae as well as in the Iridaceae subfam. Ixioideae. The septal nectaries in these groups open on to the upper surface of the ovary.

Fig. 41. Hypo- and epigyny in monocotyledons, with examples from the Liliiflorae. Hypogyny. A–D: *Milla* (Alliaceae), plant and details of flower; note the insertion of the ovary in the corolla tube. E–F: *Tillandsia pedicellata* (Bromeliaceae), belongs to a hypogynous subfamily. G: *Yucca filamentosa* (Agavaceae), belongs to the hypogynous members of this family. H–I: *Liriope graminifolia* and *Peliosanthes teta* (Convallariaceae), the former hypogynous, the latter half epigynous. Epigyny. J–K: *Rhodophiala* (Amaryllidaceae), for comparison with *Milla*. L–M: *Sisyrinchium* (Iridaceae). (A–D from Moore, 1953; E–F and J–M from Correa, 1969; G from Riley, 1892; H–I from Baillon, 1894).

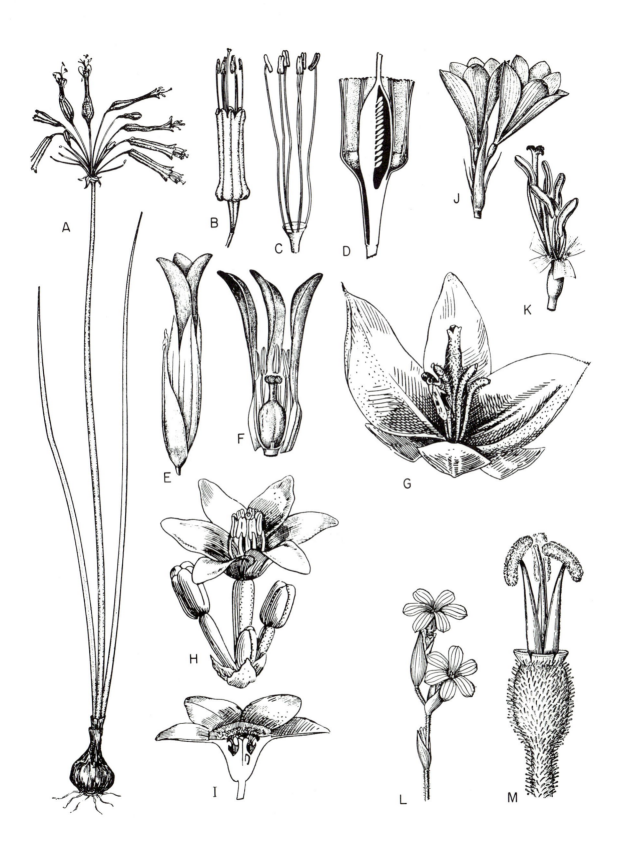

The scattered occurrence of epigyny in various orders indicates, as might be expected, that this character has evolved along several lines of evolution and must not be used in taxonomy as a character of paramount importance. However, the constancy of epigyny or hypogyny in many taxa is significant and therefore of diagnostic value making it a convenient key character.

In those dicotyledons which often possess monocotyledonous attributes epigyny has developed in the Aristolochiales, where in combination with staminal fusion, zygomorphy (in many species of *Aristolochia*) and reduction of petals it is an advanced feature. In most Nymphaeaceae the condition is closely reminiscent of that in the Hydrocharitaceae, while the Cabombaceae and the Ceratophyllaceae are hypogynous.

Base Data

ALISMATIFLORAE: HYDRO.: Hydro.: all.

LILIIFLORAE: DIOSC.: Diosc.: all; Stemo.: *Stemona* (epigynous), *Stichoneuron* (half epigynous) (but *Croomia* hypogynous).—TACCA.: all.—ASPAR.: Peter.; Conva.: *Lourya* (epigynous), *Ophiopogon* and *Peliosanthes* (half epigynous); Dorya.; Agava.: subfam. Agavoideae (*Furcraea, Beschorneria, Agave, Littaea, Polianthes* etc.); Aspho.: *Ixiolirion, Kopakowskia*; Hypox.: all; Tecoph. and Cyana. (mostly half epigynous); Amary.: all.—LILIA.: Irida.: all except *Isophysis*; Geosi.; Colch.: *Campynema* and *Campynemanthe*; Alstr.: all; Melan.: *Aletris, Nietneria* (epigynous), *Zigadenus* (half epigynous).—BURMA.: all.—ORCHI.: all.—HAEMO.: Haemo.: half of the genera.—VELLO.: all.—BROME.: Brome.: subfam. Bromelioideae.

ZINGIBERIFLORAE: ZINGI.: all, but Lowia. only half epigynous.

ARECIFLORAE: CYCLA.: all except spp. of *Sphaeradenia* and *Stelestylis*.

Associated dicotyledons:

NYMPHAEIFLORAE: NYMPH.: Nymph., gynoecium enclosed by tepals and androecium, e.g. in *Euryale* and *Victoria*, and partly in *Nymphaea*.

MAGNOLIIFLORAE: MAGNO.: Arist.: all, but *Saruma* only half epigynous.

FLORAL NECTARIES
(Fig. 42, Diagram 35)

The most extensive surveys of nectaries in monocotyledons are those of Grassmann (1884) and, in particular,

of Daumann (1970), which will be followed here. Included in this study were 425 species distributed among 196 genera.

A nectary is defined as each more or less localized part of a plant on which regularly nectar is secreted, even where the epidermis and the subepidermal tissue do not show any considerable differences in relation to the surrounding parts. It is, however, necessary that the secretion is fluid and contains sugar (Schnepf, 1964).

The floral nectaries can be divided into the following categories.

Perigonal nectaries. These are always situated on the adaxial side, and mostly on the base (rarely, as in *Galanthus*, on the apical part) of the tepals. In actinomorphic flowers they may be present and more or less equally distributed on the base of all the tepals (such as in *Disporum, Fritillaria, Iris, Luzuriaga*) or they may be present on all the tepals but more strongly developed on the base of the outer tepals only (such as in *Erythronium, Gagea, Tricyrtis, Belamcanda, Hermodactylus*) or on the base of only the inner ones (as in *Cipura, Ferraria, Gloriosa, Iris* spp., *Lilium, Smilax* male flower, *Streptopus, Veratrum, Zigadenus*). They may also be restricted either to the base of the outer tepals (as in *Iris* spp. *Neomarica, Uvularia*) or to the base of the inner tepals (as in *Aristea, Dietes, Galanthus, Moraea, Trillium*).

> Different groups among the perigonal nectaries may be distinguished according to their position:
> not, or more or less differentiated spots, or callosities (*Echinodorus, Galanthus, Iris, Libertia, Luzuriaga, Moraea, Streptopus, Trillium, Veratrum* and *Zigadenus*);
> a shallow little groove or basin or spur-like pockets (*Disporum, Erythronium, Gagea, Smilax, Tricyrtis, Uvularia*);
> a concavity or sac formed by the tepal (*Goodyera, Neottia*);
> a perigonal spur (*Angraecum, Gymnadenia, Platanthera*);
> several foldlets (*Belamcanda*);
> a medial furrow (*Gloriosa, Lilium*) etc.

The perigonal nectaries are concentrated in the following groups:

1. a single member of the Alismatiflorae (*Echinodorus* of the Alismataceae),
2. single members of the Ariflorae (e.g. *Anthurium* of the Araceae),
3. particular groups of the Liliiflorae, viz. within the Dioscoreales, in Trilliaceae (*Trillium; Paris* seems to lack nectaries), within the Asparagales, Philydraceae (*Luzuriaga, ?Lapageria*), Smilacaceae (*Smilax*), Petermanniaceae (*Petermannia*, see Clifford), and Convallariaceae (*Disporum, Streptopus, ?Clintonia*, see be-

low), and in many taxa of Liliales and Orchidales. According to Daumann, there are no nectaries in *Narthecium*, but septal nectaries in Alstroemeriaceae (*Alstroemeria* studied) as well as the genera of Iridaceae subfam. Ixioideae (*Acidanthera*, *Antholyza*, *Babiana*, *Crocus*, *Freesia*, *Gladiolus*, *Ixia*, *Lapeyrousia*, *Romulea* and *Tritonia* were studied). Most other Liliales have perigonal nectaries, viz. the Colchicaceae, Liliaceae, Tricyrtidaceae, the Iridaceae except subfamily Ixioideae, and at least a great many Melanthiaceae.

4. those members of the Orchidales which possess nectaries (by no means all taxa). In the orchids the nectar is often substituted by strongly scented substances (produced on so-called osmophores). Not even the presence of a spur is a proof for nectar secretion, because in many species of *Orchis*, which have a spur, there is no nectar.

5. in *Galanthus* of the Amaryllidaceae (Asparagales, Liliiflorae), an interesting type of weakly secreting perigonal nectary situated on the distal part of the inner tepals (in combination with nectar secretion from a disc; see below).

Androecial nectaries. These are situated on the filaments of fertile stamens or on staminodes.

Nectaries on filaments of fertile stamens are found in some scattered genera, mostly in the Liliales, for example in *Colchicum* (Colchicaceae), *Iris* (Iridaceae), *Tulipa* (?) (Liliaceae) and in the Asparagales, for example *Dianella* (Dianellaceae) and *Hessea* (= *Carpolyza*; Amaryllidaceae), and besides in the male flowers of *Chamaerops* (Arecaceae).

In *Sagittaria* (Alismataceae) there are nectaries on staminodes as well as on filaments of fertile stamens,

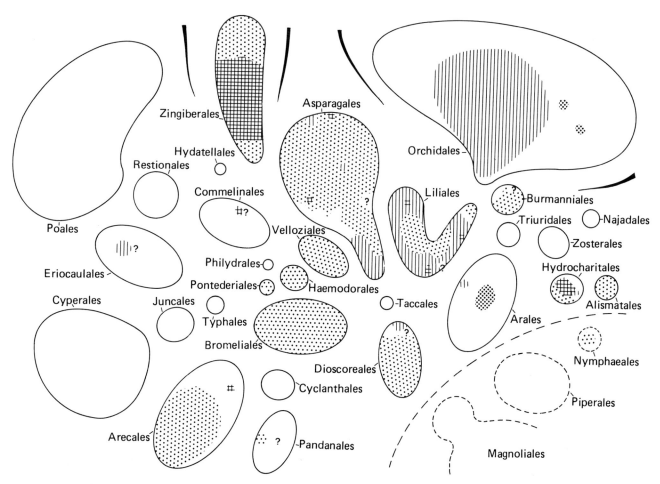

Diagram 35. Floral nectar secretion (shaded). Septal nectaries (sparse dotting) include inner as well as outer septal nectaries. Dubiously included is the nectar secretion along the carpel margins and midveins in some Burmanniales. Nectaries at tepal bases (vertical hatching) and at stamen or staminodium bases (checking). Nectar secretion from stigma papillae (dense dotting). Further details are in the text.

while nectaries on staminodes are concentrated mainly to Hydrocharitaceae (*Stratiotes*, *Hydrocharis*, *Ottelia*) and Zingiberaceae (*Brachychilum*, *Globba*, *Hedychium*, *Kaempfera*, *Roscoea*), but may also be found in the Araceae (*Aglaonema*) and Commelinaceae (*Commelina*).

Daumann estimated that only about eight per cent of

the nectariferous taxa of his material had androecial nectaries.

Septal nectaries (in a wide sense). These are the most widespread types among the monocotyledons and occur in various groups. External septal nectaries are found in

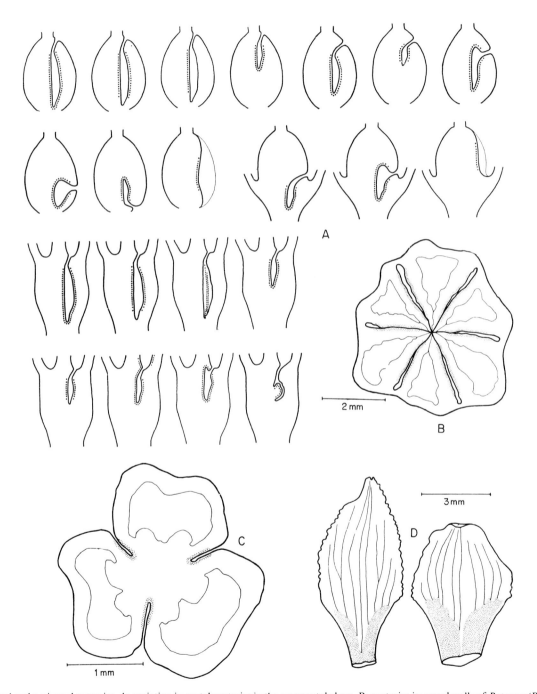

Fig. 42. Septal and perigonal nectaries. A: variation in septal nectaries in the monocotyledons. B: nectaries in carpel walls of *Butomus* (Butomaceae). C: septal nectaries in *Zigadenus* (Melanthiaceae). D: perigonal nectaries in *Veratrum* (Melanthiaceae). (All from Daumann, 1970.)

the basal parts of the carpel walls of *Luronium* (Alismataceae) and *Zigadenus* (Melanthiaceae). An external septal nectary is found on the inner (central) part of the ovary in *Tofieldia* (Melanthiaceae), where the carpels are largely free. Septal nectaries oriented towards the flower centre are also found on the inner faces of the mostly free carpels in *Trachycarpus* (Arecaceae). In some Alismataceae (*Alisma, Caldesia, Hydrocleys*) and in Aponogetonaceae "septal" nectaries are situated on the lateral walls of the free or nearly free carpels. In most other groups the septal nectaries are of the "internal" type, i.e. concealed in deep pockets or grooves in the ovary either from its upper or the lower part. Daumann (op. cit.) shows various positions and shapes of such septal nectaries in hypo- and epigynous ovaries and classifies them into a great number of types.

"Septal" nectaries are found in taxa of Butomaceae, Aponogetonaceae, and Alismataceae, viz. on the lateral walls of the free or nearly free carpels. Typical internal septal nectaries are present in most taxa of the Dioscoreales (Dioscoreaceae) and the Asparagales, although certain taxa seem to lack nectar secretion totally, such as several genera of Convallariaceae and Asparagaceae subfam. Ruscoideae. In the Liliales the Alstroemeriaceae, the Iridaceae subfam. Ixioideae and some Melanthiaceae, e.g. *Mondo, Tofieldia, Zigadenus*, have septal nectaries although they are unusual in being external (*Zigadenus* also has perigonal nectaries).

The Orchidales lack septal nectaries, agreeing thus with most Liliales. The Burmanniaceae often possess ovary nectaries, which partly take the shape of septal nectaries (*Gymnosiphon* species) partly of placental nectaries (*Gymnosiphon* species, *Cymbocarpa*); however the nectaries are independent of the septa or placentae and emerge between these on top of the ovary or are situated externally on top of the ovary (*Miersiella* and *Marthella*). Other genera may lack nectaries completely (and *Burmannia* at least partly).

The Haemodorales, Velloziales and Bromeliales have typical internal septal nectaries, and so have most taxa of Pontederiales (e.g. *Eichhornia, Pontederia*; no nectaries in *Heteranthera*) and at least some of the Velloziales (e.g. *Barbaceniopsis*), but the Philydrales lack nectaries.

The Zingiberiflorae frequently have septal nectaries. They are found in the Strelitziaceae, Heliconiaceae, Musaceae, Costaceae, Cannaceae and Marantaceae, while the Zingiberaceae have androecial nectaries.

In the Arecales nectar may be secreted (in Daumann's material six out of 11 taxa studied were devoid of nectar) from septal nectaries, viz. from external septal nectaries (*Livistonia, Sabal*), from internal ones (*Cocos*) or towards the floral centre from the partly free carpels (*Trachy-*carpus). Floral nectaries seem to be absent in the Cyclanthales and Pandanales.

Septal nectaries are probably absent in all Commeliniflorae as circumscribed here.

Gynoecial nectaries other than typical septal nectaries. These are rare. Mention should be made of *Limnocharis* (Alismataceae) and *Tupistra* (Convallariaceae) with nectar secretion on the whole of the ovary surface(s). In *Sagittaria* and *Echinodorus* of the Alismataceae nectar is secreted from the base of the free carpels, in the latter genus in combination with nectar secretion from the tepal base. (For the gynoecial nectaries of Burmanniales, see under septal nectaries above.)

Finally there remains to be noted the secretion of nectar from the stylar base in the iridaceous genera *Iris* and *Lapeyrousia* (Liliales), the former also having androecial or perigonal nectaries, the latter internal septal nectaries. However, Daumann could not verify any nectar secretion in *Tacca* (Taccales) sometimes reported to have nectar secretion from the stylar base.

In the flowers of certain monocotyledons, especially in the Arales, a sugar-rich secretion occurs from the stigmatic hairs, sometimes in combination with androecial and perigonal nectaries (*Butomus, Tulipa, Anthurium, Aglaonema*). Some orchids, such as *Aceras* and *Himanthoglossum*, for example, were thought by Daumann to possess most likely such stigmatic excretion, functioning ecologically as nectar.

Nectar-secreting discs. Discs are rarely found in monocotyledons and are morphologically of dubious nature. They occur in *Galanthus* and *Leucojum* of the Amaryllidaceae (Asparagales) and a nectar disc has also been reported in a Bolivian species of *Dioscorea* (Dioscoreales), but probably not, as sometimes stated, in the male flowers of *Hydrocharis* (Hydrocharitales).

Taxonomic evaluation

The taxonomic evaluation of nectar types must be seen against the fact that nectar is sometimes secreted on different structures in the same flower. Besides, one type of nectary may be lost in certain evolutionary lines, where nectar secretion may appear from different parts as a secondary compensatory adaptation. Thus nectary types for phylogenetic purpose must be judged in combination with other attributes and also according to their importance in relation to pollination.

The lack of nectaries, whether primary or secondary, is often combined with wind pollination. It is possible to observe groups, such as Trilliaceae, where nectaries may occur (*Trillium*) or may be lacking (*Paris*), those lacking nectar being perhaps wind pollinated. Families where some genera may be lacking, while others have nectar secretion, include Asparagaceae and several families in the Liliiflorae, such as Convallariaceae, Amaryllidaceae, Iridaceae and Orchidaceae but also Araceae, Arecaceae and Commelinaceae. Families totally devoid of floral nectaries include, for example, the Scheuchzeriaceae, Juncaginaceae and Potamogetonaceae (all wind pollinated) in the Alismatiflorae, the Taccaceae in the Liliiflorae, the Typhaceae, Sparganiaceae, Juncaceae, Cyperaceae, Poaceae etc. in the Commeliniflorae, and the Pandanaceae and Cyclanthaceae in the Areciflorae.

Staminodal nectaries seem to characterize the Hydrocharitaceae and the Zingiberaceae, the latter of which differing perhaps from all other Zingiberiflorae (having septal nectaries) in this respect.

Perigonal nectaries are concentrated to the Liliales and Orchidales, while an unspecialized gynoecial nectary type is present in the Burmanniales. The fact that *Disporum* has perigonal nectaries (like Colchicaceae and Liliaceae) indicate in combination with embryological peculiarities, for example lack of oxalate raphides etc. (Björnstad, 1970), that they may have their proper place in the Liliales, perhaps close to the *Uvularia* group of the Colchicaceae. Another genus reported by Daumann (op. cit.) to lack septal nectaries but posses perigonal nectaries is *Streptopus*, which is maybe closely allied to *Disporum*. Björnstad also reports *Clintonia* to lack septal nectaries and to have the same (liliaceous-colchicaceous) embryology as *Disporum*, but Daumann reports septal nectaries for this genus. The Convallariaceae, among which *Streptopus* and *Disporum* are normally placed, otherwise lack nectaries or have gynoecial (septal) nectaries. Within the Liliales, the Iridaceae are heterogeneous, the subfamily Ixioideae (which besides has racemose inflorescence) being possibly less closely allied to the other subfamiles than usually believed. The record of septal nectaries in *Alstroemeria* (Alstroemeriaceae) is a surprising deviation from the adjacent families in Liliales. The Melanthiaceae may have septal nectaries but these are of the external type.

Contrary to the Liliales, the Asparagales possess septal nectaries in most groups, but nectaries are often lacking in Convallariaceae, Asparagaceae (subfam. Ruscoideae), Dasypogonaceae etc. The occurrence of perigonal nectaries in *Luzuriaga* and maybe other taxa of Philesiaceae and also in some Smilacaceae and Petermanniaceae support their exclusion from the Asparagales. Judging from nectaries only, these as well as Trilliaceae agree with most Liliales, while Alstroemeriaceae, part of Iridaceae

and (maybe) some Melanthiaceae show Asparagalean character.

The role of nectar is supplemented in certain groups with cells producing strongly scented substances, for example volatile essential oils, osmophores (Esau, 1965; Vogel, 1963). These cells are located on the surface of the spathe in various taxa of Araceae (Ariflorae); they entice insects which then may become trapped and in the process of freeing themselves become infested with pollen, thereby acting as pollinators.

It is thought that many orchids also possess osmophores, only relatively few producing nectar in a floral spur.

Osmophores and nectaries are not mutually exclusive structures, for *Narcissus* (Amaryllidaceae) and *Burmannia* (Burmanniaceae) possess both.

Amongst dicotyledons with monocotyledonous affinities septal nectaries occur in Nymphaeales.

Base Data

Perigonal nectaries

ALISMATIFLORAE: ALISM.: *Echinodorus.*
ARIFLORAE: ARAL.: Arac.: *Anthurium.*
LILIIFLORAE: DIOSC.: *Trillium* spp.—ASPAR.: Smila.: *Smilax*; Peter.: *Petermannia*; Phile.: *Luzuriaga, Lapageria*; Conva.: *Disporum, Streptopus* (both probably misplaced); Amary.: *Galanthus* (perigonal apices).—LILIA.: Irida. (excl. subfam. Ixioideae): e.g. *Aristea, Belamcanda, Cipura, Dietes, Ferraria, Hermodactylus, Iris, Lapeyrousia, Libertia, Moraea, Neomarica, Tigridia*; Colch. (all or most; at tepal base in *Anguillaria, Dipidax, Gloriosa, Kreysigia, Littonia, Ornithoglossum, Sandersonia, Uvularia, Wurmbea*; Tricy.: *Tricyrtis*; Caloch.:?; Lilia.: *Erythronium, Fritillaria, Gagea, Lilium, Lloydia, Tulipa*; Melan.: several genera, e.g. *Veratrum, Zigadenus.*—ORCHI.: Cypri.: at least *Cypripedium*; Orchi.: numerous genera (but numerous also without nectaries), e.g. *Angraecum, Epipactis, Goodyera, Gymnadenia, Plathanthera.*

?COMMELINIFLORAE: ERIOC.: Erioc.:? *Eriocaulon* (glands on perianth dubiously with nectar).

Androecial nectaries

ALISMATIFLORAE: HYDRO.: Hydro. (on staminodia): *Hydrocharis, Hydrocleys, Ottelia, Stratiotes.*—ALISM.: *Echinodorus* (on stamens and staminodia).

LILIIFLORAE: ASPAR.: (on stamens): Diane.: *Dianella*; Amary.: *Carpolyza.*—LILIA. (on stamens): Irida.: *Iris*; Colch.: *Androcymbium, Colchicum*; Lilia.: ?*Tulipa.*

ZINGIBERIFLORAE (on staminodia): ZINGI.: Zingi.: *Brachychilum, Globba, Hedychium, Kaempfera, Roscoea.*

COMMELINIFLORAE: COMME.: Comme.: *Commelina* (?).

ARECIFLORAE: ARECA. (on stamens): *Chamaerops* (male flowers).

Septal nectaries

ALISMATIFLORAE: HYDRO.: Butom.: *Butomus*; Apono.: *Aponogeton.*—ALISM.: *Alisma, Caldesia, Hydrocleys, Luronium* etc.

LILIIFLORAE: DIOSC.: Diosc.: *Dioscorea*; Trill.: *Trillium* spp.—ASPAR.: Conva.: ?*Clintonia, Majanthemum, Polygonatum, Ophiopogon* (= *Mondo*), *Smilacina, Tupistra*; Aspar.: *Asparagus*; Herre.: *Herreria*; Dracae.: *Dracaena, Sansevieria*; Dorya.: *Doryanthes*; (Dasyp.: lacking in *Lomandra*;) Phorm.: *Phormium*; Xanth.: *Xanthorroea*; Agava.: *Agave, Bescorneria, Furcraea, Polianthes, Yucca*; (Hypox.: no nectaries;) Aspho.: subfam. Astelioideae: *Cordyline*; subfam. Asphodelioideae: *Aloë, Asphodeline, Asphodelus, Eremurus, Gasteria, Haworthia, Kniphofia*; subfam. Anthericoideae: *Anthericum, Chlorophytum, Paradisia*; Aphyl.: *Aphyllanthus*; Hemer.: *Hemerocallis*; Funki.: *Hosta*; Hyaci.: *Albuca, Bowiea, Camassia, Chionodoxa, Drimia, Eucomis, Galtonia, Hyacinthus, Lachenalia, Muscari, Ornithogalum, Puschkinia, Scilla, Urginia, Veltheimia*; Allia.: *Agapanthus, Allium, Nothoscordum*; Amary.: *Amaryllis, Chlidanthus, Cliva, Crinum, Cyrtanthus* (*Vallota*), *Eucharis, Griffinia, Haemanthus, Hippeastrum, Hymenocallis, Narcissus, Pancratium, Sparaxis, Sprekelia, Sternbergia, Zephyranthes.*—LILIA.: Irida. subfam. Ixioideae: e.g. *Acidanthera, Antholyza, Babiana, Crocosma, Crocus, Freesia, Gladiolus, Ixia, Lapeyrousia, Romulea, Tritonia*; Alstr.: *Alstroemeria*; Melan.: *Petrosavia, Protolirion, Tofieldia, Zigadenus* (rudimentary).—BURMA.: (*Cymbocarpa*,) *Gymnosiphon* (see also below); Thism.: *Thismia.*—PONTE.: *Eichhornia, Pontederia* (no nectary in *Heteranthera*).—HAEMO.: *Anigozanthos, Haemodorum, Wachendorfia.*—VELLO.: at least some, e.g. *Barbaceniopsis* sp. (Böhme, Hamann, Vogel; personal communication).—BROME.: *Aechmea, Ananas, Aregelia, Billbergia, Bromelia, Catopsis, Dyckia, Nidularium, Pitcairnia, Quesnelia, Tillandsia, Vriesia*.

ZINGIBERIFLORAE: ZINGI.: Helic.: *Heliconia*; Musac.: *Musa*; Strel.: *Strelitzia*; Costa.: *Costus*; Canna.: *Canna*; Maran.: *Calathea, Maranta*.

ARECIFLORAE: ARECA.: *Cocos, Livistona, Sabal*.—Panda.: *Pandanus* (dubiously present; recorded by Saunders, not found by Daumann, 1970).

Dicotyledons associated with the monocotyledons: NYMPHAEIFLORAE: NYMPH.: some taxa accord. to Emberger (1960, p. 922).

Nectar secretion at stylar base (not in the diagram)
LILIIFLORAE: LILIA.: Irid.: *Iris, Lapeyrousia*.

Nectar secretion from stigma
(ALISMATIFLORAE: HYDRO.: Butom.: *Butomus*.)
ARIFLORAE: ARAC.: *Aglaonema, Anthurium, Arum* (probably many genera).
LILIIFLORAE: LILIA.: *Tulipa.*—ORCHI.: *Aceras, Himanthoglossum*.

Other types of gynoecial nectar secretion (not in the diagram)
Basis of free carpels: ALISMATIFLORAE: ALISM.: *Echinodorus, Sagittaria*.
Entire surface of free carpels: ALISMATIFLORAE: ALISM.: *Limnocharis*.
Externally on top of ovary: LILIIFLORAE: BURMA: Burma.: *Marthella, Miersiella*.
The whole surface of the syncarpous ovary: LILIIFLORAE: ASPAR.: Conva.: *Tupistra*.
The sides of the free carpels facing the floral centre: ARECIFLORAE: ARECA.: *Trachycarpus*.

Nectar secretion from a disc
ALISMATIFLORAE: HYDRO.: Hydro.: *Hydrocharis*.
LILIIFLORAE: DIOSC.: Diosc.: *Dioscorea* sp. (Huber, 1969).

STAMEN NUMBER
(Fig. 43, Diagram 36)

The numbers of stamens per flower varies from one (bearing a solitary or half anther) as in *Canna* (Zingiberales), to numerous, as in various palms and a few grasses, e.g. up to about 150 in *Ochlandra* (Poales). The morphological basis of the variation differs from group to group, but in general numbers less than six reflect a reduction from the two whorls common in the monocotyledons, and numbers more than six result from the multiplication of stamen initials. With lower numbers interpretation of the floral morphology may render the problem of stamen number difficult. For example the flowers of *Potamogeton* may be regarded as possessing four stamens or the whole may be regarded as a synanthium with four separate unistaminate flowers. Likewise the presence of scales subtending the stamens of *Scirpodendron* and other genera of the Mapanieae (Cyperales) suggests that there the apparent flower may be a synanthium of many unistaminate flowers. Where the flower/synanthium is basically spiral in construction and the

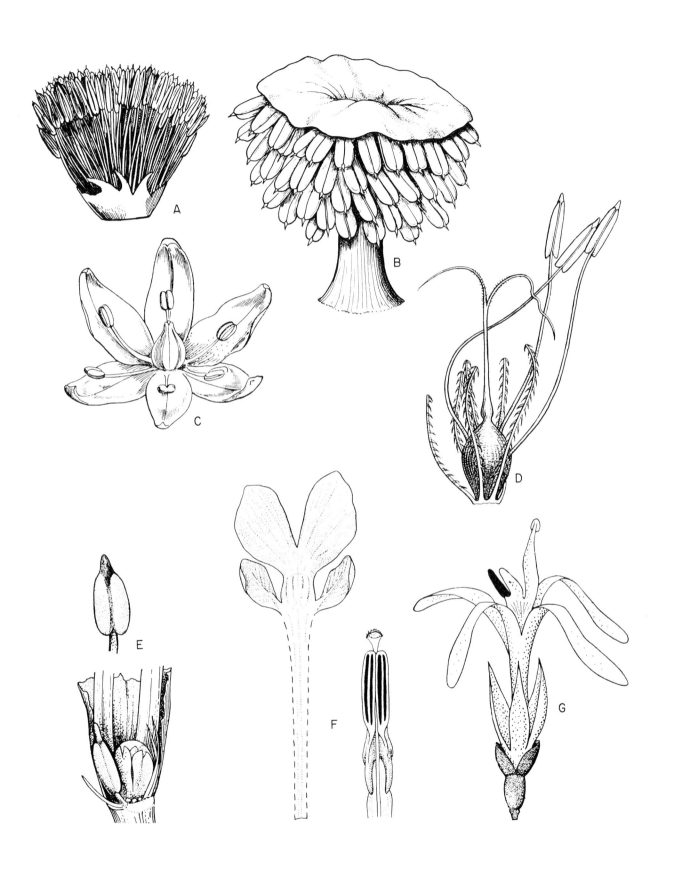

number of stamens generally in excess of six there is no obligate association between the number of stamens and the number of perianth members.

In the Alismatiflorae staminal numbers vary from numerous (Alismataceae, some Hydrocharitaceae) to solitary (Najadales) though as stated above the concept of solitary depends upon the interpretation of the floral axis (Posluszny and Tomlinson, 1977). *Triglochin* (Juncaginaceae) is, for example, mostly considered to have hexastaminate flowers, but was regarded to have dense,

lateral partial inflorescences with unistaminate, monomerous flowers of *Lilaea* type by Charlton (1976). This interpretation is not followed here in the base data table or diagram. A similar series from one to six stamens is also seen in the Ariflorae.

In the Liliiflorae, flowers with more than six stamens are rare. They occur for example in *Pleea* (Melanthiaceae) and *Pseudosmilax* (Smilacaceae), which have three trimerous whorls; *Aspidistra* (Convallariaceae) which has two tetramerous staminal whorls; and *Paris* (Trilliaceae)

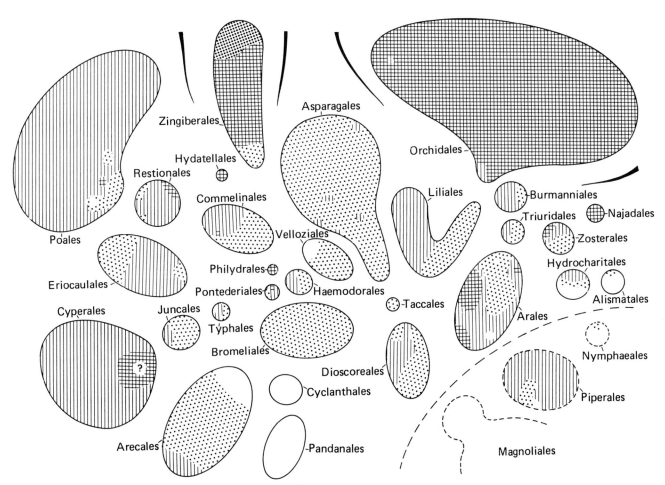

Diagram 36. Number of functional stamens. Only one theca, i.e. half of one anther functional: dense dots. One bithetic anther functional: cross-hatching. Two to three stamens functional: vertical hatching. Four to six stamens functional: dense dots. Numerous (nine or more) stamens functional: unshaded.

Fig. 43. Stamen number. A–B: numerous stamens in *Phytelephas microcarpa* (Arecaceae) and *Pandanus macrocarpus* (Pandanaceae) respectively (both male flowers). C: hexastaminate flower in *Astelia pumila* (Asphodelaceae subfam. Asteloideae). D: tristaminate flower in *Scirpus lacustris* (Cyperaceae). E: unistaminate flower in *Zannichellia palustris* (Zannichelliaceae) (note the intravaginal squamules). F: unistaminate androecium (to the right, note that the stamen encloses the style) and four petaloid staminodial lobes in *Roscoa purpurea* (Zingiberaceae). G: unistaminate flower with one half theca only (black) in *Canna* (schematic); four petaloid staminodia and a stigma may also be seen. (A from Lotsy, 1911; B from Stone, 1974; C from Pizarro, 1959; D from Schultze-Motel in Hegi, 1966; E from Vijayaraghavan and Kumari, 1974; F from Burtt and Smith, 1972; G from Koechlin, 1964.)

which has from tetra- to octomerous whorls. In the genus *Gethyllis* (Amaryllidaceae) the stamens are 6, 12, 18 or more in number. Numerous stamens also occur in *Vellozia* (Velloziales).

Many Zingiberiflorae, are unusual in that the stamens tend to be transformed into relatively large petaline staminodia readily derivable from two trimerous whorls. Thus while for example the Musaceae has 5(–6) functional stamens, Zingiberaceae has one stamen with a solitary instead of a pair of thecae and besides 3–5 staminodia.

Amongst the Commeliniflorae, single-stamened flowers are characteristic of the Centrolepidaceae, Hydatellaceae, and possibly at least the tribe Mapanieae of the Cyperaceae (depending upon the interpretation of the flower). They also occur rarely in Poaceae and Commelinaceae. Flowers with high numbers of stamens are restricted to the Poaceae.

In contrast to the Commeliniflorae, the Areciflorae are rich in flowers with many stamens, the condition applying to all Cyclanthales, most of the Pandanales and many Arecales (Uhl and Moore, 1980). Reduction to fewer than four stamens per flower occurs in relatively few Areciflorae.

Stamen number is clearly related to the pollination mechanism. Reduction of stamen number in land plants is associated with animal pollination and increasing stamen number with wind pollination. However, this is a generalization only and exceptions may be easily found because many wind pollinated plants produce both male and hermophrodite flowers and so increase pollen production by means other than multiplying stamen number.

Base Data

A single functional stamen

ALISMATIFLORAE: Hydro.: Hydro.: *Maidenia* (Kaul, 1968).—ZOSTE.: Junca. p.p.: *Lilaea*; Zoste.; Zanni.: *Althenia, Lepilaena, Vleisia, Zannichellia.*—NAJAD.: *Najas.*

ARIFLORAE: ARAL.: Arac.: several genera, e.g. *Biarum, Cryptocoryne, Gamogyne*; Lemna.: *Wolffia* etc.

LILIIFLORAE: ASPAR.: Tecoph.: *Cyanella* (rarely).—ORCHI.: Orchi. (all).—PONTE.: *Hydrothrix.*—HAEMO.: *Pyrrorhiza.*—PHILY. (all).

ZINGIBERIFLORAE: ZINGI.: all Zingi., Costa., Canna. and Maran.

COMMELINIFLORAE: COMME.: Comme.: *Callisia, Murdannia* spp., *Pseudoparis.*—TYPHA.: Typha.: *Typha* (rarely).—CYPER.: (depending on floral interpretation: in Mapanieae or more ?).—HYDAT.: Hydat. (all).—RESTI.: Centr. (all, according to pseudanthial interterpretation of flowering units).—POAL.: rarely.

Two to three functional stamens

ALISMATIFLORAE: HYDRO.: Hydro.: several genera: *Blyxa* p.p., *Enhalus, Halophila, Hydrilla, Vallisneria.*—ALISM.: *Wiesneria.*—ZOSTE. Potam.: *Ruppia*; Posid. p.p.; Zanni. p.p.: e.g. *Lepilaena*; Cymod.—TRIUR.: Triur. p.p.

ARIFLORAE: ARAL.: Arac.: many genera, perhaps c. 45%; Lemna. p.p.

LILIIFLORAE: DIOSC.: Diosc.: *Dioscorea sensu lato.* p.p. (often with 3 staminodes).—ASPAR.: Smila.: rarely; Aspar.: *Ruscus*; Hypox.: *Pauridia*; Aspho. subfam. Anthericoideae: some genera, e.g. *Anemarrhena, Arnocrinum, Hensmania, Johnsonia, Sowerbaea, Stawellia, Thysanotus* p.p.; Tecoph.: *Cyanella* p.p., *Tecophilaea, Zephyra*; Allia.: *Brevoortia, Brodiaea, Dichellostemma, Leucocoryne, Stropholirion* and some Gilliesioideae, viz. *Ancrumia, Eriuna, Gethyum, Gilliesia, Solaria* and *Trichlora*; Amary.: *Zephyra* (rarely).—LILIA.: Irida. (all); Geosi.—BURMA.: Burma. (all); Thism.: rarely.—ORCHI.: Apost.; Cypri.—PONTE.: *Heteranthera.*—HAEMO.: most taxa of subfam. Haemodoroideae: *Barberetta, Dilatris, Haemodorum, Hagenbachia, Lachnanthes, Schiekia, Wachendorfia, Xiphidium.*

COMMELINIFLORAE: COMME.: Comme.: numerous genera (but then often with three staminodia); Mayac.—ERIOC.: Xyrid. (but often with 3 staminodia); Erioc.: frequently.—TYPHA.: Sparg. and Typh. generally.—JUNCA.: rarely (some *Juncus* spp.)—CYPER.: most taxa.—RESTI.: Rest.; Anarth.; Ecdei. p.p.—POAL.: most genera.

ARECIFLORAE: ARECA.: rather rarely (perhaps c. 6%).

Four to six functional stamens

ALISMATIFLORAE: HYDRO.: Hydro.: sometimes e.g. *Ottelia* p.p.; Apono.: *Aponogeton* spp.—ALISM.: sometimes, e.g. *Baldellia, Limnophytum.*—ZOSTE.: Scheu.; Junca. except *Lilaea*; Potam. except *Ruppia.*—TRIUR.: p.p.

LILIIFLORAE: DIOSC.: Diosc. (most); Stemo.; Trill. (most; sometimes more than 6).—TACCA.: *Tacca.*—ASPAR.: most taxa except the above-mentioned with 2–3 functional stamens.—LILIA.: Colch.; Alstr.; Lilia.; Melan.—BURMA.: Thism. (most); Corsi. (—ORCHI.: none.)—PONTE.: most genera.—HAEMO.: *Phlebocarya* and subfam. Conostylidoideae.—VELLOZ.: most genera (except *Vellozia*).—BROME.: all.

ZINGIBERIFLORAE: ZINGI.: Lowia.; Helic.; Musac.; Strel.

COMMELINIFLORAE: COMME.: Comme.: more than half of the genera; Carto.—ERIOC.: Xyrid.: rarely; Rapat.: all; Erioc.: frequently.—TYPHA.: Sparg.: rarely.

—JUNCA.: Thurn.; Junca.: mostly.—RESTI.: Ecdei.: rarely; Flage.; Joinv.; Hangu.—POAL.: Poac. subfam. Bambusoideae and certain other groups; c. 5%.

ARECIFLORAE: ARECA.: maybe c. half of the genera.

Functional stamens more than six in number

ALISMATIFLORAE: HYDRO.: Butom.; Hydro.: frequently; Apono.: rarely.—ALISM.: mostly.

ARIFLORAE: ARAL.: Arac.: rarely (e.g. *Xenophya*).

LILIIFLORAE: DIOSC.: Trill.: *Paris* spp.—ASPAR.: Smila.: *Pseudosmilax* (9 stamens); Conva.: *Aspidistra* spp.; Amary.: *Gethyllis* spp.—LILIA.: Melan.: *Pleea* spp. (up to 12 stamens).—VELLO.: *Vellozia*.

COMMELINIFLORAE: POAL.: rarely in subfam. Oryzoideae (*Luziola*) and Bambusoideae (*Ochlandra*).

ARECIFLORAE: ARECA.: many, perhaps c. 40% of the genera.—PANDA.: all.—CYCLA.: all.

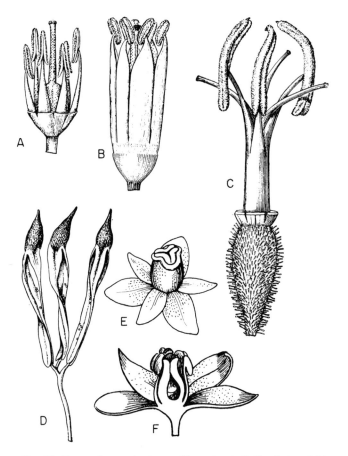

Fig. 44. Free and connate stamen filaments. A–B: free but variably broad filaments in *Nothoscordon* (Alliaceae). C: connate filaments in *Sisyrinchium arenarium* (Iridaceae). D: fused filaments in male flower of *Typha subulata* (Typhaceae). E–F: connate filaments in male flower of *Ruscus* (Asparagaceae subfam. Ruscoideae), where the filaments form a column. (A–D from Correa, 1969; E from Larsen, 1977; F from Lotsy, 1911.)

CONNATE (OR OTHERWISE FUSED) STAMEN FILAMENTS (Fig. 44, Diagram 37)

The fusion of stamen filaments to form a tube or cup may have some selective advantage related to the pollination of flowers by insects or other animals, increasing, for example, the stability of the stamens or influencing their direction of dehiscence or position in the flower. The condition thus would not be expected to show a totally random distribution among the monocotyledons.

Indeed this is so, for flowers with connate filaments occur in substantial frequencies in only three of the superorders, being absent or almost so from the Alismatiflorae, Ariflorae and Zingiberiflorae. Although many species in these superorders are almost exclusively insect-pollinated plants, they do not have fused filaments but have various other adaptations. Most Zingiberiflorae for example have but one functional stamen and in the Ariflorae the flowers are extremely small.

In the remaining three superorders connate filaments occur only sporadically and without conspicuous regularity. The fusion varies from being restricted to the very base of the filaments to resulting in a true staminal tube or corona-like structure (as in some Amaryllidaceae and in *Tulbaghia*, Alliaceae).

In the Liliiflorae, connate stamen filaments occur in a few genera of each of the Dioscoreales and Burmanniales and in quite a number of genera of each of the Asparagales and Liliales. In none of these do the connate filaments seem to be of great taxonomic importance at higher levels except that they contribute in distinguishing the Asparagaceae subfam. Ruscoideae (which might deserve family rank), in being excellent generic characters in some Hyacinthaceae and Alliaceae and in being of importance, for example, in the Amaryllidaceae, where in the *Eucharis* group the wing-like lateral flanks of the filaments fuse with each other laterally, such as in *Hymenocallis* and *Pancratium* (Traub, 1963).

Likewise, connate filaments are common in the Iridaceae, but contribute little to the recognition of divisions in the family. Also in the Orchidaceae the filaments are fused, taking part in the gynostemium.

The scattered occurrence of connate or otherwise fused filaments in the Commeliniflorae does not contribute to the taxonomy of that superorder nor does it seem to do so in the palms (Arecales), where fused filaments occur in several groups. Fused stamens besides, characterize *Pandanus* (Pandanales).

There is no doubt that filamental tubes or cups generally represent recently evolved structures. This is also obvious among the dicotyledons (mainly Magnoliales), where they occur, in, for example, Canellaceae and Myristicaceae. These have fairly specialized flowers probably having little in common with those of the hypothetical common ancestors of the dicotyledons and monocotyledons.

Base Data

LILIIFLORAE: DIOSC.: Diosc.: *Dioscorea* spp.; Stemo.: *Croomia, Stemona* spp., *Stichoneuron.*—AS-PAR.: Smila.: *Heterosmilax*; Phile. *Lapageria, Philesia*; Conva.: *Peliosanthes* (*Lourya*); Aspar.: *Danaë, Ruscus, Semele*; Aspho.: *Echeandia*; Hyaci.: at least *Androsiphon, Eucomis, Massonia, Puschkinia, Rhadamanthus, White-headia*; Allia.: at least *Androstephium, Behria, Bessera, Gethyum, Gilliesia, Miersia, Solaria, Speea, Trichlora, Tulbaghia*; Amary.: several genera, e.g. *Crocopsis, Eucrosia* and the tribe Euchariseae (incl. *Hymenocallis,*

Pancratium, Stenomesson).—LILIA.: Irida.: various genera such as *Chamelum, Ferraria, Galaxia, Gelasine, Gynandriris, Hexaglottis, Homeria, Ixia, Moraea, Patersonia, Solenostemon, Symphostemon*; Colch.: *Sandersonia*; Tricy.: *Tricyrtis.*—BURMA.: at least Thism.: *Sarcosiphon, Scaphiophora, Thismia.*—ORCHI.: all.

COMMELINIFLORAE: COMME.: Comme.: *Cochliostema.*—ERIOC.: Rapat.: p.p.—TYPHA.: *Typha* (fusion rather than connation).—RESTI.: at least in *Hopkinsia* and *Lyginia.*—POAL.: Poac.: few genera of subfam. Bambusoideae.

ARECIFLORAE: ARECA.: various genera among the Lepidocaryoid, Pseudophoenicoid, Cocosoid and Geonomoid palms.—CYCLA.: several genera (fusion of filaments).—PANDA.: *Pandanus* (fusion of filaments). The flowers are unisexual in this superorder, as in the Typhales, and there is no real staminal tube.

Dicotyledons associated with the monocotyledons:

MAGNOLIIFLORAE: MAGNO.: for example taxa in Canel., Myris. and Winte.

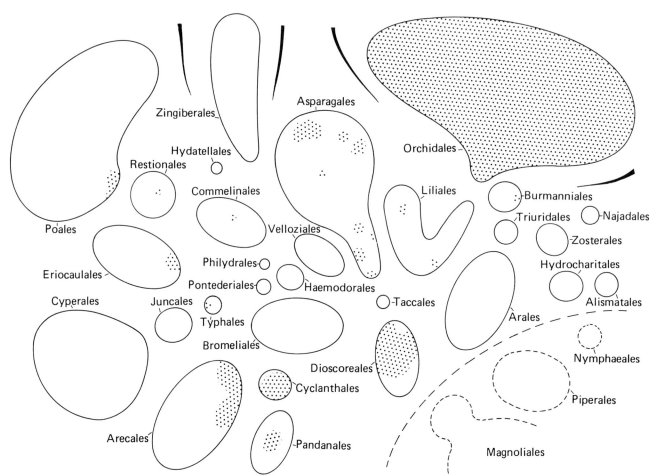

Diagram 37. Stamen filaments connate: shaded. Here are included the coherent tubular "paracorollas" connecting the stamen filaments in many Amaryllidaceae.

HAIRS ON
STAMEN FILAMENTS
(Fig. 45. Diagram 38)

Stamen filaments bearing hairs are restricted to the superorders Alismatiflorae, Commeliniflorae and Lilii-florae, amongst which they are distributed as shown in Diagram 38.

Within the Commeliniflorae only species of Commelinaceae and Cartonemataceae (Commelinales) have hairy filaments, with the exception of a species of *Rapatea* (Rapateaceae, Eriocaulales). In *Xyris* (Xyridaceae) the staminodia, not the fertile stamens, may be densely hairy.

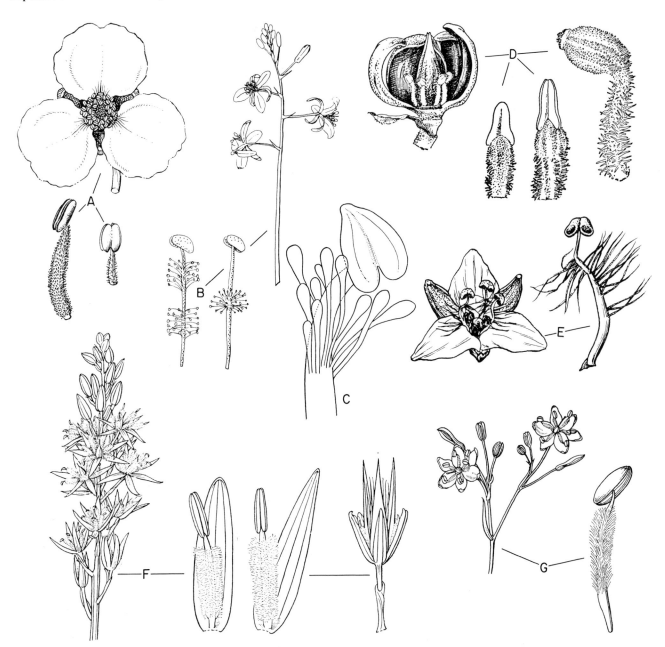

Fig. 45. Stamens with hairy filaments. A: *Sagittaria montevidensis* (Alismataceae). B–C: *Bulbine diphylla* and *B. sp.* respectively (Asphodelaceae). D: *Yucca brevifolia* (Agavaceae), flower, two stamens and a filament, the latter in detail. E: *Tripogandra grandiflora* (Commelinaceae). F: *Narthecium ossifragum* (Melanthiaceae). G: *Simethis planifolia* (Asphodelaceae). (A from Carter, 1960b; B from Nordenstam, 1964b; C from Schaeppi, 1939; D from Johnson, 1931; E from Moore, 1960; F–G from Ross-Craig, 1971.)

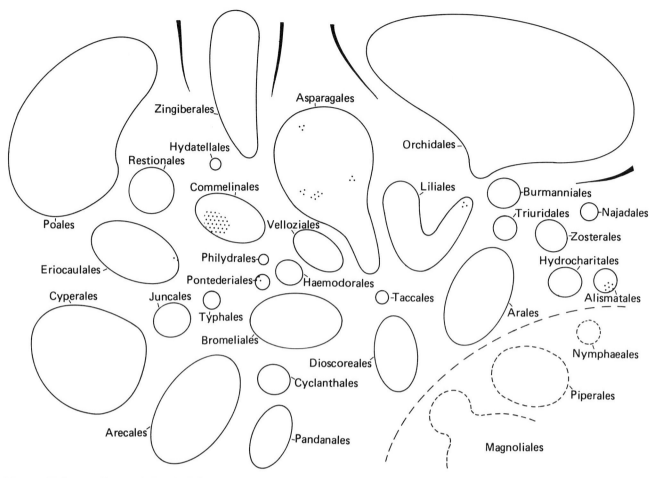

Diagram 38. Stamen filaments hairy (shaded).

Amongst the Liliiflorae hairy filaments are found in a few members of each of the Pontederiales (about half the species), the Liliales (*Narthecium* of Melanthiaceae) and the Asparagales (*Stypandra* of Dianellaceae, *Yucca* p.p. of Agavaceae, *Bulbine*, *Bulbinopsis*, *Arthropodium* and *Simethis* of Anthericaceae and *Giraldiella* of Amaryllidaceae). In *Stypandra* (Cave, 1975) the hairs are restricted to the swollen, distal part of the filament just beneath the anther. Here the trait has probably evolved independently in each of the orders, and within the Asparagales in several lines. It thus has little taxonomic significance at the high levels.

The hairs are very variable, sometimes glandular, as in some *Pontederia* species, but in most Commelinaceae and in *Heteranthera* (Pontederiaceae) of the "pearl necklace" type. In *Yucca* (Agavaceae) and, for example, *Rapatea xyridoides* (Rapateaceae) the hairs are very short, as is also the case in *Sagittaria* (Alismataceae). Their function thus may be variable, but often they may supply food for visiting insects.

Base Data

ALISMATIFLORAE: ALISM.: *Sagittaria* spp.
LILIIFLORAE: ASPAR.: Agava.: *Yucca*; Aspho.: several genera, e.g. *Arthropodium*, *Bulbine*, *Bulbinopsis*, *Glyphosperma*, *Simethis*; Diane. *Stypandra*; Amary.: *Giraldiella*.—LILIA.: Melan.: *Narthecium*.—PONTE.: *Eichhornia* (some spp.), *Heteranthera*.
COMMELINIFLORAE: COMME.: Comme.: numerous genera, e.g. *Tradescantia*, *Tinantia*, *Rhoeo* and *Cyanotis* spp.; Carto.: *Cartonema*.—ERIOC.: Rapat.: *Rapatea xyridioides*.

UNI- AND BISPORANGIATE ANTHERS
(Fig. 46, Diagram 39)

Taxa with bi- instead of the more usual tetrasporangiate anthers (Fig. 46) occur in all superorders other than the

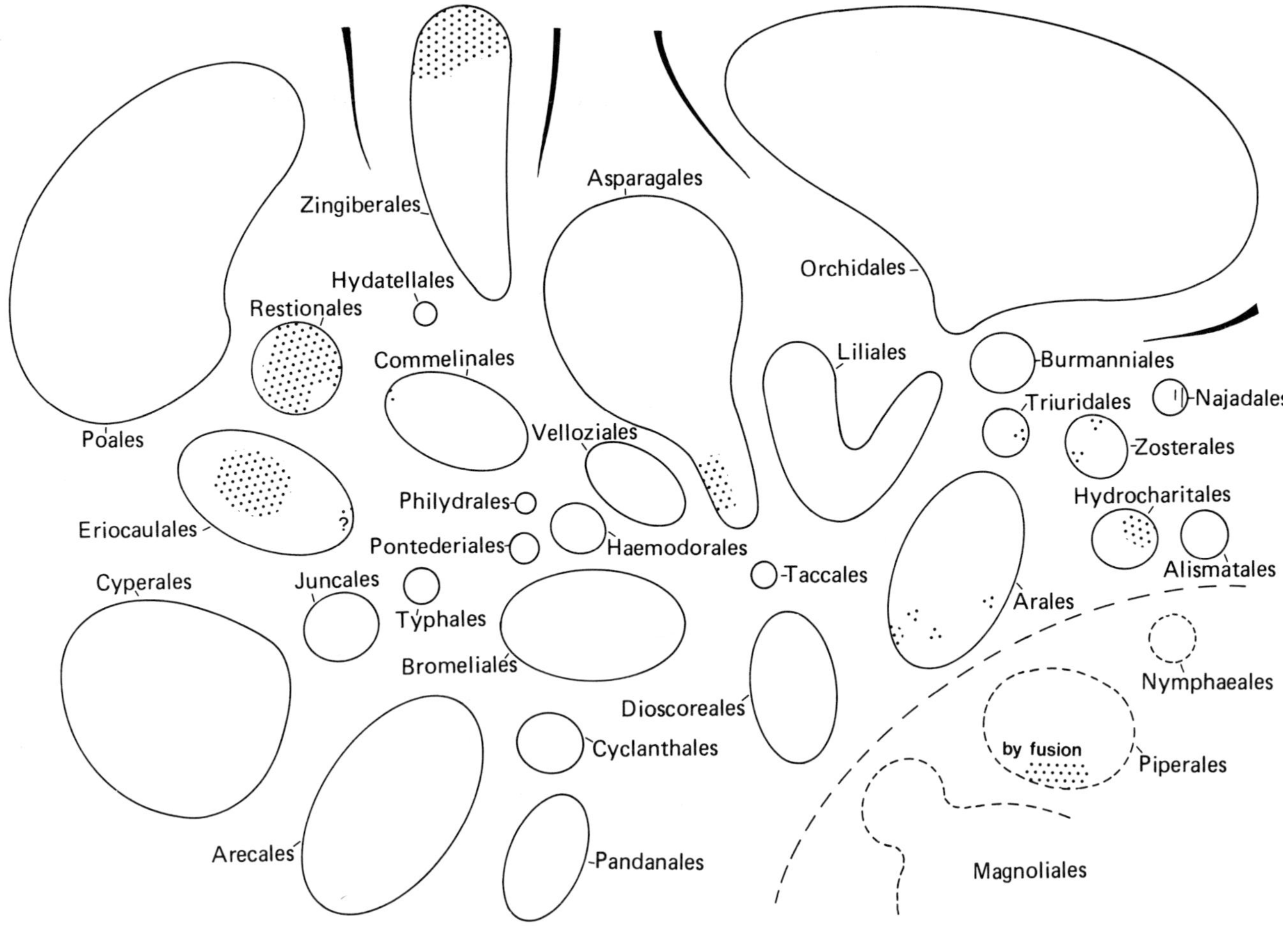

Diagram 39. Anthers bisporangiate and unisporangiate (dots and vertical hatching respectively). It should be noted that the bisporangiate anthers in the Zingiberales (Cannaceae, Marantaceae) are bisporangiate due to the development of only one of the thecae, whilst in the other groups unisporangiate thecae are developed. Unisporangiate anthers occur only in some species of *Najas*.

Areciflorae. Both conditions may occur in the same plant as with *Lilaea subulata* whose uppermost, male flowers are tetrasporangiate and whose lower, bisexual flowers are bisporangiate. Rarely as in *Najas minor* the anthers may be unisporangiate. The distribution of taxa with uni- and bisporangiate anthers is summarized in Diagram 39.

Within the Alismatiflorae bisporangiate anthers are reported from *Enhalus* and *Halophila* (Hydrocharitaceae), *Althenia* (Zannichelliaceae; Posluszny and Tomlinson, 1977), *Lilaea* (Juncaginaceae, see above), and rarely in the Triuridales. These are mostly wind- or water-pollinated plants and are further characterized by the possession of single-ovuled ovaries and hence single-seeded fruits. In the Triuridales there are occasionally bisporangiate anthers, as in *Soridium*.

The relative frequencies of bi- and tetrasporangiate anthers in the Ariflorae is uncertain. However, Davis (1966) lists two genera (*Arisaema* and *Dieffenbachia*) as possessing anthers of the former kind and three genera (*Calla, Peltandra* and *Symplocarpus*) as possessing the

latter kind. Also *Arisarum* is known to be bisporangiate. For the Lemnaceae Kandeler (1979) records *Wolffia* and *Wolffiella* as bi- and *Lemna* and *Spirodela* as tetrasporangiate. Although in the Araceae the ovule number often exceeds one per ovary there is a trend throughout the order Arales for fewer ovules and hence modest requirements of pollen.

In the Liliiflorae bisporangiate anthers are known only in the Smilacaceae (Asparagales). In the animal pollinated Zingiberiflorae there has been a reduction of the androecium to two sporangia in both the Marantaceae and Cannaceae. Again in the Marantaceae reduction of available pollen and ovule number have gone on hand in hand. Furthermore, though each locule is uniovulate as a rule, only one of the three ovules develops into a seed. The fruits of *Canna* are generally several-seeded, a situation that may be related to the pollen in that genus being shed directly onto the stigma in bud.

Within the Commeliniflorae four families seem to have bisporangiate anthers. These are the Mayacaceae (*May-*

aca, rarely; Horn af Rantzien, 1946) the Centrolepidaceae (where the condition is universal), the Restionaceae (all genera except *Lyginia* and *Hopkinsia*) (Cutler and Airy Shaw, 1965) and the Eriocaulaceae (several genera, see base data, Ruhland, 1930.). The reduction in sporangial number in most Restionaceae and Centrolepidaceae is puzzling. These are wind-pollinated plants and furthermore most Restionaceae are dioecious.

The condition of bisporangiate anthers being on the one hand so uncommon, yet their occurrence so widely scattered in the monocotyledons, suggests it has arisen on several occasions. Nonetheless the common possession of such anthers by members of the Restionaceae, Centrolepidaceae and Eriocaulaceae supports their affinity as suggested by other criteria. None of the dicotyledons commonly associated with the monocotyledons possesses bisporangiate anthers except some Piperales.

Base Data

ALISMATIFLORAE: HYDRO.: Hydro.: *Enhalus, Halophila baillonis, Hydrocharis* (often monothetic bisporangiate), *Ottelia* (incl. *Boottia*; two unisporangiate thecae).—ZOSTE.: Junca.: *Lilaea* (lateral flowers); Zanni.: *Althenia* (monothetic bisporangiate).—NAJAD.: Najad.: *Najas minor* (unisporangiate).—TRIUR.: at least *Soridium spruceanum*.

ARIFLORAE: ARAL.: Arac.: poorly known, at least in spp. of *Arisaema, Arisarum* and *Dieffenbachia*; Lemna.: *Wolffia* and *Wolffiella*.

LILIIFLORAE: ASPAR.: Smila. (all).

COMMELINIFLORAE: COMME.: Mayac. (rarely,

e.g. *M. vandelli*).—ERIOC.: Erioc.: *Blastocaulon, Lachnocaulon, Philodice, Tonina*.—RESTI.: Centr. (all); Resti.: most genera (except *Hopkinsia* and *Lyginia*).

ZINGIBERIFLORAE: ZINGI.: Cann. and Maran. (monothetic, bisporangiate).

Associated dicotyledons:

MAGNOLIIFLORAE: PIPER.: Piper. (*Peperomia*; by fusion).

ANTHER ATTACHMENT
(Fig. 47, Diagram 40)

The stamens were subjected to an extensive comparative study by Schaeppi (1939), who divided them according to appearance and attachment of the anthers into the following categories:

Impeltate (basifixed) (a) sagittate
                       (b) non-sagittate
                       (c) undifferentiated
Peltate (dorsifixed)   (a) epipeltate
                       (b) hypopeltate

This division was based on a comparison with vegetative leaves. The term peltate (dorsifixed) was used when the anther with its connective is extended beyond the point of attachment of the filament, so that this is located somewhere along the mid-line of the surface of the anther.

This contrasts to the impeltate anther attached at the base of the anther (corresponding to the commoner term basifixed.

Among the impeltate types, the sagittate and non-sagittate types of anthers are yet well set off from the filament, while the undifferentiated type has an anther continuing into the filament gradually, as in *Sparganium* (Typhales).

The part of the anther prolonged downwards beyond the attachment point of the filament faces inwards in the epipeltate type and outwards in the hypopeltate type. In some of the epipeltate anthers the upper part of the filament is enclosed in a pocket formed by the anther (and thus may appear basifixed). This condition is met with in, for example, *Kniphofia* (Fig. 47). According to Huber (1969) there is a general tendency for the Asparagales ("Asparagoide Liliifloren") to have either impeltate or hypopeltate anthers, while the Liliales ("Colchicoide Liliifloren") have either impeltate or epipeltate anthers or stamens where the anther base forms a tube around the distal part of the filaments. If this is true, *Kniphofia* and *Doryanthes*, where this also seems to be the case, form exceptions in the Asparagales.

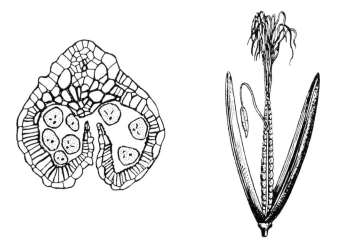

Fig. 46. Bisporangiate anther. The illustration shows a transverse section of an anther of *Centrolepis* sp. (Centrolepidaceae). Most taxa of Restionales are characterized by having bisporangiate anthers. (After Prakash, 1969, and Lotsy, 1911.)

The impeltate and "undifferentiated" anthers can be classified as basifixed, while the peltate ones are often described as dorsifixed. Schaeppi (1939), however, distinguishes between adnate, semi-adnate and dorsifixed versatile anthers, restricting the last term for those with a point-like attachment and being thus normally versatile. The term dorsifixed in this and other works does not involve any meaning as regards the ventral or dorsal side as it is used for hypo- as well as epipeltate anthers without discrimination.

As the classification of anthers according to the above is quite difficult, we shall restrict ourselves to a very approximate division of them into impeltate plus undifferentiated (basifixed) and peltate. It should be stressed that in the basifixed type with sagittate anthers, the attachment superficially may appear very like the peltate.

Basifixed (impeltate or undifferentiated) anthers have the distribution seen in Diagram 40 (shading by dots), and is the most common group. Among these the "undifferentiated" type is not common and perhaps restricted to the Typhales and maybe Smilacaceae. The basifixed, sagittate and non-sagittate, types are widely distributed, sagittate ones being found in most baccate-fruited Asparagales (also in *Dianella* and *Hypoxis*, in *Aponogeton* and many Hydrocharitaceae, in Pontederiaceae and in some Bromeliaceae). Whether the majority of the grasses belong here is uncertain (see below). Non-sagittate, basifixed anthers occur in various other monocotyledons, such as many Juncales, Cyperales, Areciflorae, but also in Dioscoreales etc. This type is no doubt the "basic" type in the monocotyledons, from which sagittate or x-shaped anthers are derivable by growth in the proximal and proximal as well as distal parts of the thecae respectively.

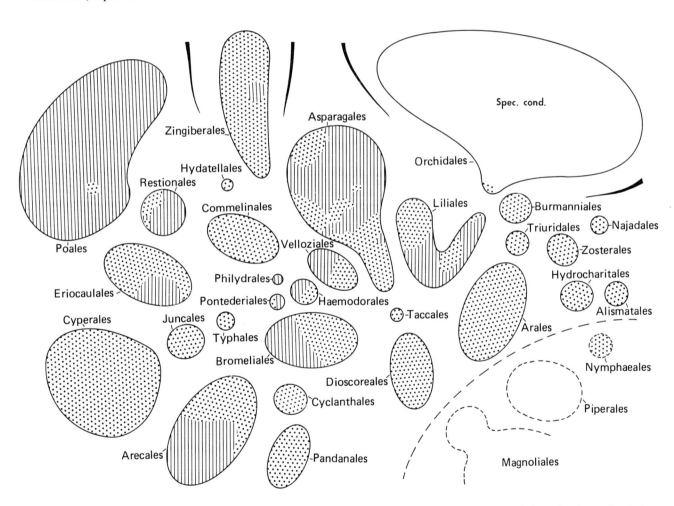

Diagram 40. Anther attachment. Anthers impeltate: dots. Anthers peltate: verticle hatching. In the orchids fusion of anther and style is conspicuous. The Magnoliales and Piperales are not classified here. In the Magnoliales the stamens are mostly flat and leaf-like with the microsporangia attached below their tips, although in some Aristolochiales they may form a staminal column. In the Piperales anther attachment varies between basifixed (Saururaceae) and broad and peltate, non-versatile (many *Peperomia* spp. etc.).

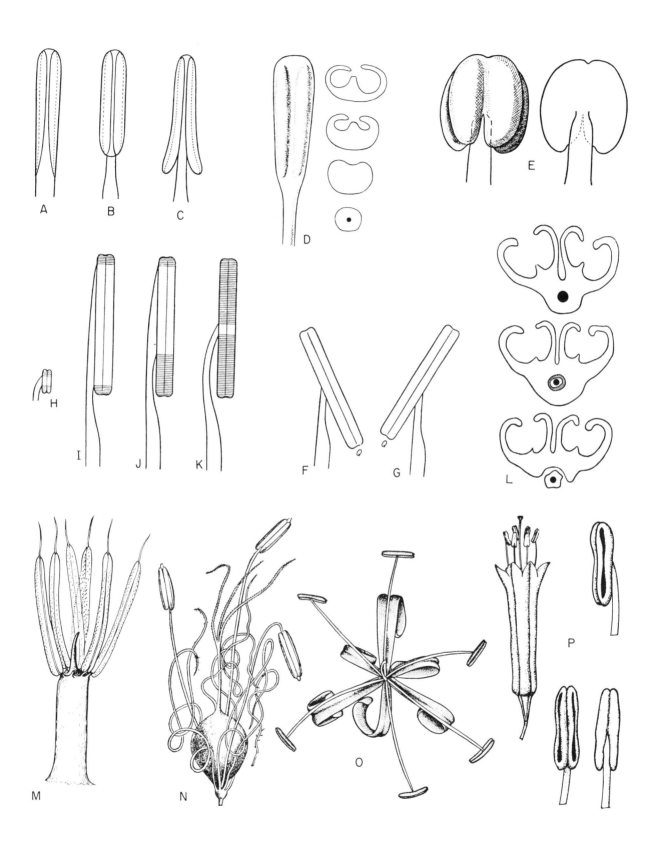

Peltate (dorsifixed) anthers are more restricted. In our diagram they are recorded for many Eriocaulales, Arecales, Restionales and most grasses, although in some of these groups (grasses and Restionales in particular) they approach the sagittate-basifixed ones. The grass anther is mostly versatile, with a narrow connective, the filament with its tapering apex being attached to it below the middle (1/4 to 1/3 from the base). The type found in the restionalean *Anarthria* (Anarthriaceae) has 4-sporangiate anthers which are very similar to those of the grasses, and so do those of *Ecdeiocolea* (Ecdeiocoleaceae), although they are described as sagittate-basifixed (Newell, 1969).

The type of anthers found in the Asparagales and Liliales are however generally very characteristically peltate.

Of the peltate anthers, the *epipeltate* ones have their main distribution in the Asparagales, where they occur in Amaryllidaceae, Alliaceae, Hyacinthaceae, Funkiaceae, Hemerocallidaceae, Asphodelaceae (subfam. Anthericoideae as well as Asphodeloideae), but also in Agavaceae (e.g. *Yucca*) and some of the other woody families. Further, this type, besides the basifixed, seems to occur in the Bromeliales (*Billbergia, Guzmania* etc.) and maybe also in the Commelinaceae (*Tinantia*).

In contrast to this, *hypopeltate* anthers occur in numerous Liliales, at least in many Iridaceae (Fig. 44C), Liliaceae (e.g. *Lilium* and *Fritillaria*) and Colchicaceae (e.g. *Colchium* and *Gloriosa*). In many other Liliiflorae the anthers are basifixed-sagittate but with the microsporangia dehiscing extrorsely as in many (most) Iridaceae (*Iris, Crocus*, and many genera on the Ixioideae mentioned by Schaeppi, 1939). In other Liliales the base of the anthers forms a tube concealing the filament tips, as in Alstroemeriaceae (Buxbaum, 1954) and *Tulipa* (Liliaceae).

The taxonomic evaluation of anther attachment is somewhat difficult. It seems that the basifixed anthers are wholly dominant in the Alismatiflorae and Ariflorae, and probably also in the Zingiberiflorae. The anthers are also basifixed in all Pandanales and Cyclanthales and in many Arecales of the Areciflorae, and in the Commeliniflorae the basifixed anthers likewise dominate (although in sagittate anthers this may be difficult to observe).

Exceptions are perhaps certain Eriocaulaceae and Xyridaceae in the Eriocaulales and various Restionales, although the conditions in this order may partly approach that in grasses, at least for the 4-sporangiate Restionaceae as well as, for example, *Anarthria*.

The most interesting pattern of variation is no doubt that in the Liliiflorae, where epipeltate anthers are fairly widespread, though often together with the presumably more basic, basifixed type. Certain patterns of distribution here may be of taxonomic significance, such as the fact that Asparagalean families with berries tend to have impeltate but families with capsules peltate (epipeltate) anthers. The fact that hypopeltate anthers are most concentrated (maybe nearly restricted) to the Liliales is also of phylogenetic interest (Huber, 1969).

Base Data

Anthers impeltate or basifixed (including subbasifixed and sessile types and also cases where the thecae are attached below stamens with a protruding "connective tip").

ALISMATIFLORAE: all.

ARIFLORAE: all.

LILIIFLORAE: DIOSC.: all.—TACCA.: all.—ASPAR.: Smila.; Peter.; Phile.; Conva.: nearly all; Aspar.; Herre.; Dorya.; Dasyp. p.p.; Hypox.; Aspho. p.p., e.g. *Astelia, Caesia* and *Chlorophytum*; Aphyl.; Diane.; Tecoph. p.p.; Cyana.; Erios.: *Eriospermum*; Amary.: perhaps *Hessea, Leucocrinum, Galanthus*.—LILIA.: Irida.: many genera; Tricy.; Lilia.: few, maybe *Gagea*; Melan. p.p. (*Veratrum*).—BURMA.: all.—ORCHI.: maybe all, though conditions specialized.—PONTE.: p.p. —HAEMO.: p.p.—VELLO.: mostly.—BROME.: p.p.

ZINGIBERIFLORAE: ZINGI.: all except perhaps Zingi.

COMMELINIFLORAE: COMME.: nearly all.—ERIOC.: Rapat.; Erioc.: mostly.—TYPHA.: all.—JUNCA.: all.—CYPER.: all.—HYDAT.: all.—RESTI.: Resti. p.p.; (Ecdei. ?) Joinv.; Hangu. (?)—POAL: at least *Coleanthus*.

ARECIFLORAE: ARECA.: c. half of the taxa.—PANDA.: all.—CYCLA.: all.

Fig. 47. Differentiation of anthers. A–C: impeltate anthers; undifferentiated (A) and differentiated non-sagittate (B) and sagittate (C) anthers. D: undifferentiated anther in *Sparganium ramosum* (Sparganiaceae). E: sagittate anther in *Maianthemum bifolium* (Convallariaceae). F–G: epi- and hypopeltate anthers respectively; the centre of the flower in both cases being to the right of the stamen, which is seen from the side. The part of the anther being prolonged (Q) beyond the attachment point faces inwards in the epipeltate, outwards in the hypopeltate type. H–K: different peltate anthers; from an undifferentiated initial (H) can be derived the adnate (I), semiadnate (J) and dorsifixed versatile anthers (K). L: three successive sections at various levels from an falsely "basifixed" anther in *Kniphofia aloides* (Asphodelaceae), where the connective is tubular and encloses the distal part of the filament. M–N: impeltate (basifixed) anthers in *Pandanus brosimos* (Pandanaceae) and *Scirpus radicans* (Cyperaceae) respectively. O: epipeltate anther type in *Sansevieria guineensis* (Dracaenaceae). P: epipeltate anther type in *Bessera tenuiflora* (Alliaceae). (A–L from Schaeppi, 1939; M from Stone, 1974; N from Hegi, 1935; O from Degener, 1936; P from Moore, 1953.)

Anthers peltate or dorsifixed

LILIIFLORAE: ASPAR.: Conva.: rarely (*Reineckea*); Dracae.; Dasyp. p.p.; Phorm.; Xanth.; Agava.; Aspho.: most genera; Tecoph. p.p. (*Tecophilaea*); Hemer.: Funki.; Hyaci.; Allia.: nearly all; Amary.: nearly all.—LILIA.: Irida.: numerous genera; Caloch.; Lilia.: most genera; Melan.: p.p.—PONTE.: p.p.—HAEMO.: p.p.—PHILY.: all.—VELLO.: p.p.—BROME.: p.p.

ZINGIBERIFLORAE: ZINGI.: Zingi.: rarely.

COMMELINIFLORAE: COMME.: Comme.: rarely (*Tinantia*).—ERIOC.: Xyrid.: all or most taxa; Erioc.: rarely.—RESTI.: Anart.; Resti. p.p.; Centr.—POAL.: nearly all.

ARECIFLORAE: ARECA.: about half of the genera.

ORIENTATION OF MICROSPORANGIA IN THE ANTHERS
(Fig. 48, Diagram 41)

The orientation of microsporangia on the anthers is generally in one of three directions. These are introrse, towards the centre of the flower; extrorse, away from the centre; and lateral, being tangential to the centre of the flower. Rarely, as with *Commelina coelestis*, both extrorse and introrse anthers may occur in the same flower.

In most cases the orientation of the microsporangia coincides with the direction of dehiscence unless secondary twisting or bending takes place. In case the orientation of the microsporangia is considered when discussing the terms introrse, latrorse and extrorse, then also the poricidal, apically dehiscent anthers can be classified in either of the three categories. Where the flower construction is in doubt or the flower is unisexual and comprises a single anther as with the Centrolepidaceae it has not been possible to assign a direction to the dehiscence of the anther. The distributions of the three anther types are summarized in Diagram 41.

From the figure it is clear that extrorse anthers occur in all superorders other than the Zingiberiflorae and Areciflorae. Amongst the Alismatiflorae and Ariflorae (Engler, 1884) the extrorse condition is by far the most common. Introrse anthers are not known in either of these groups though latrorse anthers occur infrequently

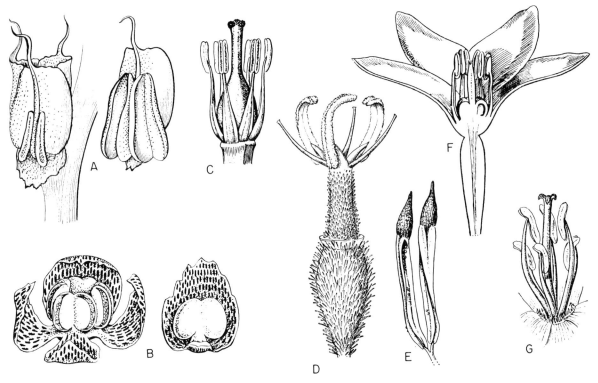

Fig. 48. Anther orientation. A–D: extrorse anthers. A: *Posidonia oceanica* (Posidoniaceae); B: *Tetroncium magellanicum* (Juncaginaceae); C: *Luzuriaga marginata* (Philesiaceae); D: *Sisyrinchium macrocarpum* (Iridaceae). E: latrorse anthers in *Typha subulata*: (Typhaceae). F–G: introrse anthers. F: *Liriope graminifolia* (Convallariaceae). G: *Rhodophiala elwesii* (Amaryllidaceae). (A from den Hartog, 1970; B from Pizarro, 1959; C, D, E and G from Correa, 1969; F from Baillon, 1894.)

in them, e.g. in *Acorus* (Araceae), *Limnocharis* (Alismataceae) and *Butomus* (Butomaceae).

Extrorse anthers occur in four of the eleven orders of the Liliiflorae but are common only in the Liliales. Here they are characteristic of the large family Iridaceae and of three small families, the Geosiridaceae, Tricyrtidaceae and Calochortaceae, but they also occur in some members of the Colchicaceae and Melanthiaceae. In contrast the Asparagales has very few members with extrorse anthers. These are the Petermanniaceae, Asparagaceae subfam. Ruscoideae, Cyanastraceae, Tecophilaeaceae and, though only partly, the Hypoxidaceae. In the Burmanniales, extrorse anthers occur in Corsiaceae.

In the Commeliniflorae only the main part of the Xyridaceae and a few species of *Commelina* have extrorse anthers. The condition in the Areciflorae is even rarer being perhaps restricted to a single palm genus (*Nypa*).

The taxonomic value of the mode of anther dehiscence is limited but it is of interest to observe that the extrorse condition is common in both the Alismatiflorae and Ariflorae, which share so many other attributes.

Base Data

Introrse anthers

LILIIFLORAE: DIOSC.: Diosc.: mostly; Stemo.; Trill.—TACCA.: *Tacca*.—ASPAR.: most groups, exceptions see under latrorse and extrorse anthers below. —LILIA.: Alstr.; Lilia.; Melan.: e.g. *Narthecium, Petrosavia, Tofieldia* etc.—BURMA.: Thism.:—ORCHI.: Apost.; also others?.—PONTE.: all.—HAEMO.: all.— PHILY.: all, (sometimes helical twisting).—VELLOZ.: all.—BROME.: all.

ZINGIBERIFLORAE: all.

COMMELINIFLORAE: COMME.: probably all or nearly all.—ERIOC.: Rapat.: where possible to classify (poricidal dehiscence); Xyrid.: (*Abolboda, Orectanthe*?); Erioc.—CYPER.: Cyper. (the distinction between introrse

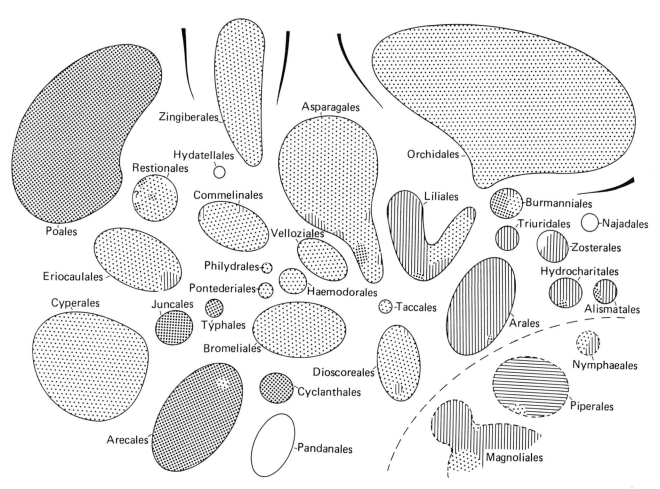

Diagram 41. Anther orientation and dehiscence, somewhat generalized. Introrse: sparse dots. Latrorse: dense dots. Extrorse: vertical hatching. Apical: horizontal hatching.

and latrorse dehiscence is difficult in many Commelini-florean orders).—RESTI.: Resti.; Ecdei.; Anart.

ARECIFLORAE: AREC.: rarely.

Dicotyledons associated with monocotyledons:

NYMPHAEIFLORAE: NYMPH.: Cabom.; Nymph. pp.

MAGNOLIIFLORAE: MAGNO.: several families.—PIPER.: Sauru.

Latrorse anthers

ALISMATIFLORAE: HYDRO.: Butom.—ALISM.: Alism.: sometimes, e.g. in *Limnocharis*.

ARIFLORAE: ARAL.: Arac.: rarely, e.g. in *Acorus*.

LILIIFLORAE: ASPAR.: Smila. except *Ripogonum*.—LILIA.: Colch.: rarely.—BURMA.: Burma.—PHILY.: Phily. p.p.

COMMELINIFLORAE: TYPHA.: mostly.—JUNCA.: Thurn.; Junca.: mostly.—RESTI.: Resti.: some genera, e.g. *Hopkinsia, Lyginia, Restio*; Joinv.; Flage. (or introrse).—POAL.: Poac.: mostly.

ARECIFLORAE: ARECA.: generally.—CYCLA.: all.—(PANDA. not classifiable.)

Extrorse anthers

ALISMATIFLORAE: HYDRO.: Apono.; Hydro.—ALISM.: mostly.—ZOSTE.: all, when classifiable.—TRIUR.: all.

ARIFLORAE: ARAL.: most (except, for example, *Acorus*), where classifiable, but sometimes dehiscing apically.

LILIIFLORAE: DIOSC.: Diosc. (rarely); Trill.; *Scoliopus*—ASPAR.: Peter.; Aspar. subfam. Ruscoideae; Hypox p.p. Tecoph.; Cyana.—LILIA.: Irida.: all, Geosi.; Colch.: generally; Tricy.; Caloc.; Melan.: several genera, e.g. *Chamaelirium, Chionographis, Helonias, Isophysis, Veratrum.*—BURMA.: Corsi.

COMMELINIFLORAE: COMME.: Comme.: *Commelina* spp.; Mayac.: *Mayaca.*—ERIOC.: Xyrid.: *Xyris*.

ARECIFLORAE: ARECA.: Areca.: *Nypa*.

Dicotyledons associated with monocotyledons:

NYMPHAEIFLORAE: NYMPH.: Nymph.: partly; Cerat.

MAGNOLIIFLORAE: MAGNO.: many, e.g. most Arist. and Annon.

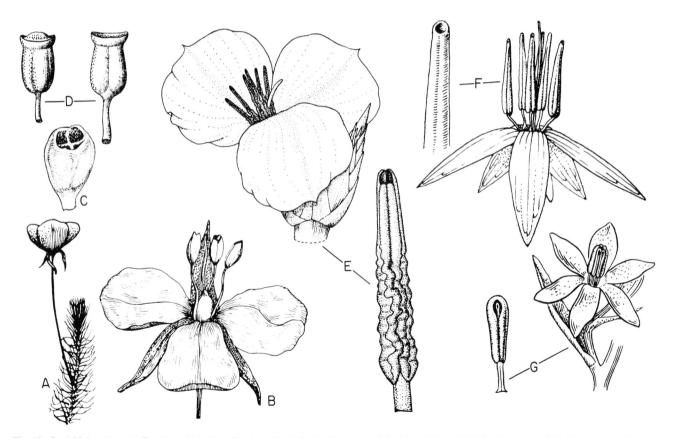

Fig. 49. Poricidal anthers. A–D: *Mayaca*. A: *M. sellowiana*, B: *M. fluviatilis*; C: *M. aubletii*; D: *M. baumii* (all in Mayacaceae). E: *Stegolepis neblinensis* (Rapateaceae). F: *Walleria mackenzii* (Tecophilaeaceae). G: *Cyanastrum hostifolium* (Cyanastraceae). (A and B from Lourteig, 1965; C from Lawrence, 1951; D from Malme, 1930; E from Maguire *et al.*, 1965; F–G from Carter, 1966.)

PORICIDAL ANTHERS
(Fig. 49, Diagram 42)

Anthers dehiscing by pores occur in only three of the superorders, being absent from the Alismatiflorae, Zingiberiflorae and Areciflorae.

The poricidal condition occurs in about half or a third of the Ariflorae possibly in response to the crowded or embedded nature of the flowers. In the Commeliniflorae dehiscence by pores is the usual condition of the Rapateaceae and the monogeneric Mayacaceae, and also occurs in a few genera of the Commelinaceae.

Amongst the Liliiflorae the attribute is limited to *Monochoria* in the Pontederiales and scattered genera of Asparagales, where several families are involved, viz. Amaryllidaceae (*Leucojum* and *Galanthus*), Hyacinthaceae (*Rhadamanthus*), Tecophilaceae (several genera), Cyanastraceae (*Cyanastrum*), Dasypogonaceae (*Dasypo-gon*), Dianellaceae (*Dianella*, *Stypandra*) and Philesiaceae (*Eustrephus* and *Geitonoplesium*).

From this distribution it would appear likely that the character has evolved repeatedly and independently along several lines and hence whilst it is of diagnostic importance, it is of little taxonomic importance.

Base Data

ARIFLORAE: ARAL.: Arac.: numerous, maybe c. a third to half of the genera.

LILIIFLORAE: ASPAR.: Phile.: *Eustrephus, Geitonoplesium*; Dasyp.: *Calectasia*; Diane.: *Dianella, Stypandra*; Tecoph.: *Conanthera, Cyanella, Odontostomum* spp., *Tecophilaea, Walleria, Zephyra*; Cyana.: *Cyanastrum*; Hyac.: *Rhadamanthus*; Amary.: *Galanthus, Leucojum.*—PONTE.: Ponte.: *Monochoria*.

COMMELINIFLORAE: COMME.: Comme.: *Dichorisandra, Porandra* (Hong, 1974), *Spironema*; Mayac.: *Mayaca.*—ERIOC.: Rapat.: all.

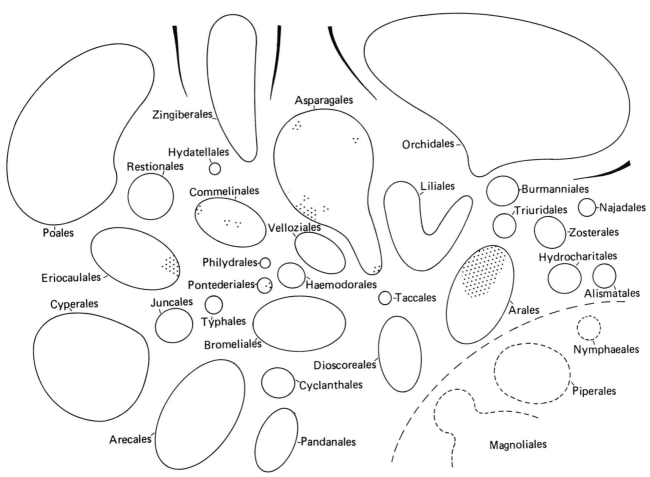

Diagram 42. Poricidal anthers (shaded).

ANTHER WALL FORMATION
(Fig. 50, Diagram 43)

The wall layers of the microsporangium formed from the two primary parietal cells and their descendants can develop in one of four principal ways as follows. The anther wall adjacent to the connective has a different origin and is not discussed here.

During the development of the **"basic type"** the two daughter cells of each primary parietal cell divide periclinally. The four resulting cell layers differentiate as the endothecium, a middle layer (two cell layers) and the tapetum. Sometimes further divisions result in more than four cell layers. This type, which is thought to be the most primitive, is not known in monocotyledons but in disparate dicotyledonous families, such as Winteraceae, Vitidaceae, Rhamnaceae, Tiliaceae, Lecythidaceae and Anacardiaceae.

The **"dicotyledonous type"** differs from the basic type by the suppression of a division in the inner secondary parietal layer, which develops directly into the tapetum. This is the most common type and according to Davis (1966) occurs in about half of the families investigated for this character, but is known among monocotyledons only in the Taccaceae. Among dicotyledonous families with this type are the Annonaceae, Aristolochiaceae and Ranunculaceae.

In the **"monocotyledonous type"** the condition is the opposite to the former, i.e. the division in the outer secondary parietal layer is suppressed, so that the middle layer and the tapetum are developed from the inner secondary parietal layer. This is the commonest type in the monocotyledons. It is known in the Alismataceae, Aponogetonaceae, Butomaceae and Hydrocharitaceae, of the Alismatiflorae, in the Dioscoreaceae, Asparagaceae (Lazarte and Palser, 1979), Agavaceae, Liliaceae *sensu lato* (?), Orchidaceae and Philydraceae in the Lilii-

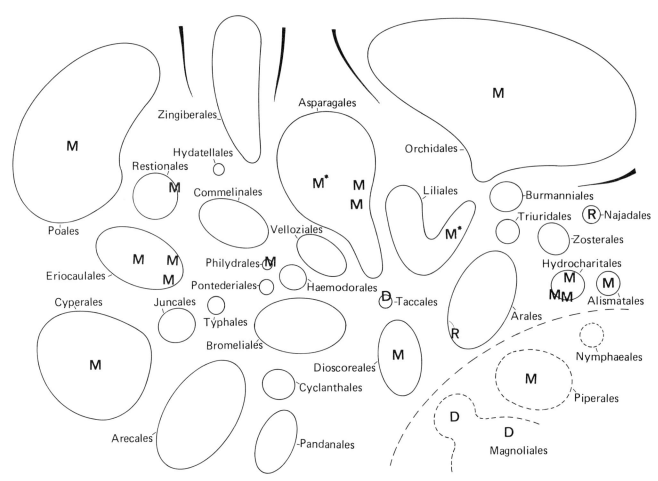

Diagram 43. Anther wall formation. Monocotyledonous type (M); dicotyledonous type (D); reduced type (R); M* refers to Liliaceae *sensu lato*, including Asparagalean as well as Lilialean taxa.

florae, and in all families of the Eriocaulales and in the Cyperaceae, Restionaceae, Centrolepidaceae and Poaceae (Mahalingappa, 1977) in the Commeliniflorae. In addition it is known in the Piperaceae and a variety of other families of dicotyledons, several of which are in the Caryophyllales (Amaranthaceae, Chenopodiaceae, Molluginaceae, Phytolaccaceae, Portulacaceae), but also the Polygonaceae, Brassicaceae, Droseraceae, Fumariaceae, Linaceae, Moraceae, Salicaceae and a few others.

The **"reduced type"** of anther wall formation is characterized by the suppression of all periclinal divisions in the secondary parietal cells, the middle layer lacking altogether, so that there are only two subepidermal layers, the endothecium and the tapetum. This type occurs in the Lemnaceae (*Lemna* and *Wolffia*) and Najadaceae, two families of extremely specialized aquatics. While in *Lemna* and *Wolffia* a middle layer is thus missing; Krajncic and Devidé (1979) found it to be present in *Spirodela* of the same family, which is thus less reduced in this, as in many other respects.

There is no reason to believe otherwise than that the reduced type has evolved independently in Lemnaceae and Najadaceae. This type is also known in *Gaultheria* (Ericaceae), and thus there is not necessarily a connection with the aquatic habitat.

The fact that 17 out of 18 sufficiently well investigated monocotyledonous families belong either to the monocotyledonous or to the reduced type, and only one to the type commonest in the dicotyledons is so far from random that it must be accorded some phylogenetic importance. The data are scarce, and further research in these matters, especially in the Taccales, Dioscoreales, Piperales and Asparagales, is desirable.

Base Data
(Mainly from Davis 1966 where basic references can be found)

Monocotyledonous type
  ALISMATIFLORAE: HYDRO.: Buto.; Apono.; Hydro.—ALISM.: Alism.
  ARIFLORAE: ARAL.: Lemna.: *Spirodela* (?)
  LILIIFLORAE: DIOSC.: Diosc.—ASPAR.: Aspar.: *Asparagus*; Agava.—LILIA.: "Liliaceae" *sensu lato* (M,* may refer to taxa of Asparagales).—ORCHI.: Orchi.—PHILY.: Phily. (Hamann, 1966).
  COMMELINIFLORAE: ERIOC.: Rapat.: *Cephalostemon* (Tiemann, personal communication); Xyrid.: *Xyris* (Hamann and E. Kircher, personal communication); Erioc.: *Eriocaulon, Lachnocaulon* (Hamann, personal communication).—CYPER.: Cyper.—RESTI.: Centr.—POAL.: Poac.
  —RESTI.: Centr.—POAL.: Poac.
  Associated dicotyledons:
  MAGNOLIIFLORAE: PIPER.: Piper.

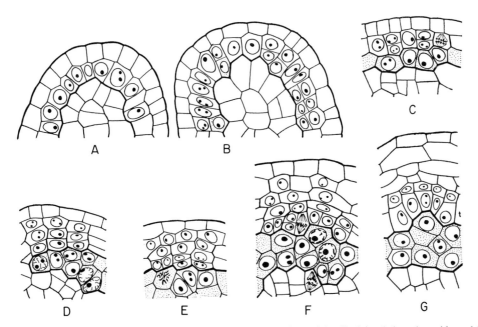

Fig. 50. Anther wall formation according to the monocotyledonous type. The archesporial cells, lying below the epidermal layer in A, cut off primary parietal cells (B), which divide into secondary parietal cells (C), the inner of which divide again to form tapetum (t) and the middle layer. (The further development of the tapetum results in a condition similar to that seen in Fig. 52 A for *Orthothylax*.) This example is *Philydrum lanuginosum* (Philydraceae) and is from Hamann (1966a).

Dicotyledonous type
   LILIIFLORAE: TACCA.: *Tacca.*
   Associated dicotyledons, e.g. MAGNOLIIFLORAE:
MAGNO.: Annon.; Arist.

Reduced type
   ALISMATIFLORAE: NAJAD.: Najad.
   ARIFLORAE: ARAL.: Lemna. (*Lemna* and *Wolffia*).

ENDOTHECIAL THICKENINGS
(Fig. 51, Diagram 44)

There is considerable diversity in the form of thickening
exhibited by endothecial cells in the anthers of monocotyledons. This diversity has been described in some
detail by Kuhn (1908) and Untawale and Bhasin (1973).
The former work was based upon serial sections and the
later on macerated tissue, which may account for discrepancies between the two reports. Much of the
contentious data has been checked and some extra added
by one of us (HTC).

Although Kuhn (1908) defined six major types of
thickening here only two are recognized, for several of
his categories intergrade and although they are useful for
detailed studies they are less useful for broad surveys. In
this regard our views approach those of Untawale and
Bhasin (1973) who recognized only two major groups,
each with two subtypes.

Girdle type
Here the endothecial thickenings are on series of finger
projections arising from a plate-like base. The projections
may be free at their tips as with *Eriocaulon* (Fig. 51A)
or fuse to form loops as with *Sagittaria* (Fig. 51B).

Spiral type
The term spiral is here employed to cover a wide variety
of thickening types from single or double spirals as with
Canna (Fig. 51C) to conditions in which the spiral
consists of incomplete loops as in *Agapanthus* (Fig. 51D–
F). The pattern of thickening amongst all these types is
complicated because of the tendency of the thickening
bands to anastomose. In the extreme, anastomosing may
lead to the endothecial walls becoming thickened all over

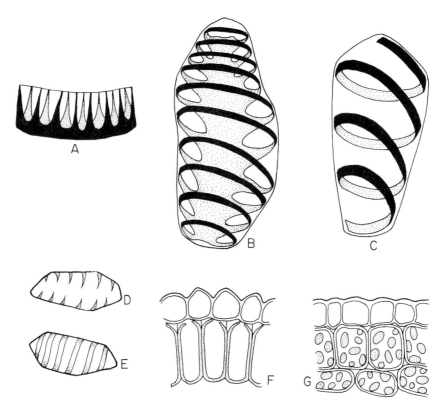

Fig. 51. Endothecial thickenings of cells in monocotyledons: girdle type, A–B; spiral type, C–G. From the following genera: A: *Eriocaulon*; B:
Sagittaria; C: *Canna*; D–F: *Agapanthus*; G: *Calectasia*. (A–C redrawn from Untawale and Bhasin, 1973; D–G redrawn from Kuhn, 1908.)

save for scattered pores as with *Calectasia* (Fig. 51G).

The distributions of the two major types are shown in Diagram 44 from which it can be seen that cells with girdle type thickening have a more restricted distribution than those with spiral type thickening. Whereas spiral type endothecial cells occur in all superorders the girdle type cells are restricted to the Alismatiflorae, Liliiflorae and Commeliniflorae, although the absence of such cells from the Areciflorae is conjectural in that no data are available from the Pandanales or Cyclanthales.

The distribution of the two types thus turns out to be more regular than expected. The pattern shows great uncertainty for the interpretation of the endothecial thickenings in the Cyperales, where Kuhn (1908) consistently reported the girdle type, whereas Untawale and Bhasin (1973) reported but spiral type, a matter that needs further consideration.

Amongst the Alismatiflorae the Alismatales appear to have only girdle type cells in the endodermis and the Hydrocharitales spiral type cells. In contrast the Zosterales are mixed in respect to this attribute.

In the Liliiflorae only the Taccales and Pontederiales have endothecial cells that are solely of the girdle type. Amongst the Asparagales, girdle type cells are rare at the moment, being known only in *Lomandra* (Dasypogonaceae). The distribution of endothecial cell types in the large order Orchidales requires further investigation as at present only a few genera have been studied and these are about equally divided as to whether they have girdle or spiral type thickening. However, within the order there is a slight indication of a pattern, the Cypripediaceae and the Orchidaceae subfam. Orchidoideae having the spiral type while the subfamily Neottioideae has the girdle type, and the subfamily Epidendroideae may be variable.

The widespread occurrence of girdle type cells in the Commeliniflorae independently of whether the species are wind or animal pollinated suggests the attribute has

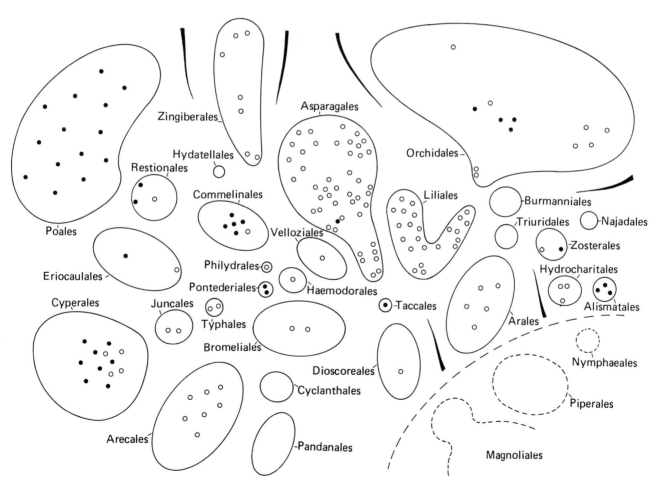

Diagram 44. Endothecial thickenings. Spiral type (hollow circles) and girdle type (solid circles). The orchids are mapped according to Diagram 2.

had a long evolutionary history in the superorder. The girdle type is the only one registered in the Poales and in the related Joinvilleaceae (Restionales). Some variation is otherwise found in the Restionales and also in the Eriocaulales, where the girdle type is known in Eriocaulaceae, but the spiral type in Rapateaceae (Tiemann, personal communication). In the Commelinaceae the girdle type seems to dominate and this is also the case in the Cyperales, where we find very few spiral thickenings to be present and these to be mixed with girdle type cells. The Juncales as well as the Typhales have spiral type thickenings as have all the palms studied; no records are available for the other Areciflorae.

Although not totally consistent this character shows a difference between the Liliiflorae (and Zingiberiflorae) and the Commeliniflorae comparable to that shown by the helobial endosperm formation. A surprising indication is that this character places the Pontederiales with the Commeliniflorae.

Base Data
(* Indicates new records supplied by H. T. Clifford)

Girdle type
ALISMATIFLORAE: ALISM.: *Alisma, Echinodorus, Sagittaria.*—ZOSTE.: Potam.: *Potamogeton.*

LILIIFLORAE: TACCA.: Tacca.: *Tacca.*—ASPAR.: Dasyp.: *Lomandra.**—ORCHI.: Orchi.: *Calanthe, Epipactis, Limodorum, Spiranthes.*—PONTE.: Ponte.: *Eichhornia, Pontederia.*

COMMELINIFLORAE: COMME.: Comme.: *Aneilema, Commelina,* Pollia, Rhoeo, Tradescantia, Zebrina.*—ERIOC.: Erioc.: *Eriocaulon.*—CYPER.: *Carex, Cladium,* Cyperus,* Kyllingia, Eleocharis, Eriophorum, Scirpus, Schoenoplectus.*—RESTI.: Anart.: *Anarthria;* Joinv.: *Joinvillea.*—POAL.: Poac.: *Andropogon, Arrhenatherum, Avena, Bromus, Cymbopogon, Cynodon, Dactylis, Digitaria,* Elymus, Eragrostis, Holcus, Phleum, Sesleria, Triticum.*

Spiral type
ALISMATIFLORAE: HYDRO.: Butom.: *Butomus;* Hydro.: *Hydrilla, Vallisneria.*—ZOSTE.: Junca.: *Triglochin.*

ARIFLORAE: ARAL.: Arac.: *Aglaonema, Anthurium, Arum, Calla, Dieffenbachia.*

LILIIFLORAE: DIOSC.: *Dioscorea.**—ASPAR.: Smila.: *Ripogonum, Smilax;* Phile.: *Eustrephus** (reticulate), *Geitonoplesium;* Conva.: *Convallaria, Maianthemum, Polygonatum;* Aspar.: *Asparagus, Ruscus;* Dracae.: *Dracaena, Sansevieria;* Dorya.: *Doryanthes;** Dasyp.: *Calectasia;* Phorm.: *Phormium;* Xanth.: *Xanthorrhoea;*

Agava.: *Furcraea;* Hypox.: *Curculigo,* Hypoxis;* Aspho. subfam. Asphodelioideae: *Aloë, Asphodeline, Caesia,* Gasteria,* Kniphofia, Sowerbaea;** subfam. Anthericoideae: *Chlorophytum;* Diane.: *Dianella;* Erios.: *Eriospermum;* Hemer.: *Hemerocallis;* Hyaci.: *Bowiaea, Chionodoxa, Galtonia, Hyacinthus, Muscari, Ornithogalum, Scilla;* Allia.: *Agapanthus, Allium;* Amary.: *Amaryllis, Clivia, Crinum, Cyrtanthus, Hippeastrum, Narcissus, Polyanthus, Sternbergia.*—LILIA.: Irida.: *Aristea, Crocosmia, Crocus, Freesia,* Libertia, Gladiolus, Iris, Moraea,* Sparaxis, Sisyrinchium, Tigridia, Watsonia;* Colch.: *Burchardia,* Gloriosa, Uvularia;* Alstr.: *Alstroemeria;* Lilia.: *Fritillaria, Gagea, Lilium, Tulipa;* Melan.: *Helonias, Tofieldia, Veratrum.*—ORCHI.: Cypri.: *Cypripedium, Selenipedium;* Orchi.: *Aceras, Eulophia, Habenaria, Pholidota, Orchis, Vanda.*—HAEMO.: *Haemodorum.**—PHILY.: *Philydrum.*—VELLO.: *Vellozia.*—BROME.: *Billbergia, Vriesia.*

ZINGIBERIFLORAE: ZINGI.: Helic.: *Heliconia;** Musac.: *Musa;* Zingi.: *Alpinia, Hedychium;* Canna.: *Canna;* Maran.: *Calathea, Maranta, Thalia.*

COMMELINIFLORAE: COMME.: Comme.: *Dichorisandra** (there is some doubt as to the interpretation of this).—ERIOC.: Rapat.—TYPHA.: Typha.: *Typha;* Sparg.: *Sparganium.*—JUNCA.: Junca.: *Juncus, Luzula.*—CYPER.: Cyper.: *Cyperus, Eleocharis, Rhynchospora, Scleria* (these all according to Untawale and Bhasin, 1973; all cyperaceous genera studied by Kuhn, 1906, were referred to the Girdle type).—RESTI.: Resti.: *Hypolaena.**

ARECIFLORAE: ARECA.: Areca.: *Areca, Caryota, Cocos, Hyphaene, Maximiliana, Phoenix, Trachycarpus.*

TAPETUM TYPES
(Fig. 52, Diagrams 45 and 46)
(in cooperation with U. Hamann)

The tapetum is the innermost layer of the anther (microsporangium) wall. It attains its maximum development at about the tetrad stage of microsporogenesis, when it normally surrounds the sporogenous tissue. Its importance is nutritive to the latter tissue, as all the nutrients pass through it on their way to the developing microspores (or pollen grains).

The anther tapetum was critically studied by Carniel (1963), who distinguished between the following principal types:
1. the cellular, multinucleate tapetum;
2. the cellular, uninucleate tapetum;
3. the true, periplasmodial tapetum.

The true periplasmodial tapetum was defined as a type

where (after disintegration of the cell walls) the formation of periplasmodium begins already in the beginning of the meiotic stage, and where there is a mitotic multiplication of the nuclei. This according to Carniel is found maybe only in the Commelinales (Commeliniflorae), the Arales (Ariflorae) and some Zosterales, as the Scheuchzeriaceae (Alismatiflorae).

The other types are characterized by retaining the cell walls for at least a somewhat longer time, and may be divided according to whether the number of nuclei are predominantly two to several, or only one.

These categories are somewhat heterogeneous. Transitional types occur between a more permanently cellular tapetum and one where the cell walls are dissolved at some stage and the "amoeboid" content is released into the anther cavity.

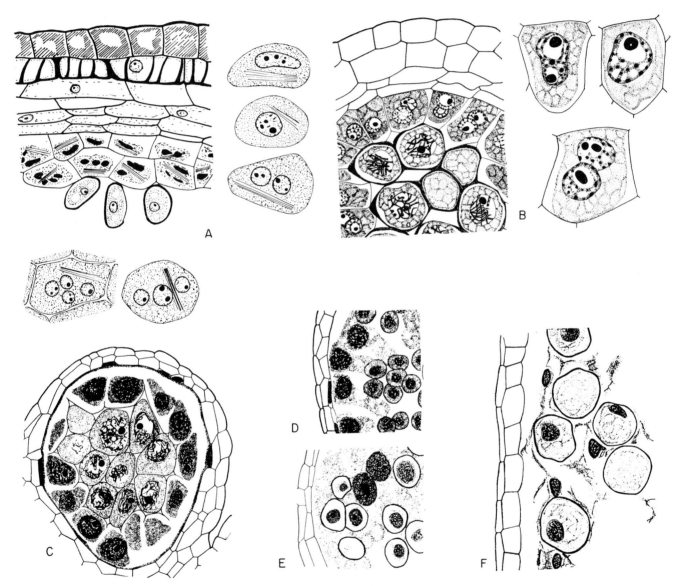

Fig. 52. Tapetum types. A–B: secretory tapetum. A: *Orthothylax glaberrimus* (Philydraceae), with a two-layered tapetum in the main figure, in the degenerating stage. Tapetal cells (illustrated separately) have variable nuclear numbers and often contain raphides. B: *Urginea indica* (Hyacinthaceae), one-layered tapetum with binucleate tapetal cells. Separately, are seen cells in the upper of which the nuclei are fusing or have fused. C–F: amoeboid tapetum in *Limnophytum obtusifolium* (Alismataceae). C: anther lobe, tapetal cells still intact and rounding up; D: tapetal cells forming projections; E: periplasmodium in process of degeneration; F: amoeboid periplasmodium formed. (A from Hamann, 1966a; B from Capoor, 1937; C–F from Johri, 1935a).

Secretory types (glandular, cellular) include those cases in which the tapetal cells finally lose their walls but the protoplasts remain separated and degenerate more or less *in situ*.

True periplasmodial type (*sensu* Carniel) represents those cases where the cell walls are dissolved already at the beginning of the meiotic stage and a periplasmodium is formed in which nuclear divisions occur.

Transitional types include late periplasmodial tapetum (often with multiplication of nuclei during the cellular phase) and other "amoeboid" types, at least partly so. Because it is often difficult to distinguish between active intrusion of tapetal protoplasts or periplasmodia and mere degeneration stages (or shrinkage by reason of bad fixation!) some tapeta reported to be "amoeboid" are likely to be in fact at the secretory type.

In the following survey we have decided to treat groups 2 and 3 together, even though being aware that there may be problems to distinguish certain cases with transitional ("amoeboid") tapetum from secretory. Davis (1966), from which part of the data has been taken, treated all non-secretory types together as "amoeboid".

The following account is thus very generalized but may cover the majority of cases.

The secretory type (incl. "glandular", "cellular", "parietal") thus is characterized by more or less stationary tapetal cells lining the inner anther cavity throughout microspore development. Secretory products containing nutrients for the sporogenous tissue are given off from the inner faces of the cells, until these break down at some stage, often not until the pollen grains are mature. Certain bodies (the "pro-Ubisch bodies"), which are spheroidal structures, penetrate the walls with the secretion developing into bodies coated with sporopollenin. The tapetal cells are usually bi-nucleate, but

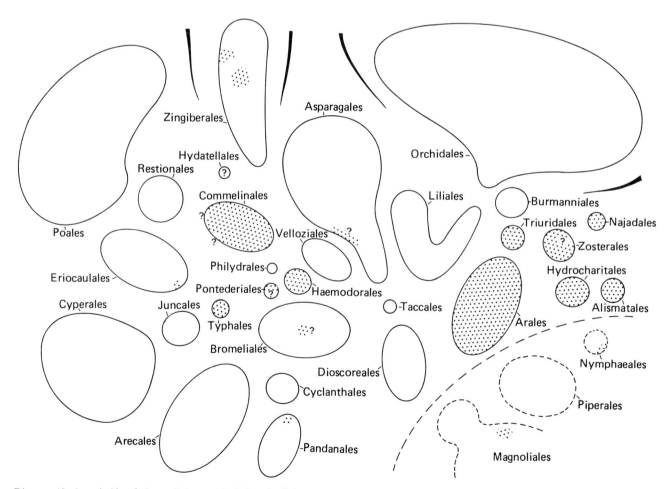

Diagram 45. Amoeboid and plasmodial types (shaded) and cellular-secretory types (unshaded) of tapetum. As the base data are sometimes restricted the diagram is generalized and should be studied along with the base data table.

may vary from uni-nucleate to multi-nucleate even in the same tapetum tissue; subsequent fusion of nuclei may reduce the nucleus number.

The periplasmodial type (incl. "true periplasmodial". "amoeboid", "invasive" and certain transitional variants) is characterized by an early break-down (probably by enzymes) of the inner and radial tapetal cell walls, the protoplasms mostly moving into the anther cavity and, at least in the true periplasmodial type, fuse to form a periplasmodium. This invests the pollen mother cells, the microspore tetrads or the free young microspores (pollen grains) depending on stage. According to Mepham and Lane (1969) the tapetal periplasmodium (of the true periplasmodial type) in the Commelinaceae seems to have an organized structure, its organelles undergoing reorganization rather than degeneration. After meiosis the callose wall around the microspores is degraded, probably also by enzymes, and the pollen is bathed in tapetal cytoplasm until shortly before anthesis (Bhojwani and Bhatnagar, 1974).

The taxonomic significance of the tapetal types, especially the periplasmodial-amoeboid types, is rarely stressed in taxonomic textbooks. As will be shown below, the taxonomic value lies not only in the distribution of tapetal types, but also to some extent in the number of nuclei in the tapetum cells, which has been discussed by Tischler (1915), Cooper (1933), Wunderlich (1954) and Carniel (1963).

The periplasmodial and amoeboid types occur in all of the investigated taxa of the Alismatiflorae, although the contents of the tapetum cells are released at a late stage in the Triuridales (Wirz, 1910). A true periplasmodial type is found at least in Scheuchzeriaceae. Further the tapetum is periplasmodial (-amoeboid?) in the Ariflorae, which is probably allied to the Alismatiflorae.

Another group of orders with amoeboid-periplasmodial tapetum is formed by some Liliiflorae, viz. the Haemodorales and maybe (part of) the Pontederiales. Whether the Hypoxidaceae (Asparagales) and the Pontederiaceae (Pontederiales), as has been stated, actually have a true periplasmodial tapetum is not quite clear. According to Stenar (1925) the cell walls in Hypoxidaceae become dissolved only when the pollen mother cells have divided and after the tapetal cells have become multinucleate.

Schnarf (1929) reported the tapetum in *Burmannia* (Burmanniales) to be transitional between secretory and amoeboid, but later (1931) he cites an old report that it does not show amoeboid character. According to Arekal and Ramaswamy (1973), the tapetum in *Burmannia*

pusilla is an ordinary, 2-nucleate secretory type, which has been confirmed for the genus (*B. stuebelii*) by Hamann and Spitmann (personal communication).

A true periplasmodium is, however, found in the Commelinaceae (Commelinales) of the Commeliniflorae, a group where the periplasmodial type reaches its most typical development. Periplasmodial tapetum also occurs in the genus *Abolboda* (but not in *Xyris*) of the Xyridaceae, Eriocaulales, this representing perhaps a link between the orders. In *Sparganium* of the Typhales the tapetal cells according to U. Müller-Doblies (1969) become up to 8-nucleate, perhaps the highest number in monocotyledons, before they degenerate, lose their cell walls and form a "false periplasmodium". Asplund (1972) reports the same kind of tapetum in *Typha*.

Also, the amoeboid type has been reported from a few orchids, and in *Canna* (Cannaceae) and *Amomum*, *Nicolaia* and *Rhoeo* (Zingiberaceae) of the Zingiberales (see, for example, Nanda and Gupta, 1977). Cheah and Stone (1975) also report amoeboid tapetum in *Pandanus parvum* (Pandanales).

In those dicotyledons which are allied to the monocotyledons, the amoeboid-periplasmodial tapetum is known only from *Ceratophyllum* (Ceratophyllaceae) and *Asimina* (Annonaceae), the remaining families of the Nymphaeiflorae and Magnoliiflorae having all or secretory types of tapetum.

The cellular or secretory types occur in most or all major groups of monocotyledons except those mentioned above.

Wunderlich (1954) made an extensive survey of the tapetum in angiosperms laying much stress on the number of nuclei in the tapetal cells. Prior to this review, Garrigues (1951) had stressed the importance of equating tapetal cells with four diploid nuclei with those having single octoploid nuclei resulting from endomitosis. Carniel (1952) drew attention to the high frequency of endomitoses in tapetal cells.

The number of nuclei is variable in tapetal cells and cannot be used indiscriminately because of endomitosis, of differences in stage when studied and because of the great variation even between different tapetal cells in the same tapetum, as shown for Philydraceae by Hamann (1966), cf. Fig. 52A.

However, as shown by Wunderlich (1954) **uni-nucleate tapetal cells** tend to dominate in the following groups:
1. Most taxa studied of the Alismatiflorae (one species of *Potamogeton* recorded for predominantly binucleate tapetal cells, however).
2. Most but not all members studied of the Ariflorae.
3. All or most taxa of the Burmanniales (Thismiaceae), most Orchidaceae (exceptions being species of

Arundina, Rao, 1967; *Spathiglossis*, Prakash and Lee-Lee, 1973) and Cypripediaceae (except a species of *Paphiopedilum*) and certain taxa of the Velloziaceae, Bromeliaceae and Pontederiaceae.

4. The taxa studied of the Juncales and probably most Cyperales.

5. Most taxa studied of the palms, although at least certain genera have two or more nuclei in the tapetal cells.

6. A few taxa of other orders where bi- or more-nucleate tapetal cells occur much more often, viz. *Canna* in the Cannaceae (Zingiberales), some species of *Allium* in the Alliaceae (Asparagales), and a few grasses, such as *Eleusine* (Mahalingappa, 1977) (Poales).

Bi- to multinucleate tapetal cells are the usual condition in the Asparagales, Liliales, Philydrales, Eriocaulales, Typhales, Poales and Zingiberales and probably also in the Restionales and Cyclanthales.

Among the dicotyledons most often associated with the

monocotyledons, there is considerable variation in the number of nuclei per tapetal cell. Uni-nucleate cells occur in *Nuphar* (Nymphaeaceae) and *Ceratophyllum* (Ceratophyllaceae), whereas most other members of the Nymphaeles have bi- to tetra-nucleate tapetal cells. In Piperales (*Piper*) the tapetal cells are, however, uninucleate. Except for the Myristicaceae which may have uni-nucleate cells, the members of the Magnoliales (Annonaceae, Magnoliaceae and Aristolochiaceae) mostly have multi-nucleate tapetal cells.

The following conclusions may be drawn from the above:

The predominance of the periplasmodial- or amoeboid tapetum with uni-nucleate tapetal cells in the Alismatiflorae and the Ariflorae supports that these superorders may be phylogenetically related;

Periplasmodial tapetum of the most characteristic type occurs in the Commelinales, but also the Typhales have periplasmodial or amoeboid tapetum, and this is also the case in Haemodoraceae and maybe the Pontederiaceae

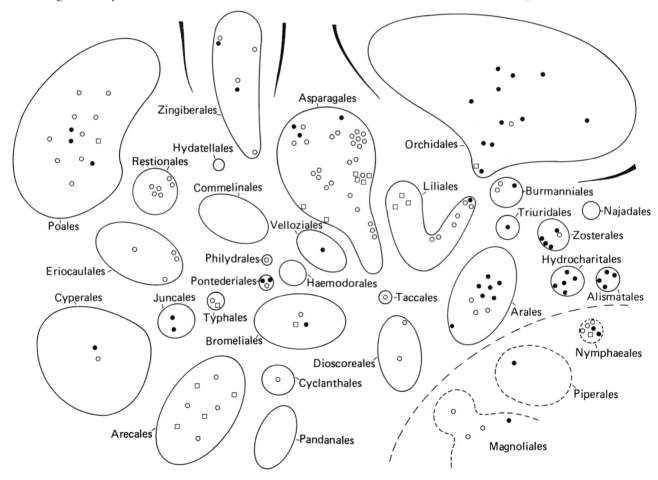

Diagram 46. Number of nuclei per tapetal cell. In the amoeboid tapetum this refers to the early stages. Mostly uni-nucleate tapetal cells: solid circles. Mostly bi-nucleate: hollow circles. Mostly with four or more nuclei: squares.

(and some Bromeliaceae?), indicating a possible connection with the Typhales. The same type where present in the Zingiberaceae and Cannaceae if correct probably has evolved independently from the preceding groups.

Base Data

Plasmodial and "amoeboid" types

ALISMATIFLORAE: HYDRO.: Butom.; Hydro.; Apono.—ALISM.: Alism.—ZOSTE.: Scheu.; Junca.; Potam. (also *Ruppia*); Zoste.; Zanni.—NAJAD.: Najad. —TRIUR.: Triur. (*Sciaphila*) (Wirz, 1910; according to Davis, 1966, "probably glandular"; probably a transitional case).

ARIFLORAE: ARAL.: Arac.; Lemn.

LILIIFLORAE: ASPAR.: Hypox. (*Curculigo, Hypoxis*) (?)—HAEMO.: Haemo.: *Anigozanthos, Dilatris, Wachendorfia, Xiphidium*.—PONTE.: Ponte.: sometimes (?).

ZINGIBERIFLORAE: ZINGI.: Zingi.: *Amomum, Nicolaia* and *Rhoeo* (but usually secretory); Canna.

COMMELINIFLORAE: COMME.: Comme. (Carto. and Mayac. not known).—ERIOC.: Xyrid.: *Abolboda* (but not *Xyris*). TYPHA.: Sparg.: Typha.

ARECIFLORAE: PANDA.: *Pandanus parvus* (Cheah and Stone, 1975).

Associated dicotyledons:

NYMPHAEIFLORAE: NYMPH.: Cerat.

MAGNOLIIFLORAE: MAGNO.: Annon.: *Asimina*.

All other groups are either incompletely known or are known to have cellular or secretory tapetum. Where no member has been investigated for tapetum type in an order or where the record is dubious question marks are given in the diagram.

Number of nuclei in tapetal cells (mainly from Wunderlich, 1954; data on Philydrales, Eriocaulales and Restionales from Hamann, personal communication.)

Predominantly uni-nucleate tapetal cells (genera confirmed)

ALISMATIFLORAE: HYDRO.: Butom.: *Butomus*; Apono.: *Aponogeton* (*Ouvirandra*); Hydro.: *Hydrocharis, Ottelia, Stratiotes*.—ALISM.: Alism.: *Butomopsis, Limnocharis, Limnophytum, Machaerocarpus*.—ZOSTE.: Scheu.: *Scheuchzeria*; Junca.: *Lilaea, Triglochin*; Potam.: *Potamogeton* p.p.—TRIUR.: Triur.: *Sciaphila*.

ARIFLORAE: ARAL.: Arac.: several genera; Lemna.: *Lemna*.

LILIIFLORAE: ASPAR.: Allia.: *Allium* spp.; Amary.: "*Amaryllis*", *Ixiolirion*.—BURMA.: Thism.: *Thismia*.— ORCHI.: Cypri.: *Cypripedium*; Orchi.: *Eria, Spiranthes*,

Vanilla, Zeuxine etc.—PONTE.: Ponte.: *Eichhornia, Pontederia*.—VELLO.: Vello.: *Vellozia* (?).—BROME.: Brome.: *Cryptanthus* (1–2-nucleate).

ZINGIBERIFLORAE: ZINGI.: Zingi.: *Elettaria*; Canna.: *Canna* (1–4 nuclei).

COMMELINIFLORAE: (COMME.: early plasmodium formation.)—JUNCA.: Junca.: *Juncus, Luzula*. —CYPER.: Cyper.: *Carex*.—POAL.: Poac.: *Eleusine, Oryza, Pennisetum*.

Predominantly bi-nucleate tapetum cells

ALISMATIFLORAE: ZOSTE.: Potam.: *Potamogeton natans*.

ARIFLORAE: ARAL.: Arac.: *Acorus* (2–4 nuclei), *Arum, Calla* (with 1, 2 or more nuclei).

LILIIFLORAE: DIOSC.: Diosc.: *Tamus*; Trill.: *Trillium*.—TACCA.: Tacca.: *Tacca*.—ASPAR.: Conva.: *Aspidistra, Convallaria, Ophiopogon, Smilacina*; Aspar.: *Asparagus*; Dracae.: *Sansevieria*; Agava.: *Yucca*; Aspho.: several genera; Hemer.: *Hemerocallis*; Funki.: *Hosta*; Hyaci.: several genera; Allia.: *Agapanthus, Tulbaghia*; Amary.: *Crinum, Galanthus, Haemanthus*.—LILIA.: Alstr.: *Alstroemeria, Bomarea*; Lilia.: *Erythronium, Lilium*; Melan.: *Amianthium, Tofieldia* (Hamann and Specht, personal communication), *Veratrum*.—BURMA.: Burma.: *Burmannia*.—ORCHI.: Orchi.: *Arundina, Spathiglossis*.—PONTE.: Ponte.: *Monochoria*.— PHILY.: Phily.: *Philydrum, Orthothylax* (1–4-nucleate). —BROME.: Brome.: *Pitcairnia* (2–4-nucleate).

ZINGIBERIFLORAE: ZINGI.: Musac.: *Musa*; Zingi.: *Nicolaia*; Costa.: *Costus*; Canna.: *Canna* (1–4 nuclei).

COMMELINIFLORAE: ERIOC.: Rapat.: *Spathanthus, Rapatea* (Tiemann, personal communication), Xyrid.: *Xyris*; Erioc.: *Eriocaulon*.—CYPER.: Cyper.: *Kyllingia*.—RESTI.: Centr.: probably all; Resti.: *Elegia, Hypodiscus, Leptocarpus, Restio*; Flage.—POAL.: Poac.: several genera.

ARECIFLORAE: ARECA.: Areca.: *Chrysalidocarpus, Martinezia, Oreodoxa, Pinanga*.—CYCLA.: Cycla.: *Carludovica*.

Predominantly 4- or more-nucleate tapetal cells

ARIFLORAE: ARAL. (see above).

LILIIFLORAE: ASPAR.: Dorya.: *Doryanthes*; Agava.: *Agave, Beschorneria, Polyanthes*; Cyana.: *Cyanastrum*; Hypox.: *Curculigo*.—LILIA.: Irida.: *Crocus, Gladiolus, Iris*; Alstr.: *Alstroemeria* (?).—ORCHI.: Cypri.: *Paphiopedilum*.—BROME.: Brome.: *Billbergia*.

COMMELINIFLORAE: TYPHA.: Sparg.: *Sparganium* (up to 8-nucleate); Typha.: *Typha* (as the preced.). —POAL.: Poac.: *Eleusine*.

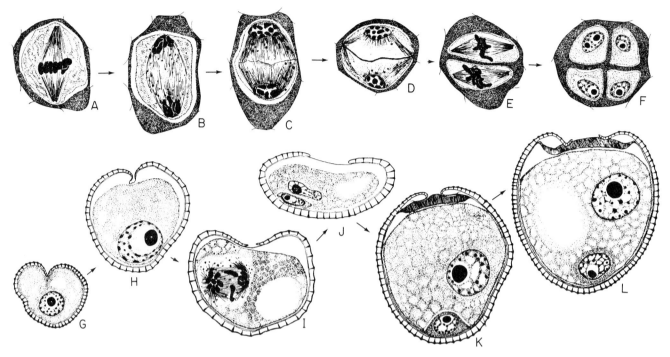

Fig. 53. Successive type of microsporogenesis (A–F) and evolution of the microspore into a ripe pollen grain (G–L) in *Urginea indica* (Hyacinthaceae) as illustrated by Capoor (1937).

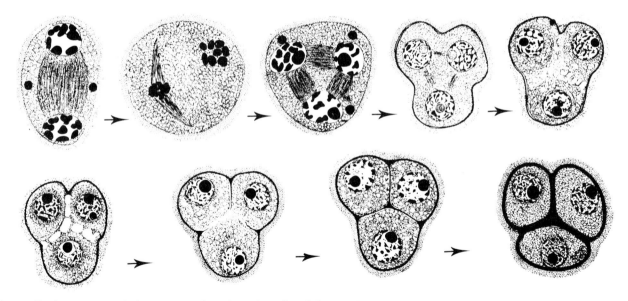

Fig. 54. Simultaneous type of microsporogenesis as shown for a dicotyledonous plant, *Melilotus alba* (Fabaceae) (from Castetter, 1925).

ARECIFLORAE: ARECA.: *Attalea, Cocos, Didymosperma, Orbignya.*

The above information is incomplete and generalized, as tapetal cells may have a very variable number of nuclei; moreover, the uni-nucleate condition may be secondary due to endomitosis of two to several separate nuclei. This is illustrated for *Orthothylax* (Fig. 52A) and for *Urginea* (Fig. 52B) respectively.

MICROSPOROGENESIS
(Figs 53–54, Diagram 47)

The tetrads of microspores are formed from microspore mother cells by two successive (meiotic) divisions. These may be either of the successive or the simultaneous type.

With the "successive type" (illustrated by Capoor, 1937; Fig. 53A-L) the cell plate is laid down immediately after the first meiotic division and another in each of the daughter cells after the second division. The four microspores are usually in one plane. With the "simultaneous type", on the other hand, no wall is laid down after the first division and the mother cell becomes separated all at once into four parts after both the meiotic divisions. While in the successive type the cell plate is laid down in the centre and then extends centrifugally, dividing the cell into two equal halves, the division in the simultaneous type usually occurs by centripetally advancing constriction furrows, which meet in the centre thereby dividing the mother cells into four parts forming a tetrahedral configuration (Maheshwari, 1950).

In the present survey information is mainly taken from Schnarf (1931), Cave (1953), Hamann (1961), Davis

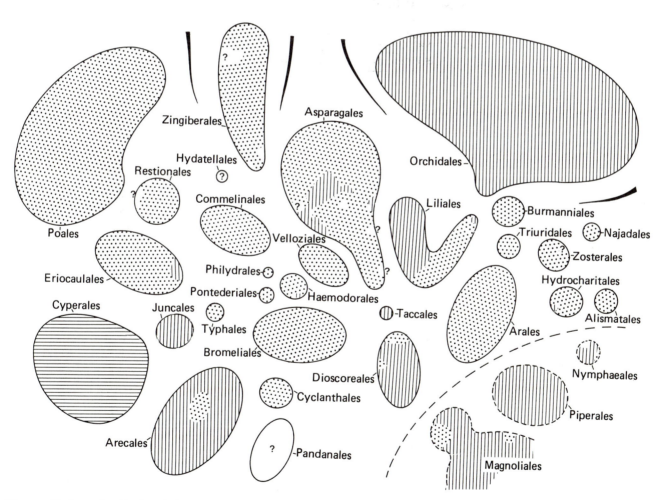

Diagram 47. Successive (dots) and simultaneous (vertical hatching) microsporogenesis in the monocotyledons. In the Cyperales the microsporogenesis is rather of the simultaneous type, but is so specialized in that three nuclei degenerate so that they are best classified separately (horizontal hatching).

(1966) and Huber (1969), where references to further publications are available.

In the monocotyledons the successive type is probably more common than the simultaneous. It has even been called the "monocotyledonous" type and was used as a criterion by van Tieghem (1891) for characterizing monocotyledons.

The distributional pattern (Diagram 47) is as follows.

In the Alismatiflorae and Ariflorae the successive type is found to occur in all families investigated other than Aponogetonaceae, this family apparently having both types of microsporogenesis.

In the Liliiflorae the microsporogenesis follows either type. The successive type is the only one recorded in the Burmanniales, Bromeliales, Velloziales, Haemodorales, Pontederiales and Philydrales, while the simultaneous type is the only one known in the Orchidales and Taccales. The Dioscoreales, Asparagales and Liliales are heterogeneous, and some features of variation, which may be of taxonomic value on the family level, are recognizable within these orders.

Thus in the Dioscoreales microsporogenesis follows the simultaneous type in the Dioscoreaceae except for *Trichopus*. In this genus and in the studied members of Stemonaceae and Trilliaceae it follows the successive type.

In most taxa of Asparagales microsporogenesis follows the successive type with the exception of the taxa studied of the Phormiaceae, (Cave, 1955; Di Fulvio and Cave, 1967), Doryanthaceae, Asphodelaceae subfam. Asphodelioideae (with a considerable number of genera studied), Dianellaceae (Raju, 1957; Cave, 1975) and Tecophilaeaceae (incl. *Lanaria*), Petermanniaceae, Herreriaceae, Xanthorrhoeaceae, Dasypogonaceae, Eriospermaceae and Asphodelaceae subfam. Astelioideae have not yet been studied in this respect. *Leucocrinum*, *Hosta* and *Hesperocallis* (Funkiaceae) all have successive microsporogenesis (Di Fulvio and Cave, 1974). It is interesting that within the Asphodelaceae the two subfamilies Asphodeloideae and Anthericoideae largely differ in this character, which thereby support their recognition, although at least *Ixiolirion* (with uncertain position here) has the simultaneous type. *Hosta*, *Leucocrinum* and *Hesperocallis*, all here referred to Funkiaceae, have successive, but *Hemerocallis* (Hemerocallidaceae) simultaneous microsporogenesis.

In the Liliales, the Iridaceae have the simultaneous type of microsporogenesis throughout, while all other members except *Tofieldia* in the Melanthiaceae possess the successive type.

All members of the superorders Zingiberiflorae and many members of Commeliniflorae investigated possess successive microsporogenesis. Simultaneous types occur in Rapateaceae (Eriocaulales) and in the Juncales and Cyperales, the latter two orders of which have tetrads. These in Juncales are tetrahedral, while in Cyperales the condition is more complicated: after the meiotic division three of the nuclei move towards one end of the cell where they gradually begin to degenerate, being finally embedded in the microspore wall of the surviving microspore cell. The tetrahedral tetrad shape found in Rapateaceae makes it probable that the microsporogenesis in this family is simultaneous as is the case in Thurniaceae (and also Juncaceae) in the Juncales, although this remains to be shown. (Generally the successive type of microsporogenesis results in tetragonal or T-shaped tetrads, although there are exceptions to this.)

The Areciflorae provide some problems and seem to exhibit a great heterogeneity. In the Cyclanthaceae microsporogenesis is successive (Harling, 1958), and this is also the case in a few palm genera investigated, though the commonest type in the palms is probably the simultaneous. The decussate and isobilateral early pollen tetrads of *Pandanus* (Pandanales) described by Cheah and Stone (1975) suggest successive microsporogenesis, although this has not been verified.

The main tendency in the dicotyledons is that the simultaneous type is dominant. The successive type is known only in certain families of Laurales, Magnoliales (Aristolochiaceae having both types) and Nymphaeales, but also in some Rafflesiales (indicating a relationship to Aristolochiaceae?), in some Podostemonaceae, and in genera of Apocynaceae and Asclepiadaceae. Within the Nymphaeales the successive type is known in the Cabombaceae and Ceratophyllaceae, whereas the simultaneous type is known in the Nymphaeaceae.

It should be noted that the above distribution pattern (Diagram 44) is based on a limited number of investigations and the actual condition may differ somewhat from that presented.

Base Data

Successive type

ALISMATIFLORAE: all cases studied (Posid.: no information), incl. TRIUR. (However, Schnarf gives both successive and simultaneous type for *Aponogeton*.)

ARIFLORAE: ARAL.: Arac.; Lemna. (all cases studied).

LILIIFLORAE: DIOSC. (heterogeneous): Diosc.: *Trichopus*; Stemo.; Trill.—ASPAR.: most taxa *except* those few groups given under simultaneous type below; (Peter., Herre., Xanth., Dasyp., Erio. and Aspho. subfam. Astelioideae: no information).—LILIA.: Colch.;

Alstr.; Lilia., Melan. (except at least *Tofieldia*).—BURMA.: Burma.: *Burmannia*.—PONTE.: all.—HAEMO.: all.—PHILY.: all.—VELLO.: all.—BROME.: all.

COMMELINIFLORAE: COMME.: Comme. (Carto. and Mayac.: no information).—ERIOC.: Xyrid.; Erioc.—TYPHA.: Sparg.; Typha.—HYDAT.: no information.—RESTI.: Resti.: *Hypodiscus* (Krupko, 1962), *Elegia, Leptocarpus, Restio* (P. Kircher, personal communication); Centr.; Flage.—POAL.: Poac.

ZINGIBERIFLORAE: ZINGI.: Helic.; Musac.; Zingi.; Costa.; Maran. (Lowia., Strel. and Canna.: no information).

ARECIFLORAE: ARECA.: at least in spp. of *Nypa* and *Pinanga*.—PANDA.: ? *Pandanus*.—CYCLA.: all.

Associated dicotyledons:

NYMPHAEIFLORAE: NYMPH.: (Cabom. not known;) Cerat.: *Ceratophyllum*.

MAGNOLIIFLORAE: MAGNO.: Arist.: several spp. of *Aristolochia*; Annon.: *Annona* and *Cananga* were reported to have both simultaneous and successive cytokinesis.

Simultaneous type

ALISMATIFLORAE: HYDRO.: Apono. (together with Successive).

LILIIFLORAE: DIOSC.: Diosc.: *Dioscorea* spp.—TACCA.: *Tacca*.—ASPAR.: Dorya.: *Doryanthes*; Phorm.: *Phormium*; Aspho.: all of subfam. Asphodeloideae, e.g. *Aloë, Apicra, Asphodeline, Asphodelus, Bulbine, Eremurus, Gasteria, Haworthia, Kniphofia* (Schnarf and Wunderlich, 1939); subfam. Anthericoideae: *Ixiolirion*; Tecoph.: *Cyanella, Lanaria, Odontostomum*; Cyana.: *Cyanastrum*; Diane.: *Dianella, Stypandra*; Hemer.: *Hemerocallis* (Cave, 1967, non Cave, 1955).—LILIA.: Irida.: all; Melan.: *Tofieldia*.—ORCHI.: all.

COMMELINIFLORAE: ERIOC.: ? Rapat. (see the text above, Tiemann, personal communication).—JUNCA.: all—CYPER. see the text above.

ARECIFLORAE: ARECA.: probably most taxa.

Associated dicotyledons:

NYMPHAEIFLORAE: NYMPH.: Nymph.

MAGNOLIIFLORAE: MAGNO most taxa except certain spp. of *Aristolochia* and certain Annonaceae.

POLLEN MORPHOLOGY, GENERAL NOTES
(in cooperation with S. Nilsson)

The terminology connected with aperture conditions has been somewhat inconsistent in the literature over the last decades. If using only the terms "colpi" and "pori", whatever their position and orientation on the pollen grain surface may be, much of the valuable information is lost and one will inevitably end up in artificial groups.

The following terms are therefore used in this book:

Sulcus, a slit-like aperture located at the distal pole. The pollen grain is sulcate.

Sulculi, slit-like apertures located at the equator and oriented along this. The pollen grain is (for example, 2- or 3-) sulculate.

Zonisulculate, pollen grains where such sulculi have fused to form a ring.

Colpi are slit-like apertures crossing the pollen equator at right angles. They are apparently lacking in monocotyledons.

Ulcus (corresponding to sulcus), a pore-like aperture located at the distal pole. The grain is ulcerate.

Foramina are (2 or more) pore-like apertures scattered over the pollen grains or at least not restricted to the distal pole or the equator. The pollen grains are (bi-, oligo-, or poly-) foraminate.

Pores are round apertures located along the equator of the pollen grain. They are perhaps lacking in monocotyledons (exception, perhaps, *Carludovica*, Cyclanthaceae).

Other types of apertures

Trichotomosulcate pollen grains have a tribrachiate slit-like aperture at the distal pole.

Spiraperturate pollen grains have one or more slit-like apertures forming a spiral.

It is generally conceived that slit-like (colpus-like) apertures in angiosperms are more primitive than round (pore-like) ones, the latter being most often associated with adaptation to wind pollination. It is also mostly considered, in monocotyledons, that distal apertures are more primitive than other types, and that sulculate types are derived; further that spiraperturate types are derived from sulcate (or 2- or oligo-sulculate) types. Trichotomosulcate types are also probably secondary in relation to simply sulcate. There are also obvious transitions between elongate and round apertures, poroid, ulceroid etc.

It must be stressed that "tetrad-ontogenetic" investigations are necessary in order to establish the position and orientation of the aperture(s) of the pollen grains. It is also essential to know whether the tetrad is square in one plane, tetrahedral, or otherwise. In case the aperture encircles the pollen grain it is sometimes taken for granted that the aperture is in the equatorial plane or nearly so, but it can also run through both poles (which is considered to be verified for *Nypa*, Arecaceae).

OCCURRENCE OF POLLEN TETRADS (Fig. 55, Diagram 48)

Pollen tetrads occur in a number of relatively large complexes of angiosperms. Many of these groups are no doubt unrelated, the tetrad formation having evolved by convergence.

This is also apparent for the monocotyledons, where tetrads are found at least in the following groups: (1) certain genera of Araceae; (2) most orchids (Orchidaceae) in Orchidales, where the tetrads are frequently agglutinated in larger aggregates: massulae or pollinia. Quite a number of taxa in Orchidales have single pollen grains (Schill and Pfeiffer, 1977), viz. Apostasiaceae, Cypripediaceae and, in Orchidaceae, the subtribes Limodorineae, Vanillinae and Pogoninae. (Vermeulen, 1966, stated that pollen in Cypripediaceae is in tetrads, but this requires verification); (3) species of *Vellozia* (Velloziales) (Ayensu and Skvarla, 1974); (4) *Philydrum* (Philydrales); (5) the genera *Hohenbergia* (Rauh, 1975) and *Cryptanthus* (both of Bromeliales); (6) some species of *Typha* (Typhales), where the tetrads are not tetrahedral, however (Plate 1A); (7) all members of Juncaceae and Thurniaceae in Juncales, where the pollen grains are united in tetrahedral tetrads; (8) all members of Cyperales, where the condition is specialized inasmuch as three of the four microspores are never fully developed but degenerate and end up as rudiments in the wall of the fourth microspore when this develops into a pollen grain (see, for example, Nijalingappa and Devaki, 1978). Besides, pollen tetrads may rarely occur in *Agave* (Agavaceae) according to M. Howard (Ph.D. study, unpublished).

The two orders (Juncales and Cyperales) are undoubtedly closely related (see also Wulff, 1939), and the Typhales are possibly rather closely allied to these. Also the orders Velloziales, Philydrales and Bromeliales show affinities, but pollen tetrads are rare in all of them. Except in the Juncales-Cyperales complex, it is likely that tetrads have evolved independently in each order. In the Araceae pollen tetrads are known in *Caladium* and *Xanthosoma* which are usually considered to be closely allied within the tribe Colocasieae. Thus tetrads may have evolved along only one line of evolution in this family.

Among the dicotyledons associated with the monocotyledons, tetrads are known, for example, within the Magnoliales in the Winteraceae and in several genera of the Annonaceae.

Base Data

(ALISMATIFLORAE: HYDRO.: Hydro. pollen free but liberated in chains resembling linear tetrads.)

ARIFLORAE: ARAL.: Arac.: at least *Caladium* and *Xanthosoma*.

LILIIFLORAE: ASPAR.: Agava.: *Agave* (rarely).—ORCHI.: Orchi. (practically all). Monads known in 17 genera: *Acianthus, Aphyllorchis, Calochilus, Cephalanthera, Chiloglottis, Cleistes, Corybas, Diuris, Epiblema, Epistephium, Glossodia, Lecanorchis, Limodorum, Pogonia, Pterostylis, Spiculaea* and *Vanilla* (Ackerman and Williams, 1980).—PHILY.: *Philydrum.*—VELLOZ.: *Vellozia* spp.—BROME.: *Cryptanthus, Hohenbergia*.

COMMELINIFLORAE: TYPHA.: Typha.: *Typha* spp.—JUNCA.: all.—CYPER.: all (with 3 microspores in each tetrad degenerating).

Associated dicotyledons:

MAGNOLIIFLORAE: MAGNO.: several genera of Annonaceae, e.g. *Monodora, Xylopia, Uvariodendron* and *Annona* (*Trigynaea* with octads) and most Winteraceae.

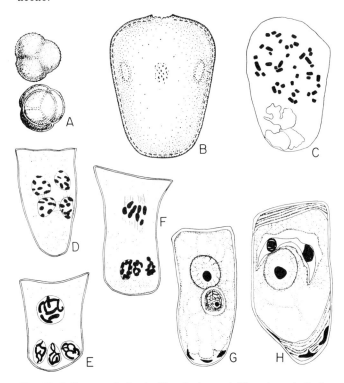

Fig. 55. Pollen tetrads in A: *Thurnia jenmani* (Thurniaceae); B: in *Lepironia mucronata* (Cyperaceae), the three degenerated microspores being incorporated in the wall of the fourth; C: in *Eleocharis uniglumis* (Cyperaceae), showing pollen mitosis in the functional pollen grain while the three degenerating microspores are seen in the lower part. (A and B from Erdtman, 1952; C from Strandhede, 1958; D–F after Piech, 1928.)

(MONO-) SULCATE POLLEN GRAINS (Fig. 56, Diagram 49)

These pollen grains have a single colpus-like aperture, a *sulcus* at the distal pole as conceived from the centre of the original tetrad (Erdtman, 1952). It is the type found in all present day gymnosperms and is also dominant in the superorders Magnoliiflorae and Nymphaeiflorae of the dicotyledons. In the monocotyledons this type and the ulcerate type (with distal pore-like aperture) are the basic and most common types beside the inaperturate, while in the dicotyledons outside the Magnoliiflorae and Nymphaeiflorae these types are wholly or nearly absent (see Walker, 1974; 1975; 1976; Dahlgren, 1977).

The distribution of the sulcate pollen grains in the monocotyledons is widespread. They dominate especially in the Liliiflorae and in the Areciflorae (except Pandanales), while they occur in a minority of the species in each of the superorders Alismatiflorae, Ariflorae and Commeliniflorae and are probably wholly absent from the Zingiberiflorae.

The following features are discernible in the distribution of sulcate pollen grains in monocotyledons:

In the Alismatiflorae they are restricted to the order Hydrocharitales: Butomaceae, Aponogetonaceae and at least some Hydrocharitaceae.

Also in the Ariflorae, this type is less common than the inaperturate type, although it occurs in genera of the Araceae. It is, however, apparently lacking in the Lemnaceae.

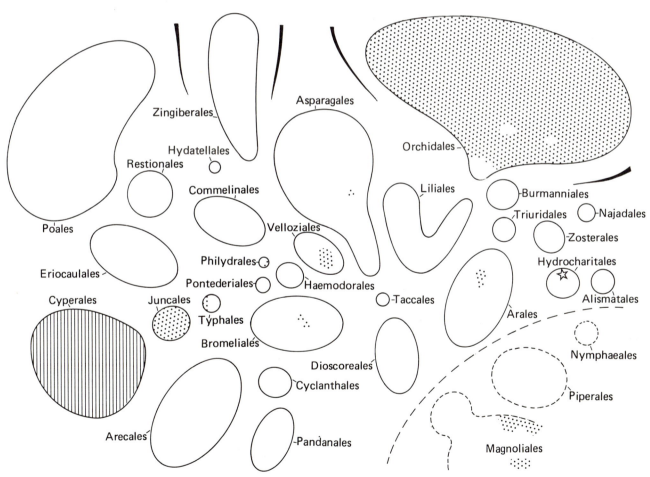

Diagram 48. Pollen tetrads (shaded). In most cases all four pollen grains develop (dots) but in the Cyperales three degenerate and become incorporated in the pollen grain wall (vertical hatching). Asterisk in Hydrocharitales indicates chain-like "threads" of pollen grains, which do not represent true tetrads.

Their occurrence in the Liliiflorae is wide and dominant (Râdulescu, 1973), and they seem to occur in all the orders except Pontederiales (having 2–3-sulculate pollen grains).

In the Burmanniales they occur in the Corsiaceae and (probably) some Burmanniaceae, while in the Orchidales their occurrence is somewhat uncertain except that they do occur in the Apostasiaceae and Cypripediaceae (Ackermann and Williams, 1980), but are probably absent in the Orchidaceae. Sulcate pollen grains dominate entirely in the Taccales, Asparagales, Liliales, Velloziales and Philydrales, some very restricted groups having either inaperturate or 2–3-ulcerate pollen grains or rarely two or more sulculi (see under these types). In the Dioscoreales sulcate pollen grains occur in the Stemonaceae, Trilliaceae and Dioscoreaceae (*Stenomeris* and some species of *Dioscorea*). Also in the Bromeliales, Haemodorales and Philydrales the pollen grains are sulcate in about half or more of the species, although two or more apertures occur in many Bromeliales and Haemodorales.

In the Commeliniflorae the sulcate type occurs in the Commelinales, Hydatellales and certain Eriocaulales, viz. in the Rapateaceae, and in some members of the Xyridaceae. In the Rapateaceae the pollen grains are divisible into four types: (1) a type with fusiform sulcus, (2) a type with a sulcus widened at each end into a "germ pore" and narrow in the middle, (3) a zonisulculate type, while (4) bisulculate pollen grains are known only in one genus (Carlquist, 1961).

Sulcate pollen grains seem to be wholly absent from the Zingiberiflorae, where the exine is thin or absent.

In the Areciflorae, finally, these pollen grains seem to occur in about 75% or more of the palm species and in most taxa of Cyclanthales, in the former group together with trichotomosulcate and bisulculate pollen grains and in the Cyclanthales together with ulcerate pollen grains.

There is no doubt that a number of taxonomic conclusions can be drawn from the distribution of this type of pollen grain, but being a basic feature ("symplesiomorphic" *sensu* Henning) it is probably wiser to draw

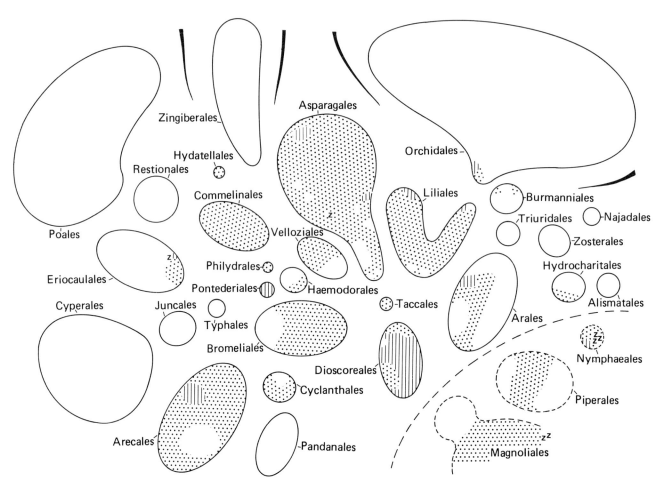

Diagram 49. Sulcate (dots) and bi- to oligosulculate (vertical hatching) pollen grains; z indicates zonisulculate pollen grains.

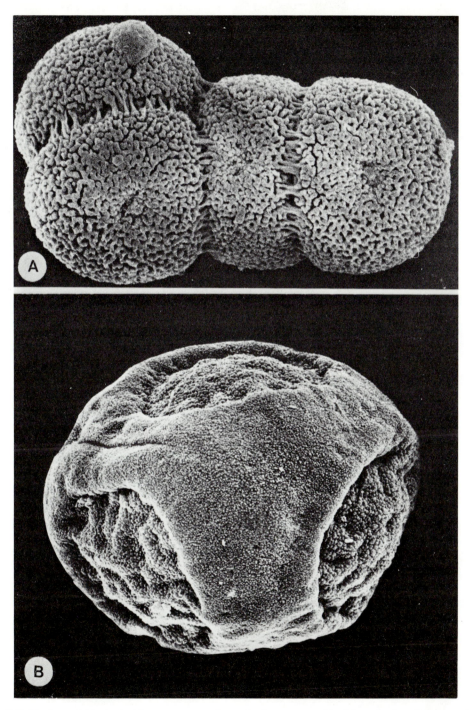

Plate 1. A: *Typha latifolia* (Typhaceae). T-tetrad with monoporate pollen grains interconnected by thin conjuctions (Nilsson *et al.*, 1977); SEM × 200. B: *Luzula pilosa* (Juncaceae). Tetrahedral tetrad with distal areas surrounded by a finely granular band-like thickening (Nilsson *et al.*, 1977); SEM × 2500.

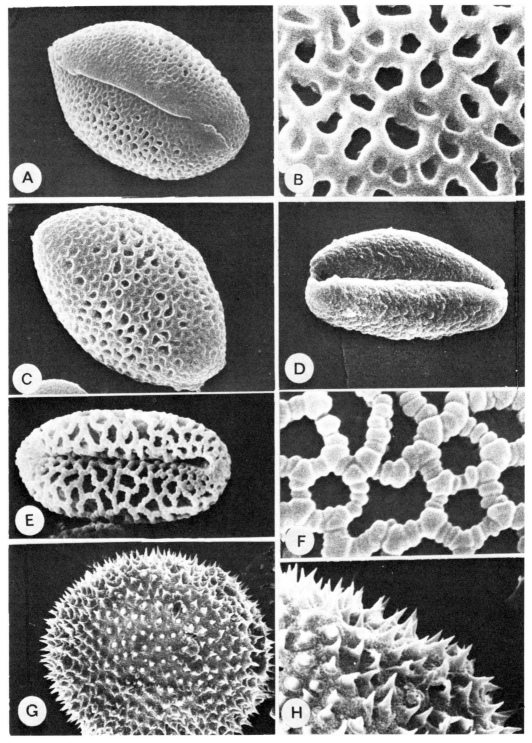

Plate 2. A–C: *Butomus umbellatus* (Butomaceae). A: 1-sulcate pollen grain showing the sulcus on the distal face; SEM × 1600. B: part of the reticulate exine; SEM × 8000. C: pollen grain in polar view, proximal face; SEM × 1600. D. *Liriope spicata* (Convallariaceae). 1-sulcate pollen grain, distal face; SEM × 1500. E–F: *Lilium tsingtauense* (Liliaceae). E: 1-sulcate pollen grain, distal face; SEM × 1250. F: part of the reticulate exine; SEM × 8000. G–H: *Ottelia ulvifolia* (Hydrocharitaceae). G: part of the inaperturate, spinulose pollen grain; SEM × 1000. H: details of the same; SEM × 3600.

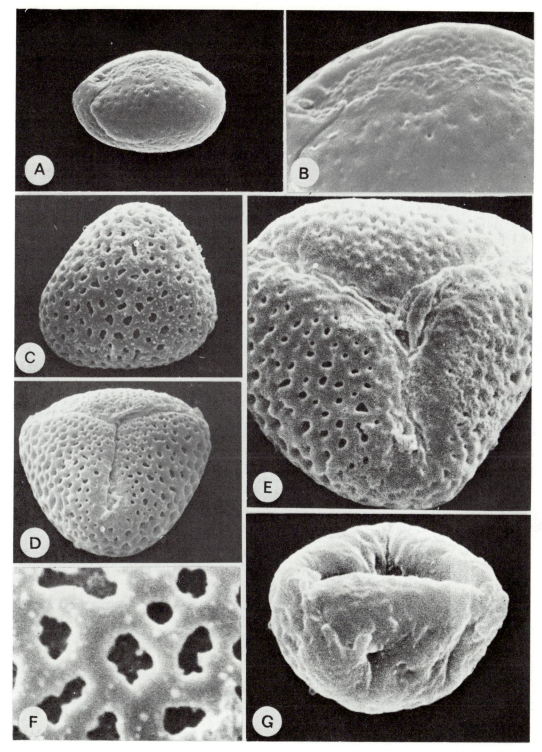

Plate 3. A–B. *Acorus calamus* (Araceae). A. 1-sulcoidate pollen grain; SEM × 1100. B: detail of the smooth, perforated exine; SEM × 3500. C–F: *Acanthorrhiza aculeata* (Arecaceae). C: 1-trichotomosulcate pollen grain in polar view, proximal face; SEM × 1500. D: pollen grain in polar view, distal face, showing the three-slit aperture; SEM × 1500. E: the same, higher magnification; SEM × 3000. F: detail of the reticulate exine; SEM × 7500. G: *Houttuynia cordata* (Saururaceae). Trichotomosulcate pollen grain in lateral view; SEM × 2800.

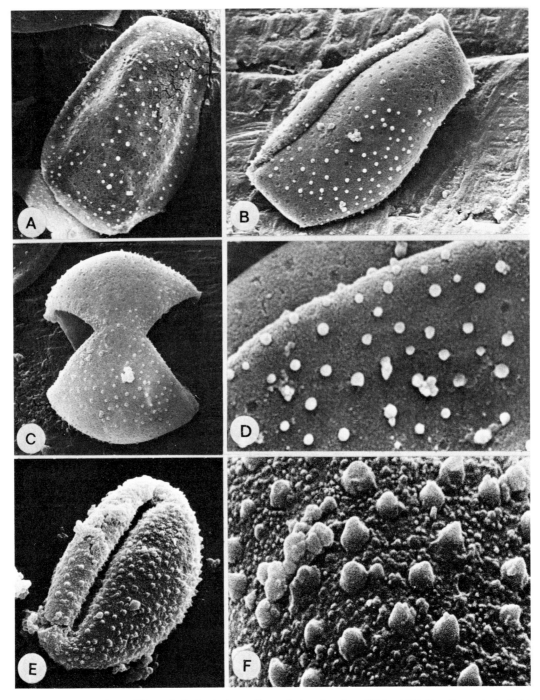

Plate 4. A–D: *Brunsvigia cooperi* (Amaryllidaceae). A: 2-sulculate pollen grain in polar view; SEM × 1650. B: pollen grain in slightly oblique polar view; SEM × 1750. C: pollen grain in equatorial view, transverse position; SEM × 1650. D: part of the spinulose exine near a sulculus; SEM × 3900. E–F: *Crinum longifolia* (Amaryllidaceae). E: 2-sulculate pollen grain showing one of the sulculi; SEM × 1300. F: part of the spinulose exine, SEM × 10 000.

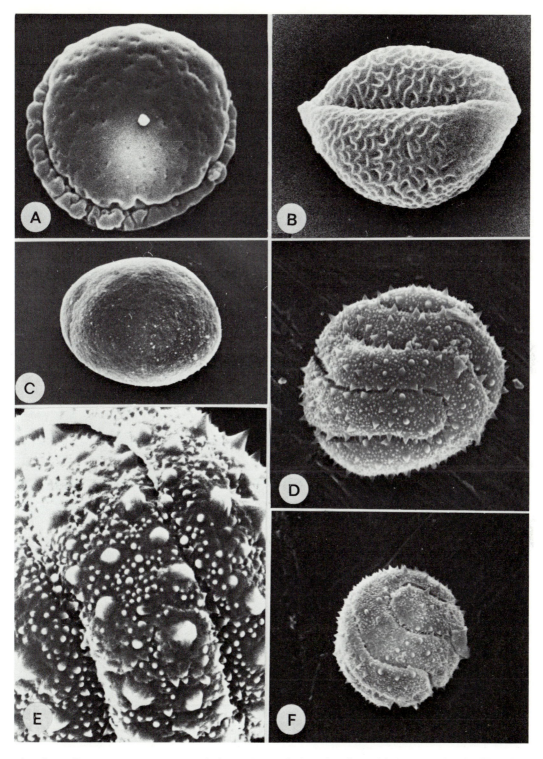

Plate 5. *Lomandra glauca* (Dasypogonaceae). A: zonisulculate pollen grain in polar view, with an equatorial ring-like sulcus; SEM × 1900. B: *Tofieldia calyculata* (Melanthiaceae). Bisulculate pollen grain, only one of the two sulculi visible; SEM × 1900. C: *Dracunculus vulgaris* (Araceae). An inaperturate pollen grain; SEM × 1800. D–F: *Eriocaulon tagawe* (Eriocaulaceae). D: a spiraperturate pollen grain; SEM × 2600. E: detail of the spinulose exine with part of the spiral aperture; SEM × 6400. F: a spiraperturate pollen grain; SEM × 1650.

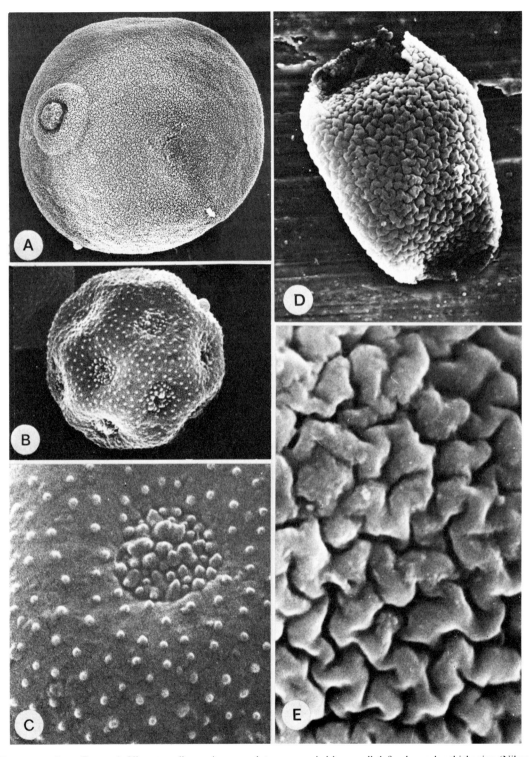

Plate 6. A *Triticum aestivum* (Poaceae). Ulcerate pollen grain, operculate, surrounded by a well defined annular thickening (Nilsson *et al.*, 1977); SEM × 1650. B–C: *Alisma plantagoaquatica* (Alismataceae). B: a polyforaminate pollen grain; SEM × 1650. C: part of the spinulose exine and two foramina. These are covered with granular opercula; SEM × 1650. D–E: *Conostylis aculeata* (Haemodoraceae). D: part of the exine; SEM × 2100. E: a biforaminate pollen grain; SEM × 10000.

conclusions from the distribution of the supposedly advanced ("apomorphic") states (see further under the inaperturate, ulcerate and other types of pollen grains).

In the dicotyledons bordering on the monocotyledons in Diagram 49, sulcate pollen grains are distributed in, for example, the Cabombaceae and Nymphaeaceae in the Nymphaeales (Nymphaeiflorae), in Saururaceae and most species of *Piper*, for example, in Piperaceae of the order Piperales, and in most families of Magnoliales (Magnoliiflorae). In the last-mentioned order inaperturate pollen grains dominate, however, in the Aristolochiaceae.

Base Data

Sulcate pollen grains

ALISMATIFLORAE: HYDRO.: Butom.; Hydro.; a few genera, e.g. *Stratiotes* (most other genera inaperturate); Apono.

ARIFLORAE: ARAL.: Arac.: maybe c. 1/3 of the genera.

LILIIFLORAE: DIOSC.: *Dioscorea* spp., *Stenomeris*; Stemo.; Trill.—TACCA.: Tacca.: *Tacca*.—ASPAR.: all except those enumerated below, with inaperturate, spiraperturate or trichotomosulcate pollen grains, and except Dasyp. p.p. and Amary. p.p. with bisulculate pollen grains.—LILIA.: Irida.: most genera (except spp. of *Rigidella* and *Tigridia*; (Geosi. not known); Colch. p.p. (except spp. of *Colchicum*, *Merendera*, *Androcymbium*); Alstr.; Tricy.; Caloch.; Lilia.; Melan. (most genera).—BURMA.: Burma.: *Apteria* (?); Corsi.—ORCHI.: at least Apost. and Cypri.—HAEMO.: Haemo.: *Dilatris*, *Haemodorum*, *Lachnanthes*, *Lanaria*, *Lophiola*, *Wachendorfia*, *Xiphidium* (mainly subfam. Haemodoroideae).—PHILY.: all.—VELLO.: Vello.: at least *Barbacenia*.—BROME.: Brome.: subfam. Pitcairnioideae and Tillandsioide and part of subfam. Bromelioideae.

COMMELINIFLORAE: COMME.: all.—ERIOC.: Rapat.: most genera; Xyrid.: *Xyris*, partly.—HYDAT.; Hydat.: *Hydatella*, *Trithuria*.

ARECIFLORAE: ARECA.: most taxa.—CYCLA.: most taxa.

Dicotyledons associated with monocotyledons:

NYMPHAEIFLORAE: NYMPH.: Cabom.; Nymph.

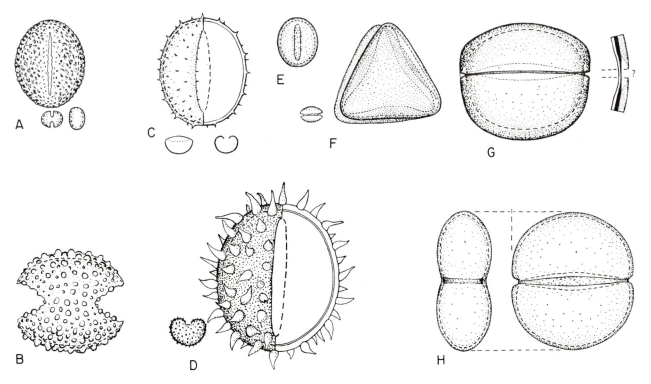

Fig. 56. Sulcate and sulculate pollen grains. A–B: bisulculate pollen grains in *Tofieldia calyculata*, Melanthiaceae (A) and in *Calamus microcarpus*, Arecaceae (B). C–E: sulcate pollen grains in *Wallichia oblongifolia* (C), in *Lepidocaryum gracile* (D), both of the Arecaceae, and in *Heckeria subpeltata*, Piperaceae (E). F: trisulculate pollen grains in *Afroradiophora africana* (Araceae). G–H: zonisulculate pollen grains in *Nymphaea zanzibariensis*, Nymphaeaceae (G) and in *Rapatea speciosa*, Rapateaceae (H). (All after Erdtman, 1952.)

p.p., e.g. *Nuphar* (Schneider and Moore, 1977) (mostly zonisulculate, e.g. in *Nymphaea*).

MAGNOLIIFLORAE: MAGNO.: many taxa in Annon.; Magn.; Degen. etc. and *Saruma* of Arist.— PIPER.: Sauru. (mostly): Piper.: *Piper* (mostly).

BI-, TRI- OR TETRASULCULATE POLLEN GRAINS (Fig. 56, Diagram 49)

The distribution of groups with two or more sulculi is scattered and the feature has no doubt evolved in several independent lines of evolution.

The following distributional pattern is observable.

Bi-, tri- or tetrasulculate pollen grains are mainly restricted to the orders Arales (Ariflorae), Dioscoreales, Asparagales, Liliales, Orchidales and Pontederiales (of the Liliiflorae) and a number of palm genera (Areciflorae). These groups are enumerated below.

In the Arales 2–4-sulculate pollen grains are relatively infrequent and occur, for example, in *Afroradiophora*, *Calla* and *Spathantheum* (Erdtman, 1952), all in the Araceae, this being apparently a relatively rare feature in the group.

In the Dioscoreaceae (Dioscoreales) *Borderea*, *Dioscorea* (often), *Rajania* and *Tamus* mostly have 2- (or 3-)- sulculate pollen grains (while *Stenomeris* and some species of *Dioscorea* have monosulcate and *Avetra* 4- foraminoidate pollen grains).

Only a few groups in Asparagales have pollen grains with two sulculi. These include, for example, several genera (largely the "Amaryllis-Crinum group") of Amaryllidaceae and Agavaceae (*Polianthes*), some genera of Dasypogonaceae, such as *Acanthocarpus*, *Chamaexeros*, and *Kingia*, while other genera of the same family (*Lomandra*, *Dasypogon* and *Calectasia*) have sulcate pollen grains (see Chanda and Gosh, 1976; Chanda *et al.*, 1979).

In the Liliales bisulculate pollen grains occur in *Rigidella* and *Tigridia* of the Iridaceae, in *Pleea* and *Tofieldia* of the Melanthiaceae, and in species of *Colchicum* of the Colchicaceae, while *Merendera* in the same family may have 3-aperturate pollen grains.

In the Orchidales bisulculate pollen grains were reported (Erdtman, 1952) to occur in Cypripediaceae, but this is probably incorrect (Schill, 1978; Ackermann and Williams, 1980); their occurrence in other Orchidales is yet incompletely known, but probably none or very restricted.

The Pontederiales (Pontederiaceae) seem to have exclusively bi- (or tri-) sulculate, elongate pollen grains.

Some single genus of the Commeliniflorae has bisulculate pollen grains, viz. at least *Spathanthus* of the Rapateaceae (Carlquist, 1961).

The palms (Arecales), finally, sometimes may have bisulculate pollen grains, e.g. species of *Calamus*, *Metroxylon*.

Among the dicotyledonous taxa bordering on the monocotyledons, one will occasionally find the pollen grains in the Nymphaeaceae (*Nelumbo* excluded) (Nymphaeales) to be 2-, 3-, 4-sulculate or zonisulculate. This may be the case in *Euryale*, *Victoria* and even in *Nymphaea*, although it is not often commented on in literature (see Erdtman, 1952).

The scattered occurrence of 2- to 4-sulculate pollen grains in the groups concerned do not permit any substantial conclusions to be drawn at the higher levels (family or order inter-relationships), but may well be of great interest at the infrafamilial level (see, for example, Chanda and Ghosh, 1976, for Xanthorrhoeaceae and Dasypogonaceae).

Base Data

Bisulculate pollen grains

ARIFLORAE: ARAL.: Arac.: rarely, e.g. in spp. of *Calla*, *Philodendron* and *Spathantheum*.

LILIIFLORAE: DIOSC.: Diosc.: frequently, e.g. in spp. of *Bordera*, *Dioscorea*, *Rajania* and *Tamus*.— ASPAR.: Dasyp.: *Acanthocarpus*, *Chamaexeros*, *Kingia*; Amary.: mainly tribus Amaryllideae.—LILIA.: Irida.: rarely, e.g. in *Rigidella*, *Tigridia*; Melan.: *Pleea*, *Tofieldia*.— ORCHI.: Cypri.: doubtful—PONTE.: most taxa (also 3-sulculate).

COMMELINIFLORAE: ERIOC.: Rapat.: *Spathanthus*.

ARECIFLORAE: ARECA.: Areca.: rarely, e.g. in spp. of *Calamus*, *Metroxylon*.

Dicotyledons associated with monocotyledons:

NYMPHAEIFLORAE: NYMPH.: Nymph.: abnormally in *Nymphaea* (?).

Tri- or tetra-sulculate pollen grains

ARIFLORAE: ARAL.: Arac.: rarely, in, for example, *Afroraphidophora*.

LILIIFLORAE: DIOSC.: Diosc.: *Dioscorea* (rarely); *Rajania* (rarely).

ULCERATE AND FORAMINATE POLLEN GRAINS
(Figs 57–58, Diagram 50)

Here we include pollen grains with one to numerous more or less circular apertures. The ulcerate pollen grains (*sensu* Erdtman, 1952) have a single pore-like aperture situated at the distal end (in relation to the tetrad centre), and the foraminate pollen grains have two to numerous, non-distal circular apertures. The data summarized in Diagram 50 are based mainly on Erdtman (1952), but also on Chanda (1966) and Chanda and Ghosh (1976).

The variation pattern is as follows:

In the Alismatales (Alismatiflorae) all members have from two (rarely) to numerous foramina scattered on the pollen surface. Such pollen grains are also rarely present in the Haemodoraceae (Haemodorales, Liliiflorae), e.g. in *Tribonanthes* (oligo-foraminoidate), and (dubiously) in Costaceae (Zingiberales, Zingiberiflorae).

In some genera of the Araceae the pollen grains may have from 3 or 4 to several foramina (and are accordingly oligoforaminate). The most common type of pollen grain in this family is the inaperturate and the next most common the sulcate; occasionally the pollen grains are ulcerate, as in species of *Stylichiton*. In Lemnaceae the pollen grains are of the ulcerate type, as is also the case in most of the following groups.

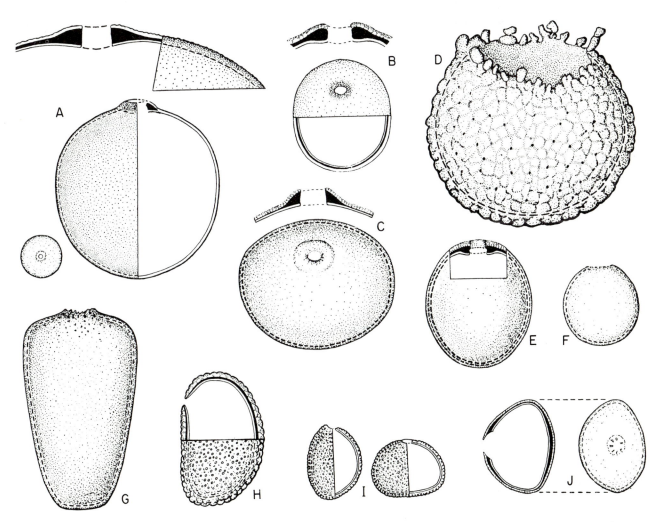

Fig. 57. Ulcerate pollen grains in A: *Dendrocalamus strictus* (Poaceae), B: *Joinvillea elegans* (Joinvilleaceae), C: *Staberhoa cernua* (Restionaceae), D: *Hypolaena lateriflora* (Restionaceae), E: *Mapania amphivaginata* (Cyperaceae), F: *Hypolytrum schraderianum* (Cyperaceae), G: *Chrysithrix capensis* (Cyperaceae), H: *Pandanus endouxia* (Pandanaceae), I: *Sararanga sinuosa* (Pandanaceae), J: *Freycinetia arborea* (Pandanaceae). (All from Erdtman, 1952.)

In the Orchidaceae (Orchidales, Liliiflorae) "3–4 poroid (or foraminoid) apertures" were reported by Erdtman (1952) for *Vanilla*, and 0–5 foraminate pollen grains by Ackerman and Williams (1980) for *Lecanorchis*. Several genera of Orchidaceae subfam. Neottioideae have more or less distinctly ulcerate pollen grains.

In the other Liliiflorae, ulcerate pollen grains are found in the Burmanniaceae and Thismiaceae (Burmanniales). At least in the former family inaperturate and sulcate types also occur. Tetraforaminoidate pollen grains are known in *Avetra* (Dioscoreaceae, Dioscoreales, Fig. 58C). In the Colchicaceae (Liliales), some genera such as *Colchicum*, *Merendera* and *Androcymbium* have pollen grains with two or three foramina or ulci situated on the distal face. *Chamaelirium* and *Chionographis* of Melanthiaceae have 4-foraminoidate pollen grains. In the Bromeliaceae (Bromeliales) two to several apertures of circular, irregular, triangular or elongate shape occur in many genera (e.g. *Acanthostachys*, *Aechmea* p.p., *Ananas*, *Gravisia*, *Hohenbergia*, *Neoregelia*, *Nidularium*, *Portea*, *Pseudananas* and *Quesnelia*), while other genera (including all genera of Pitcairnioideae and Tillandsioideae) have sulcate pollen grains. Finally in the Haemodoraceae (Haemodorales), subisopolar pollen grains with two circular apertures occur in some genera, e.g. *Anigozanthos* Fig. 58A), *Blancoa*, *Conostylis* and *Phlebocarya*, while *Tribonanthes* (Fig. 58B) has oligoforaminate and other

genera, for example, have sulcate pollen grains (see under this title).

Within the Commeliniflorae ulcerate pollen grains occur in two or three groups, viz. the Poales-Restionales complex, the Juncales-Cyperales complex and in Typhales (see tetrads, on Plate 1). In the Poales and Restionales the circular aperture sometimes has well defined, sometimes irregular margin, the latter for example in some Restionaceae. An exception is *Hanguana* (Hanguanaceae, doubtfully included in the Restionales) which has inaperturate pollen grains. In the Typhales and Juncales (*Typha* sometimes with non-tetrahedral tetrads) the pollen grains always have a circular distal aperture. In the Cyperales the tetrads are very specialized (see p. 150) having only one functional pollen grain, which may be ulcerate, or there may be in addition three minor elliptic-circular apertures (or tenuitates) laterally in the distal part or near the equatorial plane (Fig. 58E).

Finally, in the Areciflorae, all members of the Pandanales (Fig. 57 H-J), and a number of *Asplundia* species and *Cyclanthus* of the Cyclanthales, have ulcerate or ulceroidate pollen grains, while in most other genera of the Cyclanthales, including a great number of species of *Asplundia*, and in most palms, the pollen grains are sulcate or sulcoidate, in the palms often also trichotomosulcate. In *Carludovica* the single circular aperture is not distal.

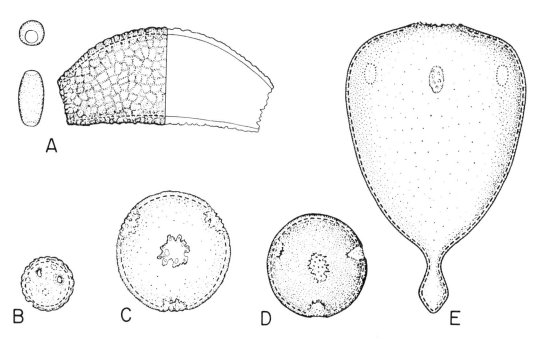

Fig. 58. Biforaminate pollen grains in A: *Anigozanthos manglesii* (Haemodoraceae) and oligoforaminate pollen grains in B: *Tribonanthes brachypetala* (Haemodoraceae), C: *Avertra sempervirens* (Dioscoreaceae), D: *Calyptrocarya glomerulata* (Cyperaceae) and E. *Cladium mariscus* (Cyperaceae), the last-mentioned perhaps best classified as ulcerate, with three elliptic tenuitates. (All from Erdtman, 1952.)

The above conditions, though variable, give useful taxonomic information. Indeed, many orders or even families possess sulcate as well as ulcerate pollen grains, and in a few families they vary from sulcate to bi- or oligosulculate or to bi- or oligoforaminate. However, it is noticeable that the Alismataceae in the present circumscription are consistently different from all other Alismatiflorae in having bi- to polyforaminate pollen grains, those in the other orders being sulcate or inaperturate.

The scattered taxa of the Liliiforae with circular apertures tend to have two or more apertures (Bromeliaceae, Haemodoraceae Colchicaceae, and some genera of Orchidaceae) while ulcerate (ulceroidate) pollen grains occur in certain Burmanniales and some genera of Orchidaceae subfam. Neottioideae. Possibly the similar conditions in the Haemodoraceae and Bromeliaceae may indicate affinity, but more likely their foraminate pollen grains have developed independently.

It is also worthy of note that the Restionales (the problematic, inaperturate *Hanguana* excluded) have ulcerate pollen grains like the Poales, indicating a close relationship between these two orders. While the Centrolepidaceae have ulcerate pollen grains, the taxa of Hydatellaceae have sulcate ones, supporting perhaps, together with other differences a view that they might be distantly related. There is also in the present character support for a relationship between the Cyperales and Juncales, and it does not contradict the suggestion that the Typhales approach these.

Finally, the Pandanales and Arecales differ conspicuously in pollen grain apertures, the former having ulcerate, the latter generally sulcate or trichotomosulcate pollen grains.

In the dicotyledonous groups adjacent to the monocotyledons, elongatedly ulcerate pollen grains are sometimes found in species of *Piper* (Piperales), and typically

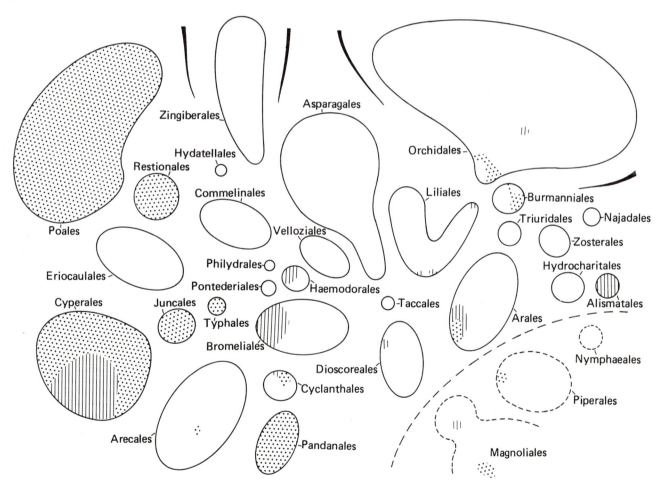

Diagram 50. Ulcerate (dots) and bi- to oligo-foraminate pollen grains in the monocotyledons. In some Cyperales there is one (distal) ulcus and a few additional pore-like apertures; this type has still been classified as ulcerate, in others there are some evenly distributed circular apertures, foramina (hatched).

ulcerate ones in tetrads occur in Winteraceae (Magnoliales).

There is some support for the theory that the circular apertures are derived from slit-like ones, and that a single aperture is more original than several. The groups with ulcerate pollen grains tend to be derived in various ways.

Base Data

Ulcerate or ulceroidate pollen grains

ARIFLORAE: ARAL.: Arac.: rarely, maybe *Stylochiton*; Lemna.

LILIIFLORAE: BURMA.: Burma. pp.: *Burmannia* spp.; Thism.: at least some *Afrothismia* sp.—ORCHI.: Orchi.: some Neottioideae with one indistinct rounded aperture.

COMMELINIFLORAE: TYPHA.: all.—JUNCA.: all.—CYPER.: mostly, some with up to 3 additional lateral poroid or elongate tenuitates.—RESTI.: Resti.; Ecdei.; Anart.; Centr.; Joinv.; Flage.—POAL.: Poac.

ARECIFLORAE: ARECA.: rarely (e.g. *Mauritia* sp.).—PANDA.: all.—CYCLA.: Cycla.: *Asplundia* subgen. *Choanopsis*, *Carludovica* (maybe more correctly porate?), *Cyclanthus*.

Dicotyledons associated with monocotyledons:

MAGNOLIIFLORAE: MAGNO.: at least Winte.—PIPER.: Piper.: *Piper* (rarely).

Bi- to oligo-foraminate pollen grains

ALISMATIFLORAE: ALISM.: Alism.: all (biforaminate in, for example, *Caldesia*).

ARIFLORAE: ARAL.: Arac.: rarely, but in, for example, *Anthurium* (3–4-foraminate); the bi-sulculate pollen grains in some genera approaching a biforaminate condition.

LILIIFLORAE: DIOSC.: Diosc.: at least *Avetra* (4-foraminoidate).—LILIA.; Colch.: *Androcymbium* sp. (one distal and a few lateral apertures), *Colchicum* spp. (2-foraminate), *Merendera* (3-foraminate); Melan.: *Chamaelirium*, *Chionographis* (4-foraminoidate).—ORCHI.: Orchi.: at least *Lechanorchis*—HAEMO.: Haemo.: mainly subfam. Conostylidoideae (2-, rarely 3- or 4-foraminate): *Angiozanthos*, *Blancoa*, *Conostylis*, *Phlebocarya*, *Tribonanthus* etc.—BROME.: Brome.: subfam. Bromelioideae p.p. (see Erdtman, 1952).

COMMENLINIFLORAE: CYPER.: Cyper.: see under ulcerate pollen grains, above; spherical 4-foraminate pollen grains occur in *Calyptrocarya*.

ARECIFLORAE: CYCLA.: Cycla.: *Thoracocarpus*.

Dicotyledons associated with monocotyledons:

MAGNOLIIFLORAE: MAGNO.: Arist.: *Asarum* spp.

SPIRAPERTURATE POLLEN GRAINS
(Fig. 59 A–B, Diagram 51)

Spiraperturate pollen grains are defined as pollen grains with one or sometimes several spiral apertures. These are known in species of *Lomandra* (Dasypogonaceae), *Aphyllanthes* (Aphyllanthaceae, Fig. 59 B), both of the Asparagales and rarely in *Crocus* (Iridaceae) of the Liliales, Liliiflorae; in the Eriocaulaceae of the Eriocaulales (Fig. 59 A), Commeliniflorae and rarely in *Costus* (Costaceae) of the Zingiberales, Zingiberiflorae.

The largest group with this character is the Eriocaulaceae, and here (as in *Aphyllanthes*) it is apparently constant. In *Crocus* and *Costus* the feature is more sporadic. No close phylogenetic connection seems to be

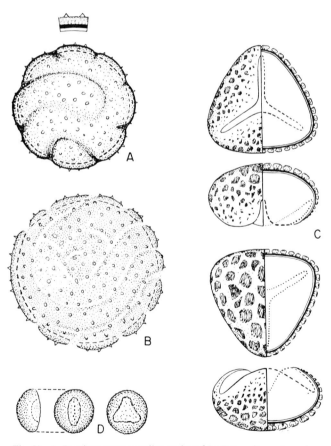

Fig. 59. A–B: spiraperturate pollen grains of A: *Eriocaulon septangulare* (Eriocaulaceae) and B: *Aphyllanthes monspeliensis* (Aphyllanthaceae). C–D: trichotomosulcate pollen grains of C: *Acanthorhiza mocinni* (Arecaceae) and D: *Piper majusculum* (Piperaceae), this having, more often sulcate pollen grains. (All after Erdtman, 1952.)

at hand between the four groups. We do not agree with Chanda and Gosh (1976) that the spirperaturate pollen grains indicate phylogenetic connections between, for example, *Lomandra* and *Aphyllanthes*, nor between either of these and the Eriocaulaceae.

Base Data

LILIIFLORAE: ASPAR.: Dasyp.: *Lomandra* sp.; Aphyl.: *Aphyllanthes*.—LILIA.: Irida.: *Crocus* (as an abberration?).

(ZINGIBERIFLORAE: ZINGI.: Cost.: *Costus* (Erdtman, 1952); the pollen grains are very dubiously aperturate.)

COMMELINIFLORAE: ERIOC.: Erioc.: all? (found in at least three genera).

TRICHOTOMOSULCATE POLLEN GRAINS
(Fig. 59 C–D, Diagram 51)

Trichotomosulcate pollen grains occur rather frequently in palms (maybe c. 15% of the species, Fig. 59 C). Sometimes they have been regarded as a primitive type of aperture, sometimes as an intermediate stage between a sulcate and a 3-colpate, equatorial aperture condition, but both hypotheses are unlikely. They are of taxonomic interest on tribal level in the palms (Punt and Wessels Boer, 1966a and b) and for the circumscription of Dianellaceae, but in the other groups of monocotyledons where they occur they are apparently of little taxonomic value (Diagram 51).

Besides in the palms trichotomosulcate pollen grains

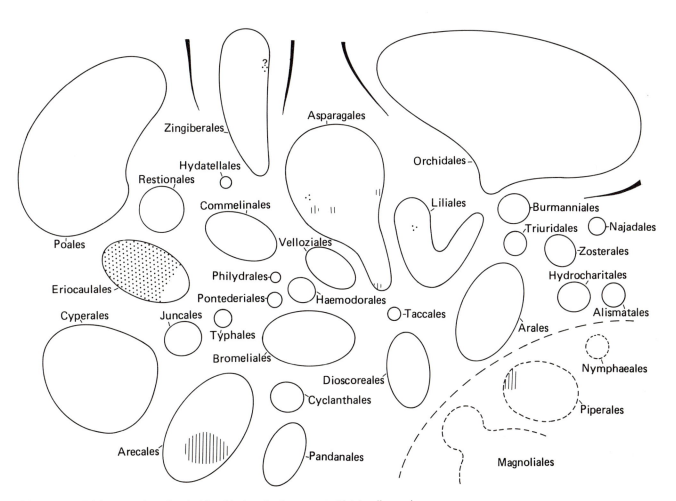

Diagram 51. Trichotomosulcate (vertical hatching) and spiraperturate (dots) pollen grains.

occur in Philesiaceae (*Geitonoplesium*), Phormiaceae (*Phormium*; di Fulvio and Cave, 1964), Dianellaceae (*Dianella, Stypandra, Excremis*) and in some Asphodelaceae (although there they are tetrachotomosulcate!), all these within the order Asparagales. Cave (1975) regarded the trichotomosulcate pollen grains, the simultaneous microsporogenesis and the absence of a parietal cell in the nucellus as diagnostic of the *Dianella* group (here: Dianellaceae).

Similar pollen grains also occur rarely in, for example, *Piper* (Piperaceae, Piperales, Fig. 59D) one of the dicotyledonous groups which shares several features with the monocotyledons.

Base Data

LILIIFLORAE: ASPAR.: Phile.: *Geitonoplesium*; Phorm.: *Phormium* (but not *Blandfordia*); Aspho. subfam. Anthericoideae: rarely tetrachotomosulcate, in at least *Arnocrinum* and *Johnsonia*; Aspho. subfam. Astelioideae: *Phormium* (Cave, 1964); Diane.: *Dianella, Excremis, Stypandra*.

ARECIFLORAE: ARECA.: Areca.: several genera, e.g. *Acanthorhiza, Astrocaryum* and *Cocos* (exceptionally).

Dicotyledons associated with monocotyledons (Walker, 1976):

NYMPHAEIFLORAE: NYMPH.: Cabom. (rarely).

MAGNOLIIFLORAE: MAGNO.: Annon. (rarely); Canel. (rarely).—PIPER.: Sauru. (rarely); Piper. (rarely).

INAPERTURATE (INCLUDING THREAD-LIKE) POLLEN GRAINS
(Fig. 60, Diagram 52)

Pollen grains without distinct apertures are scattered

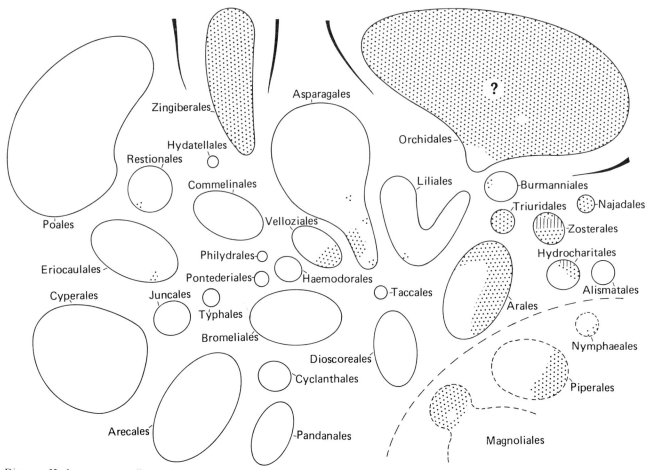

Diagram 52. Inaperturate pollen grains (shaded). In at least most Zingiberiflorae and the aquatic Alismatiflorae the inaperturate condition is associated with the lack of a substantial exine layer. The inaperturate pollen grains are often spinulose. Thread-like pollen grains indicated by vertical hatching.

throughout the angiosperm system, and so this feature probably has a somewhat restricted value as a taxonomic character at large, but may be significant at the family or even ordinal (to superordinal) level in the monocotyledons. The following data are taken mainly from Erdtman (1952). Recently new data have become available that have some impact on the pattern (especially perhaps for the orchids), as SEM and TEM have developed the possiblilities for more accurate observations in recent years (e.g. Schill, 1978).

Inaperturate pollen grains are provided with relatively to very thin exine (or none at all), apertures being then not necessary for the emergence of the pollen tube.

Inaperturate pollen grains are found in the following groups.

In the Alismatiflorae the pollen grains are constantly inaperturate in the Najadales and Zosterales; Ruppiaceae having however three thin areas (tenuitates *sensu* Erdtman, 1952). In addition inaperturate pollen grains are dominant in the Hydrocharitaceae. As in the Zingi-

beriflorae, this condition no doubt can be explained by the fact that in most or all of these taxa exine is very thin or even absent or restricted to spinulae or verrucae on the intine surface. Thus the conditions have a biological rather than phylogenetically relevant explanation. Inaperturate pollen grains with uneven exine pattern are found in *Andruris* (Triuridales) and are probably widespread in this order.

In the Ariflorae pollen grains are highly variable with 0, 1, 2 or 3 to more apertures, inaperturate pollen grains probably being found in about half of the members.

In the Liliiflorae, inaperturate pollen grains are common in the order Orchidales, where they no doubt occur in most members of Orchidaceae. These mostly have tetrads of pollen grains agglutinated in massulae or pollinia. Outside this order inaperturate pollen grains occur in some Burmanniaceae (Burmanniales), in *Campynemanthe* (here placed in Colchicaceae, Liliales) and within Asparagales in practically all members of Smilacaceae (except *Ripogonum*), in *Lapageria* and *Philesia* of the Philesiaceae, in *Aspidistra* of the Convallariaceae, *Semele* of the Asparagaceae subfam. Ruscoideae, and perhaps species of *Vellozia* ("obscure apertures" according to Ayensu and Skvarla, 1974) and other genera of Velloziaceae.

Inaperturate pollen grains also prevail or occur to exclusion in the Zingiberiflorae. However this fact is of limited value as such, as all or most members of Zingiberiflorae are virtually devoid of a conspicuous, protective exine (Kress *et al.*, 1978). Faintly discernible apertures, "tenuitates", were reported for *Heliconia* ("ulceroid aperture"; Erdtman, 1952) and members of Costaceae, the latter with "up to 15 tenuimarginate, circular or elongate apertures", and the pollen grains in the latter family also seemed to approach a spiraperturate condition. It is, however, questionable whether the conditions are comparable to the distinctly exinous pollen grains of other monocotyledons (Punt, 1968). In *Canna* the exine according to Skvarla and Rowley (1970) is restricted to isolated spinules on the intine surface as is the case in *Heliconia* (Punt, 1968).

Inaperturate pollen grains in addition occur in *Abolboda*, *Achlyphila* and *Orectanthe* of the Xyridaceae (Carlquist, 1960) and in *Hanguana* of the Hanguanaceae, the former of these families being placed in Eriocaulales and the latter (probably incorrectly) in Restionales.

The absence of species possessing inaperturate pollen grains in the Areciflorae is surprising in that this superorder includes many rain-forest species having an ecology which in part resembles that of the Zingiberiflorae.

The above distribution is less important taxonomically, than could be expected, because the inaperturate pollen

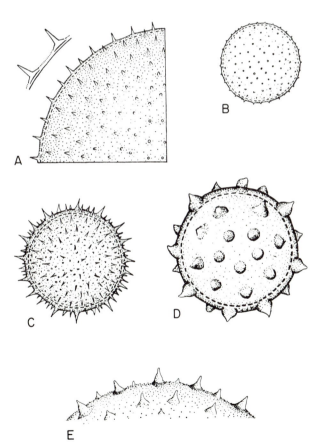

Fig. 60. Inaperturate pollen grains of A: *Ottelia alismoides* (Hydrocharitaceae), B: *Smilax aspera* (Smilacaceae), C: *Remusatia vivipara* (Araceae), D: *Synandrospadix vermitoxicum* (Araceae), and E: *Alpinia speciosa* (Zingiberaceae). (All from Erdtman, 1952.)

grains represent an adaptation to aquatic or hygrophilic habitats. Some groups consistently or nearly consistently have inaperturate pollen grains. These are the orders Zosterales—Najadales, Triuridales, and Zingiberales, and the families Orchidaceae and Smilacaceae, but the relationships of each of these groups are no doubt with groups having aperturate (and then mainly sulcate) pollen

grains. The character should be regarded more as one expressing reduction of exine than as reduction of aperture(s).

In the dicotyledons considered to be allied to the monocotyledons we find inaperturate pollen grains in various taxa, such as for example *Ceratophyllum* and (?) *Barclaya* of the Nymphaeales (Nymphaeiflorae), *Peperomia* of Piperales and most genera of Aristolochiaceae and some of Annonaceae of the Magnoliales (Magnoliiflorae).

Thread-like pollen grains are found in some marine taxa, namely *Posidonia* (Posidoniaceae), *Amphibolis*, *Cymodocea* (Fig. 61), *Diplanthera*, *Halodule* and *Syringodium* (Cymodoceaceae), in *Zostera* and *Phyllospadix* (Zosteraceae), and finally, in *Halophila* and *Thalassia* (Hydrocharitaceae). Thus they are restricted to the Zosterales and Hydrocharitales. It should be noted that these pollen grains usually start off as more or less globose but as soon as they are released gradually elongate, being totally devoid of exine (see for example, Yamashita, 1976). In *Zostera*, however, the prolongation occurs already at a stage before the first division of the meiosis (Rosenberg, 1901; Harada, 1948).

This peculiar feature of course must be seen as an interesting adaptation to water pollination (hydrogamy), the thread-like pollen grains having much greater possibility of adhering to the styles of the submersed stigmas of the female flowers. In Diagram 52 the occurrence of thread-like pollen grains is indicated separately. The submarine pollination of sea grasses has been described by Knox and Ducker (1976).

Base Data

Inaperturate pollen grains (incl. types without a coherent exine or without exine altogether).

ALISMATIFLORAE: HYDRO.: Hydro.: mostly.—ZOSTE.: all.—NAJAD.:—all.—TRIUR.:? all (see the text).

ARIFLORAE: ARAL.: Arac. (more than half of the genera).

LILIIFLORAE: ASPAR.: Smila.: *Smilax* s. lat.; Phile.: *Lapageria, Philesia*; Conva.: *Aspidistra*; Aspar. subfam. Ruscoideae (? partly).—LILIA.: Colch.: *Campynemanthe.*—BURMA.: Burma.: at least in *Burmannia.*—ORCHI.: Orchi.: poorly known, but probably the common condition—VELLO.: Vello.: at least *Vellozia* spp.

ZINGIBERIFLORAE: probably all (see the text).

COMMELINIFLORAE: ERIOC.: Xyrid.: *Abolboda, Achlyphila, Orectanthe.*—RESTI.: Hangu.

Dicotyledons associated with monocotyledons.

NYMPHAEIFLORAE: NYMPH.: Nymph.: *Bar-*

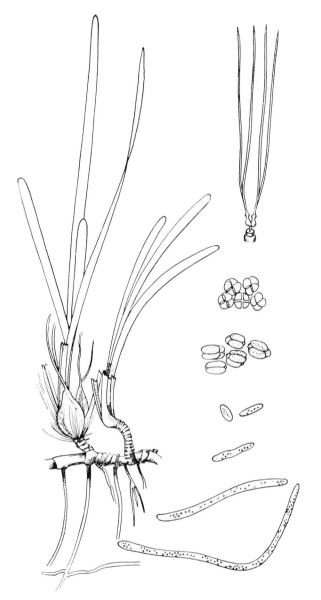

Fig. 61. *Cymodocea nodosa* (Cymodoceaceae), a marine aquatic of the order Zosterales with thread-like pollen grains. A female (fruiting) plant is seen to the left. Above and to the right is a female flower with two carpels having bibrachiate styles suitable for capturing the filiform pollen grains. The development of the pollen grains from the moment when they are formed in the tetrad to the subfiliform stage is seen in the vertical row. (From Emberger, 1960.)

claya; Cerat.: *Ceratophyllum.* (These probably are more or less devoid of exine).

MAGNOLIIFLORAE: MAGNO.: Artist.: all except *Saruma* and some spp. of *Asarum*; Annon. p.p.; Myris. p.p.—PIPER.: Sauru. p.p.; Piper.: e.g. *Peperomia.*—LAURA.: nearly all (not in the diagram).

Thread-like pollen grains

ALISMATIFLORAE: HYDRO.: Hydro.: *Halophila* and *Thalassia.*—ZOSTE.: Zoste.; Posid.; Cymod.: *Amphibolis, Cymodocea, Diplanthera, Halodule, Syringodium.*

BI- VERSUS TRINUCLEATE POLLEN GRAINS
(Fig. 62, Diagram 53)

The pollen grains of angiosperms are shed either in the binucleate or in the trinucleate state. As the generative nucleus in the binucleate grains and the sperm nuclei in the trinucleate grains are contained in cells they are more correctly described as 2- or 3-celled, as was done by Davis (1966).

An extensive survey of the pollen grains in c. 2000 species of angiosperms with respect to number of nuclei was made by Brewbaker (1967). The information presented in Diagram 53 is based mainly on this publication, supplemented by data from Davis (1966), Wunderlich (1959), Schnarf (1931) and Hamann (personal communication).

The information in Davis (1966) is largely compiled from older sources and some of the discrepancies between her statements and those of Brewbaker (for example records of 2-nucleate pollen grains in a few species of Alismatiflorae and of 3-nucleate pollen grains in Asparagales, Liliales and Commelinales and in Cannaceae, Zingiberales) may have resulted from studies of unripe or old pollen grains or other errors.

It is evident from Brewbaker's study that most families are homogeneous with respect to nucleus number in the mature pollen grain. Rarely are genera heterogeneous as indicated either by Davis or by discrepancies between her data and those of Brewbaker. These genera include *Burmannia, Canna, Ruscus, Thismia, Xyris* and *Zostera* and deserve further study.

A convenient method of studying the nucleus number in the pollen grains is by staining them in acetocarmine (Brewbaker, op. cit.) but further treatment (with Feulgen technique) may be needed to see the often weakly staining vegetative nucleus. The variable shape of the generative cell has been demonstrated by Wunderlich (1959).

One conclusion drawn by Brewbaker and others studying this character is that 3-nucleate pollen grains are derived from 2-nucleate. In no group is there indication of a reverse evolution.

Diagram 53 reveals a striking pattern of distribution, which may be summarized as follows:

Binucleate pollen grains dominate in the monocotyledons as they do in the dicotyledons. Of the 29 orders recognized herein 10 are reported to have only 2-nucleate pollen grains and 12 to have both 2- and 3-nucleate pollen grains. Binucleate pollen grains occur in about 80% of the orders.

Although usually 3-nucleate pollen grains have been reported for the Alismatiflorae there are several recent reports of 2-nucleate pollen grains in this group, e.g. for *Aponogeton distachyus* and *Triglochin striatum* (Gardner, 1976), *Lilaea* (Lakshmanan, 1970), *Ottelia* and *Blyxa* of the Hydrocharitaceae, and Zannichelliaceae (Lakshmaman, op. cit.). There is, however, probably predominance of 3-nucleate pollen grains in the superorder as a whole. The significance of this will be discussed below.

Trinucleate pollen grains are also dominant in some Commeliniflorae, in particular the grasses. While in the Poales and Juncales the pollen grains are constantly reported to be 3-nucleate there are a few exceptions reported in the Cyperales, and the few observations in the Eriocaulales indicate variation in that order, which may depend on the late division of the generative nucleus. The condition in the Restionales is also variable. Reports from Flagellariaceae and Restionaceae indicate dispersal of the pollen grains in the 2-nucleate stage, although there is a record of 3-nucleate pollen grains in *Chondropetalum* (Restionaceae) and Hamann (personal communication) reports that most, if not all, Centrolepidaceae have 3-nucleate pollen grains, and also *Floscopa* (Commelinaceae; Davis, 1966) is reported to have 3-nucleate grains.

The Ariflorae (Arales) are very variable in respect of this character and seem about equally often to have 2- as 3-nucleate pollen grains.

In the Liliiflorae, the pollen grains are mostly 2-nucleate, but records indicate that Burmanniales have both types. A few records of 3-nucleate grains also occur in other orders: *Polygonatum, Chorophytum, Ruscus*, all in the Asparagales, *Tulipa* in the Liliales (Davis, 1966) and *Schlumbergeria* in the Bromeliales (Brewbaker, 1967). These exceptions even if confirmed do not disturb the overwhelming dominance of 2-nucleate pollen in the Liliiflorae at large.

The Zingiberiflorae are almost certainly exclusively 2-nucleate, the single exceptional record of 3-nucleate pollen in *Canna* being in need of verification.

The Areciflorae according to all available reports have 2-nucleate pollen.

From the above account it is obvious that there has been a more or less complete transition from the 2-nucleate type of pollen grains (the primary state) to the 3-nucleate along a few main lines of evolution, one being the Alismatiflorae including the possibly separate line of Triuridales, one, at least, in the Ariflorae and one or more among the wind-pollinated Commeliniflorae. Exceptional cases in other groups indicate instability in this character, but possibly the changed state from 2- to 3-nucleate may be too far-reaching for the latter to be established.

Practically all groups with 3-nucleate pollen grains have dry stigmas and practically all groups with wet stigmas have 2-nucleate pollen grains (see wet and dry stigma types, p. 177).

According to the study by Brewbaker (1967), the pollen grains in the dicotyledonous orders Magnoliales, Piperales and Nymphaeales are 2-nucleate throughout, but Davis (1966) reported the pollen grains in the Nymphaeaceae to be 3-nucleate. Further study of these taxa is required.

Base Data

(*Data from Davis, 1966)

Trinucleate pollen grains

ALISMATIFLORAE: HYDRO.: Butom.: *Butomus*; Hydro.: usually; Apono.: *Aponogeton* (dubious, see below).—ALISM.: Alism.: several genera.—ZOSTE.: Scheu.: *Scheuchzeria*; Junca.: *Lilaea* (dubious, see below); Potam.: *Potamogeton, Ruppia*; Zoste.: *Zostera*; Zanni.: *Zannichellia* (dubious).—NAJAD.: Najad.: *Najas*.—TRIUR.: Triur.: *Sciaphila, Soridium* (Hamann and Weik, personal communication).

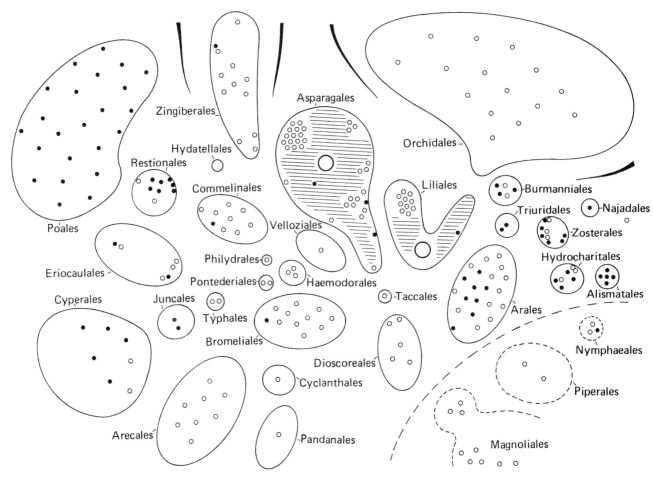

Diagram 53. Bi- and trinucleate pollen grains (hollow and solid circles respectively). The large circle surrounded by stripes in each of the orders Asparagales and Liliales indicates the predominance of binucleate pollen grains in these orders (numerous data). In the other orders each symbol represents a generic record.

ARIFLORAE: ARAL.: Arac. (heterogeneous) e.g. in *Aglaonema, Arum, Colocasia, Pinellia, Pistia, Spathicarpa, Zantedeschia*; Lemna.: *Lemna*.

LILIIFLORAE: ASPAR.: Conva.: *Polygonatum*;* Aspar. subfam. Ruscoideae: sometimes;* Aspho.: *Chlorophytum*.*—LILIA.: Lilia.: *Tulipa*.*—BURMA.: Burma.: *Apteria, Burmannia*, (e.g. *B. stuebelii*; Spitmann and Hamann, personal communication); Thism.: sometimes.—BROME.: Brome.: rarely: *Schlumbergeria*.

COMMELINIFLORAE: COMME.: Comme.: *Floscopa*.*—ERIOC.: Xyrid.: *Xyris** sp.; Erioc.: *Eriocaulon*.—JUNCA.: Junca.: *Juncus, Luzula*.—CYPER.: Cyper.: *Carex, Cyperus, Eleocharis, Rhynchospora, Schoenoplectus*.—RESTI.: Centr.: *Aphelia, Brizula, Centrolepis*; Resti.: *Elegia, Hypodiscus, Leptocarpus* (P. Kircher, personal communication), *Restio*.—POAL.: Poac.: numerous genera (all of those studied).

Dicotyledons associated with monocotyledons:

NYMPHAEIFLORAE: NYMPH.: Nymph. (cf. the text above).

Binucleate pollen grains

ALISMATIFLORAE: HYDRO.: Apono.: *Aponogeton*; Hydro.: *Blyxa, Ottelia*.—ZOSTE.: Junca.: *Triglochin, Lilaea*; Zoste.: *Zostera*; Zanni.: *Zannichellia*.

ARIFLORAE: ARAL.: Arac. (heterogeneous): e.g. *Acorus, Anthurium, Arisaema, Calla, Monstera, Philodendron, Pothos, Rhaphidiophora, Spathiphyllum, Symplocarpus, Synandrospadix, Theriophonum, Xanthosoma*.

LILIIFLORAE: DIOSC.: *Dioscorea, Rajania, Trichopus*; Stemo.: *Stemona*; Trill.: *Trillium*.—TACCA.: Tacca.: *Tacca*.—ASPAR. and LILIA.: many taxa investigated, given by Wunderlich and others in various papers (exceptions few; see above).—BURMA.: Burma.: *Burmannia*, and Thism.: both 2- and 3-nucleate.*—ORCHI.: Orchi.: all taxa studied (14 genera).—PONTE.: Ponte.:

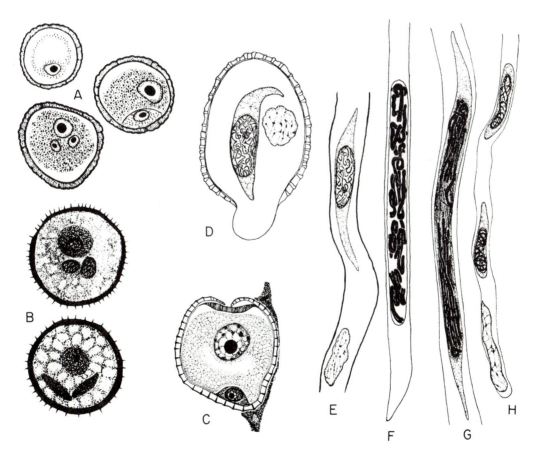

Fig. 62. Bi- and trinucleate pollen grains. A: three successive stages in the development of a trinucleate pollen grain in *Centrolepis fascicularis* (Centrolepidaceae); B: two different stages in the development of a trinucleate pollen grain in *Limnophytum obtusifolium* (Alismataceae). C: binucleate pollen grain of *Urginea indica* (Hyacinthaceae) before dehiscence (tapetal substance adhering to it). D–H: germination of a binucleate pollen grain of *Lilium regale* (Liliaceae), F–G showing the mitotic division of the sperm cell. (A from Prakash, 1969; B from Johri, 1935; C from Capoor, 1937; D–H from Cooper, 1936.)

Eichhornia, Pontederia.—HAEMO.: Haemo.: *Anigozanthos, Lachnanthes, Schiekia.*—PHILY.: Phily.: *Orthothylax, Philydrum* (Hamann, 1966).—VELLO.: Vello.: unspecified report.—BROME.: Brome.: 10 genera out of 11 studied.

ZINGIBERIFLORAE: ZINGI.: Helic.: *Heliconia*; Musac.: Musa; Strel.: *Strelitzia*; Zingi.: several genera; Canna.: *Canna*; Maran.: *Maranta*.

COMMELINIFLORAE: COMME.: Comme.: 7 genera (all of those studied by Brewbaker); Mayac.: *Mayaca.*—ERIOC.: Rapat.: *Cephalostemon, Spathanthus* (Tiemann, personal communication); Xyrid.: ? *Xyris* (cf. above); Erioc.: ? sometimes (see above).—TYPHA.: Sparg.: *Sparganium*; Typha.: *Typha,*—CYPER.: Cyper.: *Fimbristylis, Scirpus* s. lat.—RESTI.: Resti.: *Chondropetalum*; Flage.: *Flagellaria*.

ARECIFLORAE: ARECA.: Areca.: several genera.—CYCLA.: Cycla.: *Carludovica.*—PANDA.: Panda, *Pandanus.*

Dicotyledons associated with the monocotyledons:

NYMPHAEIFLORAE: NYMPH.: Nymph.: *Nuphar, Nymphaea* (cf. above, however).

MAGNOLIIFLORAE: MAGNO. and PIPER: all taxa studied by Brewbaker.

DEGREE OF CARPEL FUSION
(Fig. 63, Diagrams 54–55)

Apocarpy is generally considered to be a primitive condition in the angiosperms and from this state there is a gradual transition to a partial or entire fusion in the ovular region, but with free stylodia (the term stylodium is used here as the equivalent of a style when formed by one carpel only), then further to a partial and finally to an entire fusion of the stylodia into a style with or without apical stylodial branches.

Classifying the degree of fusion of the carpels accordingly one may distinguish between the following categories.

Apocarpy or a single carpel only (note that the single carpel in certain groups, such as in the Typhales and many Arales, may have evolved secondarily by reduction from a probably syncarpous state, while in most monocotyledons it has very likely evolved from an apocarpous state (many members of the Zosterales, the Najadales)).

Syncarpy in the ovary region but with the stylodia totally free from each other from the same ovary.

Syncarpy with a common style at the base and stylodial branches of a considerable length.

Syncarpy with one single style and an apical stigma, which may be capitate or trilobate (the lobes being however relatively short).

Syncarpy (or apocarpy) with an obsolete or no style at all, the stigmatic surface covering one or several areas on the top of the ovary.

As these categories merge gradually into each other any division must be arbitary and in several families the variation covers more than one category, the palms being without doubt that group whose variation (within basically similarly constructed carpels) is widest. Other highly variable groups are the Melanthiaceae and the Araceae.

The distribution of this character is shown in Diagrams 54–55.

Groups with apocarpy or a single carpel occur in the Alismatiflorae, where there is frequently multiplication ("dedoublement") as well as reduction in carpel number, supposing that the original carpel number in monocotyledons may have been 3. However, the stylodial branches are often two in many of these supposedly monocarpellate gynoecia, making the appearance confusing. In *Najas* the branches may also be three or more and in the Posidoniaceae they are branched and lobate. The supposition that the pistils here are not bi- or pluricarpellate is supported by the fact that the carpels in the flowers of several genera of the Hydrocharitaceae have stylodia cleft to the base. [It is interesting to note that cleft stylodia are also common in the Iridaceae and Thismiaceae (asterisks in the diagram). This condition possibly indicates a connection between the latter two families.]

Another group with apocarpy is the Triuridaceae (Triuridales) of the Alismatiflorae. In addition *Protolirion* and *Petrosavia* of the Melanthiaceae are sometimes described as apocarpous, but in fact are incompletely syncarpous in the ovary region (Sterling, 1978). They do not seem to be very closely allied to the Triuridaceae as was thought by Hutchinson (1959).

Finally, the gynoecium is apocarpous in a considerable number of palms.

In addition, the gynoecium is often monomerous by reduction in the Araceae and generally so in the Typhales. The flowers are variously interpreted in the Centrolepidaceae and Hydatellaceae.

There is a tendency in monocotyledons for apocarpy to be combined with numerous carpels. That the numerous carpels represent a primitive condition has been taken for granted more often than it has been questioned. However, according to Singh (1966), Singh and Sattler (1972, 1974, 1976) and Sattler and Singh (1978), the number of pistils in Alismataceae could represent an "amplification", i.e.

be a secondary achievement. Also *Butomus*, and the Triuridaceae have a combination of apocarpy and more than 3 carpels. Besides, certain Hydrocharitaceae should be added here in spite of their superficial syncarpy (see below). A comparison of the "primitive" dicotyledonous groups with many monocotyledonous traits (Magnoliales, Nymphaeales), will reveal apocarpy in combination with numerous or at least indefinate numbers of carpels, which are sometimes distinctly spirally set. The primitiveness of polycarpy is however not yet settled for the monocotyledons. Considering it a derived state opens new possibilities for phylogenetic roots of the aquatic and saprophytic monocotyledons.

Stylodia which are separate down to the ovary (or almost so) are found in the following groups:

The Alismatiflorae. The members of Hydrocharitaceae (Hydrocharitales) might be included in this group. The ovary part of the carpels are enclosed by the receptacle (in the epigynous flowers) are more or less tightly adnate to each other originating from a basically apocarpous condition.

The Liliiflorae. In the Dioscoreales many members of Dioscoreaceae (*Dioscorea*) and of the Trilliaceae have free stylodia (or almost so), while the probably closely related Smilacaceae (Asparagales) generally have a very short style and subsessile stigmas. In most Liliales they are fused at least basally to form a simple style, exceptions being part of the Calochortaceae, some Colchicaceae and many Melanthiaceae.

Within the Commeliniflorae free stylodia occur in some Juncaceae (e.g. *Prionium*, species of *Juncus*) and in many of the Cyperaceae (e.g. many species of *Cyperus*). The widest distribution of this state is found, however, in the Poales and Restionales. In Poales most taxa have free stylodia, although others (e.g. some genera of bamboos) have a considerable style. In the Restionales most, but not all members of Restionaceae have free stylodia or long stylodial branches as do the Flagellariaceae, Anarthriaceae and Ecdeiocoleaceae.

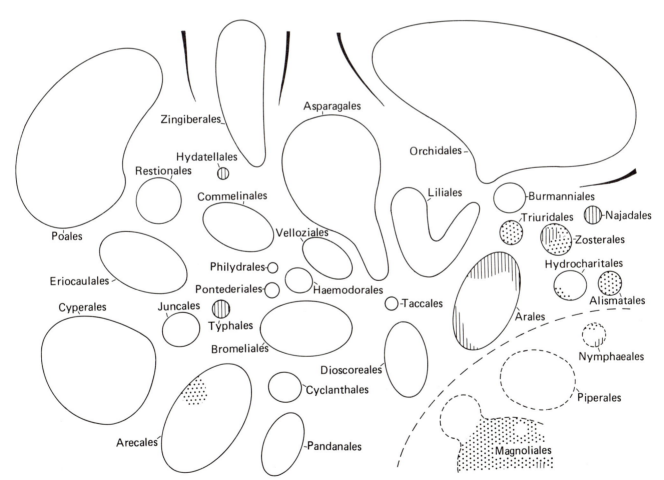

Diagram 54. Apocarpy in 2- to multicarpellate gynoecia (dots) and solitary carpels (hatching). Within the Restionales, Centrolepidaceae might partly be hatched, as the carpels here are solitary in each flower, although they are more or less fused between the flowers.

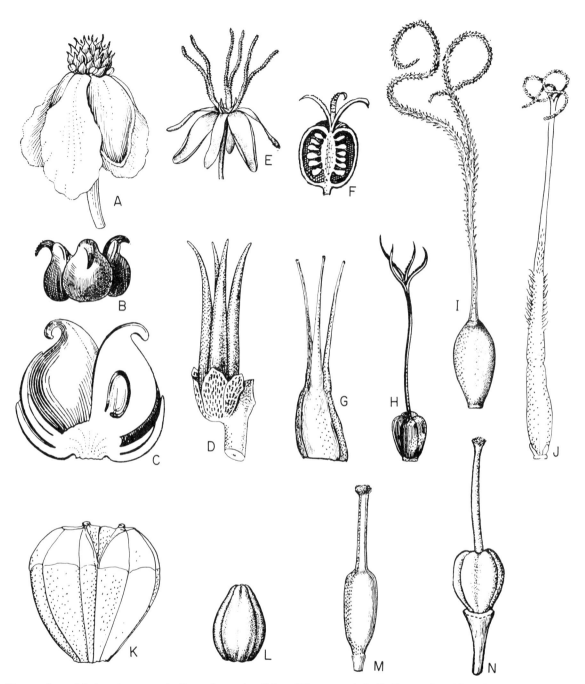

Fig. 63. Degree of carpel fusion. Apocarpy: A: *Limnophyton obtusifolium* (Alismataceae); B–C: *Phoenix dactylifera* (Arecaceae). Syncarpy but free stylodia: D: *Tetroncium magellanicum* (Juncaginaceae); E: *Elodea callitrichoides* (Hydrocharitaceae), the stylodial branches are bifurcate (asterisks in the diagram); F: *Paris quadrifolia* (Trilliaceae); G: *Wurmbea compacta* (Colchicaceae). Syncarpy, with a style which is divided into long branches: H: *Phytelephas microcarpa* (Arecaceae); I: *Scirpus spegazzinianus* (Cyperaceae); J: *Maclurolyra tecta* (Poaceae). Carpels fused with obsolete or no styles; K: *Pandanus basedowii* (Pandanaceae) with separate stigmas; L: *Astelia* sp. (Asphodelaceae), with confluent stigmas. A single style with capitate or lobate stigma: M: *Nothoscordum inodorum* (Alliaceae); N: *Allium pulchellum* (Alliaceae). (A from Carter, 1960b; B from Lotsy, 1911; C from Baillon, 1895; D from Pizarro, 1959; H from Lotsy, 1911; E and L–M from Correa, 1969; G from Nordenstam, 1964a; F from Krause, 1930; J from Calderón and Soderstrom, 1973; K from Stone, 1974; N from Moore, 1955.)

The Areciflorae are very variable, but sometimes the short stylodia are separate from the base of a syncarpous ovary.

A style which is **branched** into relatively long stylodial branches is found in the following groups.

In the Liliiflorae, many species of *Dioscorea* and of *Tacca* possess a stout lobate style and are referred to this category with some hesitation. Tribrachiate styles also occur among the Liliales, i.e. practically all taxa of the Iridaceae, Geosiridaceae and Colchicaceae and many of the Alstroemeriaceae, in the latter family of which the stylodial branches may be rather short (as in Liliaceae) and therefore are treated in the next category. Several genera of the Melanthiaceae also have a branched style. In the Asparagales, however, the style is generally simple, considerably branched only within the Asparagaceae

(*Asparagus* spp.), Dasypogonaceae and Aphyllanthaceae.

Within the Commeliniflorae a basally simple style with two or three long branches is found in most taxa of the Juncales and Cyperales and in practically all Eriocaulales except the Rapateaceae. Furthermore such styles occur in some grass genera, some genera in the Restionaceae and sometimes in *Flagellaria*.

In the Areciflorae many palms could probably be referred to this group, but where the style and style branches are often very short they may be better placed in the category of obsolete styles.

A syncarpous pistil with **simple style** that has a capitate or a distinctly trilobate (to tribrachiate) apex could be regarded the most "advanced" type in the present respect. The occurrence is concentrated especially to the Lilii-

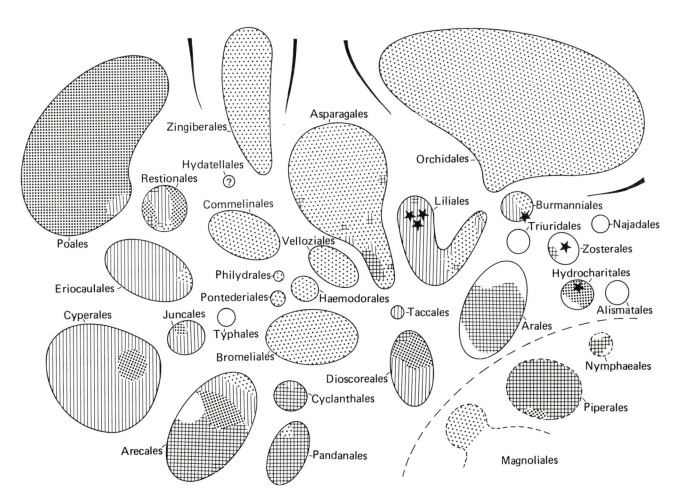

Diagram 55. Degree of carpel fusion. Apocarpy, i.e. gynoecia with wholly free carpels, as well as monocarpellate gynoecia: unshaded. Shaded parts with some degree of fusion (between two or more carpels forming a pistil). Free stylodia, sometimes also more or less separate ovary tips: dense dots. Style basally simple but tri- (or bi-) brachiate in the apical parts: vertical hatching. Style simple, although stigma sometimes trilobate: sparse dots. Styles or stylodia obsolete or extremely short, with stigma or stigma crests sessile or subsessile: cross-hatching. Bifurcate or otherwise branched styles indicated with asterisks. As regards Centrolepidaceae, see Diagram 54.

florae, the Zingiberiflorae and the Commelinales of the Commeliniflorae.

Within the Liliiflorae a simple style occurs in the majority of the Asparagales (except the Smilacaceae and few members of the Asparagaceae and Dasypogonaceae which have tribrachiate styles; and except Asparagaceae, subfam. Ruscoideae, taxa of Agavaceae subfam. Yuccoideae and Asphodelaceae subfam. Astelioideae where the style may be obsolete or lacking). In the Liliales there is a gradual transition between tribrachiate and simple but often apically trilobate styles, the simple style prevailing in the Liliaceae, Tricyrtidaceae, some Alstroemeriaceae and such genera as *Narthecium*, *Metanarthecium*, *Nietneria* and *Heloniopsis* in the Melanthiaceae. In the Orchidales the style is simple but complicated by its fusion with the androecium, while in the Burmanniales it is simple in the Corsiaceae and certain taxa of the Burmanniaceae and Thismiaceae. The pistils in the Velloziales, Haemodorales, Pontederiales, Philydrales and Bromeliales have a simple style although sometimes flat, or variously differentiated in the apical part.

The Commeliniflorae have few groups with simple style, viz. mainly the Commelinales (constant in this feature) and the Rapateaceae of the Eriocaulales, where however the apex often may be conspicuously trilobate.

Within the Areciflorae some palms have simple although short styles, and such types also occur rarely in the Cyclanthales (*Cyclanthus*) and Pandanales.

Finally, there are certain gynoecia which are totally **devoid of a style** or where the style is so short that it may be regarded to be nothing more than a slight continuation of the ovary. Here belong most of the Ariflorae, even those with a monocarpellate ovary (unshaded in the diagram). Other main groups are the three orders of the Areciflorae, where the style is often only a beaked to blunt continuation of the ovary. Finally a few more taxa should be added to this category, viz. in the Liliiflorae. They include the Stemonaceae (Dioscoreales), and the *Ruscus*, *Yucca* and *Astelia* groups of Asparagaceae, Agavaceae and Asphodelaceae respectively in the Asparagales. Perhaps *Xanthorrhoea* should be added here. Others are *Calochortus pro parte* (Calochortaceae) and the tribe Aristeae of the Iridaceae (Liliales). Also *Joinvillea* and *Hanguana* (Restionales) have sessile stigmas, a condition that seems somewhat out of place in this part of the system. In the Alismatiflorae, *Triglochin* in the Juncaginaceae, being syncarpous, belongs to this category, and among the primitively or secondarily monocarpellate pistils (unshaded in the diagram) one finds numerous examples in the Alismatiflorae, and also in the Hydatellaceae (Commeliniflorae), Lemnaceae and many Araceae (see above).

Among the dicotyledons regarded as related to the monocotyledons the Piperaceae have an obsolete style, a conspicuous similarity to the Arales and the apocarpous condition with an obsolete style or a gradually tapering pistil is held in common between many Nymphaeiflorae and Alismatiflorae.

The above pattern of variation indicates that this character can be used more extensively than hitherto to define groups and it certainly reflects phylogenetic relationships in some groups, whereas in others (Arecaceae, Melanthiaceae) it seems to be of very limited value.

Base Data

Apocarpy (2 or more carpels totally free from one another)
ALISMATIFLORAE: HYDRO.: Butom.; Apono. (Hydro. may be considered at least partly apocarpus althought the gynoecium is enclosed by the floral receptacle).—ALISM.: all.—ZOSTE.: Scheu.; Potam.; Zanni. (mostly); Cymod.—TRIUR.: all.

ARECIFLORAE: ARECA.: Areca.: several genera of the Coryphoid and Phoenicoid palms (Moore, 1973) plus *Nypa*.

Dicotyledons associated with monocotyledons:
NYMPHAEIFLORAE: NYMPH.: Cabom.; (Nymph. may also be considered apocarpous although the gynoecium is enclosed by the floral receptacle).

MAGNOLIIFLORAE: MAGNO.: most taxa (except Canel.; most Arist. and a few Annon.).

Solitary carpel (and thus no carpel fusion)
ALISMATIFLORAE: ZOSTE.: Junca.: *Lilaea*; Zoste.; Posid.; Zanni. (rarely).—NAJAD.: Najad.

ARIFLORAE: ARAL.: Arac.: many genera; Lemna.: all.

COMMELINIFLORAE: TYPHA.: mostly (occasionally 2- or 3-carpellate)—(RESTI.: Centro. the carpels although considered to belong to different, naked, unicarpellate flowers may fuse in the ovary region to form capsular structures.)—HYDAT.: Hydat. (there is doubt as to the monocarpellate nature of the pistil espec. in *Trithuria*).

(ARECIFLORAE: ARECA.: Areca.: pseudomonomerous pistils are found in some genera, but truly monomerous ones are hardly present.)

Dicotyledons associated with the monocotyledons:
NYMPHAEIFLORAE: NYMPH.: Cerat.
MAGNOLIIFLORAE: MAGNO.: rare, but found in Myris. and Degen.—LAURA.: typical condition.—(In the Piperales the seemingly simple pistils are probably pseudomonomerous and hence syncarpous throughout.)

Carpels fused only in the ovary region (or basal parts thereof)

ALISMATIFLORAE: HYDRO.: Hydro.—ZOSTE.: Junca. p.p.; stylodia mostly obscure).

LILIIFLORAE: DIOSC.: Diosc.: *Dioscorea* spp.; Trill.: spp. of *Paris* and *Trillium*.—LILIA.: Colch.: rarely, e.g. *Wurmbea*; Caloc.: partly; Melan.: several genera, e.g. *Tofieldia, Veratrum*; *Petrosavia* and *Protolirium* approaching apocarpy.

COMMELINIFLORAE: JUNCA.: Junca.: *Juncus* spp., *Prionium*.—CYPER.: ca. 15 per cent of the taxa.—RESTI.: Resti.: frequently; Anarth.; Ecdei.; (Centr.: see under monomerous gynoecia above.)—POAL.: Poac.: mostly.

ARECIFLORAE: ARECA.: Areca.: frequently, in this family there are gradual transitions in this character.

Dicotyledons associated with monocotyledons:

MAGNOLIIFLORAE: PIPER.: Sauru.: mostly.

Carpels fused in the ovary region and with a (basally) single style which is apically 2- or 3-branched

LILIIFLORAE: DIOSC.: Diosc.: spp. of *Dioscorea* and *Trichopus*; Trill.: *Trillium* spp.—TACCA.: Tacca.: *Tacca* (stout, winged).—ASPAR.: Conva.: *Drymophila, Streptopus*; Aspar.: rarely; Dasyp.: *Lomandra* spp.; Hypox.; Aphyll.: *Aphyllanthes*.—LILIA.: Irid.: nearly all; Geosi.; Colch.: mostly; Alstr.: rarely; Tricy.; Melan.: rarely.—BURMA.: Burma.: mostly; Thism.: partly.

COMMELINIFLORAE: ERIOC.: Xyrid.; Erioc.—JUNCA.: most taxa.—CYPER.: most taxa.—RESTI.: Resti. p.p.; Flage.—POAL.: Poac.: some genera esp. of subfam. Bambusoideae.

ARECIFLORAE: ARECA.: Areca.: scattered genera.—CYCLA.: Cycla.: rarely, as in *Cyclanthus*.

Dicotyledons associated with the monocotyledons:

MAGNOLIIFLORAE: e.g. in PIPER.: Sauru. p.p.

Syncarpous gynoecia with a single style bearing a terminal capitate or 3-lobate (to shortly 3-brachiate) stigma

LILIIFLORAE: DIOSC.: Stemo.: rarely.—ASPAR.: most taxa (except Smila. and for example *Ruscus, Lomandra*, Hypox.: *Aphyllanthes, Astelia* and *Clistoyucca*).—LILIA.: Alstr. (most taxa); Lilia. (mostly); Melan.: *Narthecium, Ypsilandra, Clara, Heloniopsis* and a few more genera.—BURMA.: Burma. p.p.; Thism. p.p.; Corsi.—ORCHI.: all.—PONTE.: all.—HAEMO.: all.—PHILY.: all.—VELLO.: all.—BROME.: all.

ZINGIBERIFLORAE: all.

COMMELINIFLORAE: COMME.: all.—ERIOC. Rapat.—RESTI.: Resti.: rarely, e.g. *Coleocarya, Hopkinsia, Onychosepalum*.

ARECIFLORAE: ARECA.: rarely.

Syncarpous gynoecia with very short or obsolete styles or stylodia, hence with sessile to subsessile stigmas

ALISMATIFLORAE: ZOSTE.: Junca.: mostly, e.g. *Triglochin*.

ARIFLORAE: ARAL.: Arac.: more than half of the genera.

LILIIFLORAE: DIOSC.: Stemo.: *Stemona*.—ASPAR.: Smila.; Conva.: *Rhodea*; Aspar. subfam. Ruscoideae; Dasyp.: *Lomandra* spp.; Agava.: *Clistoyucca*; Aspho.: *Astelia, Cohnia*.—LILIA.: Irida. tribus Aristeae; Caloch.

COMMELINIFLORAE: RESTI.: Restio.: rarely, e.g. *Chondropetalum*; Joinv.; Hangu.

ARECIFLORAE: ARECA.: frequently.—CYCLA.: mostly.—PANDA.: mostly.

Associated dicotyledons:

NYMPHAEIFLORAE: NYMPH.: Nymph.: all.

MAGNOLIIFLORAE: PIPER.: nearly all, except some Sauru.

DRY VERSUS WET STIGMAS
(Diagram 56)

Stigma types may be divided into two main groups, defined as "dry", that is without copious fluid secretion, and "wet", that is with surface secretions present during their receptive phase. The wetness or otherwise of a stigma being measured in terms of whether it could or could not be printed clearly onto a dry non-absorbant surface.

Amongst both monocotyledons and dicotyledons dry stigmas predominate and there is a suggestion that amongst dicotyledons especially there is an association between the nature of the stigmatic surface and the breeding system (type of self incompatibility system). Although *amongst the dicotyledons* at large there is no significant association between stigma type and the number of nuclei (cells) in the pollen grain the association is strong amongst monocotyledons. As may be seen from comparing Diagrams 53 and 56 trinucleate pollen grains in the monocotyledons are restricted to groups at least some of whose members have dry stigmas. The record of 3-nucleate pollen grains in *Canna* of the Zingiberales (Davis, 1966) requires confirmation as all other taxa in the order is reported to have 2-nucleate pollen grains.

Dry stigmas occur exclusively in the large orders Poales and Cyperales and the smaller orders Restionales, Eriocaulales, Juncales and Typhales, and also in the representatives of the Pontederiales, Philydrales and Taccales and in the two taxa studied of Arecales of the Liliiflorae; likewise all taxa known in the Hydrocharitales, Alismatales and Zosterales, forming the main part of the Alismatiflorae, have exclusively dry stigmas. An important observation is that the Poales and Juncales and part of the Restionales, Eriocaulales and Cyperales and all the Alismatiflorae have trinucleate pollen grains. Such pollen grains also occur in the Arales and very rarely in the Bromeliales, in all of which groups dry stigmas occur.

Wet stigmas are concentrated mainly in the following groups:

In the Zingiberales, Orchidales and Cyclanthales (one species only studied) where they occur to the exclusion of dry stigmas.

In many members (though in less than 30 per cent) of the Asparagales and Liliales.

In scattered taxa of the Arales, Dioscoreales and Commelinales, in the majority of the Bromeliales, and in two out of three species studied in the Haemodorales.

The occurrence of both dry and wet stigmas in the Asparagales and Liliales is somewhat surprising, but here also a certain regularity is observable. Thus the Iridaceae and Convallariaceae and some small or medium-sized families invariably seem to have dry stigmas, and in the Amaryllidaceae only few species (3 out of 17) are recorded to have wet stigmas while some families (Liliaceae, Asphodelaceae, Hyacinthaceae etc.) have a rich representation of both types. Further studies in such families will show whether the dry and wet stigmas are correlated with other character states.

In the dicotyledonous orders Nymphaeales and Piperales the few taxa studied have dry stigmas, while the condition is variable in Aristolochiaceae (*Aristolochia* having both dry and wet stigmas) as it is in Magnoliales on the whole.

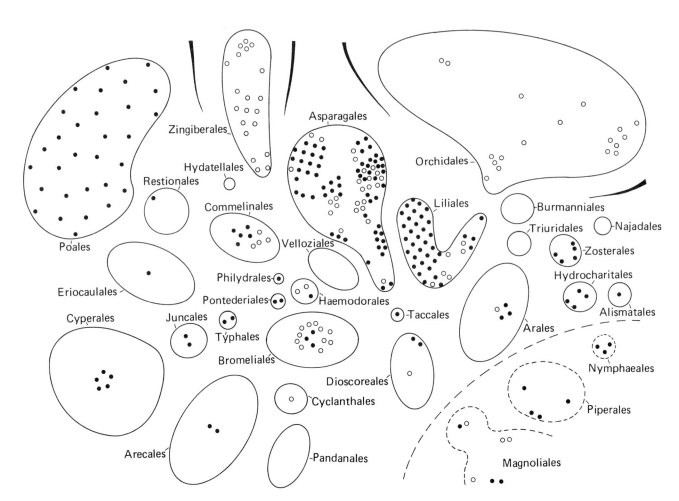

Diagram 56. Dry versus wet stigmas (solid and hollow circles respectively). Each symbol represents an investigated species.

Base Data

(Mainly according to Heslop-Harrison and Shivanna, 1977)

Dry stigmas

ALISMATIFLORAE: HYDRO.: Butom.: *Butomus*; Hydro.: *Limnobium, Thalassia*; Apono.: *Aponogeton.*—ALISM.: Alism.: *Echinodorus.*—ZOSTE.: Scheu.: *Scheuchzeria*; Junca.: *Triglochin*; Potam.: *Potamogeton*; Cymod.: *Amphibolis, Thalassodendron.*

ARIFLORAE: ARAL.: Arac.: *Arisarum, Pistia, Spathiphyllum.*

LILIIFLORAE: DIOSC.: Trill.: *Scoliopus, Trillium.*—Tacca.: *Tacca.*—ASPAR.: Smila.: *Smilax*; Phile.: *Geitonoplesium, Philesia*; Conva.: *Clintonia, Disporum, Liriope, Maianthemum, Ophiopogon, Polygonatum, Reineckea, Speirantha*; Aspar.: *Asparagus*; Agava.: *Agave, Furcraea, Yucca*; Hypox.: *Curculigo, Hypoxis, Rhodohypoxis*; Aspho.: subfam. Asphodelioideae: *Asphodeline, Asphodelus, Astroloba (Aprica), Bulbinella, Chamaealoë, Haworthia, Kniphofia*, subfam. Astelioideae: *Astelia, Cordyline*; subfam. Anthericoideae: *Anthericum, Chlorophytum, Dasystachys, Dichopogon, Echeandia, Paradisia, Pasithaea, Thysanotus*; Aphyl.: *Aphyllanthes*; Diane.: *Dianella*; Hyaci.: several genera; Allia.: *Agapanthus, Allium, Ipheion, Nothoscordum*; Amary.: 15 genera.—LILIA.: Irida.: 30 genera; Colch.: *Colchicum, Sandersonia*; Caloch.: *Calochortus*; Lilia.: *Erythronium, Notholirion, Tulipa*; Melan.: *Veratrum.*—PONTE.: Ponte.: *Eichhornia, Pontederia.*—HAEMO.: Haemo.: *Xiphidium.*—PHILY.: Phily.: *Orthothylax.*—BROME.: Brome.: *Abromeitiella, Dyckia, Neoregelia.*

COMMELINIFLORAE: COMME.: Comme.: *Aneilema, Callisia, Campelia, Commelina, Tradescantia.*—ERIOC.: Erioc.: *Eriocaulon.*—TYPHA.: Sparg.: *Sparganium*; Typha.: *Typha.*—JUNCA.: Junca.: *Juncus, Luzula.*—CYPER.: Cyper.: *Blysmus, Carex, Cyperus, Eleocharis, Scirpus.*—RESTI.: Flage.: *Flagellaria.*—POAL.: Poac.: 30 genera.

ARECIFLORAE: ARECA.: Areca.: *Chamaedorea, Ptychosperma.*

Dicotyledons associated with monocotyledons:

NYMPHAEIFLORAE: NYMPH.: Nymph.: *Nuphar, Nymphaea, Victoria.*

MAGNOLIIFLORAE: MAGNO.: Aristo.: *Aristolochia*; Magno.: *Liriodendron, Magnolia.*—PIPER. Piper.: *Peperomia, Piper*; Sauru.: *Houttuynia, Saururus.*

Wet stigmas

ARIFLORAE: ARAL.: Arac.: *Alocasia.*

LILIIFLORAE: DIOSC.: Diosc.: *Tamus.*—ASPAR.: Phile.: *Lapageria*; Aspar.: *Ruscus*; Dracae.: *Dracaena*

(= *Pleomele*); Dasyp.: *Lomandra*; Agava.: *Agave, Beschorneria*; Aspho.: subfam. Aspodeloideae: *Aloë, Eremurus, Gasteria*; subfam. Anthericoideae: *Arthropodium*; Hemer.: *Hemerocallis*; Funki.: *Hosta*; Hyaci.: *Camassia, Drimia, Galtonia, Hyacinthella, Muscari, Ornithogalum, Urginea*; Allia.: *Bloomeria, Leucocoryne*; Amary.: *Habranthus, Rhodophiala, Sprekelia.*—LILIA.: Colch.: *Iphigenia, Merendera*; Alstr.: *Alstroemeria*; Tricy.: *Tricyrtis*; Lilia.: *Fritillaria, Lilium.*—ORCHI.: Cypri.: *Paphiopedilum*; Orchi.: 17 genera.—HAEMO.: Haemo.: *Anigozanthos, Wachendorfia.*—BROME.: Brome.: 11 genera.

COMMELINIFLORAE: COMME.: Comme.: *Cyanotis, Gibasis, Thyrsanthemum, Weldenia.*

ZINGIBERIFLORAE: ZINGI.: Lowia.: *Orchidantha*; Helic.: *Heliconia*; Musac.: *Musa*; Strel.: *Strelitzia*; Zingi.: *Alpinia, Cautleya, Curcuma, Globba, Hedychium, Kaempfera, Roscoea, Zingiber*; Costa.: *Costus*; Canna.: *Canna*; Maran.: *Calathea, Maranta, Marantochloa, Stromanthe, Trachyphrynium.*

ARECIFLORAE: CYCLA.: Cycla.: *Carludovica.*

Dicotyledons associated with monocotyledons:

MAGNOLIIFLORAE: MAGNO.: Annon.: *Annona, Friesodielsia*; Arist.: *Aristolochia*; Winte.: *Drimys.*

BASAL AND LATERAL PLACENTATION
(Fig. 64, Diagram 57)

Ovaries with basal placentation are widespread and the condition is found along with both apocarpy and syncarpy, Diagrams 54–55. Although it is constant within several orders, including the Najadales, Triuridales, Pandanales and Cyperales, it is not necessarily constant within families. Thus in the Arecaceae (Areciflorae) only about 80 per cent of the species have basal placentation and within *Xyris* (Xyridaceae, Eriocaulales, Commeliniflorae) some species have basal and others parietal placentation. None the less the occurrence of basal placentation is far from randomly distributed and so has some taxonomic meaning.

In the Alismatiflorae every order has some members with basal placentation. The condition is shared by all members of the Alismatales, Najadales and Triuridales and also by at least *Aponogeton* of the Hydrocharitales and the Scheuchzeriaceae and Juncaginaceae of the Zosterales.

In the Liliiflorae relatively few taxa have basal placentation. These are restricted to the Stemonaceae (*Stemona*) in Dioscoreales and the Dracaenaceae (*Dasy-*

lirion) and Dasypogonaceae (*Calectasia* and *Dasypogon*) in Asparagales. Other members of these families have axile placentation.

In the Commeliniflorae basal placentation has its widest occurrence in Cyperaceae but is also found in a few species of *Xyris* (Xyridaceae, Eriocaulales), some Rapateaceae, and *Luzula* (Juncaceae, Juncales). In *Xyris* the placentation may be parietal, basal, or free and central (Conert, 1965).

In the Zingiberiflorae basal placentation is found in at least *Heliconia* (Heliconiaceae).

The majority of palms (Arecales) and *Pandanus* (Pandanales) have basal placentation but that condition is not shared with the third order of the Areciflorae, namely Cyclanthales.

Finally about 60 per cent of the Arales have basal placentation.

Of the dicotyledon families associated with the mono-cotyledons only the Piperaceae (Piperales) have basal placentation.

The Poaceae have lateral to apical placentation, although in early ontogeny the ovule arises from the base, at the floral axis.

Base Data

Basal placentation

In monocarpellate pistils

ALISMATIFLORAE: HYDRO.: Apono.: *Aponogeton.*—ALISM.: Alism. subfam. Alismatoideae.—ZOSTE.: Scheu.: *Scheuchzeria*; Junca.: *Lilaea.*—NAJAD.: all.—TRIUR.: all.

ARIFLORAE: ARAL.: Arac.: scattered; Lemna.: all.

ARECIFLORAE: ARECA.: Areca.: scattered (maybe in 5–10 per cent of the genera).

In syncarpous pistils

ALISMATIFLORAE: ZOSTE.: Junca.: most taxa.

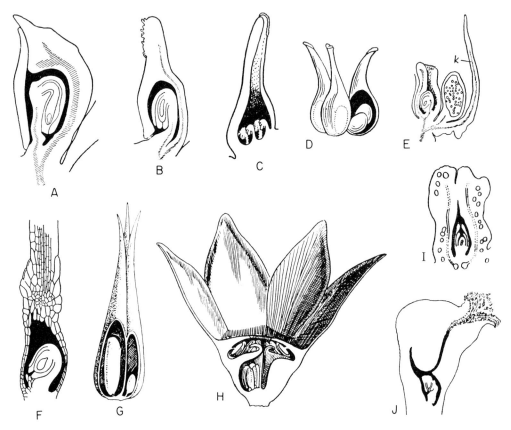

Fig. 64. Basal placentation. A–F: in monocarpellate pistils. A: *Echinodorus ranunculoides* (Alismataceae); B: *Luronium natans* (Alismataceae); C: *Aponogeton ulvaceus* (Alismataceae). In all of these cases only one carpel per flower is shown. D: gynoecium of *Aponogeton ranunculiflorus* (Aponogetonaceae, one carpel cut longitudinally). E: Flower of *Lilaea subulata* (Juncaginaceae) in longitudinal section. F: pistil of *Najas flexilis* (Najadaceae), longitudinal section. G–J: in multicarpellate gynoecia. G: *Tetroncium magellanicum* (Juncaginaceae), gynoecium with two carpels opened. H: *Lourya* sp. (Convallariaceae), flower in longitudinal section. I: *Piper medium* (Piperaceae, a dicotyledon) and J: *Pandanus polycephalus* (Pandanaceae), in longitudinal section. (A, B, C, E and J from Melchior, 1964; D from Guillarmod and Marais, 1972; F from Campbell, 1898; G from Pizarro, 1959; H from Baillon, 1894; I from Lotsy, 1911; J from Melchior, 1964.)

ARIFLORAE: ARAL.: Arac.: scattered.

LILIIFLORAE: rare; at least in: DIOSC.: Stemo.: *Stemona.*—ASPAR.: Dracae.: *Dasylirion*; Dasyp.: *Calectasia, Dasypogon.*

ZINGIBERIFLORAE: ZINGI.: at least Helic. (subbasal in some Maran.)

COMMELINIFLORAE: ERIOC.: Rapat. p.p.; Xyrid.: *Xyris* sect. *Nematopus.*—JUNCA.: Junca.: *Luzula.*—CYPER.: all.—POAL.: Poac. (rarely).

ARECIFLORAE: ARECA.: mostly (maybe in c. 70 per cent of the genera).—PANDA.: *Pandanus.*

Dicotyledons associated with the monocotyledons: MAGNOLIIFLORAE: e.g. PIPER.: Piper.

Lateral placentation (apical placentation, see pendulous ovules)

ALISMATIFLORAE: ZOSTE.: Potam. p.p. (*Potamogeton*).

ARIFLORAE: ARAL.: Arac. p.p. (maybe c. 10–15 per cent of the genera).

COMMELINIFLORAE: JUNCA.: Thurn.—POAL.: Poac. p.p.

Dicotyledons associated with the monocotyledons: NYMPHAEIFLORAE: NYMPH.: Cerat.

The above refers to unilocular pistils with ovule(s) being neither basal nor apical but inserted laterally on the ovary wall.

PARIETAL PLACENTATION IN SYNCARPOUS PISTILS
(Fig. 65, Diagram 58)

The formation of ovules in vertical rows on the ovary wall results in parietal placentation. In ontogenetic terms

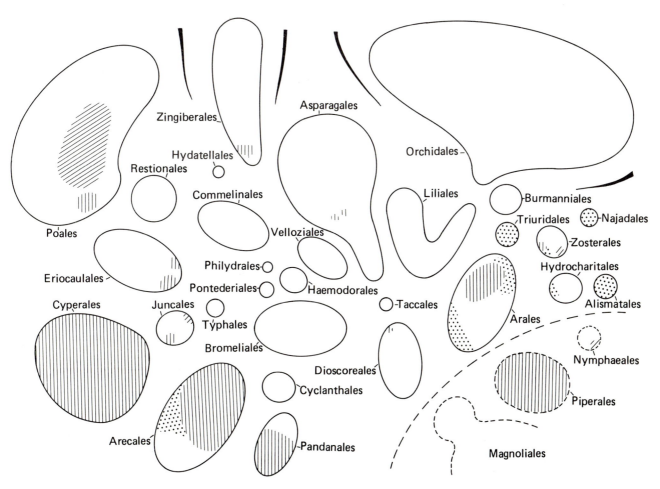

Diagram 57. Basal placentation in mono- and in bi- or tricarpellate gynoecia (dots and vertical hatching respectively). In grasses the ovule is sometimes lateral (oblique hatching) and appears to arise from the ovary wall.

this condition is clearly a forerunner of axile placentation in that in early stages of development ovaries may show parietal and in later stages axile placentation, but in phylogenetic terms it may be the reverse. Parietal placentation is a property of paracarpous (unilocular syncarpous) ovaries. Where there is only one placenta care must be exercised to determine whether one or more carpels are involved. If only one carpel is involved the placentation is generally known as marginal or lateral.

Sometimes as with some taxa of Araceae it is difficult to know the placentation type without considerable study for the ovary though often uniloculate with a single placenta may be pseudomonomerous (that is to be composed of several carpels). The evidence comes from the occasional presence of a second placenta in *Arum* (Arber, 1925) and from comparative morphological studies such as those of Engler who showed that the aroids exhibit various reduction stages from a tricarpellary ovary (Arber, 1925).

Parietal placentation is present in taxa of all super-orders except the Alismatiflorae. Proportionally, it is not well represented in any except the Orchidales, Burmanniales, Taccales and Juncales, maybe also the Philydrales of the Liliiflorae and the Cyclanthales and Pandanales (*Freycinetia*) of the Areciflorae. Its occurrence in the Burmanniales and Orchidales is sometimes considered to indicate close affinity between these groups, but this is not altogether certain as in both orders there are groups with axile placentation which could have evolved independently into parietal.

Base Data

ARIFLORAE: ARAL.: Arac.: in scattered genera, e.g. *Arum* and *Ariopsis*.

LILIIFLORAE: TACCA.: Tacca.: *Tacca.*—ASPAR.: Peter.: *Petermannia*; Philes.: *Lapageria*, *Philesia*; Hypox.: *Empodium*; Aspho.: *Astelia* spp.—LILIA.: Alstr.: *Leon-*

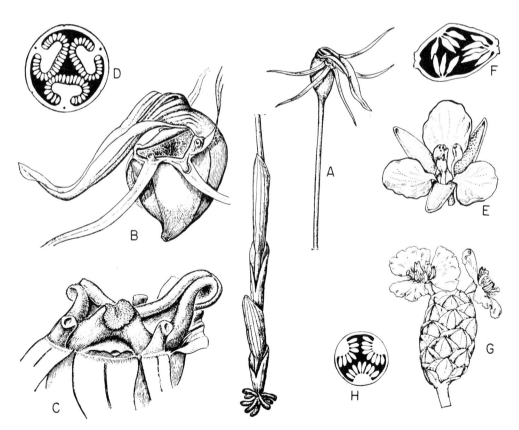

Fig. 65. Parietal placentation. A–D: *Arachnites uniflora* (Corsiaceae, Burmanniales). Plant and details of flower with a transection of the inferior ovary. E–F: *Mayaca aubletii* (Mayacaceae). G–H: *Xyris congdonii* (Xyridaceae), one of the species with parietal placentation. (A–C from Correa, 1969; D from Pizarro, 1959; E–H from Lawrence, 1951.)

tochir, Schickendantzia.—BURMA.: Burma. p.p. *Apteria, Burmannia* spp., *Cymbocarpa, Dictyostegia, Dipterosiphon, Gymnosiphon, Marthella, Miersiella;* Corsi.— ORCHI.: Cypri.: *Cypripedium, Paphiopedium* spp.; Orchi.: nearly all.—PONTE.: Ponte.: *Heteranthera, Hydrothrix.*—PHILY.: Phily.: *Philydrum.*

ZINGIBERIFLORAE: ZINGI.: Zingi.: tribus Globbeae; Costa.: *Tapeinocheilos.*

COMMELINIFLORAE: COMME.: Mayac.—ERIOC.: Xyrid.: at least in part of *Xyris.*—JUNCA.: Junca., common: many spp. of *Juncus, Marsippospermum, Rostkovia, Distichia.*

ARECIFLORAE: PANDA.: Panda.: *Freycinetia.*—CYCLA.: Cycla.: most genera: *Asplundia, Carludovica, Dicranopygium, Evodianthus, Schultesiophytum, Thoracocarpus.*

Dicotyledons associated with monocotyledons:

MAGNOLIIFLORAE: MAGNO.: Annon.: *Isolona, Monodora*; Canel.

AXILE PLACENTATION
(Fig. 66, Diagram 59)

Axile placentation is of restricted distribution being limited to three of the six superorders, Diagram 59. Within two of these, the Liliiflorae and Zingiberiflorae, the condition is widespread, but it is limited in the Commeliniflorae.

Amongst the Zingiberiflorae all taxa other than *Heliconia*, the tribe Globbeae of the Zingiberaceae, and a few members of the Costaceae have axile placentation. This is also the case with the Liliiflorae, the exceptions being *Lapageria* and *Philesia* of the Philesiaceae and the monogeneric Petermanniaceae (Asparagales), a few species of Alstroemeriaceae (Liliales), the Stemonaceae (Dioscoreales) and most taxa of Orchidales, Burmanniales and Velloziales. In the Burmanniales axile placentation is found in part of *Burmannia* and in the

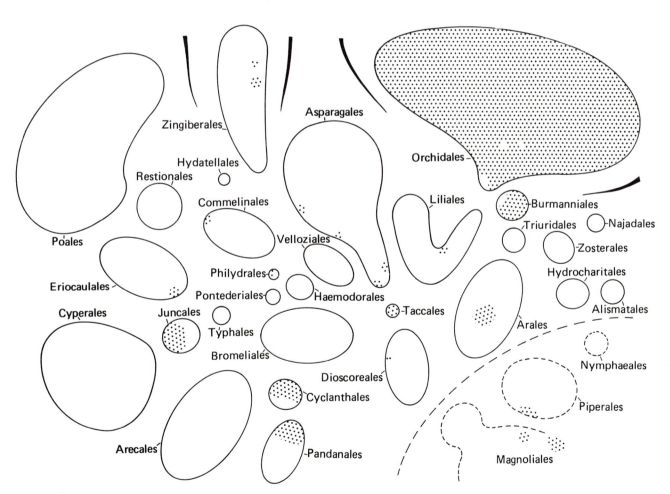

Diagram 58. Parietal placentation in bi- or tricarpellate pistils (shaded).

genera *Campylosiphon* and *Hexapterella*, and in the Orchidales the genera of Apostasiaceae, *Phragmopedium* and *Selenipedium* of Cypripediaceae and *Climatepistephium* and *Lecanorchis* of the Orchidaceae. (The placentation in most or all Velloziales is perhaps best described as laminar.)

In contrast to the above two superorders, the Commeliniflorae contain few families with axile placentation, viz. the Commelinaceae and its satellite Cartonemataceae (both in the Commelinales): most Juncaceae (Juncales), and a few members of the Rapateaceae (Eriocaulales).

The concentration of taxa with axile placentation in each of the Zingiberiflorae and Liliiflorae suggests that the character is primitive in both and indeed supports a common ancestry for the two superorders. Whether or not the Commeliniflorae show the same ancestry is uncertain but the character is still primitive in that group. Specialization to wind pollination has resulted in a reduction in ovule number within the Commeliniflorae

and a transfer of the solitary ovule to a basal, lateral or pendulous position. Perhaps here it reflects the survival of the lowest or uppermost ovule respectively from an original axile row.

The basically axile placentation of the Liliiflorae shows no obvious connections with placentation types in Magnoliiflorean dicotyledons.

Base Data

LILIIFLORAE: DIOSC.: Diosc.; Trill.—ASPAR.: all except those enumerated under parietal and basal placentations.—LILIA.: all except a few Alstr.—BURMA.: Burma.: *Burmannia* spp., *Campylosiphon*, *Hexapterella*. —ORCHI.: Apost.; Cypri.: *Selenipedium*, *Phragmopedium*; Orchi.: *Climatepistephium*, *Lecanorchis* sp.— PONTE.: mostly.—HAEMO.: all.—PHILY.: Phily.: all except *Philydrum*.—(VELLO.: see the text above.)— BROME.: all.

ZINGIBERIFLORAE: ZINGI.: Lowia.; Musac.;

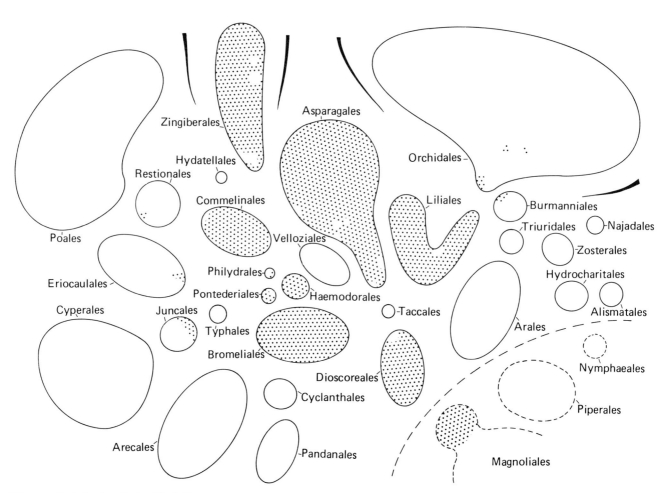

Diagram 59. Axile placentation (shaded).

Strel.; Zingi (except tribus Globbeae); Costa. (mostly); Canna.; Maran. (subbasal).

COMMELINIFLORAE: COMME.: Comme.; Carto. —ERIOC.: Rapat. subfam. Saxofridericoideae and Rapateoideae p.p.—JUNCA.: Junca.: *Juncus* p.p., *Patosia*, *Prionium*.—RESTI.: Hangu.

Dicotyledons associated with the monocotyledons: MAGNOLIIFLORAE: MAGNO.: mainly Arist.

LAMINAR (DISPERSED) PLACENTATION
(Fig. 67, Diagram 60)

The occurrence of this placentation is of particular interest in that it occurs in the Alismatiflorae of the monocotyledons and in the Nymphaeiflorae, a group of dicotyledons with which the Alismatiflorae exhibit several features in common (Diagram 60).

Besides, laminar placentation, no doubt independently, occurs in all or practically all Velloziaceae (Velloziales).

In the Alismatiflorae the laminar placentation is best developed in the Butomaceae (*Butomus*), which bears numerous ovules on the inner surface of the carpels (Fig. 67D–E). Likewise, in the Hydrocharitaceae the carpels have a placentation type referable to or derived from laminar (Eckardt, 1969). In the Aponogetonaceae, the placentation is however basal.

In the Alismataceae subfam. Limnocharitoideae (Alismatales), the carpels, as in *Butomus*, develop into follicles (Fig. 67A–C), and the placentation is typically laminar, while in the subfam. Alismatoideae there is only one ovule per carpel and this has a basal position.

In the other two orders of the Alismatiflorae (Zosterales and Najadales) the ovules are few or solitary and vary from basal to (lateral or) apical. In these orders the solitary ovules are sometimes considered to represent a reduced laminar condition.

The Nymphaeales, which comprise the Nymphaeiflorae, shows a variation comparable to that in the Alismatiflorae. Typical laminar placentation with numerous ovules is found in the Nymphaeaceae, while in the Cabombaceae the ovules are 1–3 in number occupying a laminar-median position.

It is difficult to evaluate the significance of this very marked similarity between the Nymphaeiflorae and Alismatiflorae, a similarity supplemented with numerous

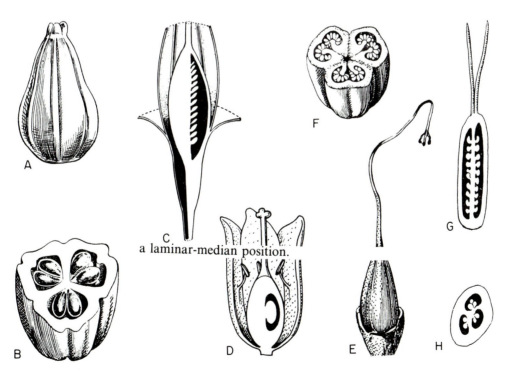

Fig. 66. Axile placentation. A–B: *Astelia pumila* (Asphodelaceae subfam. Astelioideae). C: *Bessera elegans* (Alliaceae). D: *Asparagus officinalis* (Asparagaceae). E–F: *Puya berteroniana* (Bromeliaceae). G–H: *Roscoea purpurea* (Zingiberaceae). (A–B and E–F from Pizarro, 1959; C from Moore, 1953; D from Larsen, 1977; G–H from Burtt and Smith, 1972.)

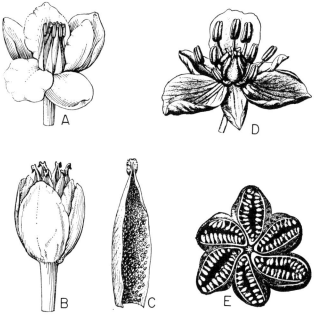

other properties such as the presence of stipules, apo-carpy, often sulcate pollen grains, common absence of vessels in the stem, and the Helobial endosperm forma-tion, which in Nymphaeales is restricted to Cabombaceae, however. (See further on p. 291.)

Base Data

ALISMATIFLORAE: HYDRO.: Butom.; Hydro. (at least in most genera).—ALISM.: Alism.: subfam. Limno-charitoideae.

LILIIFLORAE: VELLO.: Vello.

Dictoyledons associated with the monocotyledons:

NYMPHAEIFLORAE: NYMPH.: Cabom.; Nymph.

Fig. 67. Laminar placentation. A–C: *Tenagocharis latifolia* (Alisma-taceae); flower, fruit assemblage of flower and young carpel opening up. D–E: *Butomus umbellatus* (Butomaceae), flower and gynoecium in transverse section. Note that the carpels are free from each other. (A–C from Carter, 1960a; D–E from Wettstein, 1924.)

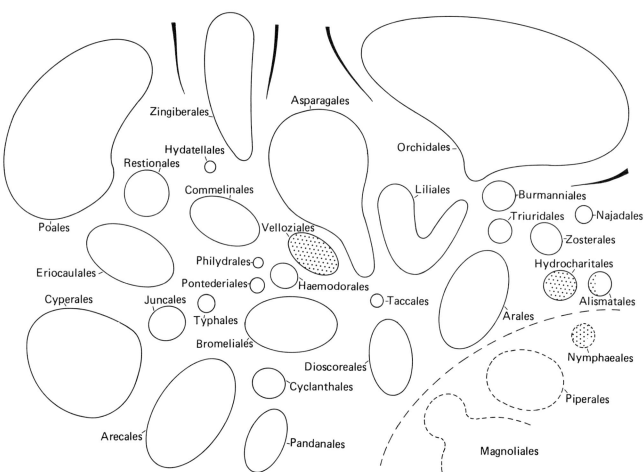

Diagram 60. Laminar or dispersed placentation (shaded).

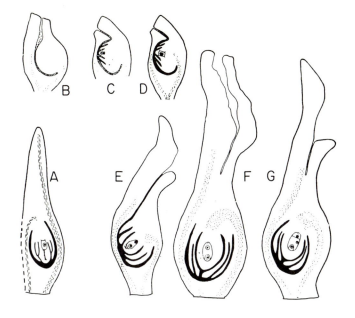

PENDULOUS (VERSUS HORIZONTAL OR ASCENDING) OVULES (Fig. 68, Diagram 61)

Ovaries bearing pendulous ovules have a rather restricted distribution, being entirely absent from the Zingiberiflorae, and occurring in relatively few Areciflorae (in Arecaceae and Cyclanthaceae) and about one-fifth of the Ariflorae.

Within the Alismatiflorae pendulous ovules are restricted to the Zosterales, five of whose seven families possess them throughout (Zosteraceae, Posidoniaceae,

Fig. 68. Pendulous ovules. A: *Centrolepis* sp. (Centrolepidaceae) and B–G in *Zannichellia* (Zannichelliaceae). It is seen that the ovule in *Zannichellia* is initiated laterally and translocated to the apex in the course of the ontogenetic development. (A from Prakash, 1969; B–G from Vijayaraghavan and Kumari, 1974.)

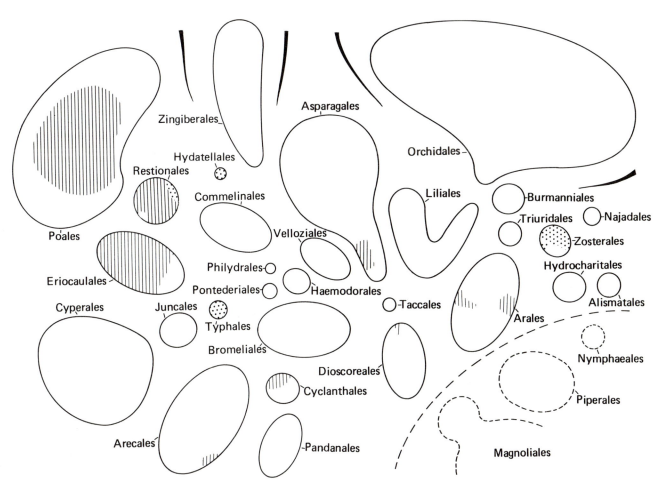

Diagram 61. Pendulous ovules (shaded). Discrimination is here made between pendulous ovules in monocarpellate (dots) and in bi- or tricarpellate ovaries (hatching).

Zannichelliaceae, Cymodoceaceae) or in part (Potamo-getonaceae).

Pendulous ovules are very rare in the Liliiflorae occurring only in Smilacaceae and two genera of Stemonaceae (*Croomia* and *Stichoneuron*), the third genus in that family (*Stemona*) having basal ovules.

Amongst the Commeliniflorae all species of the two small orders Hydatellales and Typhales have pendulous ovules. These also occur in most members of the Restionales. The axile insertion of ovules in the monogeneric family Hanguanaceae supports the view that this may be misplaced in the Restionales. In the Poales the ovules vary from apical-pendulous to lateral. In the Eriocaulales neither the Rapateaceae nor the Xyridaceae have pendulous ovules although they occur in all Eriocaulaceae.

The concentration of pendulous ovules in adjacent orders of the Commeliniflorae suggests that the character therein may or may not have a common origin, whereas its occurrence in the widely divergent Arecales, Arales and Zosterales suggests evolutionary convergence with respect to the trait.

Base Data

ALISMATIFLORAE: ZOSTE.: Junca.: *Maundia*; Potam.; Zoste.; Posid.; Zanni.

ARIFLORAE: ARAL.: Arac. p.p.

LILIIFLORAE: DIOSC.: Stemo.: *Croomia, Stichoneuron.*—ASPAR.: Smila.

COMMELINIFLORAE: COMME.: Comme.: rarely.—ERIOC.: Erioc.—TYPHA.: all.—HYDAT.: Hydat.—RESTIO.: Centr.; Resti.; Anar.; Ecdei.; Joinv.; Flage.—POAL.: Poac. (most taxa).

ARECIFLORAE: ARECA.: rarely.—CYCLA.: Cycla.: some genera.

ONE VERSUS TWO OR MORE OVULES PER PLACENTA
(Diagram 62)

The numbers of ovules per placenta varies considerably and it is convenient to distinguish two artificial groups: viz. those with one ovule per placenta and those with two or more ovules per placenta. The distribution of these categories is summarized in Diagram 62, from which it is clear that both occur, but with different frequencies, in each of the six superorders.

Thus about half the members of each of the Ariflorae and Alismatiflorae (including nearly all Zosterales and

Najadales) have placentas bearing only a single ovule. Within the Zingiberiflorae, Marantaceae and Heliconiaceae have uniovulate and the remaining families multiovulate placentas. In terms of both genera and species the Areciflorae are mostly uniovulate (exceptions being *Freycinetia* in the Pandanaceae and all members of the Cyclanthaceae).

The Liliiflorae are predominantly multiovulate, but there are few or more species with uniovulate placentas in each of Iridaceae, Smilacaceae, Asparagaceae (*Ruscus*), Dasypogonaceae, Dracaenaceae (subfamily Dracaenoideae), Asphodelaceae (subfamily Anthericoideae) and Aphyllanthaceae. In the Pontederiales, the genera *Reussia* and *Pontederia* have uniovulate and the remaining genera multiovulate placentas.

In contrast, the Commeliniflorae have mostly uniovulate placentas, the condition being universal in the Cyperales, Poales, Restionales, Typhales and Hydatellales. Elsewhere in the superorder there are indications of an evolutionary trend towards a reduction in ovule number. Thus, within the Commelinales there is a sequence of ovule numbers, the Mayacaceae having many, the Commelinaceae mostly a few and Cartonemataceae two ovules per placenta. A similar trend may be observed in the Juncales, where *Juncus* has many to few ovules per placenta, *Prionium* 6–3 and *Luzula* (unilocular) 3 ovules altogether, one per placenta. One to several ovules per placenta occur in *Thurnia*. A reduction series is also present in the Eriocaulales with the Xyridaceae being multiovulate, the Rapateaceae uni- to multiovulate and the Eriocaulaceae uniovulate.

The contrast between the Liliiflorae and Commeliniflorae is striking in that the former mainly have multiovulate and the latter mainly uniovulate placentas. An explanation for the difference may be found in their pollination mechanisms, the Commeliniflorae being largely wind- and the Liliiflorae largely animal-pollinated. See also fruits (Diagrams 70 and 73).

Base Data

Taxa with one ovule per placenta, i.e. per carpel

ALISMATIFLORAE: ALISM.: mainly subfam. Alismatoideae.—ZOSTE.: all except Scheu.—NAJAD.: all.—TRIUR.: all.

ARIFLORAE: ARAL.: Arac.: numerous genera, e.g. *Aglaonema, Amorphophallus, Biarum, Culcasia, Nephthytis, Philonotion, Pinellia, Plesmonium, Pseudodracontium, Synantherias, Thomsonia, Typhonodorum, Zamicarpella*; Lemn.

LILIIFLORAE: ASPAR.: Smila. (one or two ovules);

Aspar.: *Ruscus* (one or two ovules); Dracae.: subfam. Dracaenoideae; Dasyp. p.p.; Aspho. subfam. Anthericoideae: rarely; Aphyl.—LILIA.: Irida.: rarely.—PONTE.: Ponte.: at least *Pontederia, Reussia.*—HAEMO.: Haemo.: *Barberetta, Dilatris, Macropidia, Wachendorfia.*

ZINGIBERIFLORAE: ZINGI.: Helic.; Maran.

COMMELINIFLORAE: ERIOC.: Rapat. subfam. Rapateoideae; Erioc. (all).—TYPHA.: all.—JUNCA.: Junca.: *Luzula*; Thurn.: *Thurnia* spp.—CYPER.: all.—HYDAT.: all.—RESTI.: all.—POAL.: all.

ARECIFLORAE: ARECA.: all.—PANDA.: Panda.: *Pandanus.*

Dicotyledons associated with the monocotyledons: NYMPHAEIFLORAE: NYMPH.: Cerat.

MAGNOLIIFLORAE: MAGNO.: e.g. Myris. (all); Himan. (mostly).—PIPER.: Piper. (all).

OVULE TYPES (Fig. 69, Diagram 63)

The ovules in monocotyledons vary considerably from the anatropous type (which is believed to be the most primitive) through hemianatropous to orthotropous (= atropous) or campylotropous. The appearance of these types is illustrated in Fig. 69, but it should be noted that the terms are used somewhat differently by different authors. The account below has been compiled from various sources, the principal of which are Davis (1966) and Huber (1969). The distribution of ovule types is shown in Diagram 63.

The **anatropous** type is widespread and is found in all superorders.

Within the Alismatiflorae it is the dominant or only type in the Alismatales, Hydrocharitales, Najadales and

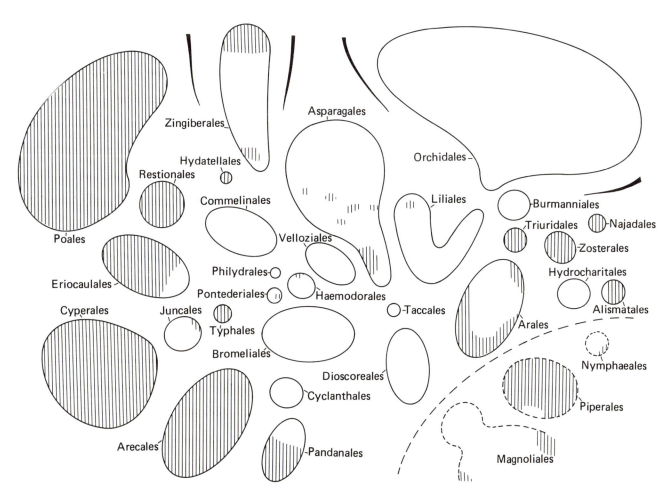

Diagram 62. Uniovulate placentas (shaded). In unilocular ovaries this coincides with uniovulate pistils (e.g. in the Cyperales, Poales, most Piperales and the apocarpous groups).

Triuridales, although *Nechamandra* and *Vallisneria* in Hydrocharitaceae are stated to have orthotropous ovules. In the Zosterales anatropous ovules occur only in the Scheuchzeriaceae and Juncaginaceae, probably the two least specialized families of that order.

In the Ariflorae anatropous and hemianatropous ovules occur in most taxa studied.

The Liliiflorae are very variable, but here the anatropous type occurs in perhaps all the members of the orders Taccales, Burmanniales, Orchidales, Velloziales, Pontederiales, Philydrales and Bromeliales. In the Dioscoreales the genus *Stemona* (Stemonaceae) has been stated to have almost orthotropous ovules, but according to Swamy (1964), they are anatropous. In the Liliales weakly campylotropous ovules may occur in the Colchicaceae. The Asparagales is highly variable in ovule conditions, anatropous ovules being however the commonest type. Exceptions (see below) occur in the Smilacaceae, Phile-

siaceae, Asparagaceae, Dasypogonaceae, Hypoxidaceae, Asphodelaceae s.lat., Hyacinthaceae, Alliaceae and Amaryllidaceae. The Haemodorales according to de Vos (1956) have orthotropous or hemianatropous ovules, whereas Davis (1966) reported anatropous ovules for this group.

In the Zingiberiflorae practically all known taxa have anatropous ovules, but in the Marantaceae there is a trend towards the campylotropous state.

In the Commeliniflorae there is a lesser dominance of anatropous ovules, although this is the general type in the Typhales, Juncales and Cyperales and in the possibly misplaced Hydatellales. In the Poales there is variation between orthotropous, hemianatropous and campylotropous ovules, and in Commelinales (Commelinaceae) the anatropous type merges into hemianatropous and orthotropous. In the Eriocaulales, the Rapateaceae have anatropous ovules, while the Xyridaceae have anatropous or

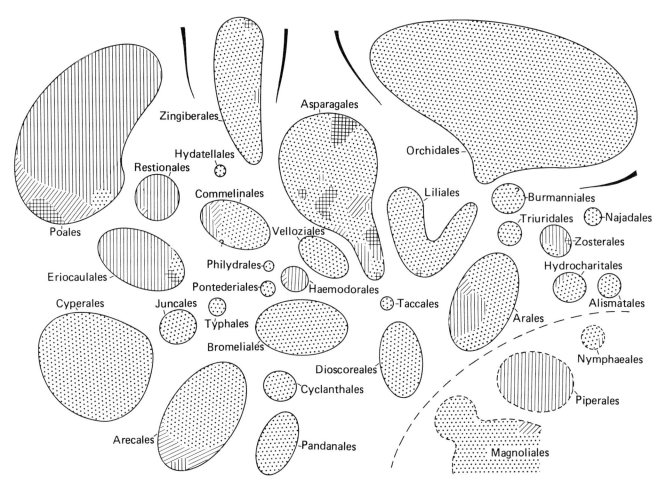

Diagram 63. Ovular morphology. Anatropous ovules: dots. Campylotropous ovules: cross-hatching. Hemianatropous ovules: oblique hatching. Orthotropous ovules: vertical hatching.

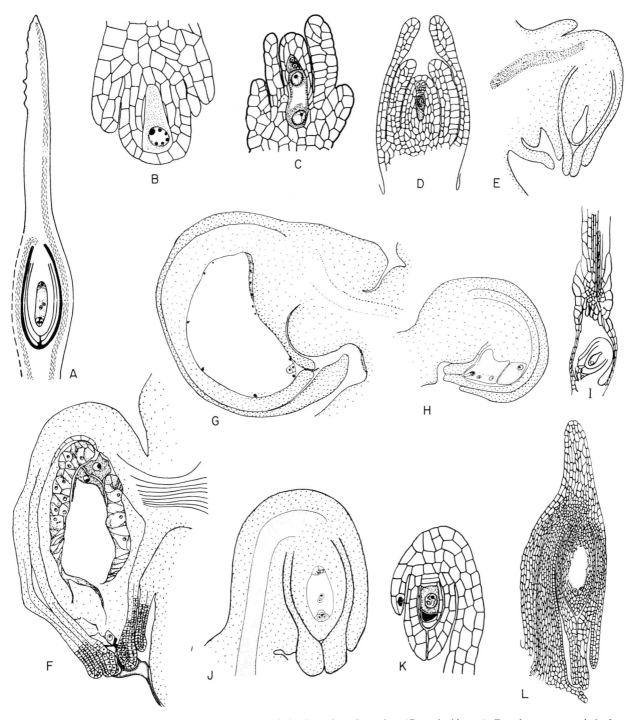

Fig. 69. Ovular morphology. A–B: pendulous orthotropous ovule in *Centrolepis fascicularis* (Centrolepidaceae). C: orthotropous ovule in *Lemna* sp. (Lemnaceae). D: orthotropous ovule in *Piper medium* (Piperaceae, a dicotyledon!). E: hemianatropous ovule in *Eremurus* sp. (Asphodelaceae subfam. Asphodeloideae). F: hemianatropous ovule in *Wachendorfia* sp. (Haemodoraceae). G–H: campylotropous ovules in *Arthropodium cirrhatum* (Asphodelaceae subfam. Anthericoideae). I: anatropous basal ovule in *Najas* sp. (Najadaceae) (cf. Fig. 68 B–G). J: anatropous ovule in *Paris quadrifolia* (Trilliaceae). K: anatropous ovule in *Arundina* (Orchidaceae), early stage. L: Anatropous ovule in *Tillandsia usneoides* (Bromeliaceae). (A–B from Prakash, 1969; C from Maheshwari and Kapil, 1963; D from Lotsy, 1911; E from Stenar, 1928a; F from Dellert, 1933; G from Stenar, 1928a; H from Schnarf and Wunderlich, 1939; I from Campbell, 1898; J from Berg, 1962b; K from Rao, 1967; L from Billings, 1904.)

incompletely campylotropous ovules except in *Xyris* where they are orthotropous, as they are also in the Eriocaulaceae. The Restionales are characterized mostly by orthotropous ovules but hemianatropous seem to occur in the Hanguanaceae; *Flagellaria* and *Joinvillea* thus agree with the Restionaceae in having orthotropous ovules (Hamann, 1974).

Finally there are mostly anatropous ovules also in all three orders of the Areciflorae, but also hemianatropous ovules occur among the palms. The record for the Pandanaceae is from Cheah and Stone (1975).

Hemianatropous ovules occur in various groups of monocotyledons; among these are scattered taxa of Arales (Araceae as well as Lemnaceae), Haemodorales (*Wachendorfia*), Asparagales (Asparagaceae: e.g. *Ruscus* and *Asparagus*, and Hypoxidaceae: *Forbesia*, *Pauridia*, perhaps some Asphodelaceae subfam. Asphodeloideae), Commelinales (genera of Commelinaceae), Poales and Arecales (a few grass and some palm genera respectively), maybe also the Hanguanaceae (Restionales).

Campylotropous ovules are not particularly common in the monocotyledons. They occur only in the Potamogetonaceae (*Ruppia*) of the Alismatiflorae and in the Philesiaceae (*Luzuriaga*), at least some members of the Convallariaceae and Dasypogonaceae, Asphodelaceae subfam. Antheriocoideae (Schnarf and Wunderlich, 1939) and *Cordyline*, and a great number of the Alliaceae, all these in the Asparagales of the Liliiflorae. In addition, weakly campylotropous ovules occur in the Colchicaceae, Liliales. In the Zingiberiflorae campylotropous ovules are known in Marantaceae, and in the Commeliniflorae they occur in some grasses (Poaceae) and rarely in Xyridaceae.

Orthotropous ovules are found in several orders of monocotyledons as discussed below.

In the Alismatiflorae orthotropous ovules are centred in the Zosterales, where they occur in the Potamogetonaceae except *Ruppia* and in the Zosteraceae, Posidoniaceae, Zannichelliaceae, and (though not quite orthotropous) in the Cymodoceaceae. In the Hydrocharitaceae, orthotropous ovules occur at least in species of *Vallisneria* and *Nechamandra*.

In the Liliiflorae, orthotropous or almost orthotropous ovules occur sporadically, viz. in Smilacaceae, some genera of Convallariaceae and a few of Asphodelaceae subfam. Asphodeloideae (Asparagales) and also in most taxa studied of Haemodoraceae (Haemodorales).

Within the Zingiberales orthotropous ovules are reported to occur in *Hitchenia caulina* of the Zingibcraccac.

Within the Commeliniflorae orthotropous ovules are known in the Eriocaulaceae and at least in *Xyris* of the

Xyridaceae (Eriocaulales), in most grasses, and in all taxa studied of the Restionales (Brongniart and Gris, 1861; Hamann, 1974) except Hanguanaceae, which has a very dubious place here. Finally orthotropous ovules occur in the Mayacaceae and in some Commelinaceae (Commelinales).

Orthotropous and pendulous ovules rarely occur in palms (Arecales), being reported for *Actinophloeus* and *Howea* and orthotropous and transverse ovules are known in *Batris* (Ventaka Rao, 1959). Mahabale and Biradar (1968) also mention ana- and orthotropous forms of ovules in *Phoenix silvestris*. According to them there is a gradual shifting of position of micropyle laterally in the course of the later stages of seed development, which forms a basis for identification of seeds.

The above pattern indicates that ovule morphology can indeed be indicative of phylogenetic relationships, although in some families the variation is considerable (Commelinaceae, Xyridaceae, Asparagaceae and especially Asphodelaceae s.lat.). Groups where attention to ovule types should be considered or have been considered for taxonomic purposes are the Restionales (note the orthotropous ovules also in the Joinvilleaceae and Flagellariaceae) and its somewhat dubious relative, the Hydatellales. Thus, Centrolepidaceae has ortho- but Hydatellaceae anatropous ovules (Hamann, 1975, 1976).

This character was also recorded as important by Huber (1969) for distinguishing the Asphodelaceae (mainly with anatropous ovules) from the Anthericaceae (with campylotropous ovules); here they are tentatively treated as subfamilies. In the Zosterales the difference between the Scheuchzeriaceae-Juncaginaceae complex (with anatropous ovules) on the one hand and the other families (with orthotropous) on the other should also be considered. The marine Zosteralean groups (orthotropous) also differ from the Najadales (anatropous) in the same regard.

The evolution from anatropous to other ovule types is often associated with other attributes which are accepted as advanced. Thus the marine Zosteralean groups may be regarded as advanced, and so are the Eriocaulalean and Restionalean-Poalean groups. However, two very specialized orders, the Orchidales and the Zingiberales have retained the basic anatropous ovule.

Among the dicotyledons associated with monocotyledons, most Nymphaeales have anatropous but the Ceratophyllaceae orthotropous ovules. The Piperales have orthotropous or almost orthotropous ovules, and most Magnoliales dealt with here (Annonaceae, Aristolochiaceae, Magnoliaceae) have anatropous ovules, although "circinotropous" ones are known in *Bragantia*

(Aristolochiaceae). Hemianatropous ovules occur in the Cannelaceae.

Base Data

Anatropous ovules

ALISMATIFLORAE: HYDRO.: Butom.; Hydro.: most genera (rarely orthotropous; see below); Apono.—ALISM.: all.—ZOSTE.: Scheu.; Junca.—NAJAD.: all.—TRIUR.: all.

ARIFLORAE: ARAL.: Arac.: most genera.; LEMNA.: at least *Spirodela*.

LILIIFLORAE: DIOSC.: Diosc.; Stemo.; Trill.—TACCA.: all.—ASPAR.: Smila. (rarely and then incompletely so); Phile.: all except *Luzuriaga*; Aspar.: subfam. Asparagoideae p.p.; Dracae. (?). (?); Phorm.; Xanth.; Agava.; Hypox.: mostly; Aspho.: subfam. Asphodelioideae and subfam. Astelioideae (except *Cordyline*); Aphyl.; Diane.; Tecoph.; Cyana.; Erios.; Hemer.; Funki. (?); Hyaci.: mostly; Allia.: partly; Amary.: mostly.—LILIA.: Irida.; Colch.: mostly; Alstr.; Lilia.; Melan.: mostly.—BURMA.: Burma.; Thism. (Corsi.: ?).—ORCHI.: probably all.—PONTE.: all.—PHILY.: all.—VELLO.: all.—BROME.: most.

COMMELINIFLORAE: COMME.: Comme.: most genera (Carto.: ?).—ERIOC.: Rapat.; Xyrid.: *Abolboda*.—TYPHA.: all.—JUNCA.: all.—CYPER.: all.—HYDAT.: all.

ZINGIBERIFLORAE: ZINGI.: all except some Maran. (Lowia.: ?).

ARECIFLORAE: ARECA.: Areca.: most genera.—PANDA.: all.—CYCLA.: all.

Dicotyledons associated with the monocotyledons: NYMPHAEIFLORAE: NYMPH.: all except Cerat. MAGNOLIIFLORAE: MAGNO.: most families except, for example, Canel.

Hemianatropous ovules (data probably incomplete)

ARIFLORAE: ARAL.: Arac.: some genera; Lemna.: sometimes in *Lemna*.

LILIIFLORAE: ASPAR.: Aspar. p.p.; Hypox.: some genera (*Forbesia*, *Pauridia*).—HAEMO.: Haemo.: at least *Wachendorfia*.

COMMELINIFLORAE: COMME.: Comme.: some genera.—RESTI.: Hangu.—POAL.: several genera.

ARECIFLORAE: ARECA.: some genera.

Dicotyledons associated with the monocotyledons: MAGNOLIIFLORAE: MAGNO.: at least Canel.

Campylotropous ovules

ALISMATIFLORAE: ZOSTE.: Potam.: *Ruppia*.

LILIIFLORAE: ASPAR.: Phile.: *Luzuriaga*; Conva.:

several genera; Dasyp. p.p.; Aspho.: most genera of subfam. Anthericoideae and *Cordyline*; Allia.: frequently.—LILIA.: Colchi.: weakly campylotropous in some taxa; Melan.: at least *Petrosavia*.—BROME.: Brome.: rarely.

COMMELINIFLORAE: ERIOC.: Xyrid.: rarely, and incompletely so.—POAL.: some genera.

ZINGIBERIFLORAE: ZINGI.: Maran.: sometimes.

Orthotropous ovules

ALISMATIFLORAE: HYDRO.: Hydro.: at least *Nechamandra* and *Vallisneria*.—ZOSTE.: Potam.: *Potamogeton* s.lat.; Zoste.; Posid.; Zanni.; Cymod.

ARIFLORAE: ARAL.: Arac.; many (c. 35) genera; Lemna.: p.p.

LILIIFLORAE: ASPAR.: Smila.: ortho- to incompletely anatropous; Conva.: p.p.; Aspar. subfam. Ruscoideae p.p.; Aspho.: rarely, some taxa of esp. subfam. Asphodeloideae nearly orthotropous, e.g. *Aloë*, *Asphodelus*, *Sowerbaea*.—HAEMO.: Haemo.: most genera, e.g. *Anigozanthos*, *Dilatris*, *Xiphidium* (de Vos, 1956).

ZINGIBERIFLORAE: ZINGI.: Zingi.: *Hitchenia* sp.

COMMELINIFLORAE: COMME.: Comme.: p.p.; Mayac.—ERIOC.: Xydid. p.p.; Erioc.—RESTI.: Resti.: Ecdei.; Anart.; Flage.; Joinv.; Centr.—POAL.: Poac.: most taxa.

ARECIFLORAE: ARECA.: rarely, e.g. *Actinophloeus*, *Batris*, *Howea*, *Phoenix* sp.

Dicotyledons associated with the monocotyledons. NYMPHAEIFLORAE: NYMPH.: Cerat. MAGNOLIIFLORAE: PIPER.: all.

PRESENCE VERSUS ABSENCE OF PARIETAL CELL
(Fig. 70, Diagram 64)

The terms crassinucellate and tenuinucellate ovules were used by Davis (1966) as synonymous with the presence and absence of a parietal cell cut off from the primary archesporial cell, and this terminology has also unfortunately been used by many Indian botanists and also by one of ourselves (Dahlgren, 1975b). However, the concepts become increasingly confused by this usage.

In most cases extremely small nucelli ("tenuinucelli"), which may become crushed by the ripe embryo sac, have no parietal cell, and thick nucelli ("crassinucelli") usually have a parietal cell that may divide to form a multicellular parietal tissue. But there are many exceptions to this. Therefore, the data of different embryological schools are not comparable in the use of the terms

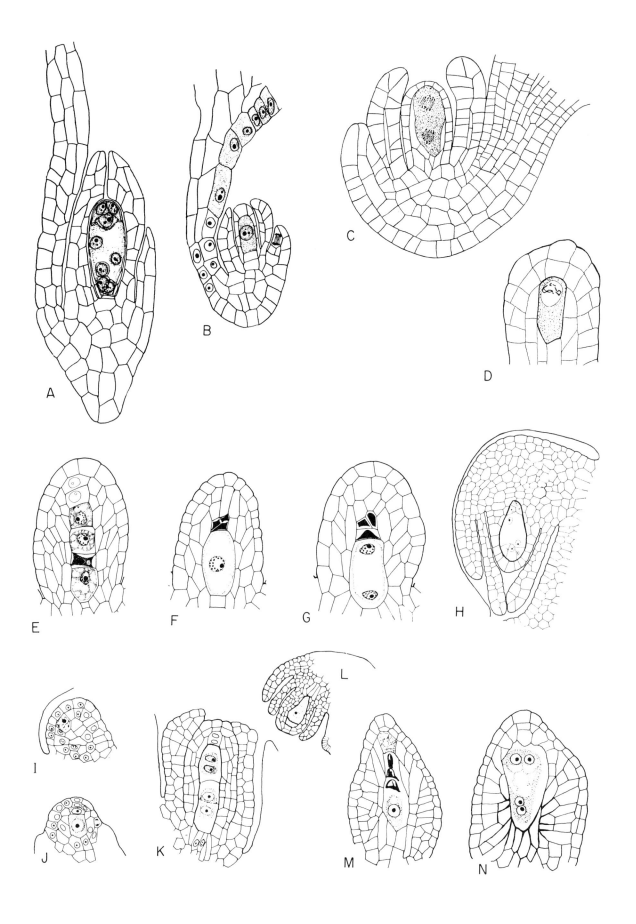

"crassinucellate" and "tenuinucellate" (Hamann, personal communication, 1966a; see also Björnstad, 1970). Thus, crassi- and tenuinucellate in the original—and preferable—definition should be used for the description of the richness in number of cells of the nucellus, including its lateral and basal parts. These terms are not sufficient to cover all the intermediate, "semicrassinucellate" forms (Hamann, 1966a), however, and therefore we have not attempted to classify the groups according to this character.

Whether a parietal cell is cut off or not, there are often periclinal divisions in the nucellar epidermis (especially at the nucellar apex). In ovules where a parietal cell is not cut off (Dahlgren, 1927) but the epidermis forms a nucellus cap; this condition is sometimes called "pseudocrassinucellate", and will be further discussed below.

The presence or absence of a parietal cell (and subsequently often a parietal tissue) is however fairly well documented for a great number of taxa of the monocotyledons and has been shown here in Diagram 64.

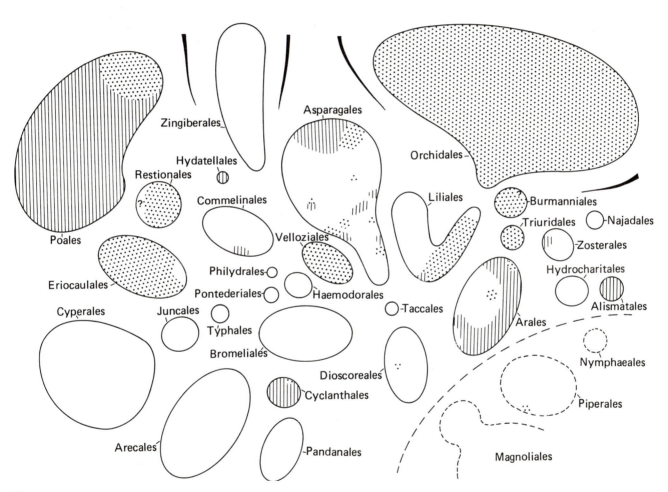

Diagram 64. Absence (shading) and presence (no shading) of parietal cell (tissue). In groups (with or without a parietal cell) the nucellar epidermis may form a cap by periclinal divisions. Where this is the case in ovules without parietal cell (tissue), vertical hatching has been used in the diagram, where parietal cell as well as periclinal divisions of the nucellus epidermis are lacking shading is by dots.

Fig. 70. Presence and absence of parietal cell or parietal tissue and of periclinal divisions of the nucellar epidermis. A–C: absence of both parietal cell and periclinal divisions of epidermis nucellus in *Arundina graminifolia*, Orchidaceae (two stages: A–B), and in *Gagea minima*, Liliaceae (C). D: absence of parietal cell (tissue) but presence of periclinal divisions of the nucellar epidermis in *Gloriosa virescens* (Colchicaceae). E–H: presence of parietal cells (tissue) in *Urginea indica*, Hyacinthaceae (different stages). I–N: as for the preceding but with occasional divisions of nucellar epidermis (K), in *Romulea rosea*, Iridaceae. (A–B from Rao, 1967; C from Stenar, 1927; D from Afzelius, 1918; E–I from Capoor, 1937; I–N from Steyn, 1973.)

In the Alismatiflorae ovules where parietal cells are cut off are dominant, but the Alismataceae (incl. Limnocharitaceae) constantly seem to lack parietal cells. Divisions may occur in the nucellar epidermis which may become irregularly 2-layered at least in the apical part. Similar conditions with a 3-layered nucellar cap formed by the epidermis also occur in the Zosteraceae. In the Triuridales parietal cells are apparently absent, and a nucellar cap is also lacking, in which regard this order differs from most other Alismatiflorae.

In the Ariflorae the ovule condition is very variable, although the Araceae are often said to lack parietal cells it may have a nucellar cap formed by the epidermis. However, a primary parietal cell is cut off from the archesporial cell in species of several genera, including *Anthurium*, *Peltandra* and *Nephthytis*, and this is also the condition in the taxa studied from the Lemnaceae. In *Arisaema*, *Acorus*, and other genera a primary parietal cell is not cut off, but the epidermis by periclinal divisions forms a nucellar cap 2–6 cell layers thick. Rarely, as in *Pistia* and *Synandrospadix* (Coccucci, 1966) the archesporial cell functions directly as the megaspore mother cell and the epidermis cells do not divide either.

The Liliiflorae are extraordinarily variable in respect of these characters. All the members of the Orchidales and Burmanniales lack parietal cells as well as nucellar cap, and within the Liliales a parietal cell is, perhaps, always lacking in the Alstroemeriaceae, Liliaceae, Colchicaceae, Tricyrtidaceae and Calochortaceae (but is present in the Iridaceae and Melanthiaceae). In the Asparagales parietal cell (tissue) is lacking in members of the Dracaenaceae (*Sansevieria*, *Dracaena*, *Dasylirion*, *Nolina*), Phormiaceae (*Phormium*), Hypoxidaceae (*Ianthe*, *Forbesia*, *Pauridia*, *Hypoxis*, *Curculigo*; de Vos, 1956), Dianellaceae (*Dianella*), most genera of the Alliaceae (*Allium*, *Tulbaghia*, *Nothoscordum*), in the Hemerocallidaceae (*Hemerocallis*), in a few genera (*Clintonia*, *Disporum*, *Theropogon*) out of the fairly numerous genera studied in the Convallariaceae, and in *Ruscus* (Asparagaceae). The absence of a parietal cell in *Clintonia* and *Disporum* is combined with differences in a number of other characters, and these genera at least seem to have a better place in or in conjunction with Colchicaceae (Liliales), as pointed out by Björnstad (1970). Also *Trichopus* (Dioscoreaceae) in Dioscoreales is stated to lack parietal tissue as well as a nucellar cap.

While the ovules in only a few of the above-mentioned taxa may have a nucellus cap formed by periclinal divisions of the epidermis, this is more frequently the case in Amaryllidaceae, in which a number of genera have parietal tissue derived from a primary parietal cell. Ovules lacking a parietal cell but having periclinal divisions in the nucellus epidermis are also known to occur in the Dracaenaceae (*Dracaena*, *Sansevieria*), Hypoxidaceae (*Pauridia*), Phormiaceae (*Phormium*) (Cave, 1955), Dianellaceae (*Stypandra*; Cave, 1975) and some genera in the Colchicaceae (*Gloriosa*, Afzelius, 1918; *Iphigenia*).

Otherwise, ovules with parietal cell (tissue) are widely distributed in the Liliiflorae, and are the only type known in the Haemodorales, Philydrales, Pontederiales, Taccales and Bromeliales, although parietal cell as well as periclinal divisions in the nucellus epidermis are lacking in Velloziales. In the Bromeliales the primary parietal cell may form a tissue up to 6 layers in thickness. Ovules with parietal tissue also dominate in the Asparagales, where they are the main or only type in families like the Agavaceae, Funkiaceae, Hyacinthaceae, Asphodelaceae (incl. *Cordyline*) and Convallariaceae. They are in addition recorded in *Aphyllanthes* (Aphyllanthaceae), *Asparagus* (Asparagaceae), *Luzuriaga* (Philesiaceae), *Cyanastrum* (Cyanastraceae), *Cyanella* and *Odontostomum* (Tecophilaeaceae), *Agapanthus* (Alliaceae), *Doryanthes* (Doryanthaceae), and *Lomandra* (Dasypogonaceae). The Amaryllidaceae may lack parietal tissue but have an epidermal nucellus cap, and also frequently have ovules with parietal tissue in which the nucellar epidermis by periclinal divisions may contribute part of the "nucellus cap". In the Liliales, the Iridaceae and Melanthiaceae with none or few exceptions have some parietal tissue formed by a parietal cell, and this is also the case in *Dioscorea*, *Stemona* and *Trillium* of the Dioscoreales.

In the Zingiberiflorae (Zingiberales) a parietal cell is consistently cut off, while there is a tendency for the nucellar epidermal cells not to divide periclinally but to elongate radially.

The Commeliniflorae are a variable complex. In the Commelinales the archesporial cell usually but not always (e.g. not in some taxa of *Cyanotis* and *Commelina*) cuts off a primary parietal cell which forms one or a few layers, in addition the epidermal cells of the nucellus may divide periclinally to form a cap of 2–3 layers. In the Restionales the few members studied seem to lack parietal cell as well as periclinal divisions in the epidermis, but according to Subramanyam and Narayana (1972), a parietal cell is formed in *Flagellaria*. The Eriocaulaceae and Xyridaceae also lack parietal tissue, while in the Rapateaceae parietal tissue is present (Hamann, personal communication). In the grasses (Poales) a parietal cell (or tissue) never seems to occur, although the nucellar epidermis may form a cap. The Typhales, Juncales and Cyperales deviate conspicuously from this pattern in having parietal tissue; the epidermal cells of the nucellus in these families do *not* divide periclinally but many elongate radially, at least in the Juncaceae (a condition similar to that in Zingiberaceae). This radial prolongation of the apical epidermal cells of the nucellus is also characteristic of the Centrolepidaceae and (as far as studied) the Restionaceae, which is of interest in regard to the nucellar cap in the grasses.

In the Areciflorae, at least most members of the Arecaceae and Pandanaceae (Cheah and Stone, 1975) have ovules where a primary parietal cell is cut off and gives rise to 2–6 cell layers, in addition to which the apical nucellar epidermis divides periclinally to contribute a nucellar cap. The condition in the Cyclanthaceae is sometimes described as similar (Davis, 1966) but a parietal cell does not seem to be cut off although a cap is formed by the nucellar epidermis (Harling, 1958; Wunderlich, 1959) except in *Cyclanthus* where even this cap is lacking.

Base Data

Presence of parietal cell (often giving rise to parietal tissue)

ALISMATIFLORAE: HYDRO.: Butom.; Hydro.; Apono.—ZOSTE.: Scheu.; Junca.; Potam.; Zanni. (Posid. and Cymod. ?).—NAJAD.: Najad.

ARIFLORAE: ARAL.: Arac.: many genera; Lemn.

LILIIFLORAE: DIOSC.: Diosc. (except *Trichopus*); Stemo.; Trill.—TACCA.: Tacca.—ASPAR.: Smila.; Phile.; Conva.: mostly; Aspar.: subfam. Asparagoideae; Dorya.; Dasyp.: at least *Lomandra*; Phorm.: *Blandfordia*; Agav. (verified in several genera); Aspho.; Aphyl.; Tecoph.; Cyana.; Funki.; Hyaci.; Alliaceae p.p., e.g. *Agapanthus*; Amary.: most genera (?).—LILIA.: Irid.; Melan.— PONTE.: all.—HAEMO.: all.—PHILY.: all.—BROME.: all.

ZINGIBERIFLORAE: perhaps all.

COMMELINIFLORAE: COMME.: Comme.: most genera.—ERIOC.: Rapat.: perhaps all.—TYPHA.: all.—JUNCA.: all.—CYPER.: all.—RESTI.: Flage.: *Flagellaria*.

ARECIFLORAE: ARECA.: ? all.—PANDA.: probably all.

Dicotyledons associated with the monocotyledons:

These with few exceptions have a parietal cell (and besides are truly crassinucellate).

Ovules mostly lacking parietal cell (tissue) but with periclinal divisions in the epidermis

ALISMATIFLORAE: ALISM.: all.—ZOSTE.: Zoste.

ARIFLORAE: ARAL.: Arac.: known in several genera, e.g. *Acorus*, *Arisaema*, *Typhonodorum*.

LILIIFLORAE: ASPAR.: Conva.: rarely as in *Smilacina*; Dracae.: *Dracaena*, *Sansevieria* (Cave, 1955); Phorm.: *Phormium*; Hypox.: *Pauridia*; Diane.: *Stypandra*;

Amary.: many, e.g. *Cooperia*, *Crinum*, *Eucharis*, *Narcissus*, *Zephyranthes*.—LILIA.: Colch.: *Androcymbium*, *Colchicum*, *Gloriosa*, *Iphigenia*.

COMMELINIFLORAE: COMME.: Comme.: sometimes; as in *Cyanotis* and *Commelina*.—HYDAT.: Hydat. (? parietal cell).—POAL.: mostly.

ARECIFLORAE: CYCLA.: mostly (illustr. in Harling, 1958).

Ovules lacking parietal cell (tissue) as well as periclinal divisions in nucellus epidermis

ALISMATIFLORAE: TRIUR.: probably all.

ARIFLORAE: ARAL.: Arac.: rarely? (at least in *Pistia* and *Synandrospadix*).

LILIIFLORAE: DIOSC.: Diosc.: *Trichopus.*—ASPAR.: Conva.: some genera, e.g. *Clintonia*, *Polygonatum*, *Theropogon*; Aspar.: subfam. Ruscoideae; Dracae.: *Dasylirion* and ? *Nolina*; Hypox.: *Curculigo*, *Forbesia*, *Hypoxis*, *Ianthe*; Diane.: *Dianella*; Hemer.: *Hemerocallis*; Allia.: p.p. at least spp. of *Allium*, *Nothoscordon* and *Tulbaghia*.—LILIA.: Colch. p.p., e.g. *Uvularia*; Alstr.; Tricy.; Caloch.; Lilia.—BURMA.: all? (Corsi.: no information).—ORCHI.: all.—VELLO.: all (Dutt, 1967).

COMMELINIFLORAE: ERIOC.: Xyrid.; Erioc.—RESTI.: Centr.; Resti. (except, perhaps *Hypodiscus*).—POAL.: many general esp. in subfam. Pooideae.

ARECIFLORAE: CYCLA.: *Cyclanthus*.

Dicotyledons associated with the monocotyledons:

MAGNOLIIFLORAE: PIPER.: Sauru.: *Houttuynia*.

EMBRYO SAC FORMATION
(Figs 71–72, Diagram 65)

In regard to the embryo sac formation there are many problems, terminological and others. It also turns out that many (? most) of the early results are so uncertain that they had better not be used as the basis of any conclusions. Thus, the following account is included with hesitation, although some features are of interest.

We have used the terminology of Maheshwari (1950, p. 86) except that the Clintonia subtype in accordance with Björnstad (1970) is treated as its own type. The bisporic Allium type (formerly called Scilla type in a wide sense) is here conceived as two types, the Allium type *sensu stricto*, in which the embryo sac develops from the lower (chalazal) dyad cell, and the Endymion type (Battaglia, 1958) with a functional upper (micropylar) dyad cell.

The Normal (or Polygonum) type of embryo sac is

derived from a single (the chalazal) megaspore and is 8-nucleate. It has the widest distribution of all types in mono- as well as dicotyledons and was probably present in the monocotyledon ancestors. It has been found in the majority of families of monocotyledons, and many cases of aberrant types have been questioned and sometimes turned out to be erroneous, the Normal type being then usually the one found to occur.

In the Alismatiflorae the Normal type occurs in most groups except the Alismatales having the Allium type and *Zannichellia* of the Zannichelliaceae having the Endymion type. In the Ariflorae the Normal type is the dominant (or only?) type in the Araceae. The Liliiflorae are more variable. In the Dioscoreales the Normal type is known in Dioscoreaceae (incl. *Trichopus*) and in *Scoliopus* of the Trilliaceae, which otherwise frequently have the Allium type of embryo sac. The Taccales are known to have the Normal type. In the Asparagales there are certain exceptions from the Normal type, in particular in the Convallariaceae, Hypoxidaceae, Alliaceae and Hyacinthaceae (Fig. 72 U–Z), but the Normal type is known to occur in most families, incl. Asparagaceae (Lazarte and Palser, 1979), Smilacaceae (Hamann and Weik, personal communication), Dianellaceae (Cave, 1955) and Cyanastraceae (Nietsch, 1941). The Liliales generally also have the Normal type of embryo sac except for the Liliaceae, where the Fritillaria type is dominant. In the Burmanniales, *Burmannia* and *Thisma* are known to have the Normal type of embryo sac; this has been confirmed for *Burmannia stuebelii*, an autotrophic species (Hamann and Spitmann, personal communication). The Orchidales have somewhat variable embryo sac formation, and exceptions from the Normal type are known in Cypripediaceae especially, but also in Orchidaceae s.str. (Wirth and Withner, 1959). The Normal type is also known from the Haemodorales (de Vos, 1956), the Velloziales (Menezes, 1973), the Bromeliales, the Ponterderiales and the Philydrales (Hamann, 1966; Fig. 71).

In the Commeliniflorae there are very few records of other types than the Normal, a few being known in the Commelinaceae, however. The Normal type has been found in the previously unknown Rapateaceae (*Spathanthus*, and is very probably present in *Cephalostemon* and *Rapatea*; Tiemann, personal communication), and has been verified for taxa of Xyridaceae (Hamann and E. Kircher, personal communication) and Restionaceae (P. Kircher, personal communication).

The Zingiberiflorae and Areciflorae have predominantly or probably exclusively the Normal type of embryo sac formation, a few exceptions being regarded as dubious in recent reviews.

Also in the dicotyledons associated with the mono-

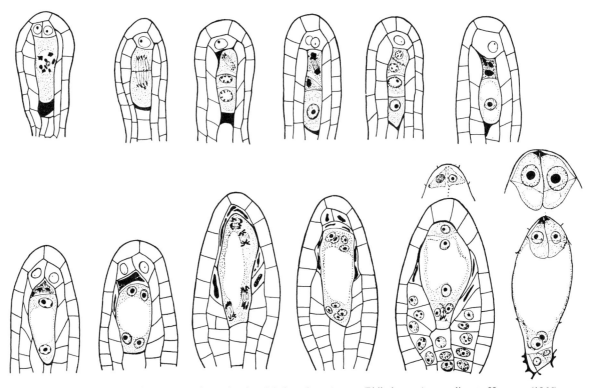

Fig. 71. Normal (Polygonum) type of embryo sac formation in *Philydrum lanuginosum* (Philydraceae) according to Hamann (1966).

cotyledons the embryo sac formation seems to be according to the Normal type, *Peperomia* of the Piperales excluded.

In connection with the Normal type of embryo sac formation should be mentioned that there is great variation in the development of antipodal cells, which may be used for taxonomic purposes. Thus, for example, there is secondary multiplication of antipodals (grasses, members of the Restionales etc.), which seems to be extremely rare in plants with Helobial type of endosperm formation.

The Allium type of embryo sac is known in the Alismatales (= Alismataceae incl. Limnocharitaceae) where it perhaps occurs to the exclusion of other types (Johri, 1935a, b, c, 1938), and in *Zannichellia* of the Zosterales (Vijayaraghavan and Kumari, 1974). In the Ariflorae it is known in the Lemnaceae (*Lemna, Wolffia*), but not in the Araceae. In the Liliiflorae the Allium type occurs in most Trilliaceae (Berg, 1962), some genera of the Convallariaceae, in *Ruscus* of the Asparagaceae, in at least part of the Hypoxidaceae, and at least in *Scilla* and *Crinum* of the Hyacinthaceae and Amaryllidaceae respectively. In the Alliaceae, the Allium type is reported for *Allium* and *Leucocoryne*, while the Normal type occurs in *Nothoscordum, Muilla* (Berg and Maze, 1966) and *Brodiaea* (Berg, 1978). The reports of Normal type in *Allium* seem to refer to apomictic species with nonfunctional embryo sacs (Berg, personal communication). The Allium type is also known in several genera of Orchidales (Swamy, 1949; Wirth and Withner, 1959; Maheshwari, 1955). Besides there are records of this type from *Rhoeo* and *Tradescantia* of the Commelinaceae and *Flagellaria* of the Flagellariaceae (Subramanyam and Narayana, 1972) in the Commeliniflorae.

The Endymion type of embryo sac is more restricted.

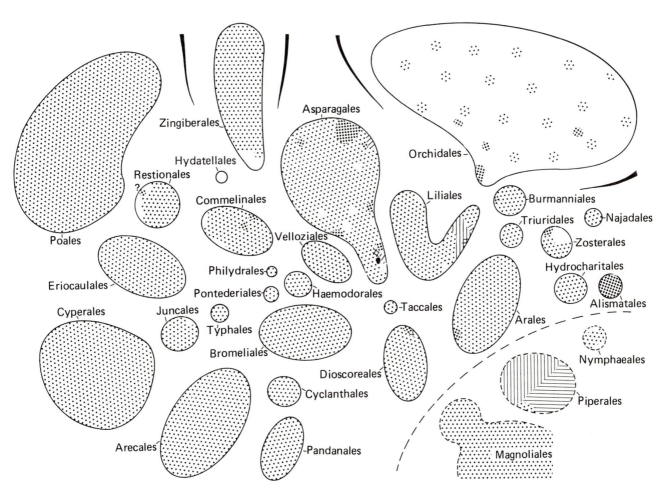

Diagram 65. Embryo sac formation (generalized). Polygonum (Normal) type: sparse dots. Allium type: dense dots. Endymion (Scilla) type: oblique hatching. Fritillaria type: vertical hatching. Clintonia type: black. Adoxa type: checking. Drusa type: oblique checking. Peperomia type: horizontal hatching. Unknown groups are left unshaded.

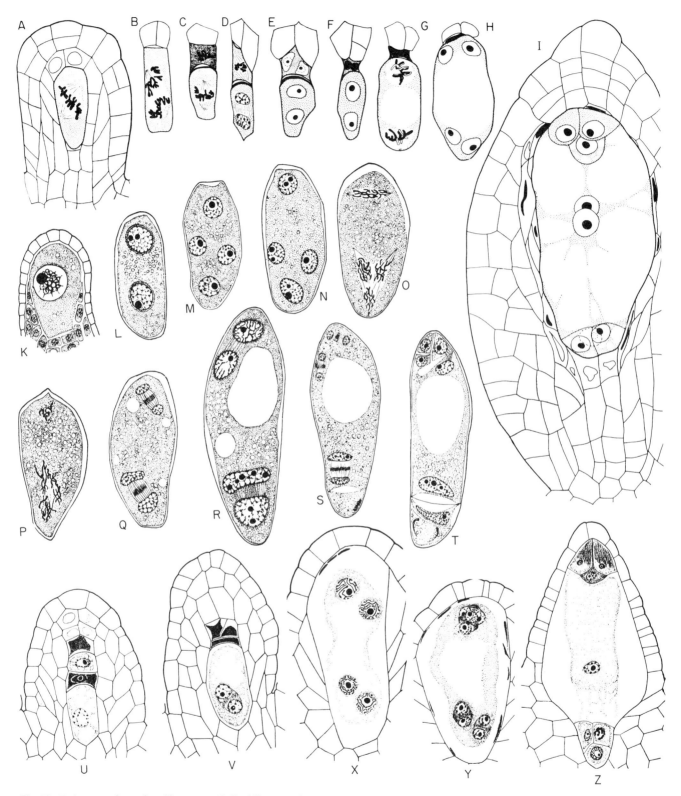

Fig. 72. Embryo sac formation. Upper row (A–I): Allium type in *Paris quadrifolia* (Trilliaceae). Centre (K–T): Fritillaria type in *Fritillaria persica* (Liliaceae), P showing fusion of three chalazal spindles ("endomitosis"). Bottom (U–Z): Polygonum (Normal) type in *Urginea indica*. (A–J from Berg, 1962; K–T from Bambacioni, 1928, U–Z from Capoor, 1937.)

It is known in a few taxa of Convallariaceae and Hyacinthaceae (Asparagales) and in *Agraphis* of Melanthiaceae (Liliales), all taxa being members of the Liliiflorae.

Of the tetrasporic types, **the Fritillaria type** is probably the most conspicuous among the monocotyledons. It is a tetrasporic, 8-nucleate type, with but two divisions following upon meiosis in a very peculiar way, partly accompanied by nuclear fusions. The nuclei of the egg cell and synergids and the upper polar nucleus are haploid; they are all derived from the micropylar megaspore nucleus, while the antipodials and the lower polar nucleus are triploid. The triploid state comes about after the 4-nucleate stage, while when the lower three megaspores divide, their spindles fuse so that two triploid nuclei are formed. The two haploid and two triploid nuclei divide once more to result in the 8-nucleate embryo sac (Bambacioni, 1928). The Fritillaria type is known mainly in the Liliaceae (Liliales, Fig. 72 K–T) where it seems to be the dominant type being known from several genera (Stenar, 1927; see also below).

In **the Clintonia type** (Björnstad, 1970), previously known as a subtype of the Fritillaria type (Maheshwari, 1946), the embryo sac is tetrasporic; after the 4-nucleate stage (meiosis), the three chalazal megaspores degenerate, while the upper one divides twice to form the nuclei of the egg cell, two synergids and a single polar nucleus, while antipodials and a second polar nucleus are not developed. This type is known in *Clintonia* of the Convallariaceae (Asparagales) and *Tulipa* of Liliaceae (Liliales).

The Adoxa type embryo sac is a regular, tetrasporic, 8-nucleate type formed after only one division of the megaspore nuclei. Its occurrence was previously thought to be scattered fairly widely in mono- and dicotyledons, but most cases upon critical reviews seem to be erroneous (Fagerlind, 1939; Maheshwari, 1946). This type may in fact be restricted among the monocotyledons to odd cases of the Liliaceae (*Erythronium, Tulipa*); a few other cases remain as dubious but not impossible.

The Drusa type is a tetrasporic, 16-nucleate embryo sac, where the megaspores undergo two divisions. Except for an egg cell, two synergids and two polar nuclei all other nuclei take part in antipodal and lateral cells. The Drusa type is known in *Smilacina* and *Maianthemum* of Convallariaceae (Asparagales) and in *Tulipa* (Liliaceae).

The many deviating types of embryo sac formation reported for *Tulipa* (Romanov, 1959) according to Berg (personal communication) may depend on the fact that apomictic taxa or non-fertile cultivars were studied, and that taxonomically, therefore, they are of no value.

Finally, **the Peperomia type** embryo sac should be mentioned—another tetrasporic, 16-nucleate type, with egg apparatus and nuclei usually fusing to form the secondary nucleus, the others becoming distributed in cells along the periphery of the embryo sac. This type is known in *Peperomia* of the Piperaceae (Piperales) among the dicotyledons closely associated with monocotyledons.

The embryo sac formation is of restricted taxonomic importance in monocotyledons. The Normal type is the dominant, but the bisporic Allium type is found in the Alismatales, where it is indeed a significant feature, in most Trilliaceae of the Dioscoreales (Berg, 1962; Fig. 72 A–I) and in some other groups, for example in certain Asparagales, where some phylogenetic importance may be attached to it. In the Lemnaceae the Normal type is present in *Spirodela*, but the Allium type in *Lemna* and *Wolffia*, which indicates an evolutionary trend comparable to the anther wall formation (p. 137).

Other types of embryo sac formation may be regarded more or less sporadic aberrations except that the Fritillaria type is the dominant in Liliaceae s.str. (Liliales), and the Peperomia type is characteristic to the genus *Peperomia* in Piperaceae (Piperales) the latter case being among the dicotyledons associated with monocotyledons.

Tetrasporic embryo sacs in monocotyledons seem to occur most frequently in Liliaceae and Convallariaceae.

Base Data

Polygonum (Normal) Type

ALISMATIFLORAE: HYDRO.: Butom.; Hydro.; Apono.—ZOSTE.: Scheu.; Junca.; Potam.; Zoste.—NAJAD.: Najad.—TRIUR.: Triur.

ARIFLORAE: ARAL.: Arac. (other types are recorded, but found doubtful or erroneous by Davis, 1966); Lemna.: rarely, as in *Spirodela*.

LILIIFLORAE: DIOSC.: Diosc. (incl. *Trichopus*); Stemo.: *Stemona*; Trill.: (*Scoliopus*).—TACCA.: *Tacca*.—ASPAR.: basically Polygonum type, but exceptions (in Conva., Hypox., Alliac., Amary. etc., see below); information available for taxa of all families, except Peter., Herre.; Dasyp.; Xanth., Eriosp.—LILIA.: Irid.; Colch. (Cave, 1941); Alstr.; Caloc.; Melan.—BURMA.: Burma.; Thism. (Corsi.: no information).—ORCHI.: Orchi.: probably in most genera.—PONTE., HAEMO., PHILY., VELLO., and BROME.: all taxa studied.

ZINGIBERIFLORAE: ZINGI.: (Lowia., Helic.: no information;) Musac.; Strel.; Zingi.; Costa.; Canna.; Maran.

COMMELINIFLORAE: COMME.: Comme.: mostly (exceptions see below); (Mayac. and Carto.: no information).—ERIOC.: Rapat.: *Cephalostemon, Rapatea, Spathanthus* (Tiemann, unpubl.); Xyrid.: *Abolboda, Xyris*

(possibly all); Erioc.—TYPHA.: all.—JUNCA.: Junca.
—CYPER.: all.—(HYDAT.: no information.)—RESTI.:
Centro.; Resti.—POAL.: Poac.

ARECIFLORAE: ARECA.: probably all (possible
exceptions, see below).—PANDA.: probably all.—
CYCLA.: all (Harling, 1958).

Dicotyledons associated with the monocotyledons:
NYMPHAEIFLORAE: NYMPH.: Nymph. (Cabom.
and Cerat.: no information).

MAGNOLIIFLORAE: MAGNO.: most taxa.—
PIPER.: Sauru. (at least some taxa).

Allium Type (mainly from S. C. Maheshwari, 1955).

ALISMATIFLORAE: Alism.: probably all, e.g. in
*Alisma, Butomopsis, Damasonium, Echinodorus, Elisma,
Hydrocleys, Limnocharis, Limnophytum, Machaero-
carpus, Sagittaria*).—ZOSTE.: Zanni.: *Zannichellia*.

ARIFLORAE: ARAL.: Lemn.: *Lemna, Wolffia*.

LILIIFLORAE: DIOSC.: Trill.: *Paris, Trillium*
(mostly).—ASPAR.; Conva.: *Convallaria, Polygonatum,
Streptopus* (?); Aspar.: subfam. Ruscoideae; Hypox.:
Forbesia, Hypoxis; Hyaci.: *Scilla* spp.; Allia.: at least
Allium (many spp.), *Leucocoryne, Tulbaghia*; Amary.:
Crinum sp.—BURMA.: Burma.: *Burmannia* sp.—
ORCHI.: Cypri.: *Cypripedium, Paphiopedilum*; Orchi.:
*Achroanthes, Corallorhiza, Cymbidium, Listera, Malaxis,
Neottia*.

COMMELINIFLORAE: COMME.: Comme.: *Rhoeo,
Tradescantia*.—RESTI.: Flage.: *Flagellaria*.

Dubious records are from *Potamogeton, Najas, Hydro-
mystris, Commelinia*, a few grass genera, *Pandanus*, and
some palm genera, but these were not found probable
by Maheshwari and are not included in the diagram.
Also *Xyris*, which was accepted by Maheshwari as probable, is dubious and has not been verified in recent studies
(Hamann and E. Kircher, personal communication).

Endymion (Scilla) Type

LILIIFLORAE: ASPAR.: Conva.: *Smilacina*; Hyaci.:
Scilla spp.—LILIA.: Melan.: *Agraphis*.

Fritillaria Type

LILIIFLORAE: LILIA.: several genera, e.g. *Cardiocrinum, Erythronium, Fritillaria, Gagea, Lilium, Lloydia,
Medeola* and mostly in *Tulipa* (also, for example, Drusa
and Adoxa Types).

Dicotyledons associated with the monocotyledons:
MAGNOLIIFLORAE: PIPER.: Piper.: at least *Piper*
spp.

Clintonia Type

LILIIFLORAE: ASPAR.: Conva.: *Clintonia* (Björnstad, 1970).—LILIA.: Lilia.: *Tulipa* (rarely).

Adoxa Type

As pointed out by Fagerlind (1939) and Maheshwari
(1946) nearly all of the records for the Adoxa type of
embryo sac are probably erroneous with the exception
of the following:
LILIIFLORAE: LILIA.: Lilia.: *Erythronium albidum,
Tulipa* (rarely).

Drusa Type

LILIIFLORAE: ASPAR.: Conva.: *Maianthemum,
Smilacina*.—LILIA.: Lilia.: *Tulipa* (rarely).

Peperomia Type

Dicotyledons associated with the monocotyledons:
MAGNOLIIFLORAE: PIPER.: Piper.: *Peperomia*.

ENDOSPERM FORMATION
(with contributions from U. Hamann)
(Figs 73 and 74, Diagram 66)

The distributions of different types of endosperm
formation and other embryological characters have long
been acknowledged as valuable taxonomic characters
although there are only few comprehensive works in
which they have been summarized, the most important
being those of Schnarf (1931), Wunderlich (1959), Swamy
and Parameswaran (1963), Davis (1966) and Swamy and
Krishnamurthy (1974). In but a few works is this character considered for the main division of angiosperms (e.g.
Dahlgren, 1975b; and for monocotyledons especially
Huber, 1969, 1977; Dahlgren, 1976). The taxa recorded
by Cave (1955) as *probably* having helobial endosperm
formation are not included here.

There are three basic types of endosperm formation:

(a) **Cellular**, in which the division of the primary endosperm nucleus is promptly followed by the formation of
a cell wall and also subsequent divisions are directly
followed by the formation of successively new cell walls;

(b) **Nuclear**, where the division of the primary endosperm nucleus and at least a number of subsequent
divisions are not initially accompanied by the formation
of cell walls;

(c) **Helobial**, the division of the primary endosperm
nucleus is followed by the formation of a transverse
cell wall, separating a larger micropylar and a smaller

chalazal chamber; free nuclear divisions follow for some time at least in the micropylar chamber, and usually more abundantly in this than in the chalazal chamber.

The data summarized in Diagram 66 are mainly taken from the compilatory works mentioned above supplemented with various articles of later date.

In contrast to dicotyledons many monocotyledons have helobial endosperm formation, a type which may have developed early in the monocotyledon ancestors (from the cellular type) and received a considerable distribution among them. The cellular endosperm development, which is fairly widely distributed in dicotyledons is rare in monocotyledons, where it occurs mainly in the Arales, but also in the small, very aberrant order Hydatellales (Hamann, 1975, 1976) and perhaps in a single species of *Thismia* (Thismiaceae), Burmanniales (Goebel and Süssenguth, 1924).

It is significant that in most orchids there is no endo-sperm formation at all. Here the central nucleus (or polar nuclei) of the embro sac may or may not fuse with the second sperm nucleus. There is generally no division of this product if it is formed or if so (in some genera of orchids) it may divide once, twice or rarely up to four times (Fig. 74), in which case no walls are formed (and endosperm formation accordingly is then nuclear).

The remaining monocotyledons either form an endo-sperm according to the helobial or the nuclear type, and it is thus crucial to discover whether or not a wall is formed after the first nuclear division. The wall is mostly formed in the lowermost (antipodal) part of the embryo sac, and the cells of the chalazal chamber can easily be confused with (multiplied) antipodal cells. Therefore helobial types have sometimes been described as nuclear (as recently discovered for both families of the Typhales by Asplund, 1968, 1972 and 1973; U. Müller-Doblies, 1968; D. Müller-Doblies, 1969). The same or reverse mis-

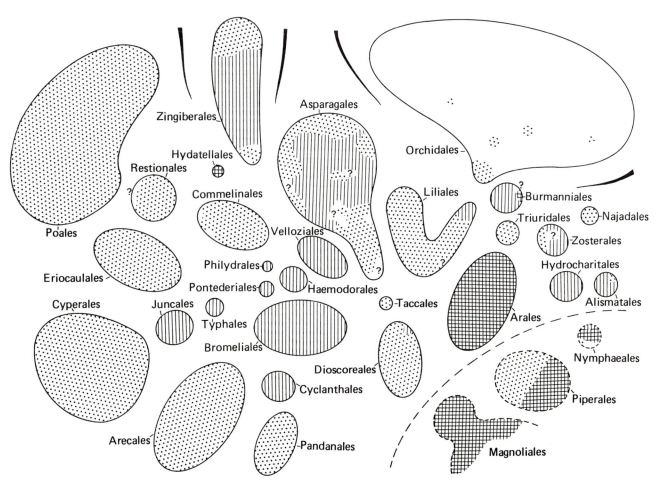

Diagram 66. Endosperm formation. Endosperm formation lacking or arrested at a binucleate stage: unshaded. Endosperm formation helobial: vertical hatching. Endosperm formation nuclear: sparse dots. Endosperm formation cellular: checking. Groups for which information is lacking are left unshaded and with question marks.

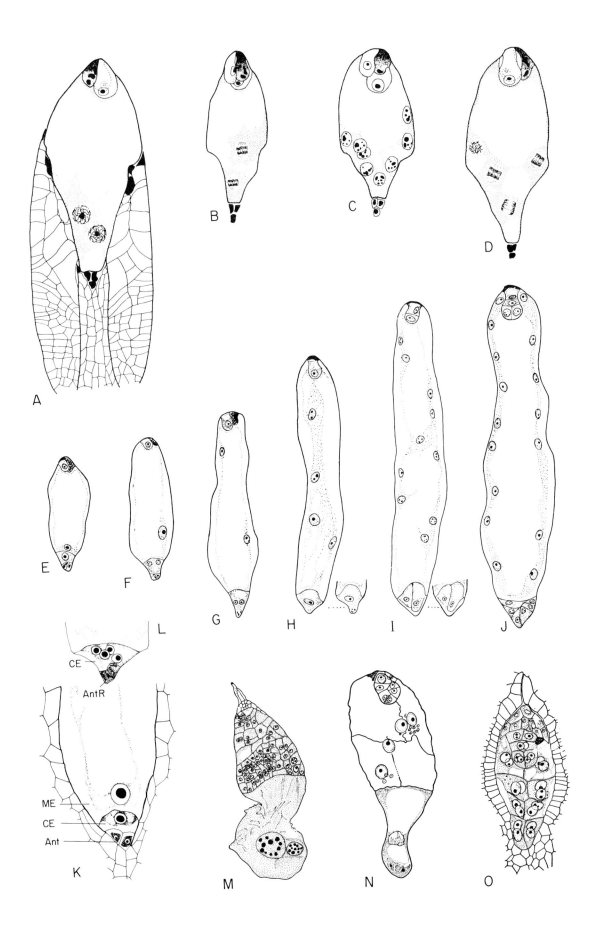

A

B

C

D

E

F

G

H

I

J

L

CE

AntR

K

ME

CE

Ant

M

N

O

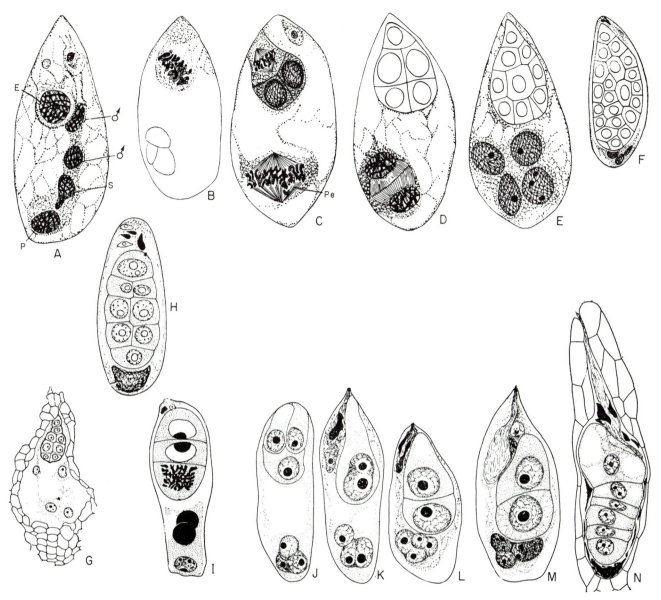

Fig. 74. Presence/absence of endosperm in Orchidales. A–F: *Cypripedium* (Cypripediaceae), development of embryo and endosperm after fertilization (A), E, the maximal 4-nucleate endosperm (P = polar nucleus, E = egg cell nucleus, Pe = primary endosperm nucleus dividing). G: *Vanilla planifolia* (Orchidaceae) showing a nuclear endosperm with several nuclei. H: *Orchis maculata* (Orchidaceae), embryo sac with embryo and a degenerating primary endosperm nucleus. I: *Spiranthes australis* (Orchidaceae), three-celled proembyro and degenerating polar nuclei (there is no endosperm formation at all!). J–L: *Corallorhiza maculata* (Orchidaceae), showing fertilization (in K) and the formation of a polyploid endosperm nucleus (in which antipodal nuclei take part). M–N: *Bletia shepherdi* (Orchidaceae) with degenerating endosperm nucleus in two successive stages. (A–F from Pace, 1909; G from Swamy, 1947; H from Hagerup, 1944; I from Maheshwari and Narayanaswami, 1952; J–N from Sharp, 1912.)

Fig. 73. Types of endosperm formation. A–D: nuclear endosperm formation in *Scoliopus bigelovii* (Trilliaceae). E–J: helobial endosperm formation (a fairly untypical variant, see p. 200) in *Philydrum lanuginosum* (Philydraceae). K–L: helobial endosperm formation, in *Sparganium erectum* (Sparganiaceae), only the chalazal chamber seen, L at a later stage than K. M: cellular endosperm in *Arisaema wallichianum* (Araceae) with a large chalazal haustorium corresponding to the chalazal chamber in the helobial type. N: cellular endosperm in *Spirodela polyrhiza* (Lemnaceae) with a similar haustorium. O: cellular endosperm in *Lemna paucicostata* (Lemnaceae), which lacks such a haustorium. (A–D from Berg, 1959; E–J after Hamann, 1966; K–L after M. Müller-Doblies, 1969; M after Maheshwari and Khanna, 1956; N after Maheshwari and Maheshwari, 1963; O after Maheshwari and Kapil, 1963.)

interpretation may apply to other taxa, and Davis (1966) warns against these sources of error.

There exist several subtypes of helobial endosperm formation. Within the Alismatiflorae the smaller chalazal chamber usually remains undivided and its nucleus becomes polyploid. In other cases the chalazal chamber becomes multinucleate and ultimately multicellular or at least pluricellular. In a rare variation (shown in Fig. 73 E–L). The chalazal chamber becomes cellular earlier than the micropylar chamber, a condition that according to Swamy *et al.* (Swamy and Parameswaran, 1963; Swamy and Krishmamurthy, 1974) occurs in dicotyledons only, but in fact does occur in particular groups of monocotyledons, viz. the Philydraceae (Hamann, 1966a), the Bromeliaceae (Hamann, personal communication; Lakshmanan, 1967), the Velloziaceae (Hamann, personal communication, based on photographs in Menezes, 1976), the Typhaceae (Asplund, 1972), the Sparganiaceae (U. Müller-Doblies, 1969; Asplund, 1973) and the Thismiaceae (*Thismia javanica*; Bernard and Ernst, 1909) (these data supplied by Hamann).

> There have been some confusions between helobial endosperm with early cellular chalazal chamber and multiplied or enlarged antipodal cells. In Wettstein's system (4th ed., 1935), multiplied antipodal cells are a characteristic feature of his order Pandanales (which included the Typhales as here conceived); but such antipodals neither occur in the Typhales (where these cells were confused with the endosperm cells of the chalazal chamber), nor in *Pandanus* of the Pandanaceae (where this tissue originates from nucellus nuclei which have invaded the chalazal part of the embryo sac; Fagerlind, 1940; Cheah and Stone, 1975).
>
> As a contrast to this it may be mentioned that the enlarged antipodal cells in the Eriocaulaceae have sometimes been confused with the chalazal part of a helobial endosperm.

The following are the main features in the distribution of endosperm formation.

Cellular endosperm formation is restricted to the Arales and Hydatellales and perhaps to a single species of *Thismia* (Burmanniales) (see below). This must not, however, be taken to indicate a close affinity between the three groups. In the Araceae (Arales, Ariflorae) the cellular endosperm is generally somewhat atypical (Maheshwari and Khanna, 1963) and different from that in most dicotyledons, the basal part of the endosperm (corresponding to the chalazal chamber in the helobial type) being generally developed as a large, mostly uninucleate haustorium (Fig. 73M). In most Lemnaceae (Maheshwari and Kapil, 1963) (*Spirodela*, Fig. 73N, somewhat transitionally; Maheshwari and Maheshwari, 1963) and some Araceae, at least in *Arisaema* and *Acorus*, the conditions are different, however, the endosperm

being of a typical cellular type without this giant uninucleate cell (Fig. 73O). It may be seriously doubted that the cellular endosperm in monocotyledons is the primary state, whereas this is probably the case in dicotyledons (Wunderlich, 1959). The type usually found in Araceae is more likely to have evolved from the helobial.

In the Alismatiflorae, endosperm formation is generally helobial (and has derived its name from the older name, Helobiae or Helobiales, for this group). However, the nuclear type has been reported to occur in the Juncaginaceae (*Triglochin* as well as *Lilaea*) and the Zosteraceae, Cymodoceaceae and part of Alismataceae. These records deserve re-examination for the reasons mentioned above. While Najadaceae (*Najas*) is sometimes reported to have helobial endosperm formation (Schnarf, 1931) further observations have shown that it is nuclear.

The Liliiflorae are extremely variable in endosperm formation with about equal dominance of nuclear and helobial types (Stenar, 1950). The nuclear type is recorded in the Taccales and prevails in the Dioscoreales (except in *Trillium*) (Fig. 73A–D) and the Liliales, where however the Melanthiaceae deviate conspicuously in having, as far as known, exclusively the helobial type (Cave, 1955). Those few orchids, which have endosperm at all (see below), show the nuclear type of endosperm formation. Asparagales is a most variable order in this respect. The nuclear endosperm formation is especially common in the Smilacaceae, Convallariaceae, and Asparagaceae (Robbins and Borthwick, 1925); it also occurs in, for example, the small families Cyanastraceae, Tecophilaeaceae and Hemerocallidaceae and in *Nolina* (Dracaenaceae), *Furcraea* (Agavaceae), *Ianthe* and *Pauridia* (Hypoxidaceae), many genera of the Amaryllidaceae, at least *Allium* and *Brodiaea* (Alliaceae) and about a third of the genera studied in the Hyacinthaceae (Buchner, 1948).

The Velloziales (see above), Bromeliales (Lakshmanan, 1967; Hamann, personal communication), Philydrales (Hamann, 1966a and b), Haemodorales (de Vos, 1956) and Pontederiales all have the helobial type of endosperm formation, the first three orders mentioned of a variation with early cellular chalazal chamber (Fig. 73E–J). Helobial endosperm formation is also no doubt found in most Burmanniales, although *Thismia luetzelburgii*, according to Goebel and Süssenguth (1924) was reported to have the cellular type.

In the Zingiberiflorae (Zingiberales), the Zingiberaceae and Costaceae have helobial endosperm formation (Panchaksharappa, 1962, 1966), while all taxa in the other families studied have the nuclear type.

The Commeliniflorae mostly have the nuclear type of endosperm formation, also the previously little known

Eriocaulaceae and Rapateaceae (Tiemann and Hamann, personal communication) and (except, most likely, *Abolboda*) the Xyridaceae (Tiemann, personal communication) with the important exceptions of the Juncales and Typhales, both (like *Abolboda*) with helobial endosperm formation (Asplund, 1968; U. Müller-Doblies, 1969). Older records for Typhales indicated that it possessed the nuclear type endosperm formation, but this resulted from a misinterpretation of the multicellular chalazal chamber as "multiplied antipodes". An exceptional group is the Hydatellales, which like the Ariflorae has cellular endosperm formation (see Hamann, 1975, 1976).

The Areciflorae have nuclear endosperm formation with the exception only of the Cyclanthales which have the helobial type (Harling, 1958).

A strong isolation of the Ariflorae is demonstrable in this character. The mode of endosperm formation does not show the great homogeneity for the Commeliniflorae otherwise so frequently displayed; nor are the Liliiflorae homogeneous. The condition is rather that the transition orders Velloziales, Bromeliales, Pontederiales, Philydrales, Typhales and Juncales consistently have some kind of helobial endosperm formation, while the most "typical" Liliiflorae (Liliales and Asparagales) are heterogeneous with respect to both types. It is also noteworthy that the Zingiberaceae and Costaceae differ from other Zingiberales in their helobial endosperm. Surprisingly, moreover, the Juncales and Cyperales differ from each other in this character, whereas otherwise they tend to agree in so many features (diffuse centromeres of chromosomes, pollen tetrads, common smut and rust genera etc.). Another marked difference is that between the Cyclanthaceae and Arecaceae.

In the dicotyledons here associated with the monocotyledons, the Nymphaeiflorae (Nymphaeales) is variable in endosperm formation. The Cabombaceae has helobial endosperm formation, which is otherwise rare in dicotyledons. Considered together with the numerous other similarities between the Cabombaceae and the Hydrocharitales-Alismatales complex, this might have some relevance in evolutionary discussions. Otherwise the cellular type is dominant in the Nymphaeales, and this type is also found, for example, in the Saururaceae and in *Peperomia* of the Piperaceae (Piperales), and in many families of Magnoliales (Bhandari, 1971), in which groups the cellular endosperm formation is believed by Wunderlich (1959) to represent a primitive state, a viewpoint we share.

Base Data

Nuclear endosperm formation

ALISMATIFLORAE: ALISM.: At least *Alisma, Damasonium, Luronium, Machaerocarpus.*—ZOSTE.: Junca.: *Lilaea, Triglochin*; Zoste.: *Zostera*; Cymod.: *Cymodocea.*—NAJAD.: Najad.: *Najas.*—TRIUR.: Triur.: *Sciaphila.*

LILIIFLORAE: DIOSC.: Diosc.: *Dioscorea, Trichopus*; Stemo.: *Stemona*; Trill.: *Paris, Scoliopus.*—TACCA.: Tacca.: *Tacca.*—ASPAR.: Smila.: *Smilax*; Conva.: *Clintonia, Liriope, Maianthemum, Polygonatum, Rhodea, Smilacina*; Aspar.: *Asparagus*; Dracae.: *Nolina*; Agava.: rarely, e.g. *Furcraea*; Hypox.: rarely, *Ianthe* p.p., *Pauridia*, Tecoph.: *Cyanella, Lanaria, Odontostomum?*; Cyana.: *Cyanastrum*; Hemer.: *Hemerocallis*; Hyaci. (Buchner, 1948: Stenar, 1950; less frequently than helobial): *Camassia, Hyacinthus, Scilla, Urginea*; Allia.: *Allium, Brodiaea*; Amary.: frequently, e.g. *Amaryllis, Calostemma, Crinum, Galanthus, Hymenocallis, Leucojum, Narcissus, Nerine, Phaedranassa.*—LILIA.: Irida.: several genera; Colch.: *Androcymbium, Colchicum, Gloriosa, Iphigenia*; Alstr.: *Alstroemeria, Bomarea*; Tricy.: *Tricyrtis*; Lilia.: all?, e.g. *Erythronium, Gagea, Fritillaria, Lilium, Medeola, Lloydia, Tulipa.*—ORCHI.: Cypri.; Orchi.: those very few forming endosperm at all. (Number of nuclei found in Orchidaceous endosperms: 16 in *Galeola*; 10–12 in *Vanilla*; 8 in *Bletilla*; "a few" in *Cephalanthera*; 4 in *Lecanorchis* and *Polystachya*; 2–4 in *Spathiglossis* and 2 in *Chamaeorchis*, *Limodorum* and *Pogonia*; in Cypripediaceae the endosperm normally becomes maximally 4-nucleate; Veyret, 1974; Prakash and Lee-Lee, 1973.)

ZINGIBERIFLORAE: ZINGI.: Musac.: *Musa*; Helic.; Canna.; Maran.

COMMELINIFLORAE: COMME.: Comme.: several genera.—ERIOC.: Rapat.: *Cephalostemon, Rapatea, Spathanthus* (Tiemann and Hamann, personal communication); Xyrid.: *Xyris*; Erioc.: *Eriocaulon, Leiothrix.*—CYPER.: Cyper.: *Carex, Cyperus, Fimbristylis.*—RESTI.: Resti.: *Chondropetalum, Hypodiscus, Leptocarpus, Restio, Thamnochortus*; Centr.: *Aphelia, Brizula, Centrolepis, Gaimardia* (Hamann, 1975).—POAL.: Poac.: numerous genera.

ARECIFLORAE: ARECA.: Areca.: probably all genera.—PANDA.: Panda.: probably all.

Dicotyledons associated with the monocotyledons:

NYMPHAEIFLORAE: NYMPH.: Nymph.: *Euryale.*

MAGNOLIIFLORAE: MAGNO.: Myris.—PIPER.: Piper.: *Piper.*

Helobial endosperm formation

ALISMATIFLORAE: HYDRO.: Butom.: *Butomus*; Apono.: *Aponogeton*; Hydro.: several genera.—ALISM.: Alism.: most taxa, e.g. *Butomopsis, Echinodorus, Limnocharis, Limnophyton, Sagittaria.*—ZOSTE.: Scheu.: *Scheuchzeria*; Potam.: *Potamogeton, Ruppia*; Zanni.: *Zannichellia.*

LILIIFLORAE: DIOSC.: Trill.: *Trillium.*—ASPAR.: Dorya.: *Doryanthes*?; Phorm.: *Blandfordia, Phormium* (Di Fulvio and Cave, 1964); Agava.: *Agave, Beschorneria, Yucca*; Hypox.: mostly, *Forbesia, Hypoxis, Ianthe* p.p.; Aspho.: subfam. Asphodelioideae: *Aloë, Asphodeline, Asphodelus, Bulbine, Caesia, Eremurus, Gasteria, Kniphofia*; subfam. Anthericoideae: *Anthericum, Arthropodium, Chlorophytum, Paradisia*; Aphyl.: *Aphyllanthes*; Diane.: *Dianella, Stypandra* (Cave, 1975); Funki.: *Hosta*; Hyaci.: known in at least 12 genera (Stenar, 1950; Cave, 1974) including *Scilla* spp., *Bowiea, Chlorogalum* and *Schoenolirion*; Allia. p.p.: "*Brodiaea*" (? = *Triteleia*), *Nothoscordum, Tulbaghia*; Amary. p.p.: *Clivia, Cooperia, Cyrtanthus* (*Vallota*), *Haemanthus, Hippeastrum, Pancratium, Zephyranthes.*—LILIA.: Melan.: recorded for at least 7 genera (incl. *Aletris*).—BURMA.: Burma.: *Burmannia* (Arekal and Ramaswamy, 1973), *Dictyostegia, Gymnosiphon, Hexapterella* (Rübsamen and Hamann, personal communication); Thism.: *Thismia* spp.—PONTE.: Ponte.: known in several genera.—HAEMO.: Haemo.: probably all, e.g. *Dilatris, Wachendorfia.*—PHILY.: Phily.: all genera (Hamann, 1966).—VELLO.: Vello.: several genera (see above).—BROME.: Brome.: maybe all genera, incl. *Aechmea, Ananas, Dyckia, Lindmania, Pitcairnia, Tillandsia, Vriesea* (largely from Hamann, personal communication).

ZINGIBERIFLORAE: ZINGI.: Zingi.: recorded in at least 6 genera; Cost.: *Costus.*

COMMELINIFLORAE: ERIOC.: Xyrid.: *Abolboda* (very probably; Tiemann, personal communication).—TYPH.: Spar.: *Sparganium*; Typh.: *Typha.*—JUNCA.: Junca.: maybe all, incl. *Distichia, Juncus, Luzula.*

ARECIFLORAE: CYCLA.: Cycla.: recorded for several genera.

Dicotyledons associated with the monocotyledons: NYMPHAEIFLORAE: NYMPH.: Cabom.

Cellular endosperm formation

LILIIFLORAE: BURMA.: Thism.: (?) *Thismia* (needs confirmation).

ARIFLORAE: ARAL.: Arac. (numerous genera; cases recorded otherwise thought to be dubious by Wunderlich, 1959); Lemn.

COMMELINIFLORAE: HYDAT.: Hydat.: *Hydatella* (Hamann, 1975).

Dicotyledons associated with the monocotyledons: NYMPHAEIFLORAE: NYMPH.: Nymph.: at least *Nymphaea*; Cerat.: *Ceratophyllum.*

MAGNOLIIFLORAE: MAGNO.: nearly all members studied incl. genera of Magno., Annon., Arist., Wint. etc.—PIPER.: Sauru.; Piper.: *Heckeria, Peperomia.*

EMBRYOGENY (Fig. 75, Diagram 67)

The types of embryo formation (embryogeny) are defined according to the first few planes of division of the zygote.

The least common condition is for the first division of the zygote to be longitudinal. Embryogenesis of this kind is known as the Piperad type and in the families considered here occurs only in the Piperaceae.

Transverse division of the zygote produces a basal and terminal cell. Further classification of embryo types depends upon whether the terminal cell divides transversely or longitudinally.

Embryos in which a transverse division of the apical cell occurs may be grouped into 3 main types.

The Caryophyllad type. In this the basal cell does not divide further but enlarges to form a vesicular suspensor cell. The terminal cell forms the embryo and a variably long neck.

As shown in Diagram 67 this type occurs in practically all families of all orders of the Alismatiflorae: viz. Alismataceae (Johri, 1935a, b and c, 1938, Aponogetonaceae, Butomaceae, Hydrocharitaceae, Scheuchzeriaceae, Juncaginaceae (Campbell, 1898), Potamogetonaceae (incl. Ruppiaceae; Yamashita, 1972), Zosteraceae, Zannichelliaceae and Najadaceae (Fig. 75). The remarkable constancy here is only rarely mentioned in the literature and should indeed deserve more attention.

In addition, this type is found in the Araceae (except at least *Pistia* and *Synandrospadix*) but not in Lemnaceae, the other family of the Ariflorae. It also occurs, though rarely, in other groups of monocotyledons, viz. in species of *Yucca* (Agavaceae, Asparagales), *Dipcadi* (Hyacinthaceae, Asparagales), *Crocus* (Iridaceae, Liliales), and probably in some members of the Zingiberaceae, Zingiberales.

In the dicotyledons it is scattered, but has been found in at least four families of Caryophyllales.

The similarity in embryogenesis between the Alismatiflorae and Arales was pointed out by von Guttenberg (1960) and, supplemented with other concordant characters, can be used as an argument for a close relationship between the groups.

The Solanad type is similar to the Caryophyllad type, but here the basal cell divides without taking part in the embryo.

It is notable that this is not recorded with certainty in the monocotyledons although suspected to occur in *Trichopus* (Dioscoreaceae, Dioscoreales), occurring also in Aristolochiaceae among the dicotyledons with several monocotyledonous features.

Similar is **the Chenopodiad type**, but here the basal cell divides and takes part in the formation of the embryo. This is reported for a few disparate monocotyledons, but further verification is sometimes needed: *Calochortus* (Calochortaceae, Liliales), *Asphodelus* (Asphodelaceae, Asphodelales), *Dipcadi* (Hyacinthaceae, Asparagales), and some Marantaceae.

Embryos in which a longitudinal division of the terminal cell takes place may be subdivided into two types.

In **the Onagrad type** the basal cell plays an unimportant part in the formation of the embryo while in **the Asterad type** the basal cell takes part substantially in the formation of embryo.

The Asterad and Onagrad types are by far the most dominant types in monocotyledons and in the Asparagales and Liliales seem to occur rather irregularly, sometimes side by side in the same families (such as in Amaryllidaceae and Alliaceae). Thismiaceae seems to have the Onagrad type of embryogenesis, the embryo being quite small. In the Burmanniaceae the embryo consists of but 4–10 cells! Similar to the embryo of the Burmanniales is that of the Orchidales, although the seed is wholly devoid of endosperm. It is generally formed either according to the Asterad or the Onagrad types. In the former, an often extensive suspensor is normally formed by the basal daughter cell of the "basal cell". As seen in Diagram 67, the Asterad type seems to be con-

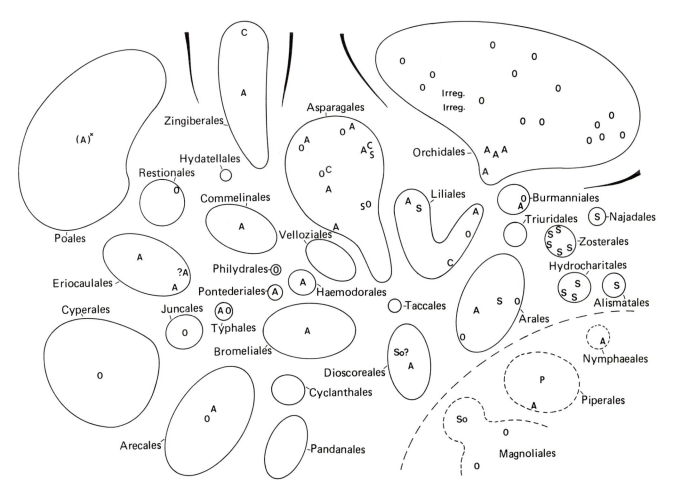

Diagram 67. Embryogenesis. Caryophyllad type: S; Solanad type: SO; Asterad type: A; Onagrad type: O. Each family is given one symbol except families where more than one type has been recorded (Agavaceae, Alliaceae, Amaryllidaceae, Araceae, Arecaceae, Asphodelaceae, Hyacinthaceae, Iridaceae, Thismiaceae) where the different types are given.

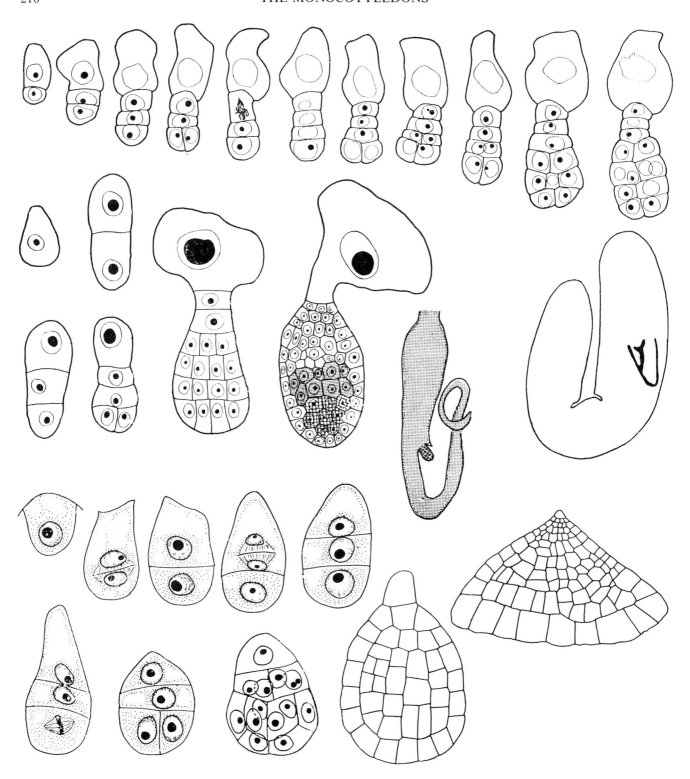

Fig. 75. Embryo formation. Upper row: embryo formation according to the Caryophyllad type, shown by *Sagittaria guayanensis* (Alismataceae), the curved embryo seen to the right. Centre: the same type of embryo formation in *Zannichellia palustris* (Zannichelliaceae). Bottom rows: showing the Onagrad type of embryo formation in *Centrolepis fascicularis* (Centrolepidaceae). (*Sagittaria* from Johri, 1935b; *Zannichellia* from Vijayaraghavan and Kumari, 1974; *Centrolepis* from Prakash, 1969.)

centrated in the Cypripediaceae and in Orchidaceae subfam. Neottioideae, but the Onagrad type in Orchidaceae subfam. Orchidoideae and Epidendroideae (Swamy, 1942). In the Commeliniflorae the Onagrad type is found in the probably related Cyperales, Juncales and Typhales (Sparganiaceae) and a similar modification also in Centrolepidaceae (Hamann, 1962a), while modifications of the Asterad type occur in most other groups.

A striking feature is the early dermatogen formation in the quadrant stage of the apical cell derivatives characteristic for the *Luzula* variation of the Onagrad type. It occurs in Juncaceae, Cyperaceae, Bromeliaceae, and Typhaceae (von Guttenberg, 1960) and in addition in the Centrolepidaceae, maybe in the Restionaceae and, in a modified form, in the Philydraceae (Hamann, personal communication). However, the base data are too incomplete for generalizations. A special condition is found in the grasses, where new walls are laid down obliquely in a very peculiar fashion (perhaps meriting recognition as a separate type). The Onagrad type is recorded for at least *Lemna* and *Pistia* (Guignard, 1963) and the Asterad type in *Synandrospadix* (Cocucci, 1966) of the Ariflorae, where the Caryophyllad type is otherwise known to occur (see below).

Few records occur in the Zingiberiflorae and the Areciflorae; these are mostly according to the Onagrad or Asterad types, although in the palms the embryogeny may follow the Onagrad type only. Haccius and Philip (1979), who described the embryogeny in *Cocos* (Arecaceae), found the palms to have a number of primitive characters, such as differentiation of the embryo proper from one cell of a pluricellular proembryo, a development of the cotyledon from a position lateral to the terminal stem tip, and a tendency to cleavage polyembryony.

The conditions in those groups of the dicotyledons which are often associated with the monocotyledons are already more variable than in monocotyledons as a whole. The Onagrad and Asterad types (dominating in the monocotyledons other than the Alismatiflorae and Ariflorae) are known in the Annonaceae, Ceratophyllaceae, Nymphaeaceae, Magnoliaceae and Saururaceae, while the Piperaceae is almost unique in having the Piperad type and the Aristolochiaceae the Solanad type.

Base Data

Here reference is given mainly to family and for the most part follows Johansen (1950) and Davis (1966), but data from many other sources are added.

Caryophyllad type

ALISMATIFLORAE: HYDRO.: Butom.; Apono.; Hydro.—ALISM.: Alism.—ZOSTE.: Scheu.; Junca.; Potam.; Zoste.; Zanni.—NAJAD.: Najad.

ARIFLORAE: ARAL.: Arac. p.p., e.g. *Arum*.

LILIIFLORAE: ASPAR.: Agava.: (*Yucca*); Hyaci. (*Dipcadi*).—LILIA.: Irida. (*Crocus*).

Solanad type

LILIIFLORAE: DIOSC.: Diosc.: *Trichopus* (?).
Dicotyledons associated with the monocotyledons: MAGNOLIIFLORAE: MAGNO.: Arist.

Onagrad type

ARIFLORAE: ARAL.: Arac.: at least *Pistia*; Lemna. (*Lemna*; *Wolffia* irregular).

LILIIFLORAE: ASPAR.: Agava.; Aspho.; Allia.; Amary.—LILIA.: Lilia.—BURMA.: Thism.—PHILY.: Phily. (specialized variation; Hamann, 1966).

COMMELINIFLORAE: TYPHA.: Spar.—JUNCA.: Junca.—CYPER.: Cyper.—RESTI.: Centr.

ARECIFLORAE: AREC.: Arec.
Dicotyledons associated with the monocotyledons: MAGNOLIIFLORAE: MAGNO: Magno.; Annon.

Asterad type

ARIFLORAE: ARAL.: Arac.: *Synandrospadix* (Cocucci, 1966).

LILIIFLORAE: DIOSC.: Diosc.—ASPAR.: Hypox.; Aspho.; Hyaci.; Allia.; Amary.—LILIA.: Irida.; Melan.—BURMA.: Thism. (?).—PONTE.: Ponte.—HAEMO.: Haemo.—BROME.: Brome.

ZINGIBERIFLORAE: ZINGI.: Zingi.

COMMELINIFLORAE: COMME.: Comme.—ERIOC.: ? Rapat.; Xyrid.; Erioc.—TYPHA.: Typha.—POAL.: Poac. (specialized type).

ARECIFLORAE: ARECA.: Areca.
Dicotyledons associated with the monocotyledons: NYMPHAEIFLORAE: NYMPH.: Cerat. MAGNOLIIFLORAE: PIPER.: Sauru.

Chenopodiad type

LILIIFLORAE: ASPAR.: Aspho.; Hyaci. (*Dipcadi*).—LILIA.: Caloch.

ZINGIBERIFLORAE: ZINGI.: Maran.

Piperad type

Dicotyledons associated with the monocotyledons: MAGNOLIIFLORAE: PIPER.: Piper.

In general, the knowledge of embryogeny is somewhat unsatisfactory for taxonomic interpretations.

Some Other Embryological Characters

Some other, possibly interesting but insufficiently known, characters that may be of taxonomic significance have been pointed out to us by Hamann (personal communication). They include the following.

The mode of division in the basal cell (in Fig. 75) of the young embryo which is usually transverse, but may be longitudinal in some groups, like Xyridaceae, Eriocaulaceae and Commelinaceae.

The shape of the suspensor.

Delayed embryo differentiation (only part of the apical quadrant giving rise to the embryo proper, the other apical cells contributing to the massive suspensor (Haccius and Philip, 1979).

The shape of the cotyledon and elongation of cotyledonary sheath.

The occurrence of a coleorhiza (rudimentary coleorhiza possibly found also in the Centrolepidaceae) and endogenous development of the primary root.

The embryo in Poaceae differs from that in most other groups of monocotyledons studied in the endogenous initiation of the first root primordium; the coleorhiza representing probably the rudiment of the radicle in other monocotyledons. This interpretation is supported by Pankow and von Guttenberg (1957), Philip and Haccius (1976) and Tillich (1977). Another monocotyledon genus, where the first root is initiated endogenously, is *Zostera* (Zosterales), according to Yamashita (1973).

FRUIT TYPES, GENERAL REMARKS

The ecological significance of various fruit types may be decided by experiment and the development of the fruit is readily determined from comparative anatomical studies. In consequence it is usually possible to determine whether similarity in fruit form is the result of evolutionary convergence or otherwise, a matter of considerable taxonomic importance. For this reason it is important, for example, to distinguish between the berries of epigynous and hypogynous flowers in that different tissues are involved in their construction; that is their phylogenetic histories are different.

In this work the distinction between the two berry types may be made by comparing Diagrams 34 and 71, a comparison that may be informative with other fruit types as well.

The terminology of fruits is very complicated and differs from country to country. That used here is very generalized. The concept of "fruit" might be improved

if restricted to "flowers in the state of maturity of seeds", in which case the German concept of "Sammelfrucht", formed from apocarpous gynoecia with several or numerous carpels—whether consisting of achenes, follicles, or single-carpel druplets—are in fact to be compared to each of the more or less syncarpous fruits. It is here a mere matter of degree of fusion of the carpels in a flower.

The classification of fruit types also poses problems in that their terminology varies somewhat according as to whether the origin or the function of the pericarp has been given weight. Furthermore, many taxonomic descriptions are difficult to interpret because such vague terms as berry-like, nut-like and drupaceous are employed.

In the following account six basic fruit types have been recognized as defined in Table 3.

TABLE 3

Fruit type	Pericarp	Carpel number	Seed number
achene	leathery, papery; indehiscent	1	1
follicle	leathery, papery; dehiscent	1	1, more than 1
nutlet, caryopsis	papery-woody; indehiscent	more than 1	1
capsule	fleshy, papery, leathery; dehiscent	more than 1	1, more than 1
berry	fleshy throughout; indehiscent	1, more than 1	1, more than 1
drupe	outer fleshy, inner woody, indehiscent	1, more than 1	1, more than 1

FOLLICLES (Fig. 76, Diagram 68)

Amongst monocotyledons follicular fruits occur most abundantly in the Alismatiflorae, where they are generally considered to represent a primitive state. They are found especially in the Alismataceae subfam. Limnocharitoideae (Fig. 76A–B) and in the Butomaceae, where they have few to numerous seeds on a laminar (dispersed) placenta, but also in the Aponogetonaceae (Fig. 76C), where the ovules are basal.

In the Typhales, follicular one-seeded fruits occur in *Typha*. As the monocarpellate condition in this order is no doubt secondarily derived from a 3- (or 2-)carpellate

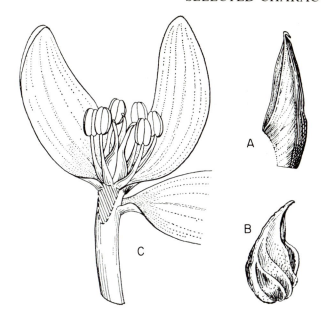

syncarpous condition, the follicular fruit represents a specialized state.

The flower in the Centrolepidaceae may be interpreted variously; accepting that the flowers are small and unisexual and contracted into synanthia, then the female flowers are monocarpellate and dehiscent (i.e. follicles by definition), but by fusion between carpels of different flowers have formed capsular units. In the Hydatellales, the pistils are dubiously monocarpellate. In *Trithuria* the fruits dehisce by three longitudinal slits, indicating a possibly tricarpellate condition, but if interpreted as

Fig. 76. Follicle and achene in the Alismatiflorae. A: follicle of *Tenagocharis latifolia* (Alismataceae). B: achene in *Limnophyton obtusifolium* (Alismataceae). C: apocarpy in *Aponogeton* which has fruits transitional between follicle and achene. (A from Carter, 1960a; B from Carter, 1960b; C from Guillarmod and Marais, 1972.)

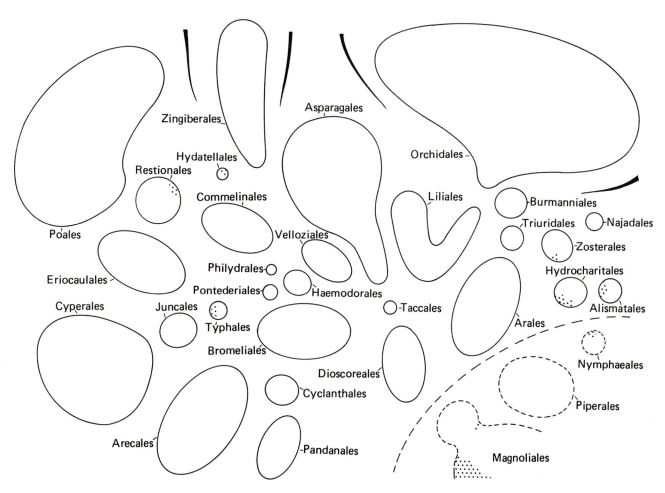

Diagram 68. Follicles (shaded).

monocarpellate the fruit might be classified as an atypical follicle.

The follicular fruits with laminar placentation in the Cabombaceae in the dicotyledonous superorder Nymphaeiflorae suggests a close affinity to the follicle-bearing Alismatiflorae. Follicular fruits are also found in certain Magnoliales.

Base Data

ALISMATIFLORAE: HYDRO.: Butom.; Apono.—ALISM.: Alism. subfam. Limnocharitoideae.—ZOSTE.: Scheu.

COMMELINIFLORAE: TYPHA.: Typha.: *Typha* (the mostly monocarpellate dry fruit in *Typha* generally seems to dehisce).—HYDAT.: Hydat.: *Trithuria* (if monocarpellate).—RESTI.: Centr. (but carpels, of dif-

ferent flowers, fused to form capsule-like fruits; see also capsules).

Dicotyledons associated with the monocotyledons: NYMPHAEIFLORAE: NYMPH.: Cabom.
MAGNOLIIFLORAE: MAGNO.: Magno.: most taxa; Degen.; Annon.: rarely; Arist.: *Saruma*; Lacto. etc.

ACHENES (MONOCARPELLATE) (Fig. 77, Diagram 69)

The distribution of achenes, if defined as monocarpellate, is very restricted (Diagram 69). They occur only in the Alismatiflorae including the Triuridales and in the Hydatellales of the Commeliniflorae, an order whose taxonomic position is unclear.

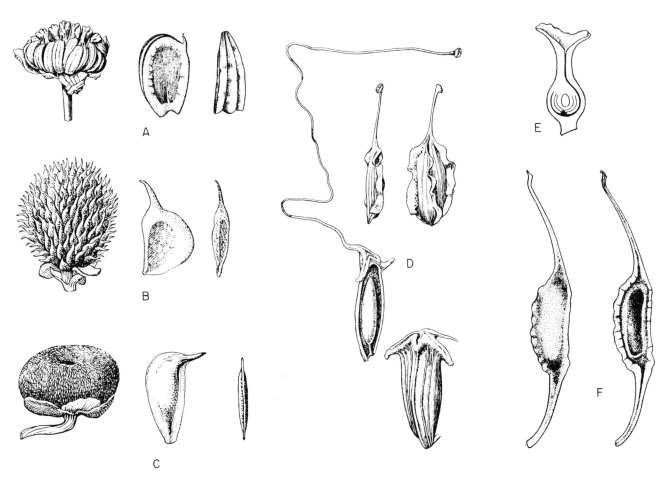

Fig. 77. Achenes ("monocarpellate nutlets"). A: *Alisma plantago-aquatica* (Alismataceae), fruit assemblage and achenes, side and back views. B: *Ranalisma humile* (Alismataceae), fruit assemblage and achenes, side and ventral views. C: *Sagittaria montevidensis* (Alismataceae), as for *Alisma* above. D: *Lilaea scilloides*, achenes from spicate flowers in different views and fruits from a basal female flower, the upper longitudinally cleft. E: *Zannichellia palustris*, female flower. F: *Zannichellia palustris*, fruits, the right one made transparant. (A from Correa, 1969; B and C from Carter, 1960b; D from Correa, 1969; E from Melchior, 1964; F from Vijayaraghavan and Kumari, 1974.)

It may be of interest to compare the taxonomic and ecological distributions of fruits with the bi- or tri-carpellate nutlets in an endeavour to determine the selection pressures that have led to the production of such morphologically similar fruits.

The single-seeded condition may in both be a response to their fundamentally similar pollination mechanisms. In the Poaceae, Cyperaceae and Restionaceae the stigmas are dry (Diagram 56) and pollination is by wind and so the pollen grains arrive only by chance but singly. Hence there is no reason to produce ovaries with many ovules and it would be a waste of reproductive energy to do so. Likewise, in the submerged aquatic families stigmas are dry and pollen will arrive mostly by chance and singly either by wind (as in *Potamogeton*) or water (as in *Zostera*).

Such an argument does not account for the evolution of the single-seeded achenes of most members of the Alismataceae, an animal pollinated group. Here advantage is taken of single pollen-loads by amassing together

several pistils. A single-seeded fruit of utricle-like construction may now have a special dispersal advantage.

In the Alismatiflorae the achenes as here defined are morphologically related to the follicles and also to druplets in some species of *Potamogeton* and the mericarps in the schizocarp of *Triglochin*, both genera being in the Zosterales.

Base Data

ALISMATIFLORAE: ALISM.: Alism.: subfam. Alismatoideae.—ZOSTE.: Junca.: *Lilaea*; Potam.: *Potamogeton* spp.; Zoste.; Zanni.; Cymod. p.p.—NAJAD.: Najad.—TRIUR.: Triur.

ARIFLORAE: ARAL.: Lemn. (interpreted here as monocarpellate, dry "utricles").

COMMELINIFLORAE: HYDAT.: *Hydatella* (? monocarpellate).

Dicotyledons associated with the monocotyledons: NYMPHAEIFLORAE: NYMPH.: Cerat.

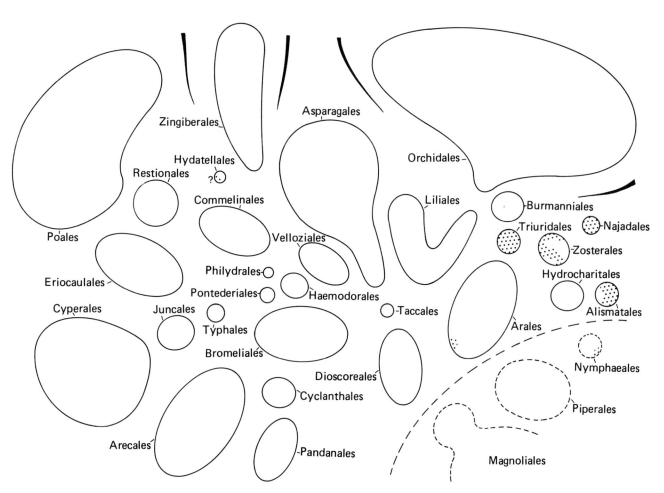

Diagram 69. Achenes ("monocarpellate nutlets"), shaded.

CAPSULES AND SCHIZOCARPS
(Fig. 78, Diagram 70)

Apart from the fruit types formed by apocarpous gynoecia the capsules are often considered the "basic" fruit type, from which baccate, drupaceous and dry indehiscent fruits can be derived and have probably evolved (cf., however, p. 219). The capsules also show a central and wide distribution in the monocotyledons. They are dominant in the Liliiflorae and Zingiberiflorae and in several orders of the Commeliniflorae: viz. in the Commelinales, Eriocaulales, Juncales and in part of Restionales.

Some capsules are formed in epigynous flowers and thus include accessory parts of the flower, such as in all Zingiberales, Orchidales, Burmanniales and Velloziales, and in various families of the Liliales (Iridaceae, Alstroemeriaceae) and Asparagales (Amaryllidaceae, Agavaceae p.p., Hypoxidaceae). In the hypogynous flowers the capsules are formed by the carpels only.

Phylogenetically the distribution of capsules (Diagram 70) is interesting in that they represent a more original state in relation to nutlets and caryopses within the

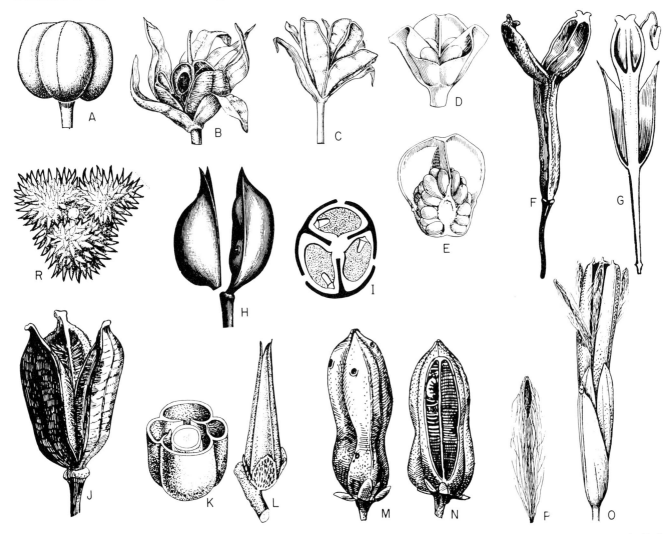

Fig. 78. Capsules, loculicidal (A–L) and septicidal (M–P), and schizocarp (R). A: *Rhodophiala elwesii* (Amaryllidaceae), unopened capsule. B–C: *Nothoscordum* spp. (Alliaceae), opened capsules, showing the septa on the centre of the valves. D–E: *Tapeina magellanica* (Iridaceae), opened capsule and valve. F–G: *Gaimardia* sp. (Centrolepidaceae), a loculicidal capsule formed by two carpels. H–I: *Commelina*, capsule and diagram. J: *Tulipa silvestris* (Liliaceae), capsule showing seeds lying as piles of coins. K–L: *Tetroncium magellanicum* (Juncaginaceae). M–N: *Yucca angustifolia* (Agavaceae). O–P: *Tillandsia pedicellata* (Bromeliaceae), septicidal capsule and seed. R: *Hypselodelphys scandens* (Marantaceae), schizocarp. (A–C and O–P from Correa, 1969; D–E and K–L from Pizarro, 1959; F–G from Baillon, 1894; H–I from Clarke in de Candolle, 1881; J from Baillon, 1894; M–N from Riley, 1892; R from Koechlin, 1964.)

Commeliniflorae and in relation to berries in most cases in the Liliiflorae (see, however, theory on pp. 219–220), while the drupes and berries in the Areciflorae and Ariflorae may not have developed from capsular fruits.

In terms of fruit type the Juncales may be considered more primitive than the Cyperales (this is also true for the perianth and pollen tetrads). Similarly there is a trend within the Restionales from capsules to nutlets. Restricted occurrences of aberrant fruit types represent similar likely transitions from capsules, such as the berries in the Musaceae and Zingiberaceae, both of the Zingiberales, and the schizocarps in the Marantaceae of the same order.

The capsule is also the basic fruit type in the Liliiflorae. In the present circumscription no complete apocarpy is met with in this superorder, although some Melanthiaceous genera (*Petrosavia, Protolirion*) represent border cases with more or less follicular fruits. Berries no doubt developed out of capsules along several separate evolutionary lines: in the Alstroemeriaceae (*Bomarea*), Amaryllidaceae, Hypoxidaceae, Asparagaceae, Dioscoreaceae, Taccaceae etc.

Capsular fruits may be further classified according to their modes of dehiscence. The most common capsule type in monocotyledons is that dehiscing by slits. Other forms such as with a circumcissile slit or dehiscing irregularly are rare. *Ophiopogon*, a member of the otherwise berry-fruited Convallariaceae, has a pistil splitting up irregularly but more or less transversely in the postfloral stage to expose the developing seeds (Rao and Kaur, 1979). Fruits dehiscing with a more regular, circumcissile slit are met with in particular in the Hypoxidaceae (Asparagales) and Thismiaceae (Burmanniales).

Capsules dehiscing by slits may be further subdivided depending upon whether they open along the midribs of the carpels, i.e. along the locules (loculicidal) or along the septae of the carpels (septicidal). Such a definition allows for comparison between septate and non-septate

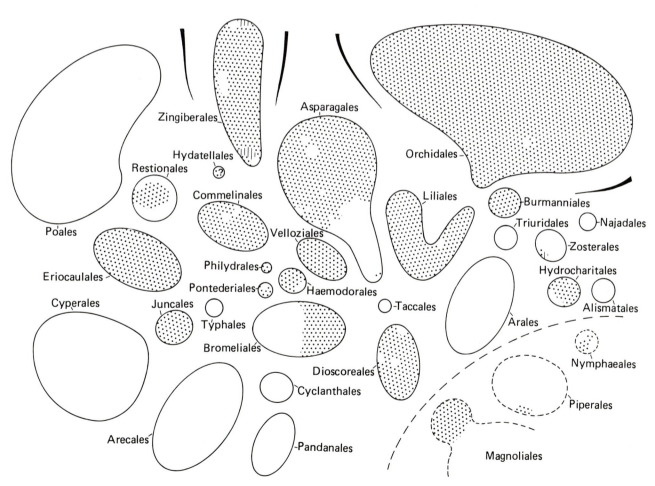

Diagram 70. Capsules (dots) and schizocarps (hatching).

capsules. Somewhat outside this classification fall most orchids, the capsules of which mostly dehisce on both sides of the midribs of the carpels. In monocotyledons loculicidal capsules are, by far, more common than the septicidal and prevail in all groups possessing capsular fruits; the Zingiberiflorae and the Commeliniflorae having exclusively loculicidal and the Liliiflorae both loculicidal and septicidal types.

Within the Liliiflorae loculicidal and septicidal capsules occur together in five orders. Each of these will be discussed in turn.

In the Bromeliales septicidal capsules are the most dominant type and are characteristic of the subfamilies Navioideae and Tillandsioideae and most part of the Pitcairnioideae (*Puya* having loculicidal, however); the Bromelioideae have mostly baccate fruits but *Karatas* has loculicidal capsules.

In the Haemodorales both capsule types occur, those in, for example, *Dilatris* being septicidal but those in *Wachendorfia* and other genera loculicidal.

Of the three families in the Dioscoreales, the Dioscoreaceae have loculicidal capsules but the Stemonaceae septicidal capsules, while in the Trilliaceae *Trillium* species may have a septicidal or irregularly dehiscent capsule.

In the Asparagales loculicidal capsules predominate totally, septicidal being reported for only a few genera, e.g. *Herreria* (Herreriaceae), *Blandfordia* (Phormiaceae), *Yucca* spp. (Agavaceae) and *Excremis* (Dianellaceae). Also in the Liliales the loculicidal capsules predominate in the Iridaceae, Liliaceae and Alstroemeriaceae, but the Calochortaceae (*Calochortus*), Tricyrtidaceae (*Tricyrtis*) and most genera of Cochicaceae (incl. *Androcymbium, Baeometra, Bulbocodium, Colchicum, Dipidax, Littonia, Merendera, Sandersonia, Synsiphon* and *Wurmbea*) and maybe also some Melanthiaceae (such as *Clara, Heloniopsis, Hewardia* and *Pleea*) have septicidal capsules. Loculicidal capsules occur in the latter two families also, however, for example in *Gloriosa, Iphigenia* and *Ornithoglossum* of the Colchicaceae and maybe in genera like *Aletris, Metanarthecium, Narthecium, Nietneria* and *Xerophyllum* of the Melanthiaceae. The Veratreae of Methanthiaceae, (e.g. *Amianthium, Melanthium, Schoenocaulon, Stenanthium, Veratrum* and *Zigadenus*) and *Tofieldia* chiefly seem to have carpels free in the apical parts and opening ventricidally from the apices and basewards, and then alongside the central column of the ovary.

From the above it may be concluded that the Liliales show a stronger tendency for forming septicidal capsules than do the Asparagales, and maybe the capsules of the Tricyrtidaceae and Calochortaceae could be taken as a support for placing these close to Colchicaceae and Melanthiaceae.

Finally, **schizocarps** form a heterogeneous assemblage of fruits and appear in different shapes within a few distantly related groups of monocotyledons. In the Alismatiflorae they occur in *Triglochin* of the Juncaginaceae (Zosterales), where the carpels are more fused than is usually the case in the hypogynous flowers of this superorder. Here the carpels at maturity separate from each other and from the central axis which remains attached to the plant.

In the Liliiflorae schizocarps occur in *Cyanastrum* of the Cyanastraceae and *Macropidia* of the Haemodoraceae (Haemodorales). Their fruit is characterized by the ovary separating into three one-seeded articles leaving behind the septa.

Finally, in the Zingiberiflorae schizocarps occur, for example, in *Heliconia* of the Heliconiaceae, where the mature ovary divides into three one-seeded cocci. In the Marantaceae, also, there are schizocarps in a few genera.

Base Data

Capsules

ALISMATIFLORAE: HYDRO.: Hydro.: most genera.—ZOSTE.: Junca.: *Maundia, Tetroncium*.

LILIIFLORAE: DIOSC.: Diosc.: mostly (except *Rajania, Tamus* and *Trichopus*); Stemo.; Trill.: *Trillium* spp.; *Scoliopus*.—TACCA.: Tacca.: *Tacca* subgen. *Schizocapsa*.—ASPAR.: Phile.: *Eustrepus* (fleshy); Herre.; Dracae.: (*Sansevieria*: see berries) *Beaucarnea, Nolina* spp.; Dorya.; Dasyp.: *Acanthocarpus, Baxteria, Chamaexeros, Lomandra* spp.; Phorm.; Xanth.; Agava.: most (except some spp. of *Yucca*); Hypox.: most (except *Curculigo, Forbesia, Molineria*); Aspho.: (except *Astelia, Lomatophyllum*); Aphyl.; Diane.: *Stypandra*; Tecoph.: (except *Walleria*); Erios.; Hemer.; Funki.; Hyaci.; Allia.; Amary.: most (except a few bacciferous genera).—LILIA.: Irida.; Geosi.; Colch.; Alstr.: (mostly); Tricy.; Caloch.; Lilia.: (mostly); Melan.—BURMA.: all.—ORCHI.: all.—PONTE.: Ponte.: most genera.—HAEMO.: Haemo.: most genera.—PHILY.: Phily.: all except *Helmholtzia*.—VELLO.: all.—BROME.: Brome.: subfam. Pitcairnioideae and Tillandsioideae.

ZINGIBERIFLORAE: ZINGI.: Lowia.; Strel.: *Ravenala, Strelitzia*; Zingi.: most genera; Costa.; Canna.; Maran.: most genera (but several exceptions).

COMMELINIFLORAE: COMME.: Comme.: most genera (except *Athyrocarpus, Palisota, Pollia*); Carto.; Mayac.—ERIOC.: Rapat.; Xyrid.; Erioc.: (?all).—JUNCA.: all.—HYDAT.: Hydat.: *Trithuria* (?).—RESTI.: Resti.: numerous genera; Anart.

Dicotyledons associated with the monocotyledons:
NYMPHAEIFLORAE: NYMPH.: Nymph.: maybe *Nuphar* and others (atypical).

MAGNOLIIFLORAE: MAGNO.: Arist.: most genera.—PIPER.: Sauru.

Schizocarps

ALISMATIFLORAE: ZOSTE.: Junca.: *Triglochin.*
LILIIFLORAE: ASPAR.: Cyana.
ZINGIBERIFLORAE: ZINGI.: Helic.: *Heliconia;* Maran.: scattered, as spp. of *Donax, Hypselodelphys, Trachyphrynium.*

BERRIES (Fig. 79K–N, Diagram 71)

Berries in monocotyledons occur in diverse orders and there is no doubt that they are of polyphyletic origin.

However, their distribution is not random but concentrated in certain orders as may be seen from Diagram 71. They abound amongst hygrophilous terrestrial plants growing in shady or semishady forests, such plants being especially concentrated in the Arales, Cyclanthales and Taccales and in some genera of the Dioscoreales and in several families in the Asparagales (Philesiaceae, Convallariaceae, Smilacaceae, Dracaenaceae etc.). Besides they are richly represented in the bromeliads and palms, the former including terrestrial plants as well as epiphytes.

Berries probably represent a derived fruit type, representing ancient to relatively recent adaptations from capsules, the recent ones being probably found in heterogeneous families. Maybe secondary adaptations from berries to capsules are feasible as well, as proposed for *Trillium* (Trilliaceae) by Berg (1962b).

According to Huber (personal communication) it is possible to interpret the capsules in most Asparagales

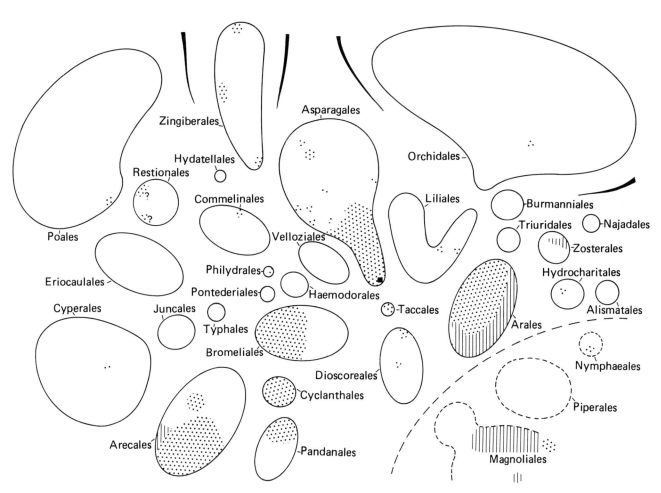

Diagram 71. Berries. Those consisting of one carpel: vertical hatching. Those consisting of two or three (or more) carpels: dots. Pome: square.

(but not in the Liliales) as secondarily derived from berries. During the baccaceous stage of the fruit, the seeds lost their ordinary type of testa, which when the fruits became capsular again was substituted by a phytomelan crust. This interpretation, which must be carefully considered, may be at variance with some considerations on fruit type phylogeny seen in current literature mentioned partly above, for example, that baccate fruits are generally derived and when having phytomelan-coated seeds are very recent in origin.

Berries may be developed from syncarpous or from monocarpellate pistils. The latter are restricted either to

taxa which are truly apocarpous or to taxa where the carpels are reduced to a single one (as in many genera of the Araceae).

Monocarpellate berries occur occasionally in the Alismatiflorae, e.g. in *Ruppia* (Potamogetonaceae). Furthermore they make up about a third of the number of genera in the almost totally bacciferous Araceae (rarely, the principally baccate fruits in Araceae become dry or dehisce irregularly). Although berries are found in c. 50% of the palm (Arecales) species, less than one per cent of them are monocarpellate.

Syncarpous berries are much more frequent, and occur

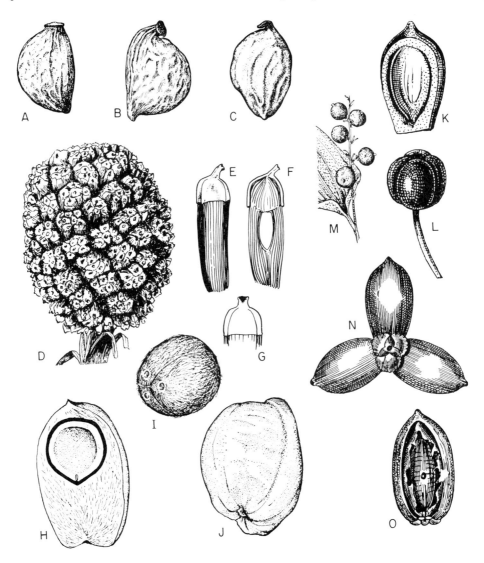

Fig. 79. Drupes (A–J) and berries (K–N). A–C: *Potamogeton species* (Potamogetonaceae), the drupes formed by separate carpels (apocarpy). D: *Pandanus* sp. (Pandanaceae) fruit assemblage; E–G: *Pandanus graminifolius*, separate drupes each with an apical cap. H–J: *Cocus nucifera* (Arecaceae), drupe-like fruit with fibrous instead of carnose mesocarp. I: showing endocarp with germination pores. K: berry of *Pothos cylindricus* (Araceae), longitudinal section. L: berry of *Paris quadrifolia* (Trilliaceae). M: bacciferous inflorescence of *Maianthemum bifolium* (Convallariaceae). N–O: berries and separate berry of a flower of *Phoenix dactylifera* (Arecaceae), each berry formed by a single carpel (apocarpy). (A–C from Correa, 1969; D–G from Stone, 1974; H–J from Lawrence, 1951; K from Engler, 1884; L from Krause, 1930; M from Turpin, in Lotsy, 1911.)

in several orders. In the Alismatiflorae they are rare, however, being perhaps mostly confined to *Hydrocharis* in Hydrocharitaceae.

In the Arales of the Ariflorae they are found in about two-thirds of the species in the Araceae (Fig. 79K), the most berry-rich order (beside Arecales and Cyclanthales) of all monocotyledons in relation to number of species.

In the Dioscoreales berries occur in *Tamus* within the Dioscoreaceae, and in *Paris* (Fig. 79L), *Scoliopus* and some *Trillium* species within the Trilliaceae. In the Taccales (*Tacca*) berries prevail, although subgenus *Schizocapsa* has capsules.

The Asparagales are rich in bacciferous families, while they are nearly lacking in the Liliales. In the Asparagales they occur in all or most taxa of the Petermanniaceae, Philesiaceae, Smilacaceae, Convallariaceae (Fig. 79M), Asparagaceae, Dracaenaceae subfam. Dracaenoideae and Dianellaceae; in addition berries are found in various genera of other families, such as in *Lomandra* (Dasypogonaceae), in some species of *Yucca* (Agavaceae), in *Astelia* (Asteliaceae), in *Molineria*, *Curculigo* and *Forbesia*, all of the Hypoxidaceae, and in a few genera of the Amaryllidaceae. Exceptional is *Behnia* in the Philesiaceae which is stated by Dyer (1976) to have a pome-fruit. In the Liliales berries are restricted to a few Alstroemeriaceae and to *Medeola* of the Liliaceae. (In case *Disporum* and a few more genera, now placed in the Convallariaceae, Asparagales, are more closely allied to the Colchicaceae tribus Uvularieae, as is indicated from embryological and other sources, then the berries are more widespread in the Liliales.)

As is mentioned under phytomelan (p. 230) berries sometimes contain seeds with a phytomelan crust, as in the Asparagaceae subfam. Asparagoideae, in *Eustrephus* and *Geitonoplesium* of the Philesiaceae and in *Dianella* of the Dianellaceae. Phytomelan-encrusted seeds are otherwise normally restricted to capsular fruits. This indicates that the baccate condition of the fruit in the mentioned taxa possibly is of very late origin and the phytomelan crust can be seen as a residual property not yet lost. If so these groups can be expected to have close affinities with taxa possessing capsular fruits.

Broadly interpreted, berries may also be the adequate designation for the fruits of *Galeola* and *Vanilla* in the Orchidales. In the Bromeliales, berries are concentrated in the epigynous subfamily Bromelioideae of the Bromeliaceae (Harms, 1930). *Helmholtzia* of the Philydrales has a berry-like fruit while the other genera of the order have loculicidal capsules.

In the Commeliniflorae berries occur only in few stray genera, such as in *Palisota* of the Commelinaceae and in *Melocanna* of the Poaceae. Also *Flagellaria* (Flagellaria-ceae), *Joinvillea* (Joinvilleaceae), and *Hanguana* (Hanguanaceae) have berries (or drupes), the seeds in the two former with a firm testa (e.g. Newell, 1969).

The Zingiberiflorae mostly have capsular fruits, but berries occur, for example, in *Musa* (Musaceae), *Amomum* and *Phaeomeria* (Zingiberaceae) and in a few genera of Marantaceae.

Finally, the Areciflorae are rich in berries (Fig. 79N–O). About half of the palm genera (including those of the Lepidocaryoidae) have berries in a wide sense. The inclusion here of the fruits of the lepidocaryoid palms (c. 25% of the palms as a whole) may be questioned but they have been accepted as berries rather than as nuts or drupes for the following reasons. Although the outer layers of the pericarp develop into a series of tough imbricate scales the thick or thin mesocarp is fleshy or spongy and the endocarp is thin in the fruits of all genera other than *Eugeissona*. Such fruits do not occur elsewhere amongst the monocotyledons. If not accepted as berries they may be treated separately or associated with another fruit type.

In the Cyclanthales the fruits are mainly baccate and among the Pandanales they are baccate at least in *Freycinetia*.

Baccate fruits occur rarely in the Nymphaeaceae, where they are formed by several, relatively loosely united carpels with some participation of the floral receptacles (cf. *Hydrocharis* in the Alismatiflorae). In the Annonaceae of the Magnoliiflorae the separate pistils of the apocarpous gynoecia usually develop into berries. Such multiple berries formed in the same flower are matched in monocotyledons only in *Cymodocea* and *Ruppia*. The Canellaceae shows paracarpous (syncarpous) berries, probably a more derived type.

Base Data

Berries

ALISMATIFLORAE: HYDRO.: Hydro.: *Hydrocharis*.—ZOSTE.: Potam.: *Ruppia*; Posid.: *Posidonia*; Cymod.: at least *Cymodocea*.

ARIFLORAE: ARAL.: Arac. (most taxa except *Pistia*; maybe c. one third with monocarpellate pistils).

LILIIFLORAE: DIOSC.: Diosc.: *Tamus*; Trilli.: *Paris*, *Trillium* spp.—TACCA.: Tacca.: *Tacca* subgen. *Tacca*.—ASPAR.: Smila.; Peter.; Phile. (except *Eustrephus* and *Behnia*); Conva.: nearly all (but *Liriope* and *Ophiopogon* with fruits rupturing and exposing the seeds); Aspar. Dracae.: subfam. Dracaenoideae (although *Sansevieria* doubtful); Dasyp.: *Lomandra* spp.; Agava.: *Yucca* spp.; Hypox.: *Curculigo*, *Forbesia*, *Molineria*; Aspho.: *Astelia*, *Lomatophyllum*; Diane.: *Dianella*; Tecoph.: *Walleria*;

Amary.: spp. of at least *Buphane*, *Choananthus*, *Clivia*, *Cryptostephanus*, *Gethyllis*, *Haemanthus*.—LILIA.: Alstr.: *Bomarea* spp.; Lilia.: *Medeola*.—ORCHI.: Orchi.: *Galeola*, *Vanilla* (atypical).—BROME.: Brome.: subfam. Bromelioideae.

ZINGIBERIFLORAE: ZINGI.: Musac.: *Ensete*, *Musa*; Zingi.: probably in several genera, at least in *Amomum* and *Phaeomeria*; Maran.: rarely e.g. *Sarcophrynium*.

COMMELINIFLORAE: COMME.: Comme.: *Palisota*.—CYPER.: Cyper.: *Carex* spp. (rarely).—RESTI.: Hangu.: *Hanguana* (maybe a drupe); Flage.: *Flagellaria* (or drupe ?); Joinv.: *Joinvillea* (or drupe ?).—POAL.: Poac.: *Melocanna*.

ARECIFLORAE: ARECA.: Areca.: many genera.— PANDA.: Panda.: *Freycinetia*.—CYCLA.: all.

Dicotyledons associated with monocotyledons:

NYMPHAEIFLORAE: NYMPH.: Nymph. (some fruits may phenetically fall in this group?).

MAGNOLIIFLORAE: MAGNO.: at least Annon., Canel. and Winte.

Pomes

LILIIFLORAE: ASPAR.: Phile.: *Behnia*.

DRUPES (Fig. 79, Diagram 72)

Other than in the Areciflorae where they comprise the fruits of about half the species, drupes are an uncommon fruit-type amongst monocotyledons.

In the Alismatiflorae there is one genus, *Potamogeton* (Fig. 79A–C), some species of which have drupes; in the Ariflorae there is likewise one genus (*Pistia*), and in the Commeliniflorae maybe two or three genera. These are *Scirpodendron* (Cyperaceae) and *Sparganium* (Sparganiaceae). From the literature it is not possible to decide for

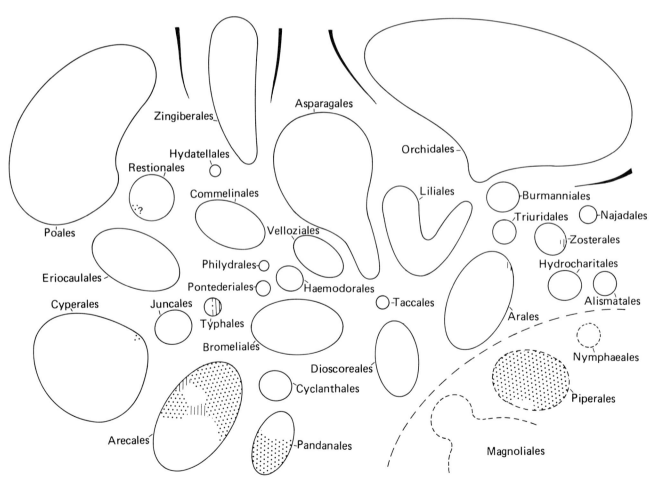

Diagram 72. Drupes. Those consisting of one carpel: vertical hatching. Those consisting of two or more carpels: dots.

certain whether the fruit of *Hanguana* (Hanguanaceae) is a berry or a drupe (in the diagrams it is included under both). Perhaps also the fruits of *Flagellaria* and *Joinvillea* (Newell, 1969) are drupes, although they are here classified as berries.

Drupes are widespread in the palms (Arecales), where they are usually tricarpellate. Unicarpellate drupes are more rare. In the Pandanales, the *Pandanus* fruits (Fig. 79D–G) in somewhat generalizing terms may be classified as drupes, although they are sometimes hard or tough.

Among the dicotyledons associated with monocotyledons, drupes occur in most or all Piperaceae (Piperales, Magnoliiflorae). Their syncarpous nature is indicated for example in *Piper* by the frequently trilobate stigma. This type of fruit is perhaps derived from a capsular or ultimately from a tri- or tetra-follicular type (as in the Saururaceae).

Base Data

ALISMATIFLORAE: ZOSTE.: Potam.: *Potamogeton* spp.

ARIFLORAE: ARAL.: Arac.: *Pistia*.

COMMELINIFLORAE: TYPHA.: Sparg.: *Sparganium.*—CYPER.: Cyper.: *Scirpodendron.*—RESTI.: (maybe Flage.: *Flagellaria*, Joinv.: *Joinvillea* and Hangu.: *Hanguana* ?).

ARECIFLORAE: ARECA.: Areca.: many genera, in the Coryphoid, Borassoid, Arecoid, Cocosoid, Geonomoid, and Phytelephantoid palms.—PANDA.: Panda.: *Pandanus* p.p., *Sararanga*.

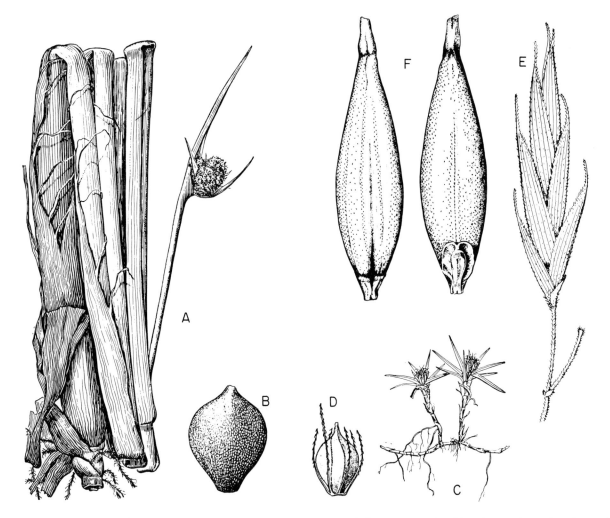

Fig. 80. Nutlets and caryopses. A–B: *Scirpus californicus* (Cyperaceae) with nutlet. C–D: *Scirpus acaulis* with nutlet carrying perigonal setae. E–F: *Arundinaria gigantea* (Poaceae-Bambusoideae), spikelet and caryopse in ventral and dorsal views. (A–D from Correa, 1969; E–F from McClure, 1973.)

Dicotyledons associated with the monocotyledons: MAGNOLIIFLORAE: PIPER.: Piper. (all).

NUTLETS
(INCLUDING CARYOPSES)
(Fig. 80, Diagram 73)

The distribution of nuts, nutlets or caryopses in monocotyledons is restricted to a few orders (Diagram 73). They are limited to three of the six superorders and within these are further restricted in distribution. Phylogenetically the nutlets represent a somewhat heterogeneous assemblage presumably assuming a common form in response to selection for dispersal. The separation of nuts from nutlets demands a criterion of size and as this is a continuous variable it has not been applied.

The apparent abundance of nutlets and caryopses in the Commeliniflorae is due to their regular occurrence in the large families Poaceae and Cyperaceae and their common occurrence in the Restionaceae, e.g. in *Alexgeorgea* (Carlquist, 1976). Within the first two of these three families the ovaries are without exception unilocular and uniovulate and so the single-seeded fruit may be regarded as the primitive condition in them. In contrast the Restionaceae include genera with capsules as well as nutlets. Some species have 2- or 3-locular ovaries each bearing a single ovule, and so here the single-seeded indehiscent fruit must be regarded as derived.

The nutlets also in the Cyperales and Poales are likely to have developed along separate but parallel lines. The Cyperales show an overwhelming number of similarities to the Juncales, which have capsular fruits and no doubt the common ancestors of the Juncales and Cyperales had capsular, several- or many-seeded fruits.

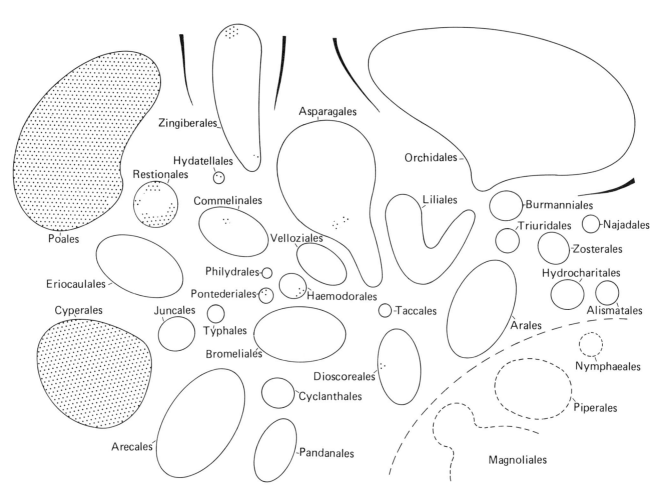

Diagram 73. Nutlets (including caryopses), here restricted to bi- to multicarpellate, dry, indehiscent fruits (shaded).

The grasses show affinities rather to *Flagellaria* and *Joinvillea* (with berries or drupes) and—although not as closely—to the Restionaceae, and thus seem to have achieved their indehiscent one-seeded fruits along this line, most likely from capsules also, if going far enough back in phylogenetic history.

Whether or not *Hydatella* in the Hydatellales has syncarpous nutlets (or utricles) is problematical in that the number of carpels is as yet not definitely known (Hamann, 1975).

Likewise the single-seeded nutlet in *Pontederia* is secondary in that most species of the Pontederiaceae have many-seeded capsular fruits.

In the Liliiflorae the occurrence of nutlets is limited to two families in the Asparagales, two of which may have strong mutual affinities. They are the Dasypogonaceae (e.g. *Dasypogon, Kingia, Calectasia*) and the Dracaenaceae subfam. Nolinoideae (*Dasylirion, Calibanus, Nolinia* spp.). Nutlets also occur in Haemodoraceae (e.g. *Barberetta, Phlebocarya*). In these families there are other genera with many-seeded, capsular or baccate fruits and so the nutlets here can also be regarded as derived fruits.

This argument is also valid for another Liliiflorean family rarely with nut fruits, viz. the Dioscoreaceae (*Rajania, Trichopus*).

In the Zingiberiflorae nuts are recorded only from the Strelitziaceae (*Phenacospermum*) and Marantaceae. They are the product of an inferior trilocular ovary in which each locule contains a single ovule only one of which matures.

Base Data

Only syncarpous, dry indehiscent fruits are included.
LILIIFLORAE: DIOSC.: Diosc.: *Trichopus, Rajania.*
—ASPAR.: Dracae.: subfam. Nolinoideae p.p.: *Dasylirion, Calibanus, Nolina* p.p.; Dasyp.: *Calectasia, Dasypogon, Kingia.*—PONTE.: Ponte.: *Pontederia, Reussia.*—HAEMO.: Haemo.: *Barberetta, Phlebocarya.*
COMMELINIFLORAE: COMME.: Comme.: *Athyrocarpus, Pollia.*—CYPER.: nearly all. (Hydat.: see under achene and follicle.)—RESTI.: Resti.: many genera; Ecdei.—POAL.: Poac.: nearly all.
ZINGIBERIFLORAE: ZINGI.: Strel.: *Phenacospermum*; Maran.: some genera, e.g. *Maranta* and *Thalia.*

ARILS, STROPHIOLES, CARUNCLES AND ANALOGOUS STRUCTURES (Fig. 81, Diagram 74)

More or less swollen, sometimes pulpy, seed appendages traditionally have been termed *arils* when developed on all sides about the hilum, *strophioles* when developed in the form of a crest along the raphe and *caruncles* when developed near the micropyle. However, intermediate structures exist and other terms and concepts have been proposed (e.g. van der Pijl, 1972). Swollen seed appendages have developed on seeds in various lines of the monocotyledons with capsular fruits. These structures are not homologous in many of the groups and therefore as considered at large are of no far-reaching taxonomic interest. Rather they have developed as means of facilitating dispersal by birds, mammals and ants. When such structures are lipoid-rich and attractive to ants they are termed *elaiosomes*.

The arillar structures have their greatest distribution in the Zingiberiflorae, formerly called "Arillatae" (Pfeiffer, 1891), where they occur in practically all families (Koch and Friedrich, 1971; Friedrich, personal communication) and are even registered in the fossil genus *Spirematospermum* (Friedrich and Koch, 1972). They are usually well developed in those members of the group which have dehiscent fruits, but rudimentary or maybe lacking in those with indehiscent fruits. In this superorder the arils always arise from the funiculus and the integument in the micropylar region. It may form a localized mass at this end or an envelope around the seed (Humphrey, 1896). Thus, in the families of Zingiberales the aril is almost universally present.

The aril consists of multicellular, woody fibres in *Strelitzia* (Strelitziaceae), trichomes in *Musa* (Musaceae) and it is absent or nearly so in *Heliconia* (Heliconiaceae); it is thin and veil-like, covering the whole seed in Zingiberaceae, but forms a bulbous mound (a "caruncula") in Costaceae; arils are also present in Cannaceae and Marantaceae, although it is reduced in *Thalia* of the latter family (Panchaksharappa, 1962).

Among the other orders seed appendages are well developed in the Taccales and some members of the Asparagales, more rarely in the Dioscoreales and Liliales. In these orders such structures are distributed in many families and appear to represent non-homologous outgrowths from different parts of the ovule.

Arillar structures are in particular well developed in the Asphodelaceae, subfam. Asphodeloideae (Fig. 81A–D), where they arise as an annular invagination at the distal part of the funicle, enveloping the seed to a variable extent during its development as if it were a "third integument" (Schnarf and Wunderlich, 1939). While generally several cell layers in thickness, the aril is but two cell layers thick in *Asphodelus, Verina, Bulbinella* and *Bulbinopsis. Bulbine* and some related genera have a richly developed arillus. In the *Aloë* group the aril is likewise well developed and in some species of *Kniphofia* supplied with a purple pigment.

Aril-like structures are also found in the Iridaceae as, for example, in the genus *Iris* sect. *Regelia*, *Pseudoregelia* and *Oncocyclus* (Fig. 81E).

Huber (1969, p. 232) classified seed appendages in the Liliiflorae in the following way:

1. arising from the chalazal region: in species of *Erythronium* (Liliaceae), *Colchicum* (*autumnale*) (Colchicaceae), *Hermodactylus* (Iridaceae) and *Galanthus* and *Leucojum* (Amaryllidaceae; Fig. 81F–G);

2. arising from chalaza and raphe: *Patersonia* (Iridaceae) and *Gagea* (e.g., *G. pusilla*, Liliaceae);

3. arising from raphe or raphe and hilum: *Croomia* (Stemonaceae), *Scoliopus* (Fig. 81I) and *Trillium* (Trilliaceae) and *Uvularia* (Colchicaceae);

4. arising from hilum: in *Caesia*, *Johnsonia*, *Simethis* and *Thysanotus* (Asphodelaceae subfam. Anthericoideae) and *Stemona japonica* (Stemonaceae). Also in *Stemona tuberosa* is there an appendage arising in the hilum region (funicle and adjacent part of the outer integument) (Swamy, 1964);

5. arising from near the micropyle: several genera of Hyacinthaceae, such as in species of *Chionodoxa* and *Scilla* (Fig. 81H).

Other groups with strophioles are to be found in the

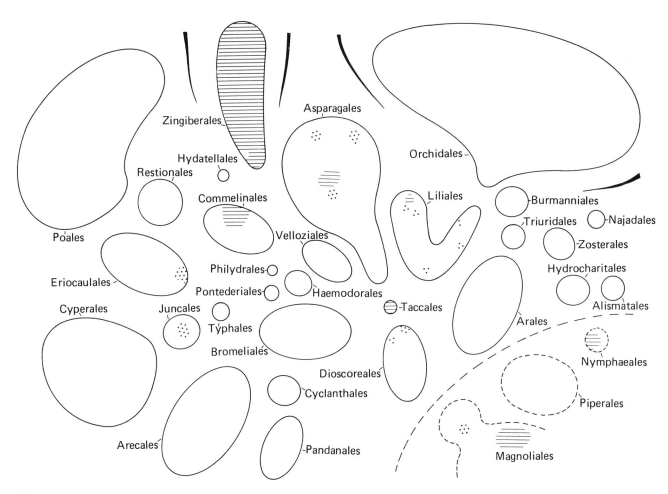

Diagram 74. Arils (hatching) and strophioles (dots).

Fig. 81. Aril and strophiole structures. A: *Asphodelus fistulosus* (Asphodelaceae), longitudinal section through ovary showing an ovule with aril, transmission tissue located above and below the ovule. B: *Bulbine annua* (Asphodelaceae), ovule. C: *Eremurus himalaicus* (Asphodelaceae), ovule surrounded by aril. D: *Kniphofia praecox* (Asphodelaceae), ar = arillus, o.i. = outer integument, i.i. = inner integument, N = nucellus. E: strophiole on seed of *Iris ruthenica* (Iridaceae). F–G: *Galanthus nivalis* (Amaryllidaceae) (F: ovule). H: *Scilla* sp. (Hyacinthaceae; development of strophiole on outer integument at the micropyle). I: strophiole on seed of *Scoliopus bigelovii* (Trilliaceae). J: strophiole on *Luzula campestris* (Juncaceae). (A–C from Stenar, 1928a; D from Schnarf and Wunderlich, 1939; E–H and J from Bresinsky, 1963; I from Berg, 1959.)

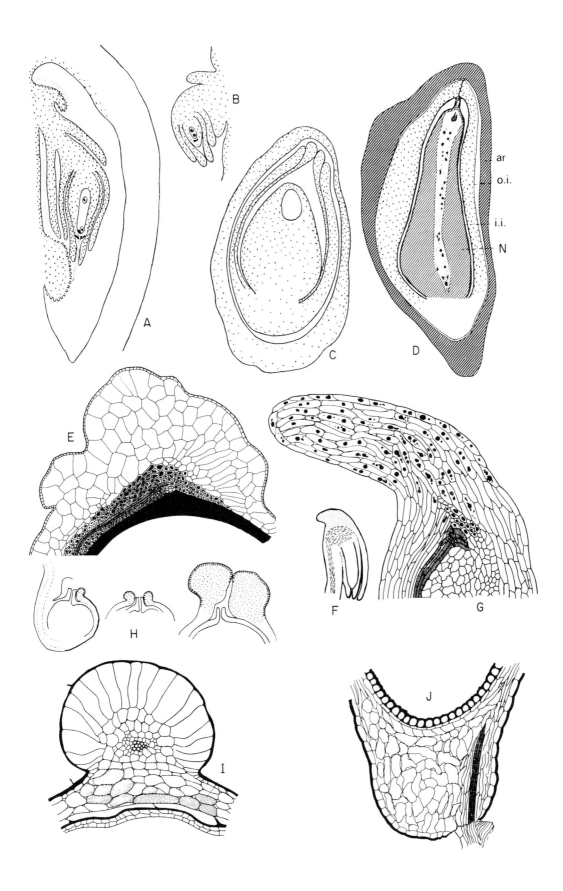

ar
o.i.
i.i.
N

Juncaceae, especially certain myrmechocorous species of *Luzula* (Fig. 81J), and in Rapateaceae.

Among the groups of dicotyledons bordering on the monocotyledons, a well developed aril is found in many Nymphaeaceae (except *Nuphar* and *Barclaya*) where it arises from the funicle. It is white, thin, saccate, non-vascular and contains stellate cells in a mucilaginous mesophyll. In *Nymphaea* it invests the whole seed; in *Euryale* and probably in *Victoria* it develops four separate outgrowths which serve as a float (Corner, 1976) and so assists in dispersal. The Cabombaceae and Ceratophyllaceae lack arils as do all the Alismatiflorae.

In the Aristolochiaceae (Magnoliales), the seeds have a strophiole at least in *Asarum* (Bresinsky, 1963, p. 41), where it arises from the apical part of the funicle. Arils occur in some genera of the Annonaceae (*Xylopia*, *Annona*, *Canangium*).

These observations do not contribute substantially to the taxonomy of the major groups of monocotyledons other than as demonstrations of the relationships of families of the order Zingiberales (though arils are rudimentary or lacking in Musaceae and Heliconiaceae) and that several genera of Asphodelaceae subfam. Asphodeloideae seem to form a natural group.

Base Data

As in the text above.

PHYTOMELAN IN SEED COATS
(Figs 82–83, Diagram 75)

The taxonomic importance of the possession of phyto-melan-encrusted seeds was in particular elucidated by Huber (1969). Phytomelan is produced in the seeds of most members of Asparagales and possibly certain Zingi-

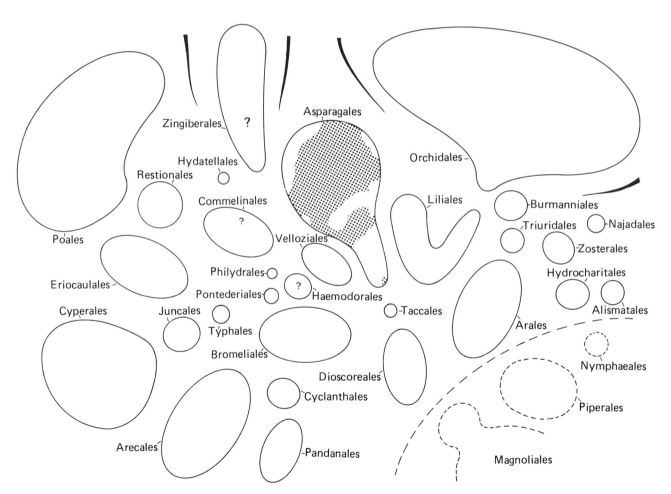

Diagram 75. Phytomelan in the outer part of the testa (shaded). The occurrence in the Zingiberales, Commelinales and Haemodorales is very dubious.

berales, and/or Haemodorales, while other groups may have phlobaphenes and other pigments in the seed coats. While tannin derivatives (such as phlobaphenes) are present in all parts of the seed coat, phytomelan production is restricted to the outer epidermis of the outer integument and of the chalazal region and raphe. Phytomelan is an opaque, brittle, charcoal-like substance which is very rich in carbon and has a hydrogen : oxygen proportion of c. 2 : 1 suggesting dehydration of carbohydrates. Phytomelan is chemically very inert.

The thickness of the phytomelan crust varies considerably, being very thick in *Astelia*, *Simethis* and *Cordyline* (Asphodelaceae) and *Geitonoplesium* (Philesiaceae). It is also thick in the Dianellaceae and Hypoxidaceae. In these groups the phytomelan crust may range between 70 μm and 180 μm in thickness.

Thinner phytomelan layers, c. 10–40 μm thick, occur in most taxa of the Alliaceae, Asphodelaceae, Aphyllanthaceae, Hemerocallidaceae, Xanthorrhoeaceae etc., which generally have capsular fruits. A few taxa with

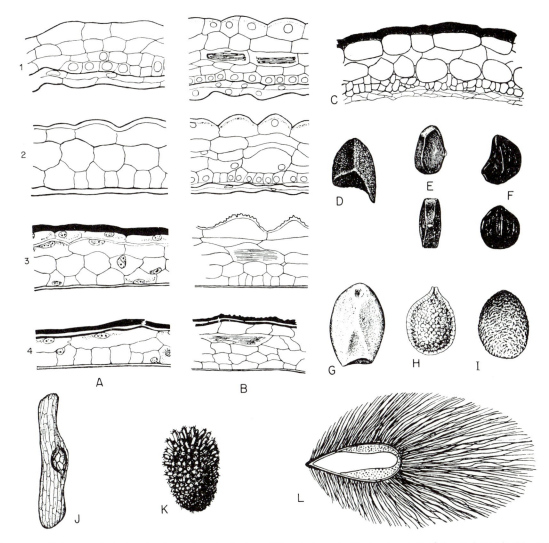

Fig. 82. Presence and absence of phytomelan in seed coats. A: four different stages in the development of the seed coat in *Muscari racemosum* (Hyacinthaceae) and B: the same development in *M. comosum*. It is seen that the inner integument disintegrates, while the epidermis of the outer becomes encrusted with phytomelan, thickly so in *M. racemosum*, more thinly and at a later stage in *M. comosum*. C: seed coat of *Camassia quamash* (Hyacinthaceae), showing a relatively thick phytomelan crust. D–E: seed of *Nothoscordum inodorum* and *N. bonariense* (Alliaceae), all black from phytomelan. F: seeds with phytomelan crust of *Rhodophiala elwesii* (Amaryllidaceae). G–L: seeds without phytomelan crust in *Luzuriaga radicans* (Philesiaceae) (G), *Sisyrinchium filifolium* (Iridaceae) (H), *Sisyrinchium macrocarpum* (Iridaceae) (I), *Arachnitis uniflora* (Corsiaceae) (J), *Walleria mackenzii* (Tecophilaeaceae) (K) and *Eriospermum* sp. (Eriospermaceae) (L). *Luzuriaga* and *Walleria* have berries; *Eriospermum* is unusual in having hairy seeds. (A–B from Wunderlich, 1937; C from Bucher, 1948; D–G and I–J from Correa, 1969; H from Pizarro, 1959; K from Carter, 1966; L from Lotsy, 1911.)

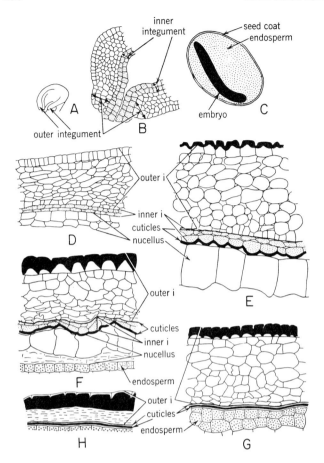

Fig. 83. Development of seed coat in *Asparagus officinalis* (Asparagaceae). In spite of the fact that the fruit is baccate the seed coat in this genus developes a phytomelan layer. A: ovule; B: integuments at the time of pollination; C: seed 44 days after pollination, schematically; D–G: successive stages of seed coat development, 8 (D), 16 (E), 20 (F) and 29 (G) days after pollination, H being taken from the mature seed. (After Robbins and Borthwick, 1925, from Esau, 1953.)

berries may also have seeds with phytomelan crusts of this thickness as, for example, species of *Asparagus* (Asparagaceae) and of the subfamily Yuccoideae of Agavaceae.

Thin phytomelan crusts of less than 10 μm occur in Alliaceae subfam. Agapanthoideae, Agavaceae subfam. Agavoideae, many Amaryllidaceae, most Hyacinthaceae, *Cyanella*, *Hosta*, *Herreria* etc. (see Wunderlich, 1937).

Huber (1969) observed that taxa in the Asparagales which were derived in various respects (e.g. by having epigynous flowers, bulbs, winged seeds) tend to have a thin crust of phytomelan, while more primitive taxa have a thicker crust.

The presence or absence of a phytomelan crust in seeds is of considerable taxonomic importance and Huber (1969) attached great importance to it in his partition of families

between the Asparagales and Liliales. Whilst there is a close connection between the presence of phytomelan and fruit type in Asparagales there is apparently a total absence of phytomelan in the capsule-fruited Liliales.

In the monocotyledons phytomelan crusts seem to be concentrated to (and perhaps restricted to) seeds of the order Asparagales. A general tendency is that it is absent from the families having berries (Fig. 82G). Thus the seeds lack phytomelan in most (but not all) members of Philesiaceae and in all known taxa of the Convallariaceae, Asparagaceae subfam. Ruscoideae and Dracaenaceae. The seeds also lack phytomelan in the Dasypogonaceae and Doryanthaceae, indicating perhaps a close relationship to bacciferous groups. However Asparagaceae subfam. Asparagoideae (*Asparagus*, *Asparagopsis*, *Myrsiphyllum*) (Robbins and Borthwick, 1925) as well as *Eustrephus* and *Geitonoplesium* (Philesiaceae) and *Dianella* (Dianellaceae) all have berries containing phytomelan-encrusted seeds. This may indicate that here the berries have been derived relatively recently from capsules.

Within the Asparagales there is some evidence that the Xanthorrhoeaceae may not be so closely related to the Dasypogonaceae as previously thought, and there is also evidence that the Ruscaceae might be distinguished as separate from the Asparagaceae *sensu stricto* (also supported by the extrorse anthers of *Ruscus* and related genera). Besides, the extremely thick phytomelan crusts of their seeds do not support the close connection between the Hypoxidaceae and, for example, the Velloziaceae as sometimes suggested, nor its close relationship to Tecophilaeaceae (with thin phytomelan layer) or *Cyanastrum*, *Eriospermum*, *Walleria* (Fig. 82K–L) and other genera without phytomelan (the latter two with hairy seeds). The phytomelan crusts of the seeds in the Hypoxidaceae also contradict any close relationships between this family and the orchids.

The possible presence of phytomelan in the seed coat of the dark seed in certain Zingiberales, in *Wachendorfia* (Haemodoraceae, Haemodorales) and in *Commelina* (Commelinaceae, Commelinales) needs verification.

Base Data

The data below have been limited only to the Asparagales (Liliiflorae). However, it is possible that phytomelan might be present in the seed coats in a few taxa— Zingiberales, Haemodorales, and Commelinales (see above).

With phytomelan crust

ASPAR.: Phile. p.p.: *Eustrephus*, *Geitonoplesium*; Aspar. p.p.: *Asparagopsis*, *Asparagus*, *Myrsiphyllum*; Herre.: *Herreria*; Phorm.: *Phormium*, *Xeronema*; Xanth.:

Xanthorrhoea; Agav.: all genera studied; Hypox.: all genera studied; Aspho.: all; Diane.: *Dianella, Stypandra*; Tecoph.: most genera (but not *Walleria*); Hemer.: *Hemerocallis*; Funki.: *Hosta*; Hyaci.: most (exceptions given below); Allia.: all; Amary.: most (exceptions given below).

Without phytomelan

ASPAR.: Smila.: all; Peter.: *Petermannia*; Phile.: *Behnia, Lapageria, Luzuriaga, Philesia*; Conva.: all; Aspar.: *Danaë, Ruscus, Semele*; Dracae.: all; Dory.: *Doryanthes*; Dasyp.: *Acanthocarpus, Lomandra* (other genera need to be verified); Tecoph.: *Walleria*; Cyana.: *Cyanastrum*; Erios.: *Eriospermum*; Hyaci.: spp. of *Chionodoxa, Puschkinia* and *Scilla*; Amary.: many tropical and South African genera; besides spp. of *Galanthus, Leucojum, Sternbergia*.

PERISPERM AND CHALAZOSPERM
(Fig. 84, Diagram 76)

Perisperm by definition represents nucellar tissue that persists and multiplies during seed development to form a nutritive tissue in the ripe seed.

In most angiosperms the nucellar tissue is consumed totally or for the most part by the developing embryo sac (where endosperm or embryo make up a variable proportion of the tissue). It is only in a few restricted groups that a considerable perisperm is developed, the most important being the members of the Caryophyllales, Piperales, Nymphaeales and Zingiberales. Sometimes a recognizable or even substantial part of the nucellar tissue remains without functioning as a copious nutritive tissue.

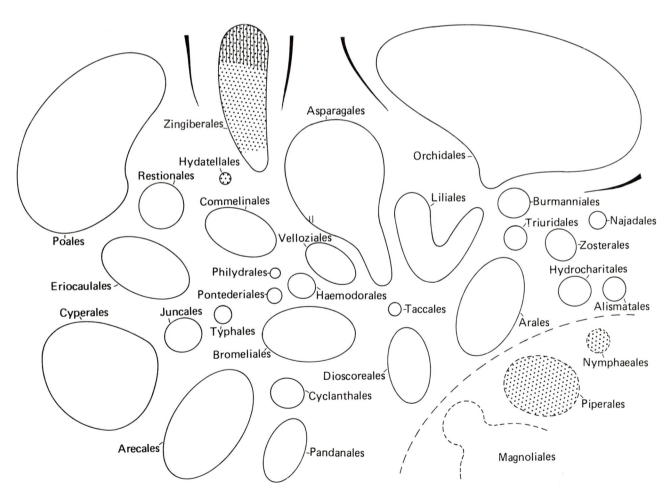

Diagram 76. Perisperm (dots) and chalazosperm (hatching) developed as main nutrient tissue of seeds.

Such nucellar tissue has not been included in the concept of perisperm employed here.

As three of the above mentioned orders are of interest here the perisperm occurrence will be treated in them separately.

Huber (1977) listed the Piperales and Nymphaeales (as superorders) among the groups connecting mono- and dicotyledons, but with "dicotyledonous features prevailing". Both groups have a richly developed perisperm.

In the Piperales this perisperm makes up the main part (c. 95%) of the seed. It is filled with starch grains, the starch increasing in quantity from the chalazal to the micropylar end of the seed. The walls between some of the perisperm cells in the Piperaceae break down and their nuclei fuse. The resulting composite cells are large, densely cytoplasmatic and store oil globules. The endosperm, which encloses the rather small, straight embryo at the micropylar end, is much less extensive than the perisperm and seems to function as an intermediary tissue between the perisperm and the embryo.

In the Nymphaeales, the perisperm is also starch-rich and much more abundant than the endosperm (Cook, 1906), which also in this order is restricted to the micropylar end, surrounding a fairly small, somewhat triangular embryo (Corner, 1976). The great resemblance between the seeds of the Nymphaeales and the Piperales (Fig. 112) indicates a closer affinity between these groups than is usually acknowledged.

In the monocotyledons perisperm is registered mainly in Zingiberales, but is also found (Hamann, 1975, 1976) in the Hydatellaceae (Hydatellales), where in *Hydatella* and *Trithuria* "the conspicuous nucellus forms a starch-containing perisperm, whereas the endosperm is restricted to a few cells which do not serve as storage tissue" (Hamann, 1976, p. 194).

Arnott (1962) questioned the endospermous nature of the nutritive tissue in the seeds of *Yucca* (Agavaceae, Asparagales) and believed it to represent a perisperm. According to Huber (1969) his arguments are not convincing, the nutritive tissue being after all endospermous.

The Zingiberales comprise the main monocotyledonous group with perisperm. Within this order there seems to be a progressive series in the development of perisperm versus endosperm. In the Musaceae, Heliconaceae and Strelitziaceae (and also probably the Lowiaceae) the endosperm tissue is abundant and starch-bearing though its outer cell layers may carry mostly aleuron (as in *Strelitzia*). In *Heliconia* there is a narrow layer of functional nucellar tissue ("perisperm") around the endosperm, while in Strelitziaceae it is reduced to a remnant (Humphrey, 1896; Decrock, 1911). In the Zingiberaceae the endosperm, though several cells in thickness in the lower part of the embryo sac, generally contains aleurone only (but starch in some cases; Panchaksharappa, 1962), while the perisperm is the most well developed, starch-bearing nutritive tissue. In the Costaceae the endosperm contains but fat, and there is a well developed perisperm but also a well developed chalazal tissue ("chalazosperm"). In the Cannaceae the endosperm is reduced to a single aleurone-layer lining the embryo sac cavity, while the perisperm is copious and a chalazal tissue fills up about half of the ovule. In the Marantaceae, finally, the condition is similar to that in the Cannaceae. A persistent perisperm of thick-walled cells with starch grains is developed, and also the chalazal tissue is at least sometimes well developed and projects into the nucellus; in *Thalia* it is bilobate but in *Maranta* reduced (Humphrey, 1896; Schachner, 1924; Panchaksharappa, 1962).

Kracauer (1930) found that in *Canna* the nucellar tissue during seed development seemed to be totally consumed, while a starch-rich nutritive tissue developed, apparently from the chalazal part of the ovule. We thus would have

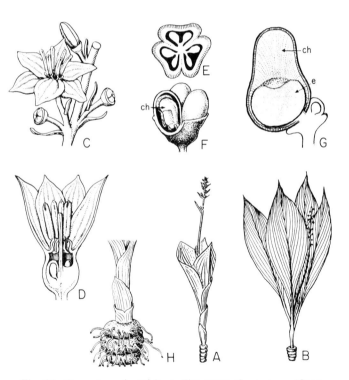

Fig. 84. *Cyanastrum hostifolium*. The genus *Cyanastrum* (Cyanastraceae) is unusual in forming a nutrient tissue from the chalazal part of the ovule: chalazosperm. A–B: plant in prefloral and fruit stage respectively. C: part of inflorescence with flower. D: flower, longitudinal section. E: ovary, transverse section. F: gynoecium in postfloral stage. G: fruit in longitudinal section (e = embryo; ch = chalazosperm), only one carpel fully developed. H: corm. (A–B from Carter, 1966; C–H from Engler, 1930.)

a *chalazosperm* here of a type corresponding to that in *Cyanastrum*.

The Cyanastraceae (Fig. 84) are unusual in the Asparagales in having no endosperm in the ripe seed, the endosperm being consumed in the course of the development. Instead, the chalazal tissue enlarges by cellular divisions to form a chalazosperm (Fries, 1919), which functionally but not histologically corresponds to a perisperm (although of course it resembles this in being diploid). This chalazosperm in *Cyanastrum* (as in *Canna*) is filled with starch grains and occupies about half of the contents of the seed, the remainder being filled up by the embryo. In spite of the apparent similarity between the Cyanastraceae and some representatives of the Zingiberales in the formation of chalazosperm these groups are not closely allied and the corresponding nutritive tissue has appeared by convergent evolution.

The common presence of a perisperm in the Piperales and the Nymphaeales, which may be relatively closely related with each other, on one hand, and the Zingiberales, on the other, could be taken as an indication of a relatively close relationship, but this must be doubted seriously in the light of their many other differences. Nor does there seem to be any close relationships between the Zingiberales and the Hydatellales.

Our conclusion is thus that perisperm most likely represents secondary, relatively recent, adaptations and has appeared separately in the Zingiberales, Hydatellales and Piperales-Nymphaeales. Similarly chalazosperm represents a recent specialization. However, the perisperm in the Piperales and Nymphaeales may have evolved originally along the same evolutionary line. In these two orders the whole seed construction exhibits certain common features otherwise difficult to explain (see Fig. 112). If this is so, then the Piperales could perhaps more easily be conceived as a transitional group between the Magnoliales and Nymphaeales, having the aethereal oils in common with the former, the herbaceous habit, the perisperm and the almost complete lack of benzylisoquinoline alkaloids in common with Nymphaeales, all representing parallel lines of evolution.

Base Data

Perisperm

ZINGIBERIFLORAE: ZINGI.: Zingi.; Costa.; Canna.: (?); Maran.

COMMELINIFLORAE: HYDAT.: Hydat.: *Hydatella, Trithuria.*

Dicotyledons associated with the monocotyledons:
NYMPHAEIFLORAE: NYMPH.: Cabom.; Nymph.; Cerat.

MAGNOLIIFLORAE: PIPER.: Sauru.; Piper.

Chalazosperm

LILIIFLORAE: ASPAR.: Cyana.: *Cyanastrum.*
ZINGIBERIFLORAE: ZINGI.: Costa.; Canna.; Maran. (? p.p.).

RUMINATE ENDOSPERM
(Fig. 86U, Fig. 100G, Diagram 77)

Ruminate endosperm, i.e. in which the margins are invaginated (Fig. 86U, *Yucca*; Fig. 100G, *Trichopus*), has a very restricted distribution in monocotyledons (Diagram 77) being known from only two superorders.

In the Liliiflorae there are three records. These are for *Yucca* of the Agavaceae (Martin, 1946) and *Avetra* and *Trichopus* of the Dioscoreaceae (Knuth, 1930). As *Yucca* and the other two genera belong to different orders it is unlikely that the endosperm in *Yucca* has any great taxonomic significance.

The situation is however quite different in the Areciflorae where ruminate endosperm is common in both the Arecaceae (Moore, 1973) and Cyclanthaceae (Harling, 1958). Here the joint possession of a ruminate endosperm supports the close affinity of these two families.

Ruminate endosperm is often considered as a primitive property. It is widely distributed in the Magnoliales where it shows a variable development from having but one or few folds to being deeply elaborately folded or invaginated. The occurrence of ruminate endosperm in *Trichopus* and *Avetra* and maybe also in palms is therefore sometimes accepted as an indication of these taxa being primitive, which seems logical, although in the Areciflorae the primitiveness is yet to be explained.

Base Data

LILIIFLORAE: DIOSC.: Diosc.: *Avetra, Trichopus.*
—ASPAR.: Agav.: *Yucca.*

ARECIFLORAE: ARECA.: Areca. (about half of the genera).—CYCLA.: Cycla.: several genera.

Dicotyledons associated with the monocotyledons:
MAGNOLIIFLORAE: MAGNO.: Annon.; Arist.: rarely; Canel.: *Cinnamosma*; Degen.; Eupom.; Magno.: sometimes, and only slightly ruminate; Myris.

STARCHY ENDOSPERM
(Fig. 85, Diagram 78)

In the monocotyledons endosperm is sometimes lacking in the mature seed (vertical hatching in Diagram 78), viz. in the Alismatiflorae, many Ariflorae, all Orchidales of the Liliiflorae and in the grass genus *Melocanna* of the Commeliniflorae. In addition a few groups, such as in Hydatellaceae and Cyanastraceae, have a very thin endosperm in which case a starch-rich perisperm or chalazosperm is fairly well developed and takes over the function of the endosperm. A similar condition is encountered in most Zingiberiflorae, where the perisperm may be much better developed than the endosperm and comprises the main nutrient storing tissue.

The endosperm when well developed may contain starch, aleuron (protein), lipoids (fat) and hemicellulose, or some of these nutrients. Starch and aleuron occur as grains and lipoids as droplets in the cell lumen, while hemicellulose is deposited in the cell walls. The occurrence of starch in the endosperm is restricted to some-what less than half of the monocotyledon species, viz. to the Commeliniflorae, certain Zingiberiflorae, and a number of orders in the Liliiflorae, an incompletely known proportion of the Ariflorae and a single genus of the Areciflorae. Especially in the Liliiflorae the starch occurs together with sometimes considerable amounts of the other kinds of nutrient substances mentioned.

The starch grains may be simple, as in most Liliiflorae, or they may be compound. In taxa of the Ponderiales (*Pontederia*), Haemodorales (both subfamilies of Haemodoraceae) and Philydrales (*Philydrum*) there are large and bean-shaped or ellipsoidal starch grains mingled with small isodiametric ones, while at least in *Xerophyta* of the Velloziales all starch grains are of the small isodiametric type (Huber, 1969). Also in *Stemona*, *Croomia* and *Paris* (all Dioscoreales) and in *Scilla* the starch grains, when present, are of both large and small size. In the Commeliniflorae the grains may be simple, as in Rapateaceae, compound, as in the Xyridaceae and Eriocaulaceae, or both types may be present, as in the Poaceae.

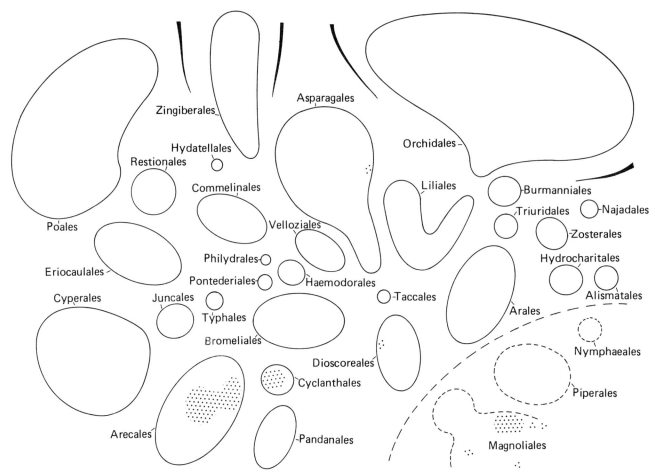

Diagram 77. Ruminate endosperm (shaded).

The following features may be observed with regard to the occurrence of a starchy endosperm:

A starchy endosperm is found in the Commeliniflorae (where it occurs in all families investigated except the Hydatellaceae), various Zingiberiflorae, many Ariflorae and Liliiflorae and a single genus of the Areciflorae.

In Commeliniflorae the starchy endosperm (except in Hydatellales) is always richly developed, while in the Liliiflorae and Zingiberiflorae where a starchy endosperm occurs, the starch varies in amount and is usually complemented with considerable amount of protein, lipoids and (sometimes) hemicellulose.

The Zingiberiflorae are variable in an interesting way in regard to the contents of endosperm, this being in some families replaced by perisperm (and sometimes chalazosperm), which is further described under that title (see p. 232). A starchy endosperm is apparently found in the families having 5 or 6 functional stamens, i.e. the Musaceae, Heliconiaceae, Strelitziaceae, and probably also the Lowiaceae, although in the last-mentioned family the conditions do not seem to be described in detail (see Humphrey, 1896), but also to some extent in Zingibera-

ceae (except Costaceae, where the endosperm contains fat only).

In the Liliiflorae starch is more or less constantly present in the endosperm of the orders Bromeliales (where it is "mealy"), Haemodorales, Pontederiales, Philydrales, Velloziales and in unripe seeds of the Burmanniales (its presence verified by Rübsamen, personal communication), while it occurs more or less sporadically in the endosperm of the Asparagales, Liliales and Dioscoreales. In the Asparagales it is most common within the Amaryllidaceae, but mainly in small amounts. In addition it is known in single species of *Scilla* and *Eucomis* (Hyacinthaceae), in *Streptopus* (Convallariaceae) and in *Ripogonum* (Smilacaceae). In the Liliales starch may be found in fairly rich amounts in the endosperm of unripe seeds but rarely in that of ripe seeds. It is, however, known to occur in considerable amounts in seeds of *Radinosiphon* (Iridaceae) and *Aletris* (Melanthiaceae). Further within the Dioscoreales, starch is known in considerable amounts in the endosperm of *Stemona* (Engler, 1892) and *Croomia* (Stemonaceae) and also in *Paris* and *Trillium* (Trilliaceae).

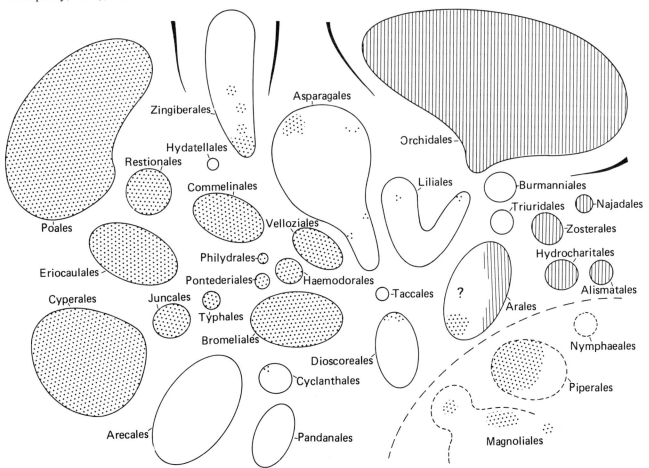

Diagram 78. Starchy endosperm (dots) and endosperm as well as perisperm and chalazosperm lacking in mature seeds (hatching).

In the Ariflorae a well developed endosperm is developed in the Lemnaceae and certain tribes of the Araceae. The endosperm at least in many of these groups contains plenty of starch grains (see Fig. 96).

In the Areciflorae the endosperm normally has a starchless endosperm, but Harling (1958) reported that starch occurs in the endosperm of several species of *Dicranopygium* (Cyclanthaceae).

According to Corner (1976) the endosperm of certain members of the Magnoliales (Magnoliiflorae) contains starch, and also the endosperm in *Piper* (Piperales) (starch grains), although much less than in the perisperm.

It can be concluded that starchy endosperm is scattered in the dicotyledons that show affinity to the mono-

cotyledons and is also scattered to a great extent in the monocotyledons (found in all superorders but the Alismatiflorae which has no endosperm except in the Triuridales). We find it likely, therefore, that the ancestors of the monocotyledons may very well have had starchy endosperm, and that starchless endosperm and no endosperm at all could represent stages of reduction.

Base Data

ARIFLORAE: ARAL.: Arac.: various genera (the extent of starchy endosperm uncertain); Lemn.

LILIIFLORAE: DIOSC.: Diosc.: *Croomia, Stemona*

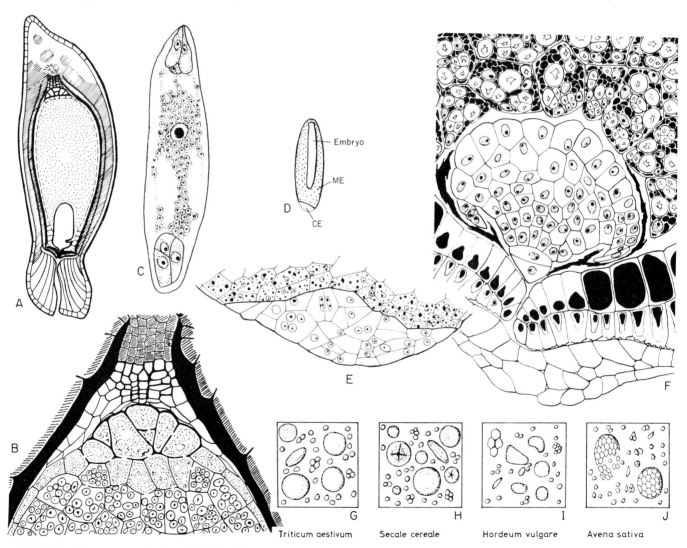

Fig. 85. Starchy endosperm. A–B: seed of *Helmholtzia acorifolia* (Philydraceae). C: embryo-sac in *Centrolepis fascicularis* (Centrolepidaceae) with starch grains formed already. D–E: chalazal endosperm (CE) and part of micropylar endosperm (ME), the latter only having starch, in *Sparganium erectum* (Sparganiaceae). F: micropylar region of seed of *Abolboda americana* (Xyridaceae) showing embryo and part of the starchy endosperm. G–J: starch grain types in some common species of cultivated grass. (A–B from Hamann, 1966a; C from Prakash, 1969; D–E from U. Müller-Doblies, 1969; F from Carlquist, 1960; G–J after Cassner, from Frohne and Jensen, 1973.)

(small amounts), *Trichopus* (small amounts); Trill.: spp. of *Paris* and *Trillium*.—ASPAR.: Smila.: *Ripogonum*; Conva.: *Streptopus* (small amounts); Hyaci.: *Eucomis* and *Scilla bifolia*; Amary.: several genera (but in small amounts).—LILIA.: Irida.: *Radinosiphon* sp.; Melan.: *Aletris* and other genera (in small amounts), otherwise only in unripe seeds.—; in unripe seeds, BURMA.: Burma.; Thism. Rübsamen, personal communication).—PONTE.: all.—HAEMO.: all.—PHILY.: all.—VELLO.: all.—BROME.: all.

ZINGIBERIFLORAE: ZINGI.: probably at least in Lowia., Musac., Helic., Strel., Zingi. (substituted partly in Zingi. and Costa. but entirely in Canna. and Maran. by starch-rich perisperm and chalazosperm).

COMMELINIFLORAE: all groups except HYDAT. which has starch-rich perisperm.

ARECIFLORAE: CYCLA.: Cycla.: only *Dicranopygium* (Harling, 1958).

Dicotyledons associated with the monocotyledons: MAGNOLIIFLORAE (according to Corner, 1976):

MAGNO.: Annon. (with or without starch); Arist. (in some cases); Myris.: sometimes: *Myristica*.—PIPER.: Piper.: *Piper* (oily-proteinaceous in *Peperomia*), smaller than the starchy perisperm.

EMBRYO TYPES
(Figs 86–87, Diagram 79)

Although embryos arrested in their growth as seen in the seed display a bewildering array of forms, they have been classified by Martin (1946) into three divisions and a number of subdivisions. Those two divisions relevant to the monocotyledons are as follows:

1. Basal division, embryo usually relatively small and restricted to lower part of seed (except in certain grasses); seeds medium to large in size.
   (a) "Rudimentary"—embryo small, globular to ovate–oblong (in fact not rudimentary in the true sense).

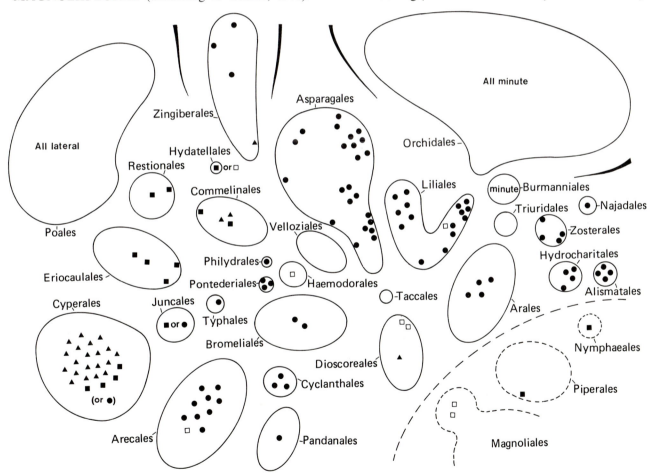

Diagram 79. Embryo types classified according to Martin (1946). "Rudimentary": hollow squares. Broad: solid squares. Capitate: solid triangles. Lateral: found only in grasses. Linear: solid circles. Minute: found only in Burmanniales and Orchidales.

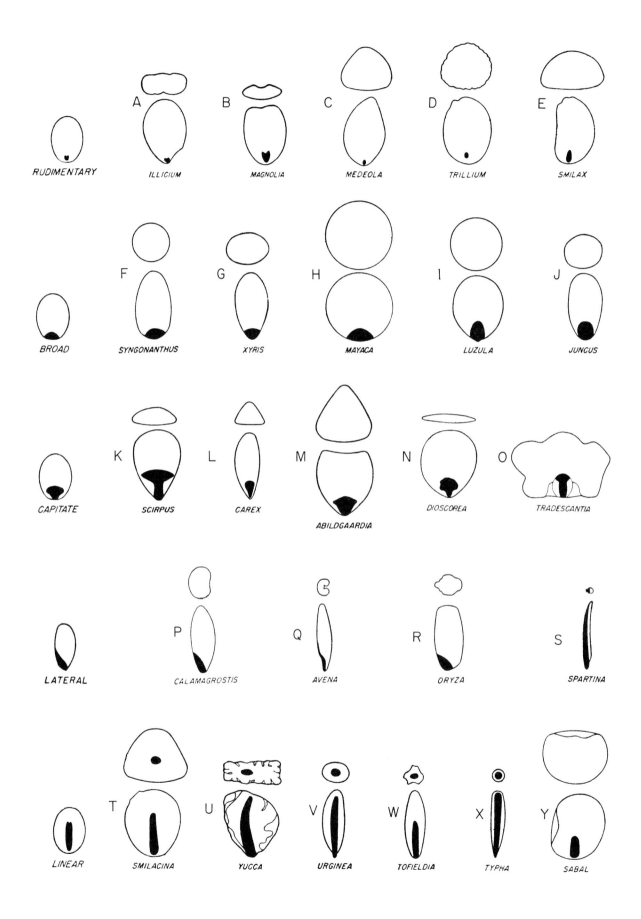

RUDIMENTARY A ILLICIUM B MAGNOLIA C MEDEOLA D TRILLIUM E SMILAX

BROAD F SYNGONANTHUS G XYRIS H MAYACA I LUZULA J JUNCUS

CAPITATE K SCIRPUS L CAREX M ABILDGAARDIA N DIOSCOREA O TRADESCANTIA

LATERAL P CALAMAGROSTIS Q AVENA R ORYZA S SPARTINA

LINEAR T SMILACINA U YUCCA V URGINEA W TOFIELDIA X TYPHA Y SABAL

(b) Broad—embryo as wide or wider than long, peripheral or nearly so.

(c) Capitate—embryo expanded distally.

(d) Lateral (in relation to endosperm and ovule symmetry)—grasses.

2. Axile division, embryo small to total, central (axile) straight or variously curved; seeds minute to large.

(a) Linear—embryo several times larger than broad, not minute.

(b) Minute—embryo minute to total, seeds usually less than 2 mm long consisting of relatively few cells.

This division may be somewhat superficial and therefore must not be the basis of far-reaching conclusions.

Most Liliiflorae (except the orchids) and most Areciflorae have linear embryos. In the Liliiflorae there is a tendency for the Dioscoreales and some supposedly primitive taxa of other orders to have small embryos which can be classified as "rudimentary" or transitional between "rudimentary" and linear. Within the Areciflorae, Harling (1958) considered the embryos of the Arecales to be transitional between "rudimentary", broad and linear, while Martin classified most as of the linear type. The embryo in the Pandanales and the Cyclanthales were classified by Harling (op. cit.) as linear.

The embryo in the Juncales was classified by Martin (op. cit.) as broad, but is transitional between the broad and the linear types, and may be better classified as "shortly linear" (Hamann, personal communication) along with those of the Sparganiaceae, Philydraceae etc. Also the embryo of part of the Cyperales may be classified as shortly linear.

From the figure it is evident that the majority of genera with "basal" and "lateral" embryos occur in the Commeliniflorae. Broad type embryos are likewise concentrated in the Commeliniflorae, where they occur in part of Commelinales, in certain (but relatively few) Cyperales and also in all families of Eriocaulales (Fig. 87), and at least in the Restionaceae and Centrolepidaceae of the Restionales. In these orders the embryos are also lateral in relation to the endosperm but medial-apical in relation to the ovule symmetry, while in the grasses the embryo is lateral in relation to the endosperm as well as the ovule symmetry (Hamann, personal communication). Outside that superorder in the sample available, basal embryos

were registered for *Dioscorea* (Dioscoreales) and *Ravenala* (Strelitziaceae). The subclass "rudimentary" intergrades with other groups and so may not represent a natural assemblage. Furthermore, within the Commeliniflorae there is only one record available for an axile type embryo (*Sparganium*, Typhales), thereby reflecting the homogeneity of the superorder and the possibly Liliiflorean affinity of the Typhales.

Of the axile embryos the subclass "minute" is restricted to the Orchidales and Burmanniales. In contrast the subclass "axile" occurs in all superorders, although as stated above it is very rare in the Commeliniflorae.

A comparison of embryo types in the monocotyledons and other presumably related dicotyledons reveals that the latter all possess basal embryos. They are either "rudimentary" as in *Magnolia* and *Liriodendron* (Magnoliales), or broad as in *Anemopsis* and *Saururus* (Piperales) and *Cabomba*, *Nuphar*, *Nymphaea* and *Brasenia* (Nymphaeales).

The "rudimentary" type thus is that found in most Magnoliales. In the Laurales there is a trend in the direction of increasing embryo volume combined with reduction of endosperm. In the Piperales the embryo is more or less "rudimentary", but the perisperm partly takes the

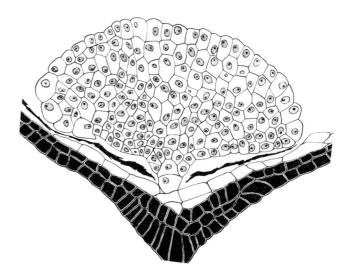

Fig. 87. An example of a "broad type" of embryo in *Abolboda americana* (Xyridaceae). The embryo is at the base of the seed and in its upper part borders on a copious starchy endosperm. (From Carlquist, 1960.)

Fig. 86. Embryo types. A–E: "rudimentary" A: *Illicium* (Schisandraceae, a dicotyledonous family); B: *Magnolia* (Magnoliaceae, also dicotyledonous), C: *Medeola* (Liliaceae); D: *Trillium* (Trilliaceae); E: *Smilax* (Smilacaceae). F–J: broad: F: *Syngonanthus* (Eriocaulaceae); G: *Xyris* (Xyridaceae); H: *Mayaca* (Mayacaceae); I: *Luzula* (Juncaceae); J: *Juncus* (Juncaceae). K–O: capitate: K: *Scirpus* (Cyperaceae); L: *Carex* (Cyperaceae); M: *Abildgaardia* (Cyperaceae); N: *Dioscorea* (Dioscoreaceae); O: *Tradescantia* (Commelinaceae). P–S: lateral (all grasses). T–Y: linear: T: *Smilacina* (Convallariaceae); U: *Yucca* (Agavaceae); V: *Urginea* (Hyacinthaceae); W: *Tofieldia* (Melanthiaceae); X: *Typha* (Typhaceae); Y: *Sabal* (Arecaceae). (All from Martin, 1946.)

place of the endosperm. It would be logical thus to conceive the "rudimentary" embryo type combined with a massive, starchy or non-starchy, endosperm as primitive in monocotyledons. This type is found most typically in the Dioscoreales (*Medeola, Trillium*, Fig. 86C–D), in certain Asparagales with baccate fruits (*Smilax*, Fig. 86E) and in certain Liliales (*Erythronium*), which is all in accordance with Huber's (1961) view that these groups are primitive amongst the Liliiflorae. However, small embryos and rich endosperm are also met with in many Commeliniflorae, Areciflorae (such as many palms) etc. On the other hand this condition is not met with in the Alismatiflorae, usually considered as primitive, nor the Orchidales or Burmanniales, the minute embryos of which are certainly derived and not combined with massive endosperm.

Base Data

(Mainly from Martin, 1946)

Linear

ALISMATIFLORAE: HYDRO.: Hydro.: several genera.—ALISM.: Alism.: several genera.—ZOSTE.: several genera.—NAJAD.: Najad.: *Najas*.

ARIFLORAE: ARAL.: Arac.: 4 genera.

LILIIFLORAE: ASPAR.: numerous genera.—LILIA.: Irid.: several genera; Colch.: *Uvularia*; Caloch.: *Calochortus*; Lilia.: *Erythronium* sp., *Lilium*; Melan.: several genera.—PONTE.: Ponte.: some genera.—PHILY.: all (Hamann, personal communication).

ZINGIBERIFLORAE: ZINGI.: Zingi., Canna. and Maran. (one genus each).

COMMELINIFLORAE: TYPHA.: *Sparganium*.—JUNCA.: Junca.: all (transitional between broad and linear, but best referred to this category).

ARECIFLORAE: ARECA.: Areca.: several genera (*Ptychosperma* classified as "rudimentary".—CYCLA.: all (Harling, 1958).—PANDA.: ? all (Harling, 1958).

Lateral

COMMELINIFLORAE: POAL.: numerous.

Capitate

LILIIFLORAE: DIOSC.: Diosc.: *Dioscorea*.
ZINGIBERIFLORAE: ZINGI.: Strel.: *Strelitzia*.
COMMELINIFLORAE: COMME.: Comme.: *Commelina, Tradescantia*.—CYPER.: most genera; see also the broad type, but often transitional between these types and the linear!

Broad

COMMELINIFLORAE: COMME.: Comme.: *Aneilema*; Mayac.: *Mayaca*.—ERIOC.: Rapat.; Xyrid.; Erioc.: probably all.[—JUNCA.: all (according to Mar-

tin, 1946, but here referred to the linear).]—CYPER.: few genera.—RESTI.: Resti.; Centr.—HYDAT.: Hydat.: transitional between broad and "rudimentary".

Dicotyledons associated with monocotyledons:
NYMPHAEIFLORAE: NYMPH.: Nymph.: *Nymphaea*.

MAGNOLIIFLORAE: PIPER.: Sauru.: *Houttuynia*.

"Rudimentary"

LILIIFLORAE: DIOSC.: Trill.: *Paris, Trillium*.—LILIA.: Lilia.: *Erythronium* sp.—HAEMO.: one genus.

ARECIFLORAE: ARECA.: *Ptychosperma* (most others linear).

Dicotyledons associated with the monocotyledons.
MAGNOLIIFLORAE: MAGNO.: e.g. Magno. and Arist.

Minute

LILIIFLORAE: BURMA.: all.—ORCHI.: all.

EMBRYOS WITH SUBTERMINAL PLUMULE

According to Huber (1969), one finds a distinctive difference between the embryo in the Dioscoreaceae (including *Stenomeris* and *Trichopus*) and the Taccaceae on one hand and that in most or all other monocotyledons on the other.

The mentioned groups have an ellipsoidal or obliquely ovoid embryo with a nearly terminal plumule and a lateral cotyledon. It is fairly well differentiated but "rudimentary" (see above) in relation to the well developed endosperm. Huber (1969) considers this type to represent a connecting link between the original, supposedly dicotyledonary embryo with terminal plumule of the monocotyledon ancestors, and the embryo of most monocotyledons having one terminal cylindrical cotyledon. This is support for considering the Dioscoreales and Taccales as "primitive" and possibly as derivatives of transitional forms between di- and monocotyledons.

MACROPODOUS EMBRYOS
(Fig. 88, Diagram 80)

Embryos with a swollen hypocotyl or radicle are referred to as macropodous (Fig. 88) and are associated with seeds lacking an endosperm. As may be seen from Diagram 80 they are restricted to two superorders.

The majority of macropodous species occur in the Alismatiflorae where they are found in the Najadales,

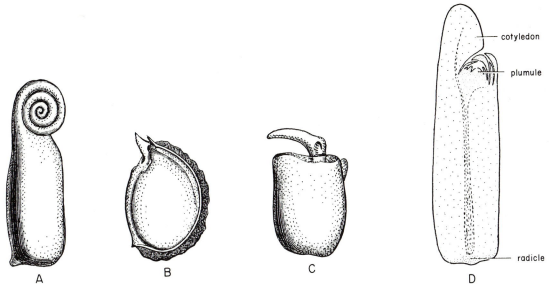

Fig. 88. Macropodous embryos in some members of the Alismatiflorae. A: *Althenia filiformis* (Zannichelliaceae). B: *Cymodocea nodosa* (Cymodoceaceae); the embryo is here made visible by splitting the nutlet longitudinally, half of the pericarp seen behind the embryo. C: *Ruppia maritima* (Potamogetonaceae). D: *Hydrilla verticillata* (Hydrocharitaceae). (A–C from Ascherson, 1889; D from Maheshwari and Johri, 1950.)

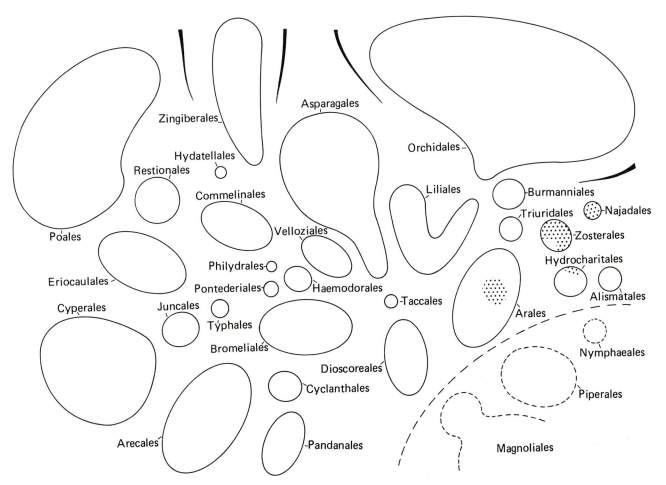

Diagram 80. Macropodous embryos (shaded).

Zosterales (except for Juncaginaceae and Scheuchzeriaceae) and Hydrocharitales (the subfamilies Thallassioideae and Halophiloideae of Hydrocharitaceae). These are all submerged aquatics.

Macropodous embryos are perhaps common in the Araceae, where they occur in *Pothos* and other genera, but need further attention.

Macropodous embryos no doubt are a secondary fairly recent adaptation. The macropodous condition is unknown in the dicotyledons.

Base Data

ALISMATIFLORAE: HYDRO.: Hydro. subfam. Halophiloideae and Thallassioideae.—ZOSTE.: Potam.; Zoste.; Posid.; Zanni.; Cymod.—NAJAD.: Najad.

ARIFLORAE: ARAL.: Arac.: at least tribus Potheae, probably more.

CURVED EMBRYOS
(Fig. 89, Diagram 81)

Embryos that are cylindrical and strongly curved in the sense of horse-shoe shaped, spiral or s-shaped occur infrequently. Less pronounced curvature of the embryo is more scattered in the monocotyledons. Curved embryos are frequently developed in the Alismatiflorae where they are universal in the Alismatales, common in the Zosterales and also present in *Halophila* (Hydrocharitaceae) in the Hydrocharitales.

Of the Ariflorae about 30% have curved embryos, the condition as in the Alismatiflorae being often accompanied by macropody (see Diagram 81), supporting the view that the Alismatiflorae and Ariflorae are possibly closely related.

Elsewhere curved embryos are rare, being found at least in the Marantaceae (Zingiberiflorae) and various genera of the Liliiflorae, such as in *Allium* (Alliaceae), several genera of the Asphodelaceae, *Cordyline* (Asteliaceae), *Xanthorrhoea* (Xanthorroeaceae) etc., all in the Asparagales. Also *Tacca* making up the Taccales has a curved embryo. In the Liliales curved embryos occur, for example, in *Calochortus* (Calochortaceae) and some genera of the Iridaceae and Liliaceae. And so in the Liliiflorae the curved embryo is the limiting condition of a common state.

The embryos of some grasses, for example *Oryza*, and most sedges have a decidedly bent embryo, the axes of the stem and root often being at right angles. These embryos are not cylindrical and so require a separate classification.

Amongst the dicotyledons with monocotyledonous affinities none have curved embryos unless the Menispermaceae is included here.

Base Data

ALISMATIFLORAE: ALISM.: all.—HYDRO.: Hydro.: at least *Halophila*.—ZOSTE.: Potam.; Zoste.; Zanni.; Cymod.

ARIFLORAE: AREC.: many genera, e.g. *Scindapsus*.

LILIIFLORAE: TACCA.: Tacca.: *Tacca* (slightly).—ASPAR.: Phile.: at least *Geitonoplesium*; Aspar.: *Asparagus* (slightly); Dasyp.: *Lomandra*; Xantho.: *Xanthorrhoea*; Agava.: at least *Agave*; Aspho.: esp. many taxa of subfam. Anthericoideae, e.g. *Anthericum*, *Arthropodium*, *Dichopogon* and *Thysanotus*, also in subfam. Astelioideae: *Cordyline*, Hyaci.: *Chlorogalum*; Allia.: at least *Allium*, *Muilla*, *Nothoscordon*.—LILIA.: Irida.: *Sparaxis* etc.; Caloch.: *Calochortus*; Lilia.: *Fritillaria* spp., *Nomocharis*. —HAEMO.: Haemo.: at least *Wachendorfia*.

COMMELINIFLORAE: poorly known but at least in most CYPER. and in POAL. (Poaceae tribus Oryzeae).

ZINGIBERIFLORAE: ZINGI.: at least in Maran.: *Donax*, *Hybophrynium*, *Ischnosiphon*, *Thalia*.

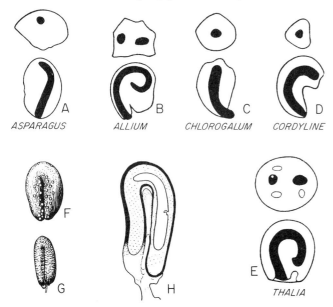

Fig. 89. Curved embryos in monocotyledons. A–E: diagrammatic representation of slightly to strongly curved, linear embryos. F: *Tenagocharis latifolia* (Alismataceae), a seed showing the strongly curved embryo. G: seed of *Alisma plantago-aquatica* in the same family. H: schematic sketch of an unripe seed of *Sagittaria sagittifolia* in the same family showing the curved embryo. The endosperm, which is dotted in the figure, is consumed during the development of the seed. (A–E from Martin, 1946; F from Carter, 1960a; G from Correa, 1969; H from Melchior, 1964.)

SEEDS WITH CHLOROPHYLLOUS EMBRYO (Fig. 90, Diagram 82)

The embryo may or may not contain chlorophyll during its development in the unripe seed. Yakovlev and Zhukova (1973, 1980; see also Dahlgren, 1980) have made an extensive study of the occurrence of chlorophyllous embryos in the angiosperms; their data on monocotyledons having been included here.

Though limited data are available, a certain pattern is discernible as may be seen from Diagram 82. Chlorophyllous embryos are common in the Araceae (Ariflorae), where they are recorded in six out of seven genera studied. They are also apparently common in the Orchidaceae *sensu stricto* (Fig. 90), where the five genera studied all

had chlorophyllous embryos. *Cypripedium* (Cypripediaceae) did not have chlorophyllous embryo and Yakovlev (personal communication) suspects the occurrence of chlorophyllous embryo in Orchidaceae not to be constant.

Netolitzky (1926) found chlorophyllous embryos in three families of the Alismatiflorae, viz. Aponogetonaceae, Najadaceae and Scheuchzeriaceae.

When critically examining the occurrence of seeds with chlorophyllous embryo it is apparent that they are localized to precisely those groups which lack endosperm (and perisperm) in the ripe seed. Moreover, in these groups the testa is not encrusted with phytomelan. Thus there is a possibility that the embryo in such cases can be exposed to a certain, albeit slight amount of light during at least some period of time in the course of their development. As light is probably essential for the formation of chlorophyll these seeds form a potentially distinct group. We will thus make some reservations for the

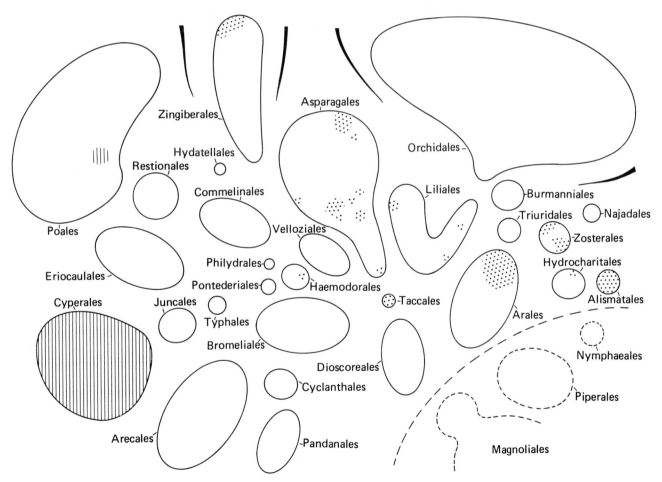

Diagram 81. Curved embryos (dots). The curvature is very variable, from the horseshoe-like type in most Alismatales to only slightly curved types in *Allium*, certain Marantaceae etc. The non-cylindrical types found in the Cyperales and certain grasses are indicated separately (hatching).

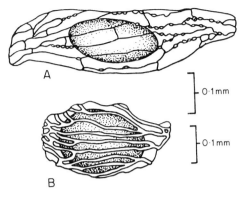

Fig. 90. Seeds of *Glossodia* (A) and *Liparis* (B), both in the Orchidaceae. The orchid seeds are minute and have a transparant or semitransparant testa. The embryos which are constantly, or at least usually chlorophyllous, are clearly discernible through the testa. (From Clifford and Smith, 1969.)

taxonomic value of this character as such. Also in dicotyledons the seeds with a chlorophyllous embryo tend to lack or have a quite thin layer of endosperm.

A few occurrences of chlorophyllous embryos in the Amaryllidaceae (*Haemanthus*) and the Iridaceae (*Tritonia*) lay outside the groups mentioned and these records need verification.

The ategmic ovules in, for example, *Crinum latifolium* (Amaryllidaceae) according to Tomita (1931), developed a naked embryo after the nucellar tissue and endosperm have been absorbed. It is likely that such an embryo could be chlorophyllous, although this is not mentioned in the literature available.

No chlorophyll-containing genera have been recorded in the dicotyledons here associated with the monocotyledons.

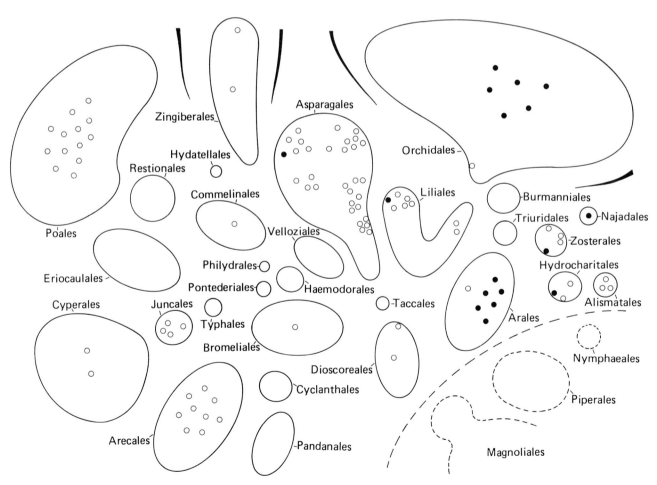

Diagram 82. Mature embryo with chlorophyll (solid circles) and lacking chlorophyll (hollow circles).

Base Data

(After Yakovlev and Zhukova, 1980)

Chlorophyllous embryo

ALISMATIFLORAE: HYDRO.: *Aponogeton.*— ZOSTE.: Scheu.: *Scheuchzeria.*—NAJAD.: Najad.: *Najas.*

ARIFLORAE: ARAC.: *Acorus, Aglaonema, Calla, Nephtytis, Pothos, Zantedeschia.*

LILIIFLORAE: ASPAR.: Amary.: *Haemanthus.*— LILIA.: Irid.: *Tritonia.*—ORCHI.: Orchi.: *Calanthe, Coelogyne, Dendrobium, Dendrochilum, Orchis.*

Non-chlorophyllous embryo

ALISMATIFLORAE: HYDRO.: Butom.: *Butomus;* Hydro.: *Stratiotes.*—ALISM.: Alism.: *Alisma, Echinodorus, Sagittaria.*—ZOSTE.: Potam.: *Potamogeton;* Zanni.: *Zannichellia.*

ARIFLORAE: ARAL.: Arac.: *Arisaema.*

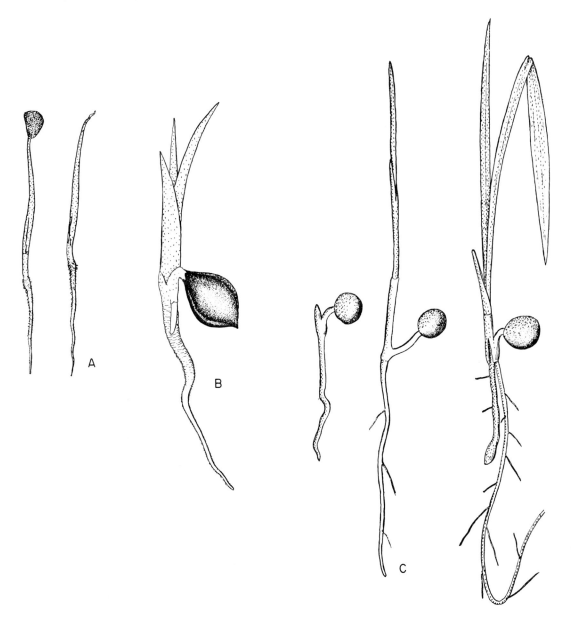

Fig. 91. Germination types. The seedlings refer to types A, B, and C in the text and bear the same letters in the illustration. A: *Agave univittata* (Agavaceae). B: *Hemerocallis fulva* (Hemerocallidaceae). C: *Nannorrhops ritchieana* (Arecaceae). (All from Holm, 1891).

LILIIFLORAE: DIOSC.: Diosc.: *Dioscorea*; Trill.: Trillium.—ASPAR.: Conva.: several genera; Aspar.: some genera; Agav.: *Agave, Yucca*; Phorm.: *Phormium*; Aspho.: *Anthericum, Eremurus, Paradisea*; Diane.: *Dianella*; Hemer.: *Hemerocallis*; Allia.: *Agapanthus, Allium*; Hyaci.: several genera; Amary.: several genera.—LILIA.: Irid.: some genera; Lilia.: *Fritillaria, Tulipa.*—ORCHI.: Cypri.: *Cypripedium.*—BROME.: Brome.: *Pitcairnia*.

COMMELINIFLORAE: COMME.: Comme.: *Tradescantia.*—TYPHA.: Sparg.: *Sparganium*; Typha.: *Typha.*—JUNCA.: Junca.: *Juncus, Luzula.*—CYPER.: Cyper.: *Bulboschoenus, Carex.*—POAL.: Poac.: several genera.

ZINGIBERIFLORAE: ZINGI.: *Zingiber*; Maran.: *Donax*.

ARECIFLORAE: ARECA.: Areca.: several genera.

SEEDLING (GERMINATION) TYPES (Fig. 91, Diagram 83)

Following Boyd (1932) three seedling types are recognized. In a few instances seedlings may be intermediate, but this situation is unusual.

Type A (Fig. 91A). Cotyledon green, plumular leaves emerging through a pore in the sheathing base of the cotyledon. The base of the cotyledon is usually at soil level and the remains of the seed may be carried up into the air. With large seeds the cotyledon may form an arch leaving the heavy seed remains on the soil or may even break off between the seed remains and the plumular bud.

Type B (Fig. 91B). Cotyledon not green, plumular leaves emerging through a pore in the sheathing base of the cotyledon. The base of the cotyledon is often at a

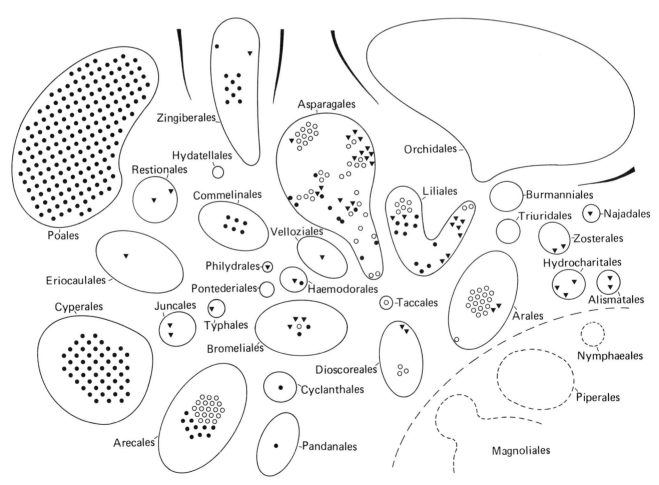

Diagram 83. Seedling types. Type A: triangles. Type B: hollow circles. Type C: solid circles. The types are shown in Fig. 89 by A, B and C respectively.

considerable distance from the seed remains and deeply buried.

Type C (Fig. 91C). As for B but the plumular leaves emerge through a collar formed about the edge of the cotyledonary pore.

The sedges can be included here and maybe also the grasses according to how their coleoptile is interpreted.

In the grasses, where the homologies of coleoptile, scutellum and epiblast have been much debated, recent German literature seems to be increasingly concordant in the following views. The coleoptile and scutellum are both parts of the cotyledon, the coleoptile representing the tubular sheath and the scutellum the "blade", which is transformed into a haustorial structure (Goebel, 1923; Troll, 1954; Pankow and von Guttenberg, 1957). In certain grasses the scutellum and coleoptile become separated from each other by the prolongation of the axis, the "mesocotyl"; thus in these grasses the scutellum and coleoptile in spite of being parts of the same structure, the cotyledon, emerge at different levels. Their connection can sometimes be seen from the course of the vascular strands in case the mesocotyl is moderately long, as in *Avena* or *Triticum* (Pankow and von Guttenberg, 1957). [The epiblast seems to be a lateral outgrowth of the root nek of the first lateral root, which represents the downwardly directed root previously thought to be the radicle, while the radicle in grasses seems to be utterly reduced (Tillich, 1977).]

The above interpretation is radically different from the "classical" interpretation, according to which the coleoptile represents the first leaf following after the cotyledon, which corresponds to the scutellum and, perhaps, the epiblast.

We have here followed the former interpretation in the diagram and base data-list.

The seedling of type C (as shown in Fig. 91 for Commelinaceae) may form a transitional stage to that of the grasses: elongation of the "collar" would lead to the coleoptile. Therefore the seedling of the Poaceae may represent a derived and modified type C seedling.

In comparing these three seedling types with those of the dicotyledons, type A may be regarded as equivalent to epigeal and types B and C except in grasses to hypogeal germination.

The distribution of seedling types amongst the monocotyledons is shown in Diagram 83, in which (except for the Poaceae and Cyperaceae for which many records exist) each symbol represents a single generic record. It is evident that type A seedlings (epigeal) are in the minority, the reverse of the situation amongst the dicotyledons.

Though the data are few and apparently diverse they are not without pattern as may be seen by considering each superorder in turn.

In the Alismatiflorae all seedlings are of type A, but in the Ariflorae the majority of seedlings of type B, only

some, e.g. *Arum* and *Philodendron*, being type A.

In the Zingiberiflorae the majority of seedlings are of type C, one exception being *Costus* (type A).

Amongst the Areciflorae there are no type A seedlings, that is they all exhibit hypogeal germination. Within the Arecaceae seedling type is of some taxonomic importance in the subdivision of the family. It was recorded for all subfamilies by Moore (1973).

Though the Commeliniflorae are diverse with respect to seedling types the orders are apparently uniform or nearly so. Thus possibly the Poales and certainly the Commelinales and Cyperales have type C seedlings and the Juncales type A. It is interesting to note that type A seedlings are typical of the studied taxa of the Eriocaulales, Restionales, Typhales and Juncales (Commeliniflorae) and the Philydrales (Liliiflorae) most of whose species grow in sites subject to periodic flooding. However, the data are as yet too insufficient for making general conclusions.

In the Liliiflorae all five orders for which more than one record is available show heterogeneity of seedlings. However, within the orders there is a considerable amount of taxonomic pattern. For example, of the two seedling types recorded for the Dioscoreales, type A is restricted to the Trilliaceae and type B to the Dioscoreaceae. In the Bromeliaceae the seedlings of species of the subfamilies Pitcairnioideae and Tillandsioideae are of type A whilst types B and C are restricted to the Bromelioideae.

Within the heterogeneous Asparagales and Liliales the distribution of seedling types is variable in many families but is constant for the Liliaceae (type A), Melanthiaceae (type B) and Colchicaceae (type C). That each of these three families may have different seedling types could be taken to support their family status.

The lack of records for the Orchidaceae is a reflection of the difficulty of assigning their seedlings to one of the seedling types since their initial stages of development are so specialized. Information on orchid seedlings is given by Veyret (1965, 1974). They usually have rudimentary cotyledons.

Base Data

Type A

ALISMATIFLORAE: HYDRO.: Butom.: *Butomus*; Hydro.: *Lagarosiphon, Stratiotes*; Apono.: *Aponogeton.* —ALISM.: Alism.: *Alisma, Sagittaria.*—ZOSTE.: Scheu.: *Scheuchzeria*; Potam.: *Potamogeton, Ruppia.*—NAJAD.: Najad.: *Najas.*

ARIFLORAE: ARAL.: Arac.: *Arum, Philodendron.*
LILIIFLORAE: DIOSC.: Trill.: *Paris, Trillium.*—

ASPAR.: Phorm.: *Blandfordia*; Aspho.: *Arthropodium, Bottinaea, Cordyline*; Hyaci.: *Albuca, Bowiea, Hyacinthus, Muscari, Ornithogalum, Scilla*; Allia.: *Allium, Bloomeria, Milla*; Amary.: *Hymenocallis.*—LILIA.: Irida.: *Diplarrhena, Sisyrinchium*; Caloch.: *Calochortus*; Tricy.: *Tricyrtis*; Lilia.: *Erythronium, Fritillaria* spp., *Lilium, Nomocharis, Tulipa.*—HAEMO.: Haemo.: *Anigozanthos.*—PHILY.: Phily.: *Philydrum.*—VELLO.: Vello.: *Vellozia.*—BROME.: Brome.: *Dyckia, Pitcairnia, Puya.*

COMMELINIFLORAE: ERIOC.: Erioc.: *Eriocaulon.* —TYPHA.: Typha.: *Typha.*—JUNCA.: Junca.: *Juncus, Luzula.*—RESTI.: Centr.: *Centrolepis*; Resti.: *Restio, Alexgeorgia.*

ZINGIBERIFLORAE: ZINGI.: Costa.: *Costus.*

Type B

ARIFLORAE: ARAL.: Arac.: 16 genera.

LILIIFLORAE: DIOSC.: Diosc.: *Dioscorea* (incl. *Testudinaria*), *Tamus.*—TACCA.: Tacca.: *Tacca.*— ASPAR.: Smila.: *Ripogonum*; Phile.: *Geitonoplesium, Lapageria*; Conva.: *Maianthemum*; Aspar.: *Asparagus, Ruscus*; Dracae.: *Dracaena, Nolina*; Phorm.: *Phormium*; Agava.: *Yucca*; Aspho.: *Anemarrhena, Anthericum, Asphodeline, Bulbine, Eremurus, Pasithea*; Diane.: *Dianella*; Hemer.: *Hemerocallis*; Funki.: *Hosta*; Hyaci.: *Chlorogalum, Dipcadi, Galtonia, Lachenalia*; Allia.: *Brodiaea, Ipheion*; Amary.: 11 genera.—LILIA.: Irida.: *Crocus, Dierama, Libertia, Melasphaerula, Neomarica, Romulea*; Melan.: *Veratrum, Zigadenus.*—BROME.: Brome.: *Aechmea.*

[COMMELINIFLORAE: (POAL.: maybe all spp. studied, but see type C).]

ARECIFLORAE: ARECA.: numerous spp.

Type C

LILIIFLORAE: ASPAR.: Smila.: *Smilax*; Conva.: *Polygonatum*; Dracae.: *Beaucarnea*; Dasyp.: *Dasypogon, Lomandra*; Aspho.: *Astelia, Bulbinopsis, Chlorophytum*; Tecoph.: *Tecophilaea.*—LILIA.: Irida.: *Aristea, Crocus, Ferraria, Hexaglottis, Lapeyrousia, Tigridia, Watsonia*; Colch.: *Colchicum, Gloriosa, Merendera*; Alstr.: *Alstroemeria.*—HAEMO.: Haemo.: *Wachendorfia.*—BROME.: Brome.: *Acanthostachys, Billbergia.*

ZINGIBERIFLORAE: ZINGI.: Zingi.: *Achasma, Amonum, Brachychilum, Cautleya, Elettaria, Hedychium, Roscoea*; Canna.: *Canna.*

COMMELINIFLORAE: COMME.: Comme. *Commelina, Cyanotis, Palisota, Rhoeo, Tinantia, Tradescantia.* —CYPER.: many.—POAL.: maybe all (see the text above).

ARECIFLORAE: ARECA.: Areca.: many spp.— PANDA.: *Pandanus.*—CYCLA.: Cycla.: *Carludovica* (?).

GEOGRAPHIC DISTRIBUTION
(Fig. 92, Diagram 84)

The past and present distribution of angiosperm families has been surveyed by Raven and Axelrod (1974) against the background of past continental connections. Many families are restricted in their distribution to the Southern Hemisphere and thus were possibly differentiated on the Gondwana continent, others are wholly or largely concentrated to the Northern Hemisphere and seem to have their past history on the Laurasian block.

Any attempt to classify families according to geographic distribution is difficult, as the centre of variation and number of taxa may not coincide with the centre of origin. Considerable migrations certainly took place in late Tertiary and later times, as is reflected in the wide and disjunct distributions of many groups (grasses, water plants etc.), and important families, like the palms, have a pantropic distribution. But even in many of the widespread families is it possible to discern a region where the greatest number of genera or tribes are concentrated, as with the Juncaceae in South America (Snogerup in Nordborg *et al.*, 1971).

Diagram 84 shows those groups which have a Southern Hemisphere concentration or a tropical concentration, and in line with the views of Raven and Axelrod (op. cit.) have a probable Gondwanaland origin. In some of these cases there is a decisive South American concentration (Bromeliaceae, Rapateaceae, Pandanaceae etc.), in others an equally Australian centre (Philydraceae, Doryanthaceae, Dasypogonaceae, Xanthorrhoeaceae), whereas in other cases there is a remarkable disjunction, e.g. in Philesiaceae (Fig. 92A), Asphodelaceae subfam. Astelioideae and Restionaceae (Fig. 92B), indicating either an early Godwanaland differentiation or successful long-distance dispersal. In some cases the dominance is less pronounced although a western (South America–Africa) or an eastern (Africa–Australia) tendency is clear.

Diagram 84 also shows an apparent tendency for some large complexes to have a pantropic concentration, some being even subcosmopolitan. To this group belong the majority of the Dioscoreales, Arales, Alismatales, Hydrocharitaceae, Triuridales, Burmanniales, Orchidales, Poales (Poaceae subfam. Panicoideae) and most Zingiberales and Commelinales (Commelinaceae). The palms (Arecaceae) also belong here (Fig. 92C), having about equal numbers of species in the New and the Old Worlds.

A second group shows a central (African) or general South Hemisphere concentration, which we may postulate has its roots in the ancestors on the central part of the

Gondwana continent. This group is very well represented in the Asparagales, Haemodorales and Liliales (Iridaceae).

A third group represents a western-concentrated South Hemisphere distribution type dominated by such orders as the Velloziales, Pontederiales, Bromeliales, Cyclanthales, Eriocaulales and to some extent also the Juncales (see above and below), the Arecales p.p. and some families of the Zingiberales (the Heliconiaceae and Cannaceae especially).

Finally, an easterly concentrated South Hemisphere distribution pattern is found in most taxa of the Restionales, Philydrales, Pandanales and in some families of Asparagales and Liliales.

The remaining (unshaded) groups either have a widely scattered, sometimes cosmopolitan or subcosmopolitan distribution or they seem to have evolved to a large extent in the Northern Hemisphere, either on a Laurasian continent or (later) on the northern continents. Here belong the Poales (Poaceae subfam. Pooideae), Cyperales, some families of Zosterales, Najadales and families in Liliales (Liliaceae, Melanthiaceae, Calochortaceae, Tricyrtidaceae and perhaps part of the Colchicaceae). In these cases the origin is often uncertain as there is a tendency for the widely distributed families to have effective means of vegetative dispersal and for various taxa to be long-distance dispersed.

A fuller account of the distributional patterns of the respective orders will now be given.

In the Alismatiflorae many of the aquatic families are widely distributed. The Aponogetonaceae, found in both Africa and Asia as well as fossilized in Patagonia, thus have perhaps a Gondwanan origin. Butomaceae are of the

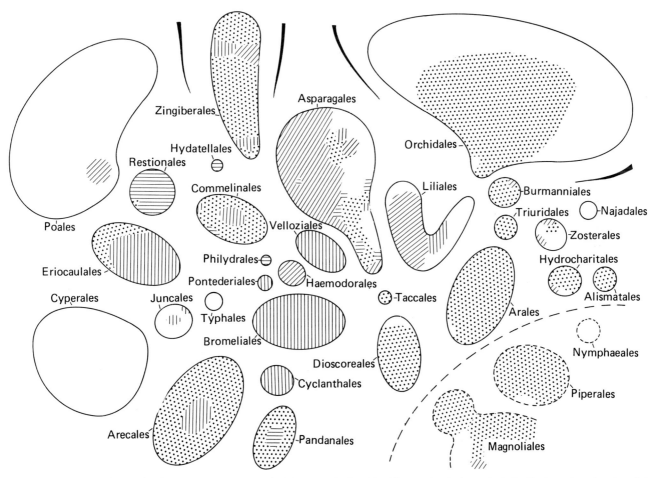

Diagram 84. Geographic distribution and concentration. Southern Hemisphere families are hatched, tropical distributions dotted and mainly Northern Hemisphere and subcosmopolitan families left unshaded. Southern Hemisphere families are divided into those with a western concentration (South American or South American-African) with vertical hatching; those with an eastern concentration (Australia or Australian-South African) with horizontal hatching; and those with a central or wide Southern Hemisphere distribution with oblique hatching. The latter distributions are of great interest in judging the past history, possibly dating back to a Gondwanaland origin.

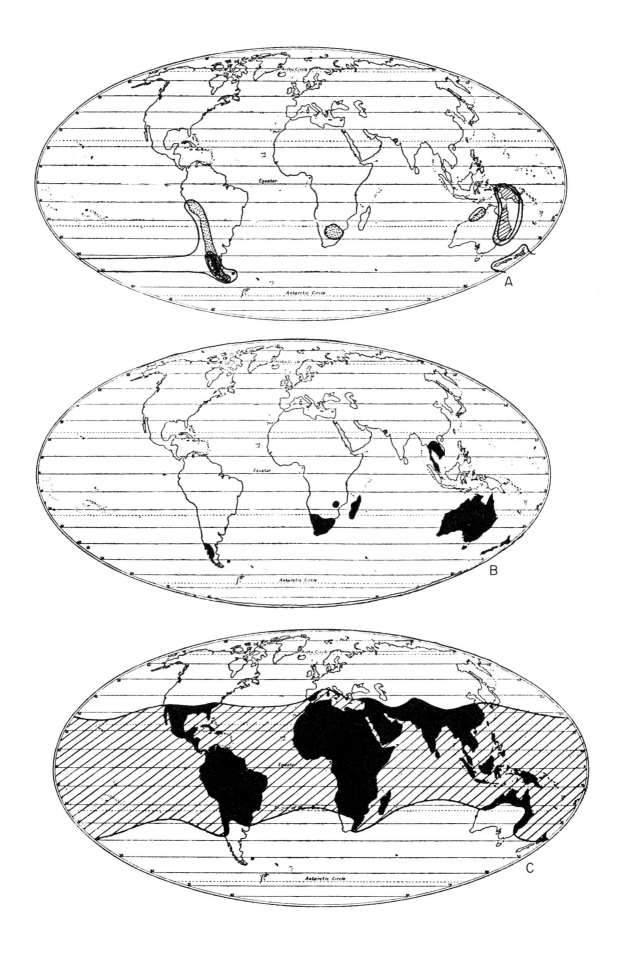

Northern Hemisphere and the Alismataceae also mainly so, although subfam. Limnocharitoideae are mainly tropical American, while the Hydrocharitaceae have a wide, mainly tropical distribution.

A similarly scattered pattern of distribution is met with in the Zosterales, the Scheuchzeriaceae being Northern Hemisphere and the Juncaginaceae, Potamogetonaceae, Zosteraceae, Zannichelliaceae and Najadaceae subcosmopolitan although one can discern a marked southern distribution pattern in the Juncaginaceae (with genera in Patagonia and Australia) as well as the Zannichelliaceae (*Althenia* in Australia). The Posidoniaceae with one species in the Atlantic–Mediterranean region and two in temperate Australian waters show a remarkably disjunct distribution. The Cymodoceaceae are tropical. (Triuridaceae, see below.)

The Arales are widely distributed with tropical concentrations showing links between America and Africa.

The Triuridales and Burmanniales are also specialized pantropical rainforest families, the Corsiaceae being found in Chile as well as New Guinea and Northern Australia, indicating perhaps long distance dispersal rather than an old southern connection.

The orchids also have a wide tropically concentrated but cosmopolitan distribution, the Apostasiaceae being found in Indomalesia and tropical Australia, and the Cypripediaceae being wide ranging including South America and Eurasia.

The Liliales comprise a phytogeographically heterogeneous order with the Iridaceae southern in its concentration, *Iris* and some other genera being conspicuous exceptions. The Alstroemeriaceae is South American, the Liliaceae, Calochortaceae, Tricyrtidaceae and Melanthiaceae are North Hemisphere families while Colchicaceae has a centre in southern Africa although *Colchicum* itself is widely distributed in Eurasia.

More southernly distributed are the Asparagales, where there are nonetheless a few North Hemisphere families: the Convallariaceae, Agavaceae, Funkiaceae, Hemerocallidaceae, possibly Dracaenaceae, and great parts of the Alliaceae and Hyacinthaceae, although even in the latter two families there are apparent centres of variation in South America and South Africa respectively. A South American ("West Gondwanaland") concentration is not so pronounced as the Australian ("East Gondwanaland") or African in this order. Australian or Australasian families are the Petermanniaceae, Dasypogonaceae,

Doryanthaceae, Xanthorrhoeaceae and Phormiaceae, African are Eriospermaceae and Cyanastraceae, while other groups are in common between two or three of the southern continents: Australasia and South America have in common Philesiaceae, Asphodelaceae subfam. Astelioideae and Dianellaceae (*Excremis* in South America); Africa and South America have in common the Herreriaceae (*Herreriopsis* on Madagascar), Hypoxidaceae and Tecophilaeaceae.

There is also a marked southern to tropical concentration for the Amaryllidaceae, Hyacinthaceae and Asphodelaceae (with the *Aloë* group in South Africa and the *Johnsonia* group in Australia), and Alliaceae has the subfamily Agapanthoideae centred in Southern Africa, but has most genera, i.e. nearly the whole subfamily of Gilliesioideae, in South America.

The Velloziaceae though occurring also in Africa have their centre in South America (Brazil) and the numerous Bromeliaceae have an even stronger concentration there. The Pontederiales also belong to this group although *Monochoria* is an Old World genus. The Haemodorales have a South Hemisphere distribution with centres in Australia and South Africa, but genera such as *Xiphidium* and *Lachnanthes* are American. Finally the Philydraceae range from southern Australia to Japan.

The Zingiberiflorae have a pantropical distribution through the large families Zingiberaceae and Marantaceae, but the Heliconiaceae are mainly, and the Cannaceae exclusively, South American. The Strelitziaceae are South American–African (*Ravenala* on Madagascar) and Musaceae *sensu stricto* palaeotropic. This suggests an old South Hemisphere ("Gondwanaland") origin for the whole order.

The Commeliniflorae include some widely distributed groups (Poales, Typhales, Cyperales, Juncales) which are in fact subcosmopolitan and well represented in the Northern Hemisphere. At least the Juncales have however a South American centre of variation, the Thurniaceae and most genera of Juncaceae (and even most sections of *Juncus*) being concentrated here.

The Eriocaulales are South American in concentration with almost the whole of the Rapateaceae (except the African *Maschalocephalus*), most Xyridaceae (including *Abolboda*, *Orectanthe* and *Achlyphila*) and also most Eriocaulaceae in this continent. The last mentioned family has a few genera also in Africa but only *Eriocaulon* in Eurasia.

Fig. 92. Distribution of the families Philesiaceae (A), Restionaceae (B) and Arecaceae (C). Philesiaceae has a general "Gondwana continent" distribution with representation in all three southern continents and concentrations in the East as well as West. Restionaceae also has a Gondwanaland distribution but with a pronounced (central-) eastern concentration. Arecaceae, finally, has a wide pantropical distribution (From Good, 1947.)

The Commelinales are a widely distributed tropical order with a marked South American concentration, most commelinaceous genera and most species of *Mayaca* being found here, while the probably derived *Cartonema* is Australian.

The Restionales are decidedly Southern Hemisphere with an eastern Gondwana concentration. Most genera of Restionaceae are Australian and South African. One genus, *Leptocarpus*, extends from Australia and the Indo-China Peninsula to South America. The Centrolepidaceae occur in Australia, Tasmania, New Zealand, Malaysia and part of South-eastern Asia, with but one species, *Gaimardia australis* (cf. *Leptocarpus*), in South America. The Hydatellales are also Australian.

The Areciflorae are mostly tropical groups. The palms doubtless have their greatest concentration in South America, with subcentres on Madagascar and in Asia, but not very many genera in Australia; the Cyclanthaceae are wholly South American. In great contrast to these two groups the Pandanales are palaeotropic.

From the above it is clear that many monocotyledonous groups have a Southern Hemisphere concentration (variously striated in the diagram), which implies that much of the earliest differentiation of the subclass took place on the "Gondwana continent" partly before, partly after, this broke up and the daughter continents separated from each other. The fact that certain families (like the Bromeliaceae and Cyclanthaceae) are restricted, for example, to one continent (e.g. America) in itself does not necessarily mean that they "appeared" after the period with an Antarctic connection; their ancestors might well have occurred in a restricted part of Gondwanaland, and thus had a narrow distribution.

The dicotyledons associated with monocotyledons are largely tropical, and except for the Magnoliaceae and a few more families are possibly South Hemisphere in origin. The Nymphaeales comprise a widely distributed group showing obvious signs of being advanced ecologically as well as in regard to dispersal means.

Base Data

General tropical distribution (or at least with some tropical concentration).

ALISMATIFLORAE: HYDRO.: Hydro.—ALISM.: Alism.—ZOSTE.: Cymod.—TRIUR.: Triur.

ARIFLORAE: ARAL.: Arac.

LILIIFLORAE: DIOSC.: Diosc.—TACCA.: Tacca.—ASPAR.: Smila.: *Smilax*; Dracae. (incl. Nolin.); Hyaci. p.p.; Allia. p.p.—BURMA.: Burma.; Thism.—ORCHI.: Apost.; Orchi.

ZINGIBERIFLORAE: ZINGI.: Lowia.; Musac.; Zingi.; Maran.

COMMELINIFLORAE: COMME.: Comme. (a western South Hemisphere centre may be discerned). —ERIOC.: Erioc. p.p.—RESTI.: Hangu.

ARECIFLORAE: ARECA.: Areca.—PANDA.: Panda. (in these orders there is a western and eastern concentration respectively).

Dicotyledons associated with the monocotyledons:

MAGNOLIIFLORAE: MAGNO.: Annon., Aristo., Myrist. etc.—PIPER.: Piper.

Distribution centred in the western parts of the Southern Hemisphere ("West-Gondwanaland" distribution)

LILIIFLORAE: ASPAR.: Herre.; Allia. subfam. Gilliesioideae.—LILIA.: Alstr.—PONTE.: Ponte.—VELLO.: Vello.—BROME.: Brome.

ZINGIBERIFLORAE: ZINGI.: Helico.; (Strel.) Canna.

COMMELINIFLORAE: COMME.: Comme. p.p.; Mayac.—ERIOC.: Rapat.; Xyrid.; Erioc.—JUNCA.: Thurn.; Junca. (main centre in S. America).

ARECIFLORAE: ARECA.: Areca. (a certain concentration in the western hemisphere).—CYCLA.: Cycla.

Distribution centred either in the African part of the Southern Hemisphere or evenly Southern Hemisphere ("Central or General Gondwanaland" distribution)

ALISMATIFLORAE: HYDRO.: Apono.—ZOSTE.: Junca.; Zanni.

LILIIFLORAE: ASPAR.: Phile.; Hypox.; Aspho. (all its three subfamilies Asphodelioideae, Anthericoideae and Astelioideae); Diane. p.p.; Tecoph. (incl. *Lanaria*); Cyana.; Erios.; Hyaci. p.p.; Amary.—LILIA.: Irida.; Geosi.; Colch.—BURMA.: Corsi.—HAEMO.: Haemo.

ZINGIBERIFLORAE: ZINGI.: Strel.; Costa (?).

COMMELINIFLORAE: possibly POAL. only (?).

Dicotyledons associated with the monocotyledons:

MAGNOLIIFLORAE: MAGNO.: Winter.

Distribution centred in the eastern parts of the Southern Hemisphere ("East-Gondwanaland" distribution)

LILIIFLORAE: ASPAR.: Smila. p.p. (*Ripogonum*); Peter.; Dorya.; Dasyp.; Phorm.; Xanth.; Diane.; p.p.—PHILY.: Phily.

COMMELINIFLORAE: COMME.: Carto.—HYDAT.: Hydat.—RESTI.: Restio.; Anart.; Ecdei.; Centr.; Flage. (to tropical); Joinv.

ARECIFLORAE: PANDA.: Panda.

The families not enumerated either have a subcosmopolitan distribution or a North Hemisphere ("Laurasian") distribution, with (arctic-) temperate-subtropical range.

DIFFUSE CENTROMERE
(Diagram 85)

Chromosomes with "diffuse centromeres" have been described in the Cyperaceae (Tanaka, 1939 etc.; Håkansson, 1954; Strandhede, 1958 etc.) and in the Juncaceae (Nordenskiöld, 1951, 1956). The "diffuse" (non-localized) centromere was first described by Schrader (1935). It is not always an adequate term, as centromeric activity may be localized to several points on the chromosome (being polycentric). It is known in several genera of the Cyperaceae, but whether it is consistent in this family is not known. It also occurs in *Luzula* (Juncaceae) where the chromosomes may be few ($2n = 6$) and relatively large or more numerous and small, the number of nucleoli

being however the same number. At metaphase the chromosomes lie flat with their whole length in the equatorial plane and in anaphase the daughter chromosomes move apart parallel to each other (Håkansson, 1954). In the Juncaceae the condition seems to be variable, however, and within *Juncus* chromosome numbers are often constant and centromere activity concentrated to definite parts of the chromosomes.

There is no doubt, however, that this type of chromosome, apparently unknown outside the Cyperales and Juncales among the seed plants, has developed in the common ancestors of these orders and supports their close relationship.

An additional important circumstance is that precisely these two orders have post-reductional meiosis (Battaglia and Boyes, 1955).

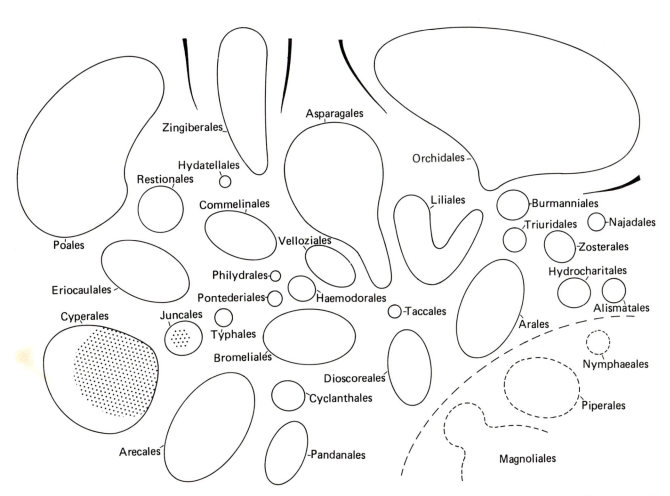

Diagram 85. Chromosomes with "diffuse centromere" (shaded). This feature forms a strong argument for the close connection between the Juncales and Cyperales.

SALT TOLERANCE (Diagram 86)

In the absence of experimental data the ability of plants to grow in saline soils cannot be predicted. However, the natural distribution of salt tolerant species indicated certain families possess this characteristic. The distribution of such families is given in Diagram 86, from which it can be seen that they occur in three of the superorders.

The truly aquatic marine angiosperms comprise a great part of the Zosterales and some genera of the Hydrocharitales (Alismatiflorae) and no doubt represent a few distinct evolutionary lines.

The marine aquatics of the Zosterales include all taxa of the Zosteraceae, Posidoniaceae and Cymodoceaceae, and within the Hydrocharitaceae, the genera *Enhalus*, *Thalassia* and *Halophila*. The genera of Zannichelliaceae and Potamogetonaceae occur in fresh or brackish water,

but *Ruppia* sometimes in salt-ponds with quite high salinity (Setchell, 1924; Davis and Tomlinson, 1974).

Species of *Triglochin* (Juncaginaceae) are often rooted in saline soil.

Among the Areciflorae the nypa palm (*Nypa*) grows along the banks of tidal rivers and the date palm (*Phoenix*) is highly tolerant of saline soil. In addition, several species of *Pandanus* (Pandanales) are sea-shore plants.

In the Commeliniflorae five of the families have genera which may live in saline or brackish environments. These are the Poaceae, with many genera involved (mostly Eragrostoideae), the Cyperaceae (many genera), the Juncaceae (e.g. *Juncus maritimus*), the Restionaceae (*Leptocarpus*) and the Typhaceae (*Typha* spp.). In spite of being somewhat related families it is unlikely that this property has a common origin.

In the Liliiflorae there appears to be mainly one genus,

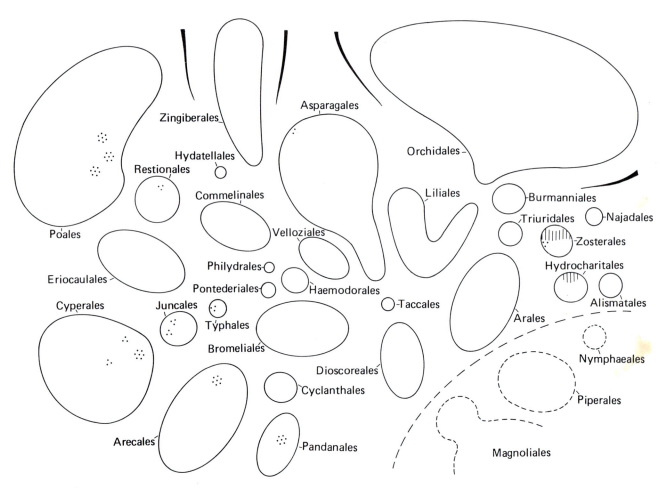

Diagram 86. Salt tolerance (shaded). The halophytic monocotyledons are divisible into those rooted in salt water only (dots; mainly in the left part of the diagram plus species of *Triglochin*, Juncaginaceae) and the submerged marine aquatics (hatching) which all belong to the Alismatiflorae.

Crinum (possibly also *Pancratium*; both of the Amaryllidaceae, Asparagales) which grows in saline environments.

Base Data

Submersed marine aquatics
ALISMATIFLORAE: HYDRO.: *Enhalus, Halophila, Thalassia.*—ZOSTE.: (Potam.: *Potamogeton* and *Ruppia* in brackish waters;) Zoste.: all; Posid.: *Posidonia*; Cymod.: all. (—NAJAD.: Najad.: *Najas marina* in brackish water.)

Plants rooted in salt water
ALISMATIFLORAE: ZOSTE.: Junca.: *Triglochin* spp.

LILIIFLORAE: ASPAR.: Amary.: *Crinum* spp. (maybe also spp. of *Pancratium*).

COMMELINIFLORAE: TYPHA.: Typha.: *Typha* spp.—JUNCA.: Junca.: *Juncus* spp.—CYPER.: Cyper.: spp. of several genera, e.g. *Scirpus* and *Schoenoplectus.*—RESTI.: Resti.: *Leptocarpus.*—POAL.: Poac.: several genera.

ARECIFLORAE: ARECA.: Areca.: certain species.—PANDA.: Panda.: *Pandanus* spp.

ALUMINIUM ACCUMULATION
(Diagram 87)

Species accumulating aluminium though reasonably common amongst dicotyledons are rare amongst monocotyledons. Such plants may sometimes be recognized amongst herbarium specimens because their dried foliage is a bright yellowish-green. When reporting upon his survey of aluminium in the plant kingdom, Chenery (1950) defined as accumulators plants possessing in excess of 1000 p.p.m. dry weight of the element in oven-dried material (105° C).

According to this criterion most members of the Rapateaceae, several species of *Aletris* (Melanthiaceae) and one species of *Eleocharis* (Cyperaceae) are accumulators. From his tables of data two other species came near to qualifying. These were *Xyris indica* (832 p.p.m.) and *Centrolepis novoguineensis* (850 p.p.m.).

The above results indicate that the capacity to absorb high amounts of aluminium and to survive is a property possessed by few families. The distribution of these is shown in Diagram 87.

From the figure it is evident that most aluminium accumulators are in the Commeliniflorae, the single exception being *Aletris* in Liliiflorae.

Base Data

See Chenery (1950) for full data. Mention is made above only for figures either above or approaching 1000 p.p.m. dry weight.

SOME CHEMICAL CHARACTERS
(Fig. 93)
(in cooperation with
S. Rosendal Jensen)

General Remarks

Chemical characters in the last decades have played an increasingly important role in taxonomy at the higher as well as lower levels. This has given a great impetus to new aspects, and one may be inclined to pay more attention to these characters than to morphological ones. However, one must be cautious when handling chemical data for several reasons.

Often, as in the accounts to follow, presence versus absence of particular compounds or groups of compounds have been registered. This is a dubious approach, as in such a case no attention is paid to the relative quantities of the substances. What one is noting is that either of a number of possible pathways in the synthesis of a class of compounds occurs in the plant species. As the methods used have their limitations, and often not the same methods and at least not the same degree of exactness were at hand behind the different data, there are also other sources of error. Besides it is known that different parts of the plant accumulate different substances (in particular seeds versus vegetative parts) and that there are often seasonal changes in the accumulation of compounds in the same parts of the plants.

When a number of chemical similarities occur between different plant groups (such as the presence of certain flavonoids in grasses and palms below) then there is also the problem as to whether these can be regarded as independent similarities weighing heavily in judging relationships, or are side branches or steps in fundamentally the same biosynthetic pathway, a pathway which may even have evolved independently in different plant groups.

Finally, the occurrence of particular groups of compounds must be judged with regard to their possible functions as poisons or repellant substances in relation to insects or higher animal herbivores. Thus there is a tendency that steroid saponins are absent where rich amounts of alkaloids are present, a condition observable for Amaryllidaceae and Colchicaceae (Diagram 89). It seems that the plant in the course of its evolution (by mutation and selection) ends up with, under the circumstances, the least energy-consuming combination of character states.

In spite of all this: where a definite pattern of variation in the possession of chemical constituents is present, this no doubt can reflect relationships. This is perhaps confirmed most reliably through its correlation with a number of independent morphological characters.

For the purpose of this study only a very limited set of chemical characters has been chosen. These have been selected by virtue of their more common presence in at least certain orders of monocotyledons, their more definite distribution pattern indicating possible use for phylogenetic conclusions, and their being fairly well known in the orders of monocotyledons.

No attempt has been made here to give detailed

references to the literature for these groups of compounds. Instead, reference has usually been given to one or a few sources where the distribution of the compounds concerned are more fully treated.

For saponins, chelidonic acid and cyanogenic compounds data from Professor R. Hegnauer's personal card indexes have been placed at the disposal of one of the contributors (S. R. Jensen). Therefore, there has not been considered a great need for further references.

Besides the groups of compounds included here, substantial taxonomic and phylogenetic contributions have been obtained from, for example, the occurrence of non-protein amino acids and from the occurrence of diverse alkaloids, the latter of which are systematic markers for the Amaryllidaceae, the Colchicaceae and other groups. The taxonomic value of these compounds, however, lie on a lower taxonomic level than is being surveyed and/or are too little known over the whole monocotyledon system and so have been neglected here.

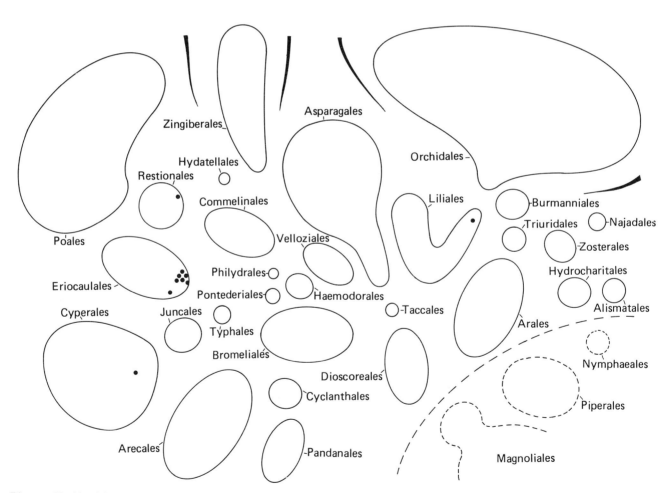

Diagram 87. Aluminium accumulation. The dots refer to the tabulated taxa with aluminium concentrations in excess of 850 p.p.m.

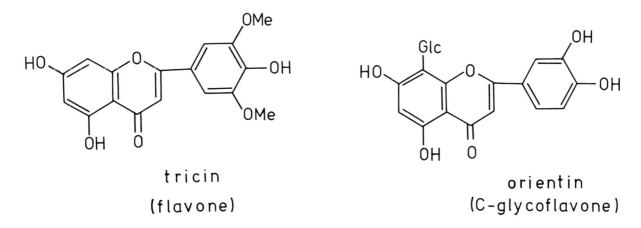

Fig. 93. Chemical formulas of some substances belonging to groups of compounds surveyed in this study.

Saponins: General Occurrence (Diagram 88)

Saponins include two main groups, viz. the steroid and triterpene saponins.

They comprise a heterogenous group of glycosides with terpenoid aglyca. In water solutions they are characterized by the ability to foam and many of them have a haemolytic effect.

Steroid saponins have aglyca with furostanol and spirostanol structures while triterpene saponins are derivatives of oleanans and related compounds. These kinds of saponins have widely different distributions, and as will be shown below, the steroid saponins are the ones most widely distributed in monocotyledons, while the triterpene saponins are the commonest type in dicotyledons.

Apparently no true triterpene saponins have been isolated as such from the monocotyledons, but their aglucones, the triterpenes, are rather common, particularly among the grasses, where they are known to occur in the free form in most species.

The verified occurrences for the steroid saponins will be dealt with below, but as they are expected to have a rather incomplete coverage, we have here included the records of saponins at large, based mainly on the accounts in Hegnauer (1963) and Gibbs (1974, pp. 1854–1980). These data, most of which refer to older literature, are based on qualitative tests and some are probably unreliable. Yet the overall distribution has a certain interest in comparative discussions.

In the Alismatiflorae there is just one positive (but doubtful) record (*Sagittaria*), while in the Ariflorae 10 species have been recorded as saponine-positive, 13 as negative, the latter including the two records for *Lemna*.

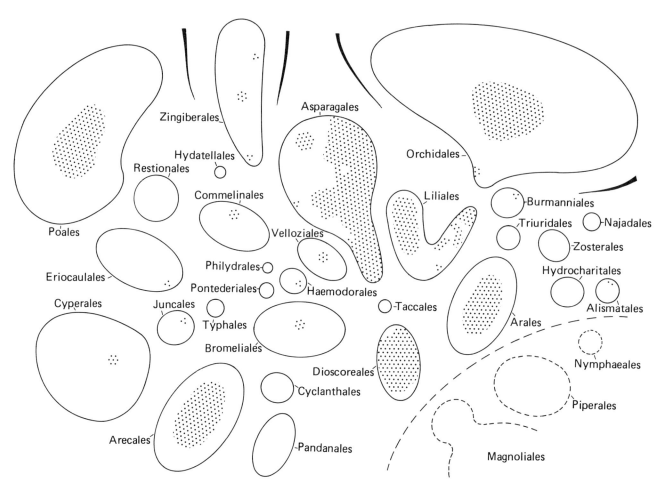

Diagram 88. Saponins (unspecified); the diagram presents the distribution of plants giving a saponin reaction. It is approximate and includes unverified cases but shows a better coverage than Diagram 89.

Saponins are frequent in the Dioscoreales (the Taccales being as yet unknown?) and most Asparagales and are also known in many Liliales and taxa of the Orchidales and Burmanniales. Within the Asparagales there are positive records in all families studied except the Hypoxidaceae, Cyanastraceae, Tecophilaeaceae, Dianellaceae and Hemerocallidaceae, in all of which only few taxa have been studied. (Yet a certain tendency may be discerned?) For the Amaryllidaceae Gibbs (1974) quotes several saponin records (nine genera), which is surprising against the common impression that at least steroid saponins are here substituted by alkaloids.

The Liliales seem to be poorer in saponins than are the Asparagales, and the compounds are apparently lacking in the Colchicaceae, where they are again substituted by alkaloids. Saponins are reported to occur in more than half of the genera of the Iridaceae investigated, in species of *Alstroemeria* and *Bomarea* (Alstroemeriaceae), and are possibly proportionally rare in the Liliaceae. In the Melanthiaceae they are stated to occur in many (11) genera, and there is also a positive record for the Tricyrtidaceae. One species is given as positive for the Burmanniales (Corsiaceae) and some for the Orchidales, where they seem to occur sometimes in both Cypripediaceae and Orchidaceae.

While saponins mostly seem to be lacking in the orders of Liliiflorae with starchy endosperm there are positive records for each of the Haemodorales (*Conanthera*) and Bromeliales (*Ananas*, *Hechtea*). There are also single positive records in the Commelinales (*Tradescantia*) and Eriocaulales (*Xyris*) as well as Juncales (seeds of *Marsippospermum*) and Cyperales (*Carex*). The grasses and palms include many plants positive for saponins, and both steroid saponins and free triterpenes are known from these orders.

A few new records of saponins are from the Zingiberiflorae (Heliconiaceae, Zingiberaceae and Costaceae); in two of these at least they represent steroid saponins.

Steroid Saponins (Diagram 89)

In the present compilation these compounds comprise several groups (steroid amines, bufadienolides, cardenolides and spiroketal-saponins), all having a steroid skeleton. The compounds presumably form a homologous biosynthetic group, derived from triterpenes by a specific series of reactions whereby the steroid skeleton has been formed. The compounds are usually constituted by a steroid aglucone attached to a mono- or (more often) a di- or oligosaccharide.

The occurrence of these compounds has been rather intensively investigated due to their interesting pharmacological properties or because some of the compounds are precursors of economically important pharmaceuticals (diosgenin etc.).

Apparently triterpene and steroid saponins have a tendency to be vicarious (mutually exclusive) although this is by no means absolute.

In some monocotyledons the steroid saponins have not been verified with certainty as with a record in *Sagittaria* (Alismataceae). Saponin compounds recorded in the Ariflorae (see general occurrence) may comprise both steroid and triterpene saponins, but the occurrence of steroid saponins is certain for a couple of genera (*Pinellia*, *Montrichardia*).

On the other hand the steroid saponins occur rather generally in the Dioscoreales, Asparagales and Liliales, although they are perhaps absent in the alkaloid-rich Amaryllidaceae and probably also so in the Colchicaceae. Only a single verified record is seen in our list for Orchidales (*Dendrobium*), but this group has probably been investigated insufficiently. The single record for a saponin in Velloziales (*Barbacenia*) is a triterpene saponin and so does not indicate any relationship to the Asparagales. Likewise this type of saponin occurs in the Bromeliales (*Ananas*), but *Hechtea* in the same order is reported to have steroid saponins. The saponins of the Haemodorales are not known.

Surprisingly, saponins in the Commelinaceae (*Cyanotis*), Zingiberaceae (*Alpinia*) and Costaceae (*Costus*) of the Commeliniflorae and the Zingiberiflorae respectively are of steroid type, while in grasses (Poaceae) the saponins are probably mainly of the triterpene type. Two exceptions are given, viz. *Avena* and *Sorghum*.

In the palms (Arecales) the saponins may be both of triterpene and steroid types.

Steroid saponins are generally poisonous (blood haemolysing) and thus seem to serve mainly as a means of protecting the plants. This may explain their probable absence in families having rich amounts of poisonous alkaloids. These include the Amaryllidaceae and Colchicaceae, where they would otherwise be expected from a phylogenetic point of view. On the whole the concentration of the steroid saponins are in the orders of the Liliiflorae without or with starchless endosperm, and the occurrence outside these orders is—as far as is known—only sporadic. The (few) verified occurrences in the Arales, Zingiberales, Poales, Commelinales and Arecales may serve as a warning against drawing far-reaching conclusions solely from the occurrence of steroid saponins.

Steroid saponins in the above general sense are found

rather widely scattered in the dicotyledons, where they have so far been found in the Anacardiaceae, Apocynaceae, Asclepiadaceae, Brassicaceae, Cactaceae, Celastraceae, Euphorbiaceae, Lamiaceae, Melianthaceae, Moraceae, Ranunculaceae, Scrophulariaceae, Solanaceae, Sterculiaceae, Tiliaceae, Zygophyllaceae and probably more families. These families do not form any natural groups and thus these saponins are of little taxonomic significance at this level of dicotyledons.

Base Data

ALISMATIFLORAE: ALISM.: Alism.: *Sagittaria* (?).
ARIFLORAE: ARAL.: Arac.: *Montrichardia, Pinellia.*
LILIIFLORAE: DIOSC.: Diosc.: *Dioscorea, Tamus*; Trill.: *Paris, Trillium.*—ASPAR.: Smila.: *Smilax*; Phile.: *Lapageria, Philesia*; Conva.: *Aspidistra, Clintonia, Convallaria, Liriope, Maianthemum, Ophiopogon, Polygona-* *tum, Reineckea, Rhodea, Smilacina*; Aspar.: *Asparagus, Ruscus, Semele*; Herre.: *Herreria*; Dracae.: *Calibanus, Dasylirion, Dracaena, Nolina, Sansevieria*; Dorya.: *Doryanthes*; Agava.: *Agave, Beschorneria, Clistoyucca, Furcraea, Hesperaloë, Polianthes, Samulea, Yucca*; Aphyl.: *Aphyllanthes*; Aspho.: *Anemarrhena, Asphodelus, Bottinaea, Chlorophytum, Cordyline, Debesia, Eremurus, Pasithea*; Funki.: *Hosta*; Hyaci.: *Albuca, Bowiea, Camassia, Chlorogalum, Dipcadi, Hyacinthus, Muscari, Ornithogalum, Puschkinia, Schizobasis, Scilla, Urginea*; Allia.: *Agapanthus, Allium, Brodiaea, Gilliesia, Leucocoryne, Nothoscordum, Tristagma.*—LILIA.: Irida.: *Crocus, Eleutherine, Gladiolus, Homeria, Moraea, Romulea, Tritonia*; Lilia.: *Erythronium, Fritillaria, Korolkowia, Lilium*; Melan.: *Aletris, Chamaelirium, Chionographis, Heloniopsis, Metanarthecium, Narthecium, Schoenocaulon, Tofieldia, Veratrum, Zigadenus.*—BURMA.: Burma.: *Arachnites* (?).—ORCHI.: Orchi.: *Dendrobium.*—BROME.: Brome.: *Hechtea.*

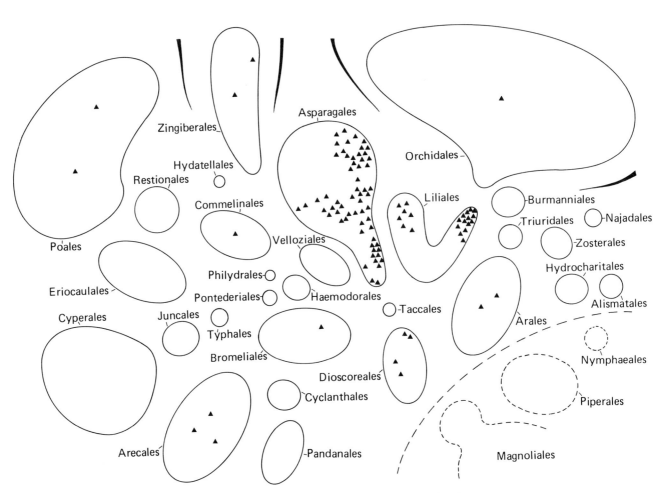

Diagram 89. Steroid saponins. Only verified cases of the respective groups are included. Each triangle represents a genus.

ZINGIBERIFLORAE: ZINGI.: *Alpinia*; Costa.: *Costus.*

COMMELINIFLORAE: COMME.: Comme.: *Cyanotis.*—POAL.: Poac.: *Avena, Sorghum.*

ARECIFLORAE: ARECA.: Areca.: *Chamaedorea, Chamaerops, Pseudophoenix.*

Chelidonic Acid (Diagram 90)

Knowledge of the distribution of this compound stems mostly from works of Molisch (around 1900) and Ramstad (1953) who investigated a number of species for the occurrence of chelidonic acid. The data are considered to be fairly representative.

Chelidonic acid in monocotyledons is mainly concentrated in the Liliiflorae, where it has its richest occurrence in the Dioscoreales, Asparagales, Liliales and Haemo-

dorales. Its occurrence in the Orchidales and other liliiflorous orders is probably not well enough known for it to be considered as rare or absent in these groups. A number of records are also known in the grasses (Poaceae) and one in *Canna* (Cannaceae).

The Haemodorales agree well with the Asparagales in the richness of chelidonic acid. The Iridaceae, however, are surprisingly poor and the Liliaceae apparently deficient in chelidonic acid, contrasting with the Colchicaceae and Melanthiaceae which have many records. In the Asparagales the distribution is more even.

The records in Cannaceae and the grasses are difficult to evaluate at present.

In dicotyledons, chelidonic acid occurs in the Ranunculaceae, Papaveraceae, Rhamnaceae, Thymelaeaceae and Lobeliaceae. The three first-mentioned families are also characterized by their general ability to synthesize benzylisoquinoline alkaloids, and of these at least the Ranunculaceae and Papaveraceae are decidedly related.

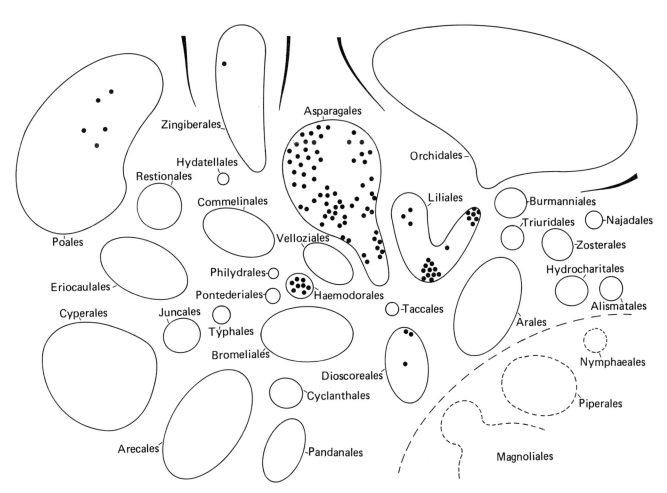

Diagram 90. Chelidonic acid. Each solid circle represents a genus shown to contain this substance.

There is no evidence that any of the dicotyledonous families producing chelidonic acid are particularly closely allied to the Liliiflorae, although a relationship between Ranunculaceae and Liliiflorae was at one time a popular hypothesis.

Base Data

LILIFLORAE: DIOSC.: Diosc.: *Dioscorea*; Trill.: *Paris, Trillium.* ASPAR.: Smila.: *Ripogonum, Smilax*; Conva.: *Aspidistra, Convallaria, Liriope, Ophiopogon, Reineckea, Tupistra*; Aspar.: *Asparagus, Ruscus, Semele*; Dracae.: *Dasylirion, Dracaena*; Dasyp.: *Acanthocarpus, Dasypogon, Lomandra*; Phorm.: *Blandfordia, Phormium*; Agava.: *Agave, Yucca*; Hypox.: *Hypoxis*; Aspho.: subfam. Asphodeloideae: *Aloë, Asphodelus, Bulbine, Bulbinella, Eremurus, Kniphofia*; subfam. Anthericoideae: *Anthericum, Borya, Bottinaea, Chlorophytum, Dichopogon, John-*

sonia; subfam. Astelioideae: *Astelia, Cordyline*; Tecoph.: *Conanthera (Cummingia), Lanaria*; Hyaci.: *Hyacinthus, Lachenalia, Muscari, Ornithogalum, Scilla*; Allia.: *Allium, Brodiaea*; Amary.: 22 genera (c. 60% of the species studied by Ramstad, 1953). LILIA.: Irida.: *Lapeyrousia, Romulea, Trimezia*; Colch.: *Androcymbium, Anguillaria, Baeometra, Colchicum, Gloriosa, Iphigenia, Merendera, Ornithoglossum, Sandersonia, Uvularia, Wurmbea*; Alstr.: *Alstroemeria*; Melan.: *Heloniopsis (Sugerokia), Melanthium, Narthecium, Schoenocaulon, Tofieldia, Veratrum, Zigadenus.* HAEMO.: Haemo.: *Anigozanthos, Blancoa, Conostylis, Dilatris, Lachnanthes, Lophiola, Tribonanthes, Wachendorfia.*

ZINGIBERIFLORAE: ZINGI.: Canna.: *Canna.*

COMMELINIFLORAE: POAL.: Poac.: *Lolium, Phalaris, Phleum, Sorghum, Zea.*

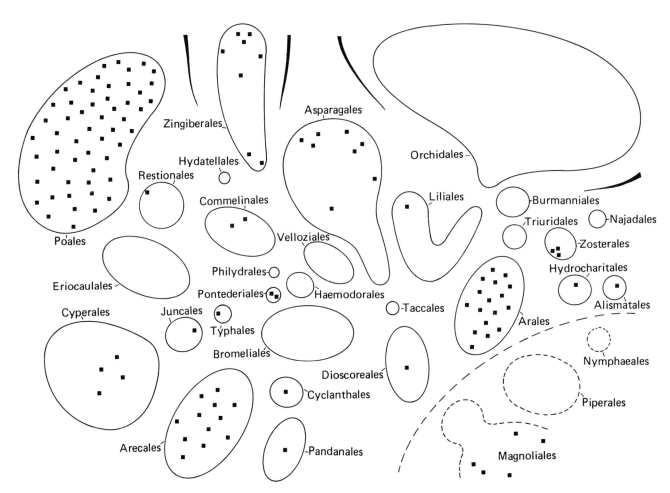

Diagram 91. Cyanogenic compound. Each square represents a genus shown to contain cyanogenic substances.

Cyanogenic Compounds (Diagram 91)

The presence of cyanogenic compounds is easily detected from minute amounts of fresh plant material or herbarium material by a simple test, and a large number of species have been examined.

Cyanogenic compounds are usually found in the plants as glycosides. They are derived from the amino acids from the primary metabolism, and except for a single example, the glycosides known from the monocotyledons and from the Magnoliiflorae are all derived from tyrosine (see below).

The data presented here are mainly from Hegnauer (1963) and Gibbs (1974).

Cyanogenic compounds are known from various orders in the monocotyledons and show no definite pattern, although certain families are particularly rich (Poaceae, Araceae, Arecaceae). The absence in certain large orders (Eriocaulales, Bromeliales, Orchidales, nearly so in the Liliales) may partly depend on lacking investigations.

Conspicuous is the richness of cyanogenic compounds in *Scheuchzeria*, *Triglochin* and *Lilaea* of the Zosterales, which indicates a relatively close relationship. A certain tendency for concentration of cyanogenic compounds in the Poaceae, Flagellariaceae, Arecaceae and partly Cyperaceae on the one side and the Ariflorae on the other is perhaps discernible (cf. flavonoid pattern), while the Liliiflorae seem to be poor in these compounds, but this impression may be false as more investigations are made. All orders of the Areciflorae have representatives with cyanogenic compounds.

While cyanogenic compounds seem to be missing in the Nymphaeiflorae and in the Piperales of the Magnoliiflorae, they occur in many families in the Magnoliales and Laurales of the latter superorder (van Valen, 1978b).

Two recent papers have been published by Hegnauer (1973, 1977) on the importance of the distribution of cyanogenic compounds; the former deals with monocotyledons in particular.

The cyanogenesis in nearly all monocotyledons proceeds along the tyrosine pathway, exceptions being, for example, *Chlorophytum capense* and *C. comosum* (Asphodelaceae, Asparagales), with holocalin, i.e. the phenylalanine pathway (van Valen, 1978a).

Cyanogenic compounds formed along the tyrosine pathway are also known in *Liriodendron* (Magnoliaceae), *Calycanthus* and *Chimonanthus* (Calycanthaceae), *Trochodendron* (Trochodendraceae), some Ranunculaceae, *Nandina* (Berberidaceae), *Papaver* and *Eschscholtzia* (Papaveraceae), *Dicentra* (Fumariceae), *Andrachne*, *Bridelia*, *Poranthera* and *Securinega* (Euphorbiaceae), certain Fabaceae, certain Proteaceae, *Borago* (Boraginaceae) and some Campanulaceae.

From this it is uncertain whether the tyrosine-derived cyanogenic compounds (with few exceptions) in the monocotyledons have any great significance, but perhaps they could be compared with the *P*-type plastids in the sieve tubes, which are consistent in the monocotyledons but have a somewhat scattered distribution in the dicotyledons.

Base Data

ALISMATIFLORAE: ALISM.: Alism.: *Alisma*.—HYDRO.: Hydro.: *Vallisneria*.—ZOSTE.: Scheu.: *Scheuchzeria*; Junca.: *Lilaea, Triglochin*.

ARIFLORAE: ARAL.: Arac.: *Alocasia, Amorphophallus, Anthurium, Arisarum, Arum, Calla, Colocasia, Cyrtosperma, Dieffenbachia, Dracontium, Eminium, Lasia, Lysichiton, Pothoidium, Schismatoglottis, Schizocasia, Scindapsus, Xanthosoma, Zantedeschia*.

LILIIFLORAE: DIOSC.: Diosc.: *Dioscorea*.—ASPAR.: Agav.: *Yucca*; Aspho.: *Chlorophytum* (see above); Hyaci.: *Albuca, Urginea*; Allia.: *Allium* (?); Amary.: *Amaryllis, Cyrtanthus* (*Vallota*), *Leucojum*.—LILIA.: Irida. *Aristea*.—PONTE.: Ponte.: *Eichhornia, Monochoria*.

ZINGIBERIFLORAE: ZINGI.: Musac.: *Musa*; Helic.: *Heliconia*; Zingi.: *Hedychium*; Costa.: *Costus*; Canna.: *Canna*; Maran.: *Maranta, Phrynium, Thalia*.

COMMELINIFLORAE: COMME.: Comme.: *Commelina, Tinantia*.—TYPHA.: Typha.: *Typha*.—JUNCA.: Junca.: *Juncus*.—CYPER.: Cyper.: *Cyperus, Fimbristylis, Kyllingia, Schoenus*.—RESTI.: Flage.: *Flagellaria*.—POAL.: Poac.: 53 genera (neg. only in 8 genera studied!).

ARECIFLORAE: ARECA.: Areca.: *Adonidia, Areca, Arenga, Calamus, Caryota, Cocos, Corypha, Dypsis, Livistonia, Phytelephas, Ptychosperma, Roystonea*.—PANDA.: Panda.: *Pandanus*.—CYCLA.: Cycla.: *Carludovica*.

Dicotyledons associated with the monocotyledons:

MAGNOLIIFLORAE: MAGNO.: Annon., Canel., Magno., Degen., Winte.—LAURA.; Calyc.; Laura.

FLAVONOID COMPOUNDS

(By J. B. Harborne)
(Diagrams 92–98)

Introduction

The flavonoids are a large group of phenolic plant constituents, which are united in their biosynthetic origin. All are derived from the same 15-carbon skeleton, as in flavone, the parent compound of the series, and commonly contain hydroxyl, methoxyl and sugar substituents. Visually, the most important classes of flavonoids are the anthocyanins (red to blue pigments) and the chalcones and aurones (yellow pigments). However, the largely colourless flavones and flavonols are the most widely distributed classes in plants. Not only do they occur in almost all flowers, but they are also always present in leaves, where their presence is masked by the ubiquitous chlorophyll.

Because of their universal distribution in vascular plants, their immense structural variation, their ease of detection and their chemical stability, the flavonoids are among the most valuable chemical markers for systematic purposes. Following the pioneering surveys of Bate-Smith (1962, 1968), their distribution within the angiosperms has been extensively investigated. As a result, it

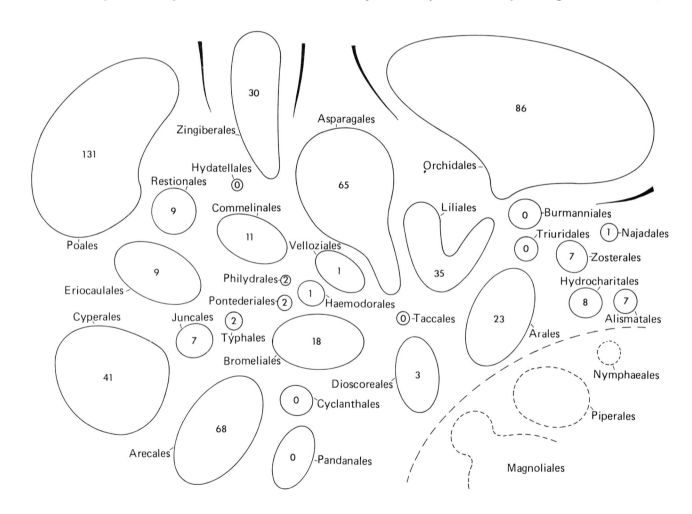

Diagram 92. Number of genera investigated for flavonoids. This diagram is of relevance for the interpretation of Diagrams 93–98 inasmuch as it shows the maximum number of genera of each order investigated for flavonoids. Thus the symbols in any of Diagrams 93–98 cannot exceed the numbers in each ordinal bubble. For example, a negative score for the Pandanales does not indicate absence and ten positive scores for the grasses may still indicate relative rareness of a flavonoid group in that family. (The data on which these diagrams are based are from Gornall *et al.*, 1979.)

is clear that in dicotyledons flavonoid patterns are correlated closely with the woody-herbaceous dichotomy in habit. Thus there is a significant changeover in the type of flavonoids found in the leaves of herbaceous families as compared to those of woody plant families. By contrast, in the monocotyledons, although the same structures are present the situation is very different; the flavonoid patterns encountered are complex and reticulate. There are thus difficulties in interpreting the significance of varying flavonoid patterns within the monocotyledons. In the present review, only a few tentative systematic conclusions can be drawn.

The present account of monocotyledon flavonoids is based mainly on the summary provided by Gornall *et al.* (1979) for all angiosperm taxa investigated to that date. Diagram 92 shows the number of plant species of each order of monocotyledons that have been investigated for their flavonoids. In addition, most of the major families of the monocotyledons have been systematically analysed for flavonoids, with representative sampling at the generic level, and some of these results are shown in Table 4 (from Williams *et al.*, 1981). The available data refer largely to analyses of leaf tissue, although in some cases other plant organs, (e.g. flowers, fruits, roots) have also been analysed. There is sometimes significant variation in pattern between different organs so that results on different tissues cannot always be equated with each other. With this proviso, the distribution of the major classes of flavonoid in each of the monocotyledon groups will now be considered. Subsequently, the few systematic conclusions that can be elicited from the available data will be briefly discussed.

The Flavonols

Two flavonols, kaempferol and quercetin, are extremely widespread in plants so it is not surprising to find that in the monocotyledons they have been detected in every family that has been investigated in depth. In general, therefore, little significance can be attached to their presence/absence although the frequency of their occurrence does vary. It may be noted, for example, that they are significantly uncommon in the families of the Alismatiflorae and that they are also rare in the Poaceae and Juncaceae (Table 4). In all other families investigated, they occur in at least 10% of taxa.

There is a third common flavonol of plants, namely myricetin. This is an interesting taxonomic marker in the dicotyledons, since its occurrence as a leaf constituent is closely associated with the woody habit (Bate-Smith, 1962). In the monocotyledons, myricetin is generally less frequent. It has a very scattered distribution, being

TABLE 4
A Summary of the Leaf Flavonoid Patterns found in the Major Monocotyledonous Families and Orders

Order or family	Flavone C-glycosides	Flavonols	Flavones	Proantho-cyanidins	Tricin	6- and 8-hydroxy-flavonoids	Flavonoid sulphates
Araceae	+ + +	+ +	(+)	+ +	—	—	(+)
Bromeliaceae	+	+ +	+	—	—	+	—
Commelinaceae	+ + +	+	+	nd	(+)	(+)	—
Cyperaceae	+ +	+	+	+ +	+	+	(+)
Alismatiflorae*	+ +	(+)	+	(+)	—	—	+
Poaceae	+ + +	(+)	+	(+)	+ + +	—	+
Iridaceae	+ +	+	nd	+	+	—	nd
Juncaceae	(+)	(+)	+ + +	+	—	—	+
Liliaceae	(+)	+ +	+	+	(+)	—	(+)
Orchidaceae	+ +	+ +	(+)	(+)	(+)	+	(+)
Restionaceae	+	+ + +	+	+ + +	—	+	(+)
Arecaceae	+ + +	+	+	+ + +	+ +	—	+ + +
Zingiberiflorae†	+	+ +	+	+ + +	—	—	(+)

Key:– = absent; (+) = in < 10% of species surveyed: + = in 10–25% of species; + + = in 25–50% of species; + + + = in > 50% of species; nd = not determined
* based on an analysis of Butomaceae, Hydrocharitaceae, Alismataceae, Juncaginaceae, Potamogetonaceae, Posidoniaceae, Zosteraceae, Zannichelliaceae, Ruppiaceae, Najadaceae.
† based on an analysis of Zingiberaceae, Marantaceae, Strelitziaceae, Cannaceae.

recorded in at least five families: Iridaceae (*Iris*), Marantaceae (*Calathea*), Restionaceae (*Chondropetalum*, *Elegia*), Sparganiaceae (*Sparganium*) and Zingiberaceae (*Brachychilum*, *Hedychium*). While it is a useful marker at tribal level within the Zingiberaceae, its overall distribution is so sporadic that no general significance can as yet be attached to its presence in these plants.

Flavonols carrying extra *O*-methyl substituents are of more limited occurrence in plants than the above three substances and thus they have more potential as taxonomic markers. Such *O*-methylation may affect particularly the 7-, 3'-, 4'- or 5'-positions. Two examples are isorhamnetin, the 3'-methyl ether of quercetin, which is of fairly regular occurrence in several monocotyledon groups (e.g. in the genus *Roscoea*, Zingiberaceae) and syringetin, the 3', 5'-dimethyl ether of myricetin, which is

a rarer substance. Syringetin is at present only known from *Philydrum lanuginosum* (Philydraceae), *Hedychium stenopetalum* (Zingiberaceae) and *Chondropetalum* and *Elegia* spp. (Restionaceae).

The distribution of flavonol derivatives with extra *O*-methylation is shown in Diagram 93 (solid circles). The twenty records are divided between 13 families in 10 orders and thus seem to show no particular pattern. The apparent absence of these methyl ethers from the Ariflorae and Alismatiflorae probably only reflects the fact that flavonols as a class are rare in these two particular orders.

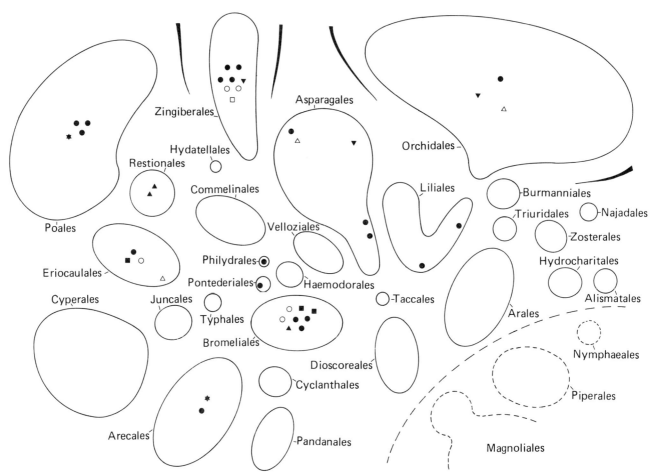

Diagram 93. Flavonols (cf. Diagram 92 and its legend). 7-3'-4' or 5'-*O*-Me-flavonols: solid circles. 3-, 6- or 8-*O*-Me-flavonols: hollow circles. 8-oxy-flavonols: solid triangles. 6-oxy-flavonols: solid squares. 5-*O*-Me-flavonols: hollow squares. *O*-acyl-flavonols: hollow triangles. Flavonol sulphates: asterisks. Dihydroflavonols: solid inverted triangles.

The Flavones

The distribution of the various flavones known in the monocotyledons is illustrated in Diagrams 94–98. The most distinctive of these is tricin, the 3′, 5′-dimethyl ether of tricetin (5, 7, 3′, 5′-pentahydroxyflavone) which is almost entirely restricted in its distribution to these plants, although there are a few rare records in the dicotyledons. Tricin is almost universal in the grasses, having been detected in 93% of 274 species surveyed representing 121 genera. It is also very common in the Arecaceae (in 51% of 125 spp.) and the Cyperaceae (in 45% of 121 spp.). These three families apart, it is otherwise uncommon. It is known from a few species of the Hyacinthaceae (Asparagales), Orchidaceae (Orchidales), Colchicaceae and Iridaceae (Liliales), all in the Liliiflorae.

By contrast, in spite of representative surveys, it has not been detected at all in the Alismatiflorae, Zingiberiflorae and the Ariflorae; it is also absent from the Bromeliaceae, Juncaceae and Restionaceae. At present, therefore, tricin appears to have a very discrete distribution and its major occurrence in Arecaceae, Cyperaceae and Poaceae would seem to link these three families chemically (see also concluding section).

The two common flavones of angiosperm plants are apigenin (5, 7, 4′-trihydroxyflavone) and luteolin (5, 7, 3′, 4′-tetrahydroxyflavone). They are of an expectedly wide occurrence both in the free state and in *O*-glycosidic combination throughout monocotyledon families (Diagram 96, solid circles). Their frequency varies considerably from 95% in Juncaceae to as low as 6% in Araceae and 1% in Orchidaceae. The majority of families fall between these two extremes, with most having luteolin or apigenin in about 15% of the species (Table 4). Taxonomically, the

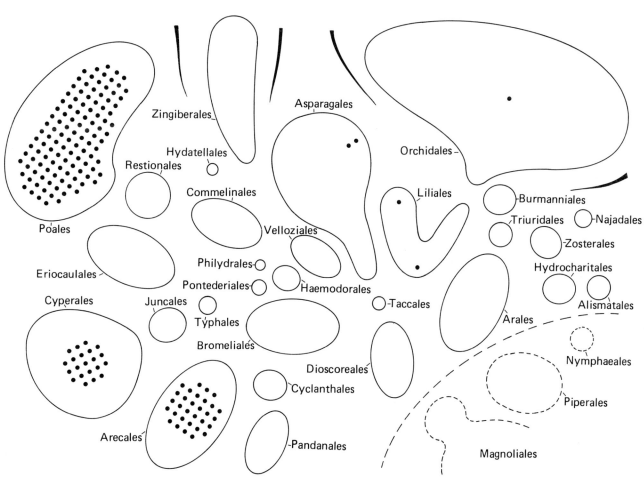

Diagram 94. Tricin (cf. Diagram 92 and its legend).

unusually high concentration of luteolin in members of the Juncaceae sets this family and the order Juncales apart from other members of the Commeliniflorae. Otherwise, the two common flavones do not appear to be of especial systematic importance.

The flavones apigenin and luteolin, besides being present as *O*-glycosides (with sugar linked through phenolic groups), also occur in plants in *C*-glycosidic combination. Such glycoflavones, in which sugar is linked directly to the carbon nucleus at the 6- or 8-position, are readily distinguished from other flavone derivatives during surveys because of their resistance to acid treatment. A considerable range of such glycoflavones have been encountered in the monocotyledons; many (e.g. vitexin, orientin etc.) are similar to those reported in dicotyledons but some (e.g. 8-galactosylapigenin of *Briza*) are unique to these plants. In the dicotyledons, glyco-

flavone distribution is rather heterogenous. They are well represented in many orders (e.g. Caryophyllales, Rosales) but not in many others (e.g. Araliales, Lamiales). By contrast, glycoflavones are more evenly distributed in monocotyledon groups and have been recorded in practically every family so far surveyed (Diagram 95).

In many monocotyledon families, glycoflavones are the most common class of flavonoid encountered. This is true in the Araceae, Arecaceae, Commelinaceae and Poaceae (Table 4). Also, these substances are well represented in most other families that have been studied. In only two families are they comparatively rare: the Juncaceae, where their presence is subsumed by free luteolin, and the Liliaceae, where the major flavonoid type is flavonol.

Glycoflavones are of some systematic interest because there is a geographical element in their frequency of occurrence. For example, in the Orchidaceae, they are

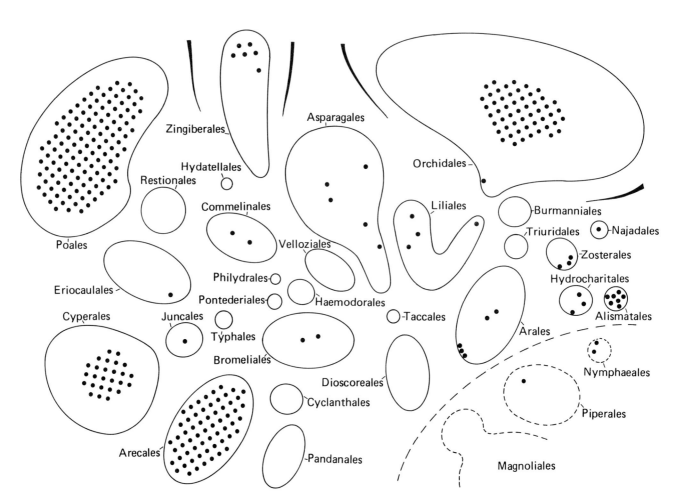

Diagram 95. C-glycoflavones (cf. Diagram 92 and its legend).

much more common in tropical and subtropical than in temperate species. Again, in the Restionaceae, they are restricted to South African members (not in the diagram), and have not been recorded in any Australasian taxa.

Finally, under flavones, mention must be made of *O*-methylated derivatives. The distribution of one such compound tricin, a 3′, 5′-dimethyl ether, has already been annotated. Another common methyl derivative is chryso-eriol, luteolin 3′-methyl ether, which is occasionally found in those families where luteolin is a common constituent. More rarely, *O*-methylation may take place at the 5-position of luteolin, producing luteolin 5-methyl ether. This latter substance is unique to two families of the monocotyledons, the Cyperaceae (Cyperales) and the Juncaceae (Juncales). This rare compound thus links together two families which for many other reasons are regarded as closely allied.

6- and 8-Hydroxyflavonoids

The introduction of a hydroxyl at the 6- or 8-position in flavones and flavonols causes a significant shift in colour so that such compounds are yellow instead of cream. This is true of quercetagetin or 6-hydroxyquercetin and of gossypetin or 8-hydroxyquercetin, which besides occurring as yellow flower pigments are present in leaves and other plant parts as well. The restricted distribution of such 6- and 8-hydroxyflavonoids means that they are of some interest as taxonomic markers.

Gossypetin, quercetagetin and the related flavone derivatives are mainly known from dicotyledons, but they have been recently found with increasing frequency in

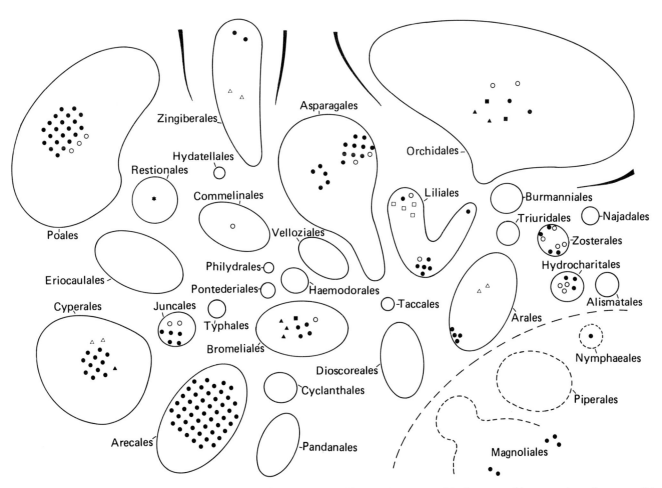

Diagram 96. Luteolin/apigenin (dots) and other flavones (cf. Diagram 92 and its legend). 6- or 8-*O*-Me-flavones: solid squares. 6-oxy-flavones: solid triangles. Dihydroflavones: hollow triangles. 7-, 3′, 4′-, or 5′-*O*-Me-flavones: hollow circles. 8-oxy-flavones: asterisk. Isoflavones: hollow squares.

monocotyledon families. 8-Hydroxyflavonoids are only reported in two families, the Restionaceae and Bromeliaceae. Gossypetin, together with the related 8-hydroxyluteolin (or hypolaetin), are characteristic of the Restionaceae and occur widely in Australian members of this family. By contrast, there is a single record of gossypetin in the Bromeliaceae, in one of 61 species surveyed.

6-Hydroxyflavonoids are regular components of several families: the Eriocaulaceae (with quercetagetin and its 6-methyl ether patuletin), the Cyperaceae and Commelinaceae (with 6-hydroxyluteolin), the Orchidaceae (with 6-hydroxyapigenin methyl ethers) and the Bromeliaceae. The latter family is particularly rich in derivatives of both 6-hydroxyluteolin and 6-hydroxymyricetin; many of the 6-hydroxy compounds here are partly O-methylated.

The scattered occurrence of these derivatives in six families each belonging to a different order suggests that the ability to introduce a 6- or 8-hydroxyl group into the flavonoid nucleus arose more than once during the evolution of the monocotyledons. The fact that three of the six families fall into one superorder, the Commeliniflorae (Cyperaceae, Commelinaceae, Restionaceae) may be purely coincidental but this does suggest that surveys of families in this group which have not yet been examined for these yellow pigments might be rewarding.

Sulphated Flavonoids

In addition to occurring in glycosidic combination, flavones and flavonols may exist in plant tissues covalently linked with inorganic sulphate. Such con-

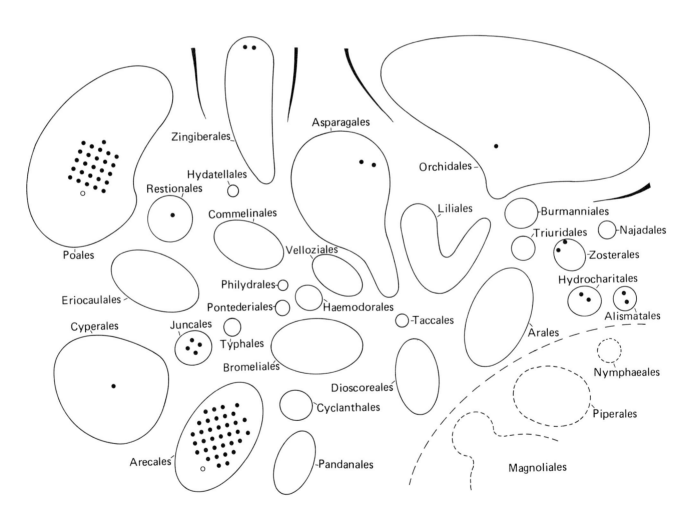

Diagram 97. Flavonol sulphates (hollow circles) and flavone sulphates (solid circles). (Cf. Diagram 92 and its legend.)

jugates are anionic and can be distinguished from the more common glycosides by their ability to migrate during electrophoresis in an acidic buffer. Surveys (Harborne, 1977) indicate that they are present in several hundred species representing at least 25 angiosperm families. While their occurrence may sometimes be associated with a fresh water or saline habitat, they are nevertheless of some interest as taxonomic markers.

In the monocotyledons, flavone sulphates occur (Diagram 97) abundantly in the Arecaceae (in 52% of spp. surveyed), Juncaceae (in 35%) and Poaceae (in 16%). They are also recorded in the Cyperaceae (in 2 of 120 spp.) and the Restionaceae (in 2 of 47 spp.). Furthermore, they are accompanied by flavonol sulphates in two of the above families, namely the grasses and the palms. This pattern of distribution may be of some significance in as much as the affinities between the palms, grasses and related families of the Juncales-Cyperales alliance has been pointed out earlier (Clifford, 1967).

A second centre of distribution of flavone sulphates is in the Alismatiflorae, where they are known in the Alismataceae, Hydrocharitaceae, Zosteraceae and Zannichelliaceae. Interestingly, the sulphates found in these water plants are slightly different chemically from those of the grasses and palms; also these families are distinctive in containing sulphates of simpler phenolic acids, such as caffeic acid.

The ability to sulphate flavonoids is also scattered among several other monocotyledon groups—there are isolated records of sulphates in the Orchidaceae (Orchidales), Hyacinthaceae (Asparagales) and Marantaceae (Zingiberales).

Other Flavonoid Classes

Data on other flavonoids besides those mentioned above are generally too sparse for it to be useful to draw any taxonomic conclusions. Anthocyanin pigments, for example, have only been systematically investigated in some families (see Stirton and Harborne, 1980) and the results (Table 5) only indicate as yet that each family tends to have its own individual pattern of anthocyanin. Besides these common anthocyanins, a few rarer structures have been recorded. The recent report of 5-methylcyanidin in *Egeria* (Hydrocharitaceae) is of some interest, since 5-*O*-methylation has also been observed

TABLE 5
Summary of Anthocyanin Patterns of Eight Monocotyledonous Families

Family	Anthocyanidins present*				Acylation	Major glycosidic patterns
	Pg	Cy	Dp	Methyl derivatives		
Commelinaceae	—	+ +	+ +	—	+ +	3, 5-diglucoside, 3, 7, 3'-triglucoside
Araceae	+	+ +	(+)	—	—	3-rutinoside
Bromeliaceae	+	+	+	+ +	—	3, 5-diglucoside, 3-rutinoside-5-glucoside
Poaceae	—	+ +	(+)	+	(+)	3-arabinoside, 3-galactoside
Iridaceae	+	+	+	+	(+)	3, 5-diglucoside, 3-rutinoside-5-glucoside
Marantaceae	—	+ +	+ +	+	—	3-rutinoside, 3-sophoroside-5-glucoside
Orchidaceae	+	+ +	+	+	(+)	3-glucoside, 3, 5-diglucoside
Zingiberaceae	—	+ +	—	+	—	3-glucoside, 3-rutinoside

*Key: Pg, pelargonidin; Cy, cyanidin; Dp, delphinidin; methyl derivatives, peonidin, petunidin, and malvidin; + +, very common; +, frequent; (+), infrequent; —, absent.

TABLE 6
Some Possible Groupings of Monocotyledon Families based on Major Leaf Flavonoid Patterns

Families	*Flavonoid constituents*
Sparganiaceae Typhaceae Flagellariaceae Anarthriaceae	Flavonols, mainly quercetin and kaempferol
Eriocaulaceae Restionaceae Orchidaceae Bromeliaceae	Flavonols dominant, together with 6- or 8-hydroxyflavonoids
Commelinaceae Araceae Lemnaceae	Flavonols and Glycoflavones in quantity
Poaceae Cyperaceae Juncaceae Arecaceae	Luteolin, glycoflavones and flavone 5-glucosides in all four. Sulphates common, except in Cyperaceae. Tricin in all except Juncaceae.

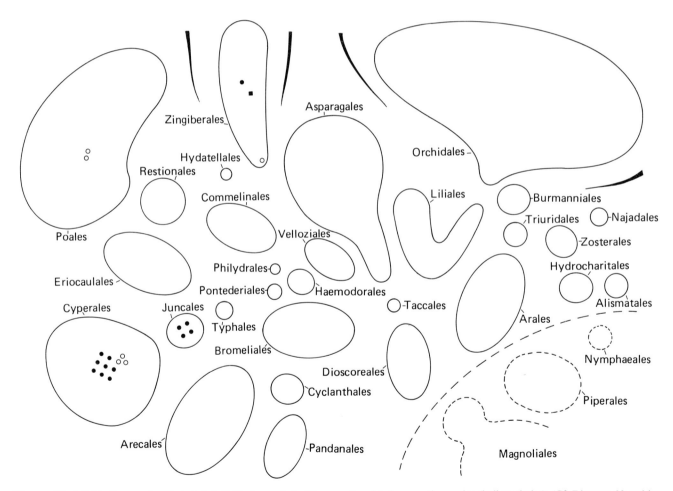

Diagram 98. 5-*O*-Me-flavones (solid circles), 5-*O*-Me-flavonols (solid squares) and 3-deoxy-anthocyanins (hollow circles). (Cf. Diagram 92 and its legend.)

in the flavone series in Juncaceae and Cyperaceae (see above). Other rare pigments, based this time on 3-deoxyanthocyanin structures, have been reported variously in grasses (Poales), sedges (Cyperales) and Musaceae (Zingiberales), but the records are far too few to be meaningful.

Proanthocyanidins are polymeric plant tannins, which although structurally and functionally unrelated to the anthocyanins, are similar to these pigments in that they yield anthocyanidins when heated with acid. These materials are extremely widely present in leaves of monocotyledons (Table 4) and their almost universal occurrence robs them of any systematic significance.

Four other classes of flavonoids are recorded in monocotyledons, but these are all known from only a very few families. Chalcones are present in Cyperaceae, Lilaceae, Zanthorrhoeaceae and Zingiberaceae; flavanones in Cyperaceae, Liliaceae, Zingiberaceae; aurones in Cyperaceae; and isoflavones in Iridaceae.

Systematic Conclusions

From the above data, it is possible to make three generalizations about the flavonoids found in monocots:
(1) The pattern is reasonably consistent and representative at the family level. A good example of such consistency is the Poaceae, where an extensive survey of 274 spp. representing 121 genera and all tribes has confirmed that the pattern is very uniform. Geographical variation may upset this pattern to a minor extent. In another cosmopolitan family, the Cyperaceae, one interesting marker, luteolin 5-methyl ether, which links it to the Juncaceae, is only found

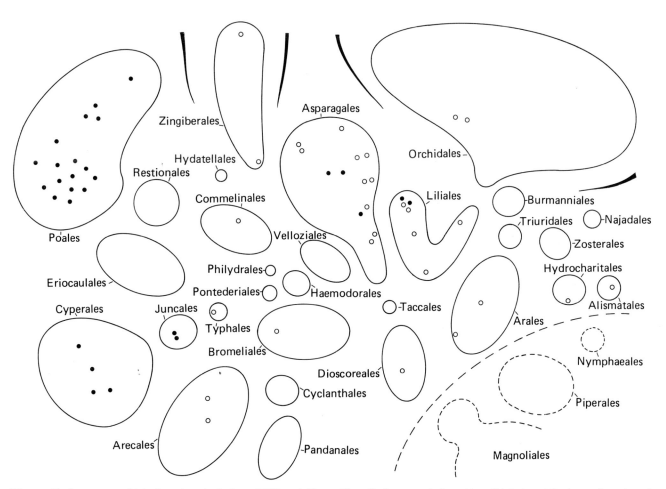

Diagram 99. Isoenzymes of dehydroquinate hydrolyase (DHQ-ase). Taxa with verified presence indicated by solid circles, while absence (or extremely small amounts?) are indicated with hollow circles.

in tropical members;

(2) Both qualitative and quantitative variations are significant and should be used in making comparisons. The data in Table 4 could usefully be analysed by computer for similarities between families using five concentration values for each flavonoid character. In some cases, however, further surveys are needed to be sure of the correctness of the frequencies of occurrence;

(3) Families can be placed into groupings according to the major flavonoids recorded in the leaves (Table 6). It is important to stress that these groupings do not necessarily indicate any relationship with systematic groupings. Indeed, it is apparent from a cursory comparison with the monocotyledon system presented on p. 22 that there are a few correlations with morphological groupings.

The difficulty of drawing conclusions from the present results is that data for certain key families are still incomplete (e.g. for the Eriocaulaceae) or are absent. This is largely due to the inaccessibility of many of these plants to phytochemists. For the purposes of the present discussion, it is perhaps only worthwhile looking at one of the groupings in Table 6 and determining its systematic possibilities. This is the suggested alliance based on five flavonoid characters between the Poaceae, Cyperaceae, Juncaceae and Arecaceae. That a close relationship exists between the first three families of this alliance is hardly disputed although currently they are each placed in separate orders (Poales, Cyperales, Juncales) within the superorder Commeliniflorae.

The inclusion of the Arecaceae into this threesome is perhaps not so clearcut and indeed such an alliance has been strongly disputed on morphological grounds (see Bendz and Santesson, 1973). However, the possibility that the grasses, in particular, and the palms should be placed nearer each other in systems of classification emerged about the same time from a computer generated classification of these and related families (Clifford, 1970). In the final chapter of this book this point will be considered in some detail. The similarities in flavonoid pattern are counterbalanced by a conspicuous number of differences in other characters. It is therefore possible that the occurrence of a remarkably similar pattern of flavonoids in these two groups may be a result of convergent evolution. A final point may be made that the comparative analyses of flavonoid patterns in the grasses, palms and related plants has served at least one useful purpose; it has stimulated a much needed reassessment of taxonomic relationships between these large and economically important monocotyledon groups.

ISOENZYMES OF DEHYDROQUINATE HYDROLASE (DHQ-ASE) (Diagram 100)

Boudet et al. (1975) have identified two dehydroquinate dehydrolyases ("DHQ-ases"), the third enzyme of the shikimate pathway. One isoenzyme is associated with the shikimate: $NADP^+$ oxidoreductase ("SH.OR-ase") in a complex which is at least bifunctional and which seems widely distributed in plants (Boudet and Lecussan, 1974). The other isoenzyme is a free form specifically activated by shikimic acid. The distribution of this isoenzyme was investigated by Boudet et al. (1977). The investigation was performed through a simple diagnostic test connected with the regulatory properties of the enzyme. It turned out that this DHQ-ase isoenzyme activated by shikimic acid was present only in monocotyledons, and essentially in the Juncaceae, Cyperaceae and Poaceae among the taxa studied. It was also present in a few more species, viz. in the Liliiflorae (Asparagales and Liliales): Asparagus officinales (Asparagaceae), Asphodelus aestivus (Asphodelaceae) and Hemerocallis sp. (Hemerocallidaceae) and two species of Iris (I. germanica and I. pumila) (Iridaceae).

The taxonomic distribution of DHQ-ase shows that the methodology seems to be reliable, as there is a great taxonomic regularity in its distribution. Its presence in the Juncales, Cyperales and Poales is of interest, for example, because of the occurrence in these groups of certain flavonoid compounds (see pp. 261–267). (However, although these flavonoids are also as a rule present in palms, the two palms studied did not possess this DHQ-ase.) The extension of these studies to taxa of the Restionales, Eriocaulales and other orders is highly desirable.

The few scattered occurrences of DHQ-ase in the Asparagales and Liliales are interesting but can hardly be used for drawing taxonomic conclusions in these orders, as two species of Iris studied proved to have them and two other species did not. The same irregularity was observed in the Asparagales, where DHQ-ase was present in one species of Asparagus but not in one of Ruscus, and in one species of Aspodelus and one of Hemerocallis but not in other Asparagalean taxa of the Hyacinthaceae, Alliaceae or Amaryllidaceae.

Base data

See Boudet et al. (1975).

HOST SPECIFICITY OF PARASITIZING FUNGI

Attempts to correlate the host preferences of parasitizing fungi with angiosperm taxonomy and phylogeny have been made only in the last years, for example by Nannfeldt (1968), Holm (1969), Savile (1954, 1971) and El Gazzar and Badawi (personal communication). The topic presents many problems relating to morphology, taxonomy and interrelations of the fungi as well to the factors deciding the host specificity. It is an acknowledged fact that many fungal species occur on a variety of quite unrelated host plants, while others show a strong host specificity, being restricted to one or a few closely allied species of the same genus of host plants. Problems regarding the possibility for a normally host specific fungus to "jump" to an unrelated host and establish there are crucial.

In relation to phylogeny and taxonomy on the family, order and superorder levels dealt with in this book, the problems are particularly acute, but have been focused by Savile, and need being presented briefly and related to other data.

Savile (1979) recently presented extensive data on host relationships of parasitizing fungi. His work covers all higher host plants. For the main features of the monocotyledon system his conclusions are noteworthy with regard to the following features.

The Typhales because of the host distribution of Uromyces spargani are regarded to be closely allied to the Arales, notably Acorus. The Commeliniflorae are considered a relatively homogeneous group provided the Typhales are excluded. The Juncales are regarded as derived from ancestral Cyperales as judged from their Puccinia and Uromyces rusts, some of the uredinospores of the juncalean parasites being considered more derived than many that occur on the cyperalean taxa. The Liliiflorae are also considered phylogenetically younger than the commeliniflorean core groups (Poales, Cyperales, Juncales) and unrelated to them. In the Asparagales and Liliales certain groups of genera (in our book considered families) are attacked by particular species of rusts.

Uredinales (by L. Holm) (Diagrams 100–101)

This order, the rusts, includes c. 5000 species distributed amongst about 100 genera, most of which are relatively small. The rusts occur exclusively on ferns, gymnosperms and angiosperms.

The following genera are known to occur on monocotyledons (see Diagram 102).

Puccinastrum, with c. 35 species. The aecial stage (0 + I) is restricted to Pinaceae and the dicaryophase (II + III) to dicotyledons, except for two species which attack Goodyera (Orchidaceae). These are apparently closely allied to P. pyrolae, and Savile (1971) has presented the very likely hypothesis that Goodyera has been colonized by the Pyrola rust secondarily.

Melampsora, with c. 75 species, no doubt has primarily the 0 + I stages on Pinaceae and II + III on taxa of the dicotyledonous families Salicaceae and Flacourtiaceae. In some species the aecial stage has moved to angiosperms, mainly dicotyledons, but in a few cases to monocotyledonous families, including the Araceae (Arum), Hyacinthaceae (Muscari), Alliaceae (Allium), Amaryllidaceae (Galanthus, Leucojum) and Orchidaceae subfam. Orchidoideae (Gymnadenia, Ophrys, Orchis, Platanthera). These host groups, separately, show mutual affinities, such as those between Galanthus and Leucojum, between the genera of the Orchidoideae, and between the three Asparagalean families Hyacinthaceae, Alliaceae and Amaryllidaceae, but the secondary "jumps" to monocotyledons may have happened separately and conclusions must not be strained (see Diagram 101).

Coleosporium, with c. 80 species, as far as known has stages 0 + I on Pinus only and states II + III on diverse groups of angiosperms though mostly on dicotyledons and with only three species recorded from monocotyledons, these all belonging to Orchidaceae.

Phakopsora, with c. 50 species, likewise attacks mainly dicotyledons, but one species has stages II + III on Commelina (Commelinaceae) and 5 species on grasses. The aecial stage is yet unknown for most species of Phakopsora including all the species attacking monocotyledons.

Physopella has c. 20 species, 13 of which occur on grasses (Poaceae). The aecial stage is unknown.

Dasturella, 3 species, with stages II + III parasitizes members of Poaceae subfam. Bambusoideae. Stages 0 + I are known for one species only, viz. on *Randia* (Rubiaceae), a dicotyledon.

Tunicopsora and *Stereostratum* are two monotypic genera both with stages II + III on Poaceae subfam. Bambusoideae, while the other stages are unknown.

Blastospora, has 2 species. Stages II + III occur on *Smilax* (Smilacaceae), while the other stages are yet unknown.

Goplana, has 6 species. Stages II + III are known on dicotyledons except for one species on *Dioscorea*; stages 0 + I are not known.

Hemileia, with c. 40 species, is known only in the II + III stages which attack mainly dicotyledons, but 4 species are recorded on taxa of Orchidaceae and one on Dioscoreaceae.

Cerradoa is a monotypic genus with the only known stages, II + III, occurring on *Attalea* (Arecaceae). This is a recent find and the only rust known to occur on a palm!

Puccinia sensu lato (including *Uromyces*) is a vast genus with perhaps about 4500 species. Many are host-alternating, this being probably the original condition.

The different hostal stages are treated separately here:

Stages 0 + I generally occur on dicotyledonous plants, though there are some exceptions, e.g. *P. sessilis* (stages II + III on grasses): the aecial stage never occurs on dicotyledons, but is reported on a number of monocotyledonous genera, viz. *Arum* (Araceae), *Paris* (Trilliaceae), *Convallaria*, *Maianthemum*, *Polygonatum*, *Smilacina* (Convallariaceae), *Hosta* (Funkiaceae), *Dipcadi*, *Ornithogalum*, *Scilla* (Hyacinthaceae), *Allium* (Alliaceae), *Leucojum* (Amaryllidaceae), *Iris* (Iridaceae), *Uvularia* (Colchicaceae), *Plantanthera* and *Orchis* (Orchidaceae).

The host spectrum is reminiscent of that for *Melampsora*, and certainly is not random amongst the monocotyledons. Thus the species seems to prefer the Liliiflorean core groups (those without starchy endosperm) and, rarely, the Araceae.

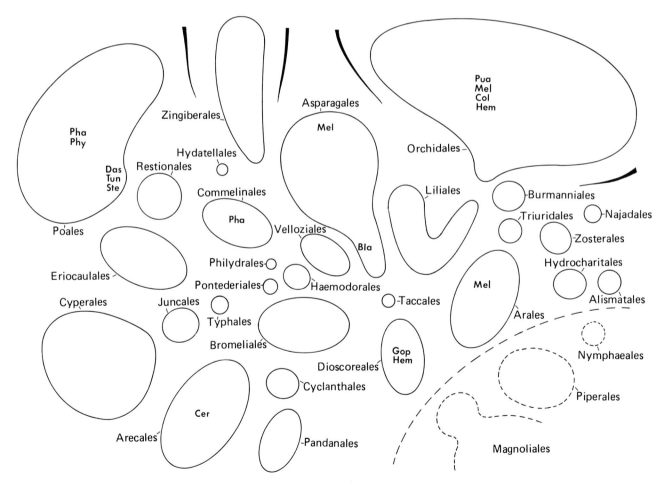

Diagram 100. Genera of Uredinalean fungi on groups of monocotyledons (orders only). Abbreviations: Bla, *Blastospora*; Cer, *Cerradoa*; Col, *Coleosporium*; Das, *Dasturella*; Gop, *Goplana*; Hem, *Hemileia*; Mel, *Melampsora*; Pha, *Phacopsora*; Phy, *Physopella*; Pua, *Puccinastrum*; Ste, *Stereostratum*; Tun, *Tunicopsora*. The genera attacking bamboos are in the lower right area of the Poales bubble.

P. amphigena (stages II + III on *Calamovilfa*, Poaceae) has stages 0 + I on *Leucocrinum* (Funkiaceae) and *Smilax* (Smilacaceae), which are not closely allied.

Stages II + III. A very great number of *Puccinia* species develop their dicaryophase on monocotyledons, in particular on grasses (more than 350 species), but numerous species also on members of Cyperaceae (c. 70 species) and fairly many on members of the Juncaceae and of the Asparagales and Liliales, while only single species are recorded to occur on members of the Araceae, Bromeliaceae, Cannaceae, Commelinaceae, Dioscoreaceae, Haemodoraceae, Marantaceae, Musaceae, Orchidaceae, Pontederiaceae, Sparganiaceae and Zingiberaceae. One highly interesting example among the latter is *Uromyces spargani* with 0 + I on *Hypericum* (Hypericaceae) and II + III on *Sparganium* (Sparganiaceae) and *Acorus* (Araceae) which was referred to by Thorne (1976) as indicating affinity between the Typhales and Arales. If the *Puccinia* (incl. *Uromyces*) species are classified according to their supposed primitiveness, judged from the sculpturing patterns of the spores (Savile, 1979) it turns out that the Poales, Cyperales and Juncales are attacked to the largest proportion by "primitive" species, while the hosts of the Liliiflorae are more often attacked by species referable to the "advanced" types.

As a consequence of the widely distributed cases of parallel and convergent evolution within *Puccinia sensu lato* it is difficult to present a phylogenetically plausible grouping of the species, such as would be desirable for our purpose. One probably natural species group is the section *Caricinae* (see Holm, 1966) which is characterized by aecial spores with a granular surface, and which have cells in the aecial peridium which are rhomboid in longitudinal section. This group includes a great number of species occurring on taxa of Cyperaceae, fairly many species on taxa of Juncaceae, and at least two species, *Puccinia atropuncta* and *Uromyces veratri*, on genera of the Melanthiaceae (*Amianthium, Melanthium, Schoeno-*

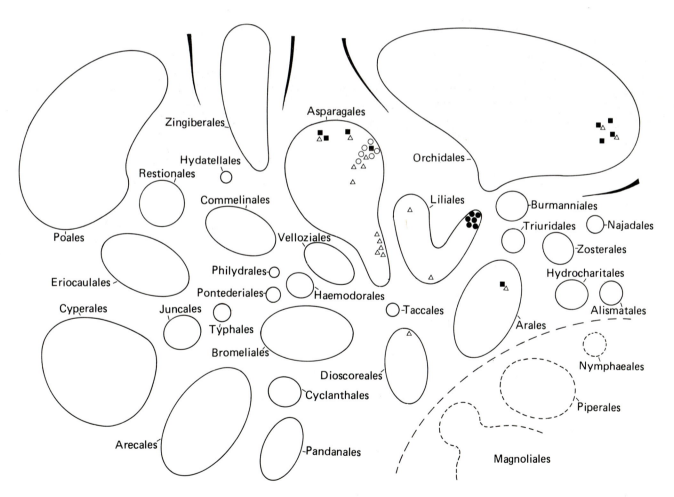

Diagram 101. Some Uredinalean parasitic taxa. Solid squares: *Melampsora*; hollow triangle: *Puccinia sessilis*, aerial stage; solid circles: the two supposedly related species *Puccinia atropuncta* and *Uromyces veratri*; hollow circles: *Uromyces muscari*.

caulon, Stenanthium, Veratrum, Xerophyllum and *Zigadenus*). The aecial stage is found on diverse dicotyledons especially among the Asteraceae.

Savile (1962) has also pointed out a group of "microforms" (i.e. with only the stages 0 + III) occurring on certain Liliiflorae. They probably form a natural group of species judging from some shared unusual spore characters. To this group belong some species attacking *Allium*, e.g. *Uromyces ambiguus* and *U. aemulus*. Savile referred to this fact as an indication that *Allium* may not be closely allied to Amaryllidaceae, as the members of the latter family are not attacked by these species. *U. muscari* and closely related species attack species of *Muscari, Bellevalia, Scilla, Dipcadi, Hyacinthus, Lachenalia* and *Urginea*, all in Hyacinthaceae.

The host choices of *U. muscari* and the above-mentioned two species attacking only genera of Melanthiaceae demonstrate that further research in *Puccinia* could contribute substantially to the taxonomy of the angiospermous hosts.

Ustilaginales (by J. A. Nannfeldt) (Diagrams 102–103)

This group is natural and easily circumscribed. The host plants are restricted to angiosperms (records among gymnosperms, vascular cryptogams and mosses being either wholly erroneous or at least very dubious). One feature not yet satisfactorily explained is that in the temperate flora no woody host plants are known at all, and in the tropics only two small genera, it appears, of smuts, on *Grewia* (Tiliaceae) and *Cissus* (Vitidaceae). The host plants include various dicotyledonous families and also a number of families in the monocotyledons. About half of the smut genera are restricted to monocotyledonous hosts.

The generally recognized genera of Ustilaginales amount to c. 30 in number. Most of them are probably reasonably homogeneous, though this is not the case with, for example, the giant *Ustilago*, a genus characterized by several "trivial", i.e. common attributes. Some monotypic genera are probably not worthy of their rank.

The occurrence among the monocotyledons is shown in Diagrams 102–103. Far-reaching conclusions should not be drawn about relationships among these hostal taxa.

The genus *Doassansia* (c. 40 species) together with the satellite genera *Burrillia* (c. 7 species) and *Tracya* (2 species) comprises a homogenous group of smut fungi. This group of genera is obviously adapted to hydrochory and therefore possibly of restricted interest in this context. Its host genera are to be found in the following monocotyledonous genera:

Alisma, Echinodorus, Elisma, Limnocharis, Sagittaria (Alismataceae), *Butomus* (Butomaceae), *Hydrocharis* (Hydrocharitaceae), *Potamogeton* (Potamogetonaceae), *Lilea* (Juncaginaceae), *Spirodela* (Lemnaceae), *Eichhornia* (Pontederiaceae), and such dicotyledons as *Callitriche* (Callitrichaceae), *Nymphoides* (Menyanthaceae), *Epilobium* (Onagraceae), *Hottonia* (Primulaceae), *Ranunculus* (Ranunculaceae) and *Limosella* (Scrophulariaceae).

Entorrhiza, with c. 10 species, is a very natural and isolated genus, which induces galls on roots. It is known as a parasite on the Juncaceae (at least 3 species of *Juncus*) and on the Cyperaceae (*Carex limosa, Cyperus* sp., *Eleogiton fluitans* and *Eleocharis pauciflora*).

This distribution supports the previously well demonstrated relationship between Cyperales and Juncales.

Farysia, with c. 20 species, likewise is a natural and isolated genus, which is found mainly in warmer parts of the world. It parasitizes genera of Cyperaceae only (mainly species of *Carex* including the subgenus *Indocarex*, and of *Gahnia*).

Anthracoidea, with at least 50 species, represents a homogeneous group, which is not closely allied to *Cintractia* in which it was included until 1963. Its species parasitize members of the Cyperaceae, notably members of the subfamily Caricoideae tribus Cariceae (*Kobresia, Uncinia, Carex*, however not *Carex* subgenus *Indocarex*) and also *Trichophorum caespitosum*.

Planetella (one species), which is possibly closely allied to *Anthracoidea*, is likewise found on cyperaceous hosts, viz. certain species of *Carex*.

Cintractia sensu stricto, with c. 30 species, may include some alien elements, but is wholly restricted in its hostal choice to Cyperaceae, with most species on *Rhynchospora* and *Fimbristylis* but some on *Cyperus, Carpha* and other genera.

Schizonella (2 species) is a natural genus parasitizing the tribe Cariceae (*Carex, Kobresia*) of the Cyperaceae.

Orphanomyces (3 species) has recently been distinguished from *Ustilago* and is probably a natural group. It attacks *Carex* (Cyperaceae).

Cintractiella (one species) is known as a parasite on *Hypolythrum* (Cyperaceae).

Testicularia (2 species) has the type species on *Rhynchospora* (Cyperaceae) and a second species on *Leersia* (Poaceae). It is probably heterogenous.

Tilletia (c. 80 species) is a natural genus characterized by, for example, its odour of trimethyl amine ("stink smut"). It is restricted in its host choice to various groups of grasses (Poaceae).

Neovossia (2 species) is a satellite genus of *Tilletia* and likewise is restricted to the Poaceae (*Phragmites* and the probably closely allied *Molinia*).

Jamesdicksonia (one species) attacks grasses.

Moesziomyces (4 species) recently distinguished for good reasons from the heterogeneous *Tolyposporium* is restricted in its host choice to grasses (Poaceae) where it is known on *Echinochloa*, *Leersia*, *Paspalum* and *Pennisetum*.

Dermatosorus (one species) was likewise recently distinguished from *Tolyposporium*. It is known on *Eriocaulon* (Eriocaulaceae).

Urocystis (at least 80 species) for the most part seems to represent a natural genus and is known to attack a number of disparate dicotyledonous families, such as Araliaceae, Brassicaceae, Polemoniaceae, Primulaceae, Ranunculaceae, Rosaceae and Violaceae. The monocotyledonous hosts are also somewhat scattered in our system and include the following.

Dioscoreales: Dioscoreaceae (*Dioscorea*), Trilliaceae (*Paris*, *Trillium*).

Asparagales: Convallariaceae (*Convallaria*, *Polygonatum*), Hypoxidaceae (*Hypoxis*), Asphodelaceae (*Ixiolirion*), Eriospermaceae (*Eriospermum*), Hyacinthaceae (*Muscari*, *Ornithogalum*, *Scilla*), Alliaceae (*Allium*), Amaryllidaceae (*Galanthus*, *Leucojum*, *Sternbergia*).

Liliales: Iridaceae (*Gladiolus*), Colchicaceae (*Colchicum*), Alstroemeriaceae (*Bomarea*), Liliaceae s.str. (*Erythronium*).

Juncales: Juncaceae (*Juncus*, *Luzula*).

Cyperales: Cyperaceae (*Carex*).

Poales: Poaceae (several genera of different groups).

The host preferences in this case give little phylogenetic evidence except that it indicates that the three Liliiflorae orders attacked probably form a coherent group.

Ustilago (more than 300 species) is a collective genus

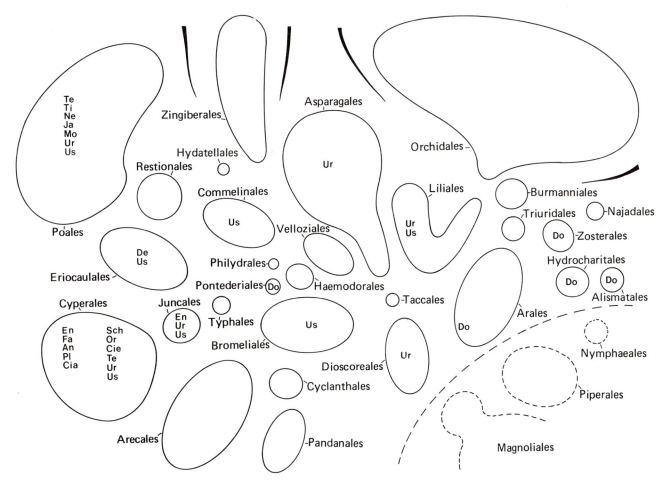

Diagram 102. Occurrence of genera of parasitizing Ustilaginalean fungi in groups of monocotyledons (orders only). Abbreviations: An, *Anthracoidea*; Cia, *Cintractia*; Cie, *Cintractiella*; Do, *Doassansia*; En, *Enthorrhiza*; Fa, *Farysia*; Ja, *Jamesdicksonia*; Mo, *Moesziomyces*; Ne, *Neovossia*; Or, *Orphanomyces*; Pl, *Planetella*; Sch, *Schizonella*; Te, *Testicularia*; Ti, *Tilletia*; Ur, *Urocystis*; Us, *Ustilago*.

including, no doubt, disparate elements. Some "natural" groups of species in *Ustilago* can be discerned however. The genus attacks members among series of families, most of them dicotyledonous (many distantly related) but also several monocotyledonous families of different orders among which should be mentioned the following.

Asparagales: Hyacinthaceae (*Albuca*, *Bellevalia*, *Eucomis*, *Hyacinthus*, *Muscari*, *Ornithogalum*, *Urginea*), Alliaceae (*Allium*); Liliales: Liliaceae (*Erythronium*, *Gagea*, *Tulipa*); Bromeliales: Bromeliaceae (*Tillandsia*); Commelinales: Commelinaceae (*Aneilema*, *Commelina*, *Pollia*); Juncales: Juncaceae, *Juncus*, *Luzula*; Cyperales: Cyperaceae (*Cyperus*, *Dichronema*, *Diplasia*, *Mariscus*, *Rhynchospora*, *Scleria*); Eriocaulales: Eriocaulaceae (*Eriocaulon*); Poales: Poaceae (at least 130 host species distributed among disparate subfamilies).

Most of the graminicolous species of *Ustilago*, including all pests on economically important cereals, belong to a natural group of species, which seem to be restricted to grasses. It is obvious that also a number of

species now placed in the genera *Sphacelotheca* and *Sorosporium* belong to this group of species.

Of interest, apart from this, is the marked host preference in the Liliiflorae to the families Hyacinthaceae and Liliaceae *sensu stricto*. This is a far from random distribution and deserves further study.

Tolyposporium and the related genus *Tecaphora*, which may be derived from *Tolyposporium* according to Savile (1979), are reported from the following plant host groups:

Tolyposporium: Restionaceae (1 sp.), Eriocaulaceae (2 spp.), Juncaceae (2 spp.), Cyperaceae (6 spp.), Poaceae (15 spp.).

Thecaphora: Restionaceae (1 sp.), Cyperaceae (4 spp.), Poaceae (4 spp.), Liliaceae *sensu lato* (1 sp.). Although *Tolyposporium* (in which *Thecaphora* might be included) may rather represent a generic complex than a single genus and secondary jumps have occurred to ecologically associated dicotyledons, there is an indication that these smuts have a significant phylogenetic interest and may have been present on the ancestors of these groups, which

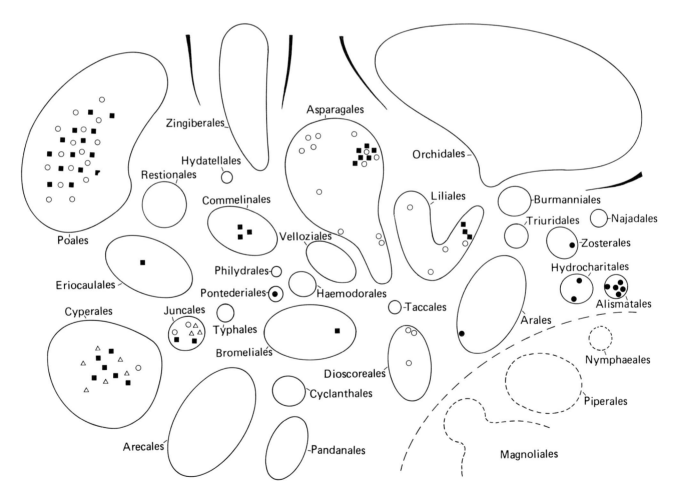

Diagram 103. Some Ustilaginalean genera. Solid circles: *Doassansia*; hollow circles: *Urocystis*; solid squares: *Ustilago*; hollow triangles: *Enthorrhiza*.

are all members of the Commeliniflorae, with the exception of the exceptional liliaceous(?) host.

Of the remaining smut genera *Entyloma*, *Melanotaenium*, *Sorosporium* and *Sphacelotheca* are still too heterogeneous or imperfectly known to be discussed here.

Conclusions. There are obviously too scarce records for parasites on tropical and South Hemisphere angiosperms to get a clear picture of the host distribution of the smut genera. Notable is the absence of any records of smut parasites on orchids. In this connection it should be pointed out that absence of records does not imply evidence for absence of fungal parasites. The total picture indicates, however, that smuts seem to be concentrated in temperate regions.

Regularity in host choice in smuts is much greater in monocotyledons than in dicotyledons and is largely concentrated to the Poales and Cyperales, but it is notable that these two orders of monocotyledons do not share any other smut genera (except for the very dubious *Testicularia*) than those having a wide hostal spectrum. Genera concentrated on the Cyperales (plus rarely the Juncales) are *Enthorrhiza*, *Farysia*, *Anthracoidea*, *Planetella*, *Cintractia*, *Schizonella*, *Orphanomyces* and *Cintractiella*, while those attacking grasses only are *Tilletia*, *Neovossia*, *Jamesdicksonia* and *Moesziomyces*. It is likely that host specificity in these genera may be useful in assessing generic affinities in the Cyperaceae and Poaceae respectively.

Erysiphales: Erysiphaceae (by J. A. Nannfeldt) (Diagram 104)

This is a natural group of fungi which like the Ustilaginales parasitizes only angiosperms. Unlike the smuts, they often attack the green parts of woody plants, and the family is apparently concentrated in temperate regions. It consists of about ten genera and several hundred species.

More than 150 families (and 1200 genera) of dicotyledons are attacked by erysiphaceous fungi, while only a few are able to parasitize monocotyledons. *Erysiphe graminis*, however, occurs on a great number of grass genera and thus no doubt exhibits a physiological specialization. The species moreover shows a number of distinct morphological features and it has even been proposed that it can be regarded as a separate genus (*Blumeria*).

Furthermore, according to Hirata (unpublished) *Leveillula taurica* attacks *Allium* (*A. cepa, porrum, sativum*) in Asparagales, Liliiflorae, and also *Zanted-* *eschia* in Arales, Ariflorae. The species occurs on a variety of host plants, mainly in regions with a mediterranean climate. Other records of mildews on monocotyledons (for Amaryllidaceae, Iridaceae, Commelinaceae, Cyperaceae and Orchidaceae) must be interpreted with caution as they are probably largely erroneous or at least very dubious. They are not included in the diagram.

Among the dicotyledons associated with monocotyledons there are several attacked by mildews, including *Piper betle* of Piperaceae, two (dubious) records on Annonaceae, *Aristolochia bracteata* and *A. indica* and *Asarum canadense* of Aristolochiaceae, and *Magnolia* (several species) and *Liriodendron* of Magnoliaceae. None are known from any species of Nymphaeales.

Taphrinales (by J. A. Nannfeldt) (Diagram 104)

This is another easily and well circumscribed order which comprises a single genus, *Taphrina*, with at least 100 species parasitizing ferns and angiosperms. It has a rather peculiar host pattern: c. 25 species of ferns, including members of the Polypodiaceae *sensu lato* as well as of the Osmundaceae, are parasitized by this genus, and so are c. 25 genera of dicotyledons, viz. members of the Fagaceae, Betulaceae, Corylaceae, Ulmaceae, Rosaceae, Malaceae, Anacardiaceae, Aceraceae, Hippocastanaceae and Asteraceae. The host specificity of some *Taphrina* species seems to give indication of phylogenetic (chemical) affinity between the hosts.

Two species of *Taphrina* attack monocotyledons, viz. members of Zingiberaceae (*Curcuma, Globba, Hedychium, Zingiber*). This fact gives no indication of relationships between monocotyledonous families but may serve as an example of good host specificity on the family level.

HOST SPECIFICITY OF INSECTS

It is to be expected that valuable contributions for phylogeny can be achieved by a profound study of the host faithfulness of insects in their feeding and oviposition behaviour.

Basically it is a matter of chemical characters, viz. a dependence on:
(1) the amount of available food material of the plant;
(2) its chemical composition as suitable to the particular group of insect;
(3) the lack of repellant, deterrent of poisonous substances;
(4) the possible occurrence of attracting substances.

Unlike the dicotyledons where there are many more base data from which to draw conclusions, so far there are apparently few examples of such host specificity fully worked out for monocotyledons. Only two very illuminating examples are selected here.

Hosts of Plant-lice (*Sternorrhyncha*)

Eastop (1979) reports about the hosts of certain plant-lice (Stenorrhyncha), that several groups show striking examples of host specificity. Thus taxa of Juncaceae and Cyperaceae are the hosts of the psyllid subfam. Liviinae and the aphid tribe Saltusaphidini, both groups probably containing mostly host specific species. Poaceae, according to Eastop, are not colonized by psyllids, but by many aleyrodids, aphids of the Siphini, Rhopalosiphini, Macrosiphini, Cerataphidini, Eriosomatini and Fordini; and many Coccoidea belonging to different

families also occur on grasses. Moreover, Arecales and Pandanales have a characteristic fauna of aphids (Cerataphidini), aleyrodids and coccids, but no clear distribution pattern was recognizable. The aphids on palms often have relatives on large grasses, in particular of the Bambusoideae. Finally, the Orchidaceae are colonized by relatively few Stenorrhyncha.

Hosts of Lygaeidae subfam. Blissinae (*Hemiptera*), a Group of Chinch Bugs (*Fig. 94, Diagram 105*)

In an article by Slater (1976) the host plant relationships of the lygaeid subfam. Blissinae, a group of chinch bugs, were analysed. The members of this subfamily breed only on monocotyledonous plants, which is *a priori* very promising for taxonomic conclusions.

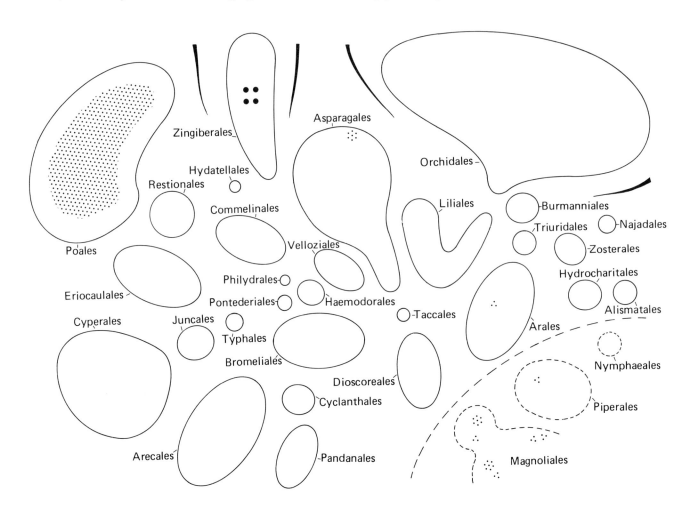

Diagram 104. Erysiphalean and Taphrinalean parasites on monocotyledons. Small dots: records of Erysiphaceae; main dotted area in Poales represents *Erysiphe graminis*. Large dots: *Taphrina*.

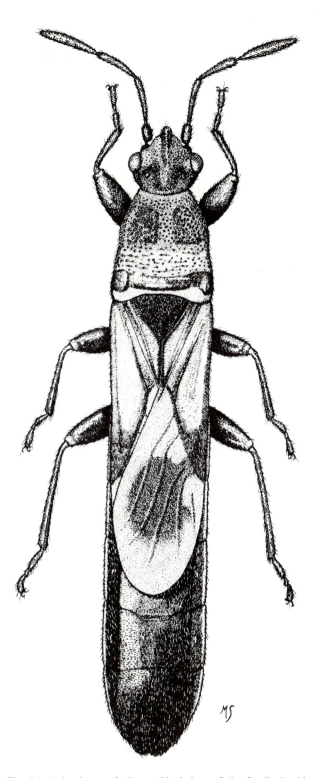

Fig. 94. *Ischnodemus tibialis*, a chinch bug of the family Lygidae subfam. Blissinae. This species is known to breed on *Paspalum virgatum*, a grass. It is an example of these more or less host specific insects. (From Slater, 1976.)

In Diagram 105 are shown the records enumerated by Slater for different orders of monocotyledons. Slater supplies data for 21 genera and 111 species of chinch bugs. The breeding insects are indicated separately from mere finds on plants. It is seen that the monocotyledons on which these chinch bugs breed are restricted to the Poales (where the records are very numerous), Restionales, Cyperales, Juncales, Typhales, Zingiberales, Haemodorales and (perhaps) Bromeliales. Visits are also known on a few species of Araceae, Orchidaceae and Iridaceae.

The chinch bug genus *Ischnodemus* (Fig. 94) is that which has the greatest range outside the grasses, occurring besides on species of Cyperaceae, Juncaceae, Typhaceae, Sparganiaceae, Zingiberaceae, Cannaceae, Marantaceae, Haemodoraceae and Bromeliaceae (and also Araceae). Species of *Macchiademus* are known for Poaceae, Restionaceae and Juncaceae (and besides on Iridaceae: *Moraea*), species of *Capodemus* for Poaceae, Restionaceae and Cyperaceae, species of *Spalacocoris* for Zingiberaceae, and species of *Extrademus* for Araceae and Orchidaceae; more casual finds being placed in parentheses.

In evaluation the relationships of these groups—as judged by the feeding chinch bugs—it may be noted that all genera found on other monocotyledons than grasses (breeding or not) are also found breeding on grasses, except *Spalacocoris sulcatus* feeding on roots in Zingiberaceae, and *Praeblissus albopictus* intercepted on an orchid.

Thus the following pairs of orders have two genera of chinch bugs in common: Poales-Restionales, Poales-Cyperales and Poales-Juncales. Poales otherwise has the genus *Ischnodemus* in common with the Typhales, Haemodorales, Zingiberales and Bromeliales, and these four orders agree in being visited exclusively by this genus, except for Zingiberales with one or two more genera. The data may slightly support affinity between the Haemodoraceae, Bromeliaceae, Typhales and Zingiberales.

One possibility may be that the diverse groups outside the Poales-Cyperales are but secondary hosts for the Blissinae chinch bugs, although this is partly contradicted by the fact that there seems to be no host plants for them among dicotyledons, nor among, for example, the Asparagales.

The rich data for grasses may give valuable suggestions about infrafamilial relationships, though the results are far from clear.

Some conclusions may be drawn from the specificity of the individual species of chinch bugs. These turn out to be restricted quite often to one or a few host genera in the same family, but there are some conspicuous exceptions. *Ischnodemus fulvipes* is known from *Canna* (Cannaceae), *Thalia* (Marantaceae), *Musa* (Musaceae) and on a "Bromeliad" species, which may be taken to support a close relationship between the three families in the Zingiberales *inter se* and perhaps also between the Zingiberales and Bromeliales. *Ischnodemus brinckii*, a South African species, attacks *Wachendorfia paniculata* and *Dilatris viscosa* (Haemodoraceae), *Tetraria ustulata* (Cyperaceae) and besides has been collected on a species of *Zantedeschia* (Araceae), a somewhat scrambled group of plants. A close relationship between the host groups of *I. brinckii* is thus more difficult to believe. Slater (op. cit.) regards this species as a possible relict, feeding on dry seeds, which would suggest that the early hosts may have been liliiflorean. The poalean and other hosts

would then be secondary, the parasites subsequently having diversified under co-evolution with them to evolve sap-feeding forms often adapted to concealed habitats in leaf sheaths such as in grasses. Although this seems somewhat speculative taking into account the few finds on liliiflorous plants, the theory is interesting in discussing the evolution of the Commeliniflorae.

Base Data

Abbreviations: I: Species of *Ischnodemus*; C: *Capodemus*; M: *Macchiademus*; S: *Spalacoris*; A: *Atrademus*; P: *Praeblissus*; E. *Extrademus*.

ALISMATIFLORAE: none.

ARIFLORAE: ARAL.: Arac.: *Zantedeschia* (I), *Philodendron* (E).

LILIIFLORAE: LILIA.: Irid.: *Moraea* (M). ORCHI.:

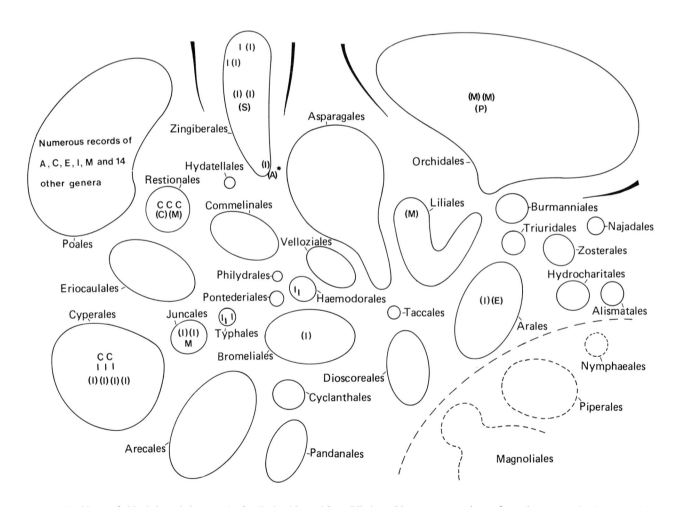

Diagram 105. Hosts of chinch bugs belong to the family Lygidae subfam. Blissinae. Numerous records are from the grasses; in the other orders all records are inserted. Each symbol represents a species attached. Abbreviation in parenthesis denotes visits (but not breedings) others indicate breeding insects. Abbreviations: I, Species of *Ischnodemus*; C, of *Capodemus*; M, of *Macchiademus*; S, of *Spalacoris*; A, of *Atradenus*; P, of *Praeblissus*; E, of *Extrademus*.

Orchi.: *"Chrysis"* (P); unspec. (E). HAEMO.: Haemo.: *Dilatris* (I), *Wachendorfia* (I). BROME.: Brom.: unspec. (I).

ZINGIBERIFLORAE: ZINGI.: Musac.: *Musa* (I, ?A); Zingi.: *Alpinia* (I), *Elettaria* (I), unspec. (S); Canna.: *Canna* (I); Maran.: *Thalia* (I).

COMMELINIFLORAE: TYPHA.: Sparg.: *Sparganium* (I); Typha.: *Typha* (I). JUNCA.: Junca.: *Juncus* (I, M). CYPER.: Cyper.: *Cladium* (I), *Cyperus* (I), *Ficinia* (C), *Gahnia* (I), *Scirpus* (C, I), *Tetraria* (I). RESTI.: Resti.: *Elegia* (C), *Thamnochortus* (C), unspec. (M). POAL.: Poac.: numerous genera of grasses attacked by a considerable number of chinch bug genera, these including I, C, M, A, and E (see above).

Evaluation, General Procedures

The data summarized in the diagrams given above vary from nearly complete to very incomplete depending upon the character considered. Accordingly, there is no doubt that the variation in some taxa is greater than we have recorded. Nonetheless there are supposedly sufficient data in most instances to estimate the affinities between groups by using numerous characters in combination. This approach should reveal affinities between groups even though selection by environmental influences has caused them to differ morphologically from each other and their less diversified ancestors.

In judging the probable affinities between taxa from the distributions of character patterns it is clear that not all characters are of equal importance. Some quite variable in the monocotyledons are, even so, quite constant over whole orders and superorders. Others are variable within the one family or even genus. Thus embryo sac formation is quite variable in families such as the Convallariaceae, Alliaceae and Trilliaceae. In contrast it is fairly constant in the large family Poaceae.

Because we have accepted that characters differ in taxonomic importance from part to part of the system and are not all equally important in the same part of the system we have not subjected the data to a formal numerical analysis.

In collecting the data our attention has been drawn to the paucity of information available for several small families, such as the Geosiridaceae, Hanguanaceae and Thurniaceae, and also for certain large families some of whose characters are poorly known.

Bearing in mind these limitations we have attempted to estimate the degree of similarity between groups from the level of family (sometimes also genus) to superorder. These comparisons will deal firstly with problems at the ordinal to superordinal level. Where appropriate, discussions will be given about the family and generic groupings as conceived in the preliminary diagram on which the data have been mapped.

Superorders Alismatiflorae and Triuridiflorae

The Alismatiflorae have long been considered a natural entity at either the ordinal or, in recent years, on the superordinal level, but there remain some problems associated with them such as:

(1) their affinity in regard to other monocotyledons has been very uncertain. Mostly they have been suggested as having affinities lying in the direction of the Liliiflorae or Areciflorae. However, a closer agreement with the Ariflorae has become obvious in the course of this study;

(2) the acknowledged similarity, with respect to numerous characters, to the waterlilies and allied plants (Nymphaeiflorae = Nymphaeales) has caused taxonomists either to believe in a true and rather close phylogenetic relationship (even to including the Nymphaeales in the monocotyledons close to the Alismatiflorae) or to suppose a far-reaching convergence due to aquatic adaptations;

(3) the order Triuridales has either been referred to the Alismatiflorae or to the Liliiflorae;

(4) the number of orders acknowledged within the Alismatiflorae as well as their circumscription has been variable.

These problems will now be discussed.

AFFINITIES WITH THE ARIFLORAE

Early in this project we found a number of similarities between the Alismatiflorae and Ariflorae and the evidence of agreement between these two superorders has increased with the amount of data, as has also the evidence for a more distant relationship between the Ariflorae and the Areciflorae than is usually expressed in literature.

Some similarities between the Ariflorae and the Alismatiflorae are given below.

Both groups are almost completely herbaceous.
Secondary growth in the stem is always absent.
The leaves are conspicuously similar in the non-aquatic and non-saprophytic taxa of the two groups: petiolate, sheathing at base, and with a well developed, often cordate lamina.
Ptyxis in such leaves is supervolute.
Corms and true bulbs are lacking.
Silica bodies are absent.
Both are mainly glabrous.
Intravaginal squamules are restricted to these two superorders, being found in most Alismatiflorae except Triuridales and in a few taxa (?) of the Arales.
Vessels are constantly lacking in the stems and leaves of both superorders; they are sometimes also lacking in the roots as well.
Laticifers occur in both superorders.
Both groups may have an inflorescence with the shape of a spadix or spike (consider, for example, the inflorescence in Juncaginaceae and Zosteraceae, both of the Zosterales, or in Aponogetonaceae of the Hydrocharitales, Alismatiflorae). Furthermore, a sheathing spathe as found in the Ariflorae has its counterpart in some Alismatiflorae.
Anther attachment is basal.
Anther dehiscence is often extrorse.
There is a (weak) tendency for the anthers in both superorders to be bisporangiate.
Tapetum in both superorders is amoeboid, with as a rule uninucleate tapetal cells, a condition uncommon elsewhere in the monocotyledons.
Microsporogenesis is almost constantly successive in both superorders.
A similar variability in the aperture conditions of pollen grains occurs in both superorders.
Both groups frequently have trinucleate pollen grains, the Alismatiflorae relatively more often so than the Ariflorae.
The stigmas in both superorders usually have a dry surface.
Axile placentas are lacking in both superorders, and the ovules per placenta are often reduced to one.
Perisperm is lacking in both superorders.
In all Alismatiflorae and maybe about one third of the Ariflorae the endosperm, though formed, is consumed during seed development, these two superorders being

the only monocotyledons with these conditions.

Embryo formation proceeds according to the Caryophyllad type in taxa of both superorders, a condition that is otherwise rare in the monocotyledons. The data on the Ariflorae are, however, scarce.

The embryo is linear in both superorders and frequently macropodous in the Alismatiflorae, less commonly so in the Ariflorae; still the agreement is a conspicuous one.

Curved embryos occur in a considerable number of taxa in both superorders, but are otherwise rare in monocotyledons.

Chlorophyll often occurs in the embryos of the seeds in both superorders, a condition otherwise rare in monocotyledons except amongst the orchids.

Flavonols are rare in both groups.

Sattler and Singh (1978) found great similarity in floral organogenesis between *Aponogeton* (Aponogetonaceae, Hydrocharitales) and, for example, *Acorus* (Araceae, Arales) and, besides, in *Triglochin* (Juncaginaceae, Zosterales), which may be taken to support further the affinity between the Alismatiflorae and Ariflorae.

Among the differences that distinguish the two superorders from each other the following should be mentioned especially:

The abundance of oxalate raphides in the Ariflorae, and their total absence from the Alismatiflorae.

Stomata are normally tetracytic (to hexacytic) in the Ariflorae and either missing or paracytic in the Alismatiflorae.

The gynoecia in the Alismatiflorae are frequently apocarpous.

The absence of petaloid tepals in the Ariflorae.

The absence of laminal placentation in the Ariflorae.

The constantly cellular endosperm formation in the Ariflorae in which this group deviates from practically all other monocotyledons (Hydatellaceae and perhaps some single species of Thismiaceae excepted; other *Thismia* species investigated have helobial endosperm formation).

The baccate fruit of most Ariflorae.

These differences can be complemented with, and are also to some extent connected with, differences in growth form and habitats: while most Ariflorae are shade herbs which grow on the floor of forests or scrubs and have baccate, animal dispersed fruits, the Alismatiflorae are inhabitants of moist ground or are aquatics, some of them comprising the only wholly marine aquatics among the angiosperms. The fruit dispersal thus is mainly by water, and pollination in some taxa is also hydrogamic.

AFFINITIES WITH THE NYMPHAEIFLORAE

Similarities between the Alismatiflorae and the Nymphaeiflorae have been enumerated previously, e.g. by Dahlgren (1974, 1976), Haines and Lye (1975), El Gazzar and Hamza (1975) and Huber (1977), most of whom were inclined, like Takhtajan (1969), to consider the Nymphaeales as closely related to, or to be included in, the monocotyledons near the Alismatiflorae.

In evaluating the data on the previous pages, it should be noted that Haines and Lye (1975) found the two cotyledons of some Nymphaeaceae to be basally connate at one side and thus that they might be interpreted as a single cotyledon.

The similarities between the Alismatiflorae and the Nymphaeiflorae include the following.

Both are primarily aquatic herbs, with an atactostele and a radicle, the growth of which is arrested at an early stage.

Not previously mentioned by us is the observation made by van Tieghem (1891) that the root caps of the Nymphaeales are of the monocotyledonous type.

Secondary growth is absent.

The leaves are basically of a similar shape, petiolate, sheathing and with a well developed cordate-oval lamina.

Stipules are present in taxa of both superorders, their presence being otherwise very rare in the monocotyledons

Inclusions of oxalate and silica are lacking in both superorders.

Vessels are lacking in stems of all the Alismatiflorae and probably in many Nymphaeales (Kosakai *et al.*, 1970), but in *Cabomba* at least vessels are present in the rhizomes and young aerial stem (according to Inamdar and Aleykutty, 1979).

Laticifers are present in both superorders, viz. in the Alismataceae and Nymphaeaceae.

The flowers have a basically trimerous perianth in some representatives of the Nymphaeiflorae, viz. in the Cabombaceae, and in *Cabomba* also the androecium is trimerous as in a few Alismatiflorae (*Butomus* etc.).

The anthers are basifixed in both superorders.

Anther dehiscence is extrorse in some Nymphaeiflorae as in most Alismatiflorae.

Although the tapetum is secretory in most Nymphaeiflorae it is amoeboid in the Ceratophyllaceae as in the Alismatiflorae.

Microsporogenesis is succesive in the Cabombaceae and

Ceratophyllaceae (a very rare condition in dicotyledons), as in the Alismatiflorae (and most other monocotyledons).

Pollen grains are usually either sulcate, zonisulculate or inaperturate in the Nymphaeiflorae; they are mostly sulcate or inaperturate in the Alismatiflorae (except Alismatales, where they are foraminate) and about half of the other monocotyledons. Sulcate pollen grains are found among the dicotyledons only in the Nymphaeiflorae and Magnoliiflorae (including Piperales).

Apocarpy is widespread in both superorders, sometimes (as in most Nymphaeales and Hydrocharitales) the carpels being enclosed by the floral receptacle.

The placentation in both superorders is laminal or lateral laminal to basal or possibly derived from these types.

A parietal cell is usually cut off from the archesporial cell in both groups.

Endosperm formation is reported to be helobial in taxa of Cabombaceae, a very rare condition in dicotyledons but one which is in agreement with most Alismatiflorae.

Follicular fruits are found (though rarely) in both groups. Achenes occur in the Ceratophyllaceae and numerous Alismatiflorae.

As pointed out by Huber (1977) and others these similarities are counterbalanced by a number of conspicuous differences, some of which are so striking that one may be inclined to guess that several (but by no means all) of the above-mentioned similarities could have evolved by convergence. Among the differences to be stressed particularly are the following.

The Nymphaeales have two and the Alismatiflorae one cotyledon; Haines and Lye's (1975) observations (see above) do not reduce this difference substantially.

The leaves of the Nymphaeales have reticulate venation and involute ptyxis, while venation is not reticulate in the Alismatiflorae where ptyxis is also supervolute.

Intravaginal squamules are found in most or all Alismatiflorae, but are absent from the Nymphaeales.

Sieve tube plastids constantly have triangular protein bodies in the Alismatiflorae as in other monocotyledons, but protein bodies are missing in the Nymphaeales, where the sieve tube plastids accumulate starch grains.

Stomata, when present, tend to be paracytic in the Alismatiflorae but are anomocytic in the Nymphaeales.

The perianth members are not spirally inserted in the flowers of the Alismatiflorae, nor are stamens or carpels, but distinctly spiral insertion is present in many Nymphaeales.

The ripe seeds in the Alismatiflorae lack perisperm and mostly endosperm as well, while in the Nymphaeales there is copious perisperm enclosing near its top an endosperm tissue which encloses the embryo (Fig. 112).

Chemistry is different: ellagi-tannins found in at least some Nymphaeales are not known to occur in any monocotyledons at all (Bate-Smith, 1973).

Our conclusions are that a rather distant relationship may exist between the superorders Alismatiflorae and Nymphaeiflorae. However, it is not altogether apparent which of the similarities are from an old residual relationship (but apocarpy and sulcate pollen grains should be such attributes) and which have developed by convergence. This is further discussed on p. 323. The Alismatiflorae and the Nymphaeiflorae thus are probably less closely allied than are the former and the Ariflorae.

THE POSITION OF THE TRIURIDALES

The Triuridales (Fig. 95) comprise a group of terrestrial, chlorophyll-less mycotrophic plants growing on the forest floor, thus differing considerably from any of the conventional Alismatiflorae. The apocarpy and sometimes gynobasic stylodia of the carpels have a superficial similarity to those of, for example, *Alisma*, which has previously been one of the reasons for placing the Triuridales in the alismatifloraean assemblage. Although these similarities are probably by parallel evolution, there are other reasons for placing the Triuridales in the Alismatiflorae. The more conspicuous of these are as follows.

Both taxa comprise herbaceous plants lacking vessels in the stems and leaves (the Triuridales and many other Alismatiflorae also lack vessels in the roots).

Secondary thickening growth does not occur.

Silica bodies and oxalate raphides are lacking.

Hairs are largely lacking.

Stomata are lacking in the Triuridales and many other Alismatiflorae.

The anthers are basifixed and extrorse in the Triuridales as in most of the other Alismatiflorae.

Tapetum is stated to be of the amoeboid type and the tapetal cells are uninucleate. Although being a poorly investigated attribute, this may be one of those pointing most definitely in favour of placing the Triuridales with the Alismatiflorae.

The pollen grains are inaperturate, a feature common also in certain non-aquatic Zosterales members (e.g. *Scheuchzeria*).

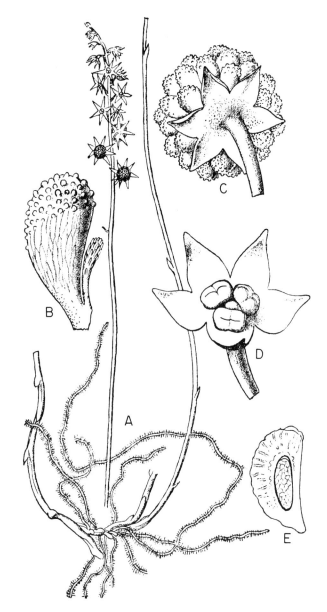

Fig. 95. Triuridales; Triuridaceae: *Sciaphila major* and (the details) *Sc. sumatrana*. A, habit; B, female flower; C, male flower; D, carpel; E, seed. (After Beccari, from Wettstein, 1924).

Microsporogenesis is successive.

The pollen grains of the Triuridales are trinucleate as in most other Alismatiflorae (but in contrast with most Liliiflorae).

Placentation is basal in the Triuridales as in the Alismatales and some taxa of the other orders.

The ovules are anatropous and have the Polygonum type of embryo sac formation.

The fruits are monocarpellate achenes as in many other Alismatiflorae (this being basically a consequence of the uniovulate apocarpous condition).

There is a lack of information with respect to many characters in the Triuridales, and notwithstanding the above similarities there are some differences which make the phylogenetic position of the Triuridales most uncertain. The following differences are notable.

The plants are saprophytic herbs growing in non-aquatic and non-marshy habitats.

Intravaginal squamules have not been reported.

The ovules are tenuinucellate and lack parietal cell; the nucellus epidermis does not divide periclinally.

The endosperm formation is nuclear in Triuridales, helobial in most Alismatiflorae.

The seeds are endospermous, unlike those of the Alismatiflorae.

These attributes appear to be very significant as they form exceptions from most or all other Alismatiflorae. They are, however, found in some Liliiflorae, and there is indeed a need for further research on the Triuridales to establish their real affinities. We are therefore not inclined to judge which alternative is to be preferred: placing the Triuridales in the Alismatiflorae, or placing the order in the Liliiflorae, and then probably close to the Melanthiaceae (Liliales). Another possiblility is to treat the Triuridales as a separate superorder, Triuridiflorae, as done by Thorne (1968). In the light of the information available, each of these treatments has merit. We have here chosen the last alternative.

THE CIRCUMSCRIPTION AND JUSTIFICATION OF THE ORDERS ALISMATALES, HYDRO-CHARITALES, ZOSTERALES AND NAJADALES

The Alismatales according to our circumscription are distinguishable from the remaining three orders in terms of the following attributes.

They are constantly non-marine plants being more often marsh plants than submerged aquatics, although the latter category is not lacking.

Vessels with simple perforation plates are present in the roots of the Alismatales; in other Alismatiflorae (except *Butomus*) vessels are either lacking or have scalariform perforation plates.

The Alismatales have laticifers, such are lacking in other Alismatiflorae.

The floral parts are frequently more numerous than in other Alismatiflorae.

From the available evidence it seems that the Alismatales have endothecial thickenings of the girdle type in the anthers, while most other Alismatiflorae have the spiral type.

The pollen grains have more than one aperture, being bi- to polyforaminate, unlike the remaining Alismatiflorae.

The carpels are usually numerous and frequently in superimposed whorls unlike other Alismatiflorae.

There is no parietal cell formed in the nucellus, as in most other Alismatiflorae (except *Zostera*), but there are mostly periclinal divisions in the epidermis of the nucellus in the Alismatales, which seems to be rare in other Alismatiflorae.

The embryo sac formation is constantly of the Allium type.

The embryo is horse-shoe shaped, which is not the case in the Hydrocharitales, although curved embryos are known in *Halophila* of Hydrocharitaceae and several zosteralean genera.

Anthocyanin pseudobases occur in the Hydrocharitales but are absent in the Alismatales.

Singh and Sattler (1977) found that the floral organogenesis of *Sagittaria* and other Alismataceae differ from, for example, that in *Aponogeton*, *Scheuchzeria* and other Alismatiflorae with trimerous whorls of stamens and carpels, in that secondary primordia are superimposed on the primary (trimerous) ones. Thus, it seems, the polymerous conditions in most Alismataceae are likely to be secondary, and thus to be considered an advanced feature. These authors found no sign that the androecia or gynoecia in the Alismataceae have a spiral organization.

There is no doubt that the formerly acknowledged Limnocharitaceae, sometimes included in the Butomaceae, are in closest agreement with the Alismataceae in practically all attributes except the ovaries and fruits which are reminiscent of those in *Butomus* (Butomaceae, Hydrocharitales). The Limnocharitaceae are therefore included in the Alismataceae as a subfamily. Though containing only one family, the Alismatales seem to be justified as an order in the present circumscription.

The Butomaceae, the Aponogetonaceae and the main part of the Hydrocharitaceae (see p. 25) form a fairly well definable order. The marine genera *Thalassia* and *Halophila* in their filiform pollen grains, lack of stomata and other specialized features approach (although probably by convergent evolution) the taxa of Zosterales.

The Najadales are sometimes treated as an order separate from, and at other times are included in, the Zosterales. In the course of the present study it has appeared as increasingly obvious that the Najadaceae are in close agreement with the Zosteralean families. The so-called verticillate leaves seem to be in fact opposite and by transposition pseudoverticillate. The anther wall was found to be of the reduced type (p. 137) a derived condition of little use for taxonomic purposes. The basal and anatropous ovule represents a conspicuous difference from all aquatic Zosterales which tend to have apical or lateral ("parietal") orthotropous-campylotropous, pendulous ovules. In this the Najadaceae agree with most taxa of the Juncaginaceae and Scheuchzeriaceae; these also seem to have nuclear endosperm formation as have the Najadaceae (Davis, 1966). The embryo is also chlorophyllous as in *Scheuchzeria*.

From the above evidence there does not seem to be any justification for upholding the Najadales. The Najadaceae are probably best included in the Zosterales in sequence with the Scheuchzeriaceae and Juncaginaceae as was done by Stebbins (1974).

Superorder Ariflorae

The Ariflorae with the single order Arales have often been associated with the Areciflorae (Arecales, Pandanales, and Cyclanthales), or the Typhales. Our results do not particularly support these affinities. The superficial similarities between these groups have presumably evolved by convergence.

AFFINITIES WITH THE DIOSCOREALES

There is some evidence that the Arales may have affinity with members of the Liliiflorae, notably the Dioscoreales. Vegetative similarities include the petiolate leaves with a broad, often cordate, sometimes compound lamina with the same ptyxis and though rarely so in Arales, with reticulate venation. Both groups accumulate oxalate raphides. The inflorescences (spikes) agree principally and the fairly small or moderate-sized often greenish flowers of many Dioscoreales approach the minute ones in the Arales. In both groups we find some primary attributes such as basifixed anthers, spiral type thickenings in the endothecium, mostly dry stigmas, mostly anatropous ovules, mostly Polygonum type embryo sacs etc.

There is a tendency for the Arales to show advancement, however, and in a number of attributes we find great variation in the direction of specializations. Thus, we (may) find root hairs growing out from specialized, short cells; velamen; laticifers; tetracytic stomata; reduction in tepal, stamen and carpel number; extrorse anthers often with apical poricidal dehiscence; amoeboid tapetum; inaperturate and/or 3-nucleate pollen grains; macropodous and/or curved embryo; basal rather than axile placentation; orthotropous ovules; lack of parietal cell; cellular endosperm formation; Caryophyllad type of embryogenesis; baccate fruit; frequent loss of endosperm; loss of vessels in the stem. Septal nectaries (if these are primitive ?) are also lost in the Arales. In the Arales the embryo is larger in relation to the endosperm than in the Dioscoreales, and the baccate fruit (rarer in the Dioscoreales) no doubt represents a specialization, while on the other hand the epigynous flowers in most Dioscoreales represent a specialization in that order.

AFFINITIES WITH THE ARECIFLORAE

It has been demonstrated above that the Ariflorae most likely have their closest affinity with the Alismatiflorae. This affinity is considerably closer than that with the Areciflorae as illustrated by the following differences.

The Areciflorae are mostly large woody plants or lianas, more rarely (as in Cyclanthales) giant herbs; the Ariflorae are small to giant herbs.

The leaves have a fundamentally different construction, those of the Areciflorae having a plicate or (Pandanales) conduplicate ptyxis, those of the Ariflorae a supervolute ptyxis. In those Areciflorae which are petiolate as in most Ariflorae, the lamina tends to be dissoluted ("compound") or bifurcate.

The leaf bases in the two groups are different, those of the petiolate Areciflorae being of the Bambusa type (p. 67), those of the Ariflorae being of the Alisma type.

Vessels are constantly present in the leaves of all the Areciflorae, in the Pandanales and in addition in the Arecales also constantly in the stem. In the Ariflorae vessels are always absent from the whole shoot. (The Cyclanthales in this feature are intermediate, lacking, as it seems, vessels in the stems.)

Laticifers often found in the Ariflorae are lacking in the Areciflorae, the mucilage canals in the Cyclanthales being probably of a different construction.

Stomata, though tetracytic in both superorders, tend to be dissimilar, the terminal subsidiary cells being generally different from the lateral in the Areciflorae.

Silica bodies are often (though not constantly) found in the Areciflorae (Arecales) but never in the Ariflorae.

Septal nectaries often found in the Areciflorae are lacking in the Ariflorae.

Anther dehiscence is latrorse (or introrse) in the Areciflorae, extrorse in the Ariflorae.

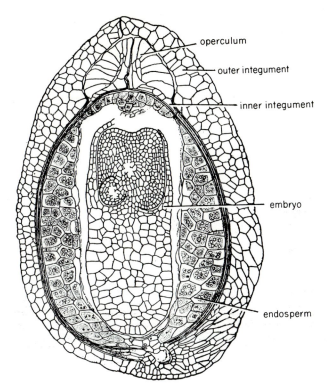

Fig. 96. Mature seed of *Lemna paucicostata* in longitudinal section. The inner integument forms a prominent operculum about the micropyle. Note the copious endosperm containing starch grains. (After Maheshwari and Kapil, 1963.)

the Areciflorae and the Ariflorae, the Cyclanthales showing less difference to the Arales than the other orders. The following similarities between the superorders should however be noted.

The common presence of oxalate raphides in both groups.
The tetracytic stomata (cf., however, above). This similarity may be superficial (see Diagram 30).
The frequently very short or obsolete styles or stylodia of both superorders.
The absence of axile placentation in both groups.
The joint presence of so-called "pseudocrassinucellate" ovules in the Ariflorae and the Cyclanthales of the Areciflorae.
The high proportion of baccate fruits in the Areciflorae and Ariflorae respectively, this being perhaps the most conspicuous similarity, and one which assumed great importance in some earlier classifications.

Besides these, the frequently spadix-like inflorescence and the spathe-like inflorescence bracts in the Areciflorae have been referred to as similarities with the Ariflorae. Though several in number, the above similarities do not seem to be significant, each either having a wide distribution outside these superorders or being explained as a simple biological adaptation arising by convergence.

Tapetum where known is secretary in the Areciflorae, amoeboid in the Ariflorae.
Endosperm formation in the Areciflorae is mostly nuclear (in the Cyclanthales helobial); it is cellular in the Ariflorae.
The embryogenesis seems (data are incomplete) to be of the Onagrad or Asterad type in the Areciflorae, of the Caryophyllad type (in Lemnaceae, however, of Onagrad type) in the Ariflorae.
The endosperm when present is at least frequently starchy in Ariflorae, starchless in Areciflorae except the cyclanthaceous genus *Dicranopygium*, which has a starchy endosperm (Harling, 1958).
In the Ariflorae but not the Areciflorae embryos are generally chlorophyllous.
The Areciflorae (at least the palms) seem to have a totally different set of flavonoids from the Ariflorae: including rich amounts of luteolin/apigenin (found in the Ariflorae only in the Lemnaceae) and tricin (lacking in the Ariflorae) and flavone sulphates.

We judge that these differences are significant enough to reject theories of a particularly close affinity between

AFFINITIES WITH THE TYPHALES

The frequently mentioned affinity between the Ariflorae and the Typhales of the Commeliniflorae should also be considered in the light of the data presented. Among the notable similarities are the superficially similar (but, as shown by D. Müller-Doblies, 1968, and U. Müller-Doblies, 1969, basically different) inflorescences; the common susceptibility to the rust *Uromyces spargani*; the absence of silica bodies but presence of oxalate raphides, the non-petaline tepals, the basifixed stamens, the amoeboid tapetum, the successive microsporogenesis and the rare occurrence in each order of pollen tetrads. However, in the light of the large number of significant differences (see Dahlgren, 1968) the close relationship is not very probable.

The fact that *Uromyces spargani* attacks *Acorus calamus* (Araceae) as well as species of *Sparaganium* (Sparganiaceae) contributed to the treatment of Typhaceae, Sparganiaceae and Araceae in the same order, Arales, by Thorne (1968, 1976). Savile (1979) supports this view in considering the Typhales and Arales closely related for the same reason. He noted that he "cannot believe that it is mere chance that the one susceptible species of *Sparganium* is the species that most resembles *Acorus* in leaf morphology and anatomy and in the morphology of the individual fruits, that most closely duplicates its ecology, and that has almost the same North American range." "If *Acorus* is a true aroid (it is unusual but not unique in lacking alkaloids and raphides) it must surely be among the most primitive members of Araceae", Savile concludes, and "it is also reasonable to assume that the distigmatic *Sparganium eurycarpum* is more primitive than the remaining, monostigmatic species".

However, considering the highly significant differences between the Arales and the Typhales (see also p. 300), these conclusions seem dubious (especially as they imply a *very* close relationship); similarity in distribution, habitat, life form etc. would rather provide arguments in favour of a "successful jump" from one of the hosts to the other. These fairly superficial similarities could hardly outweigh the fundamental differences between the two orders, although there are, admittedly, a number of similarities between them.

AFFINITIES WITH THE PIPERALES

Finally, several workers have pointed out the conspicuous similarity between the monocotyledonous Ariflorae (Arales) and the dicotyledonous Piperales, which are here provisionally included in the superorder Magnoliiflorae. This similarity impressed Lotsy (1911) and Emberger (1960) in constructing their phylogenetic diagrams. Monocotyledonous attributes of the Piperales have been mentioned by Huber (1976) and one of us (Dahlgren, 1974, 1976) and were in particular stressed by Burger (1977), who was struck by the monocotyledonous features in Saururaceae and suggested a close connection with monocotyledonous ancestors (discussed on pp. 330–332). There is no doubt that the Piperales belong among the hemi-monocotyledons (cf. Huber, 1977), but a close connection with in particular the Ariflorae has been doubted. The following conspicuous similarities between the Ariflorae and Piperales should be noted:

Both consist of herbs or herbaceous vines.
Both have scattered vascular bundles in the stem.
Both groups have petiolate leaves with well developed cordate lamina.
Rarely reticulate venation may occur in the Ariflorae (*Pothos, Arum*) as in the Piperales.
The stomata in both groups tend to be tetracytic.
The inflorescence is a spadix, in most members of both

groups and the spathe which is typical of the Ariflorae is matched by similar leaves in *Houttuynia* and *Anemopsis* (Saururaceae), although they are four to numerous in the latter genera.
The flowers in both groups are trimerous, often lack a perianth and undergo the same types of reductions.
The Piperales may have the Monocotyledonous type of anther wall formation, a condition which is otherwise rare in dicotyledons.
The pollen grains in both groups are often either sulcate or inaperturate.
In both groups the pistils tend to have an extremely short or obsolete style.
The stigma generally has a dry surface in both groups.
Placentation is basal in many Ariflorae and in practically all Piperales, and the ovules in the Ariflorae are often solitary as in the Piperales.
Endosperm formation in the Ariflorae is cellular, which is also, for example, the case in *Peperomia* (Piperaceae) and in the Saururaceae. However, there are differences in endosperm appearance (a big unicellular haustorium in most Ariflorae).
The fruit is a berry in many Piperales (e.g. in *Peperomia*) and in most Ariflorae.
Essential oils as found in the Piperales are also found in *Acorus* (Ariflorae).

This list may appear impressive but it is counterbalanced by a number of equally conspicuous differences which include the following:

The stems of the Piperales have vessels which are, moreover, of advanced types, with simple or scalariform perforation plates with few bars. In contrast the stems of the Ariflorae are constantly devoid of vessels.
The ptyxis is normally supervolute in the Ariflorae, but very variable in the Piperales, being more often involute or conduplicative.
Stomata have no preferred orientation in the Piperales, but are parallel to the leaf axis in the Ariflorae.
Oxalate raphides, which are of almost universal occurrence in the Ariflorae, are missing in the Piperales (though oxalate is found in other crystal forms).
Laticifers which are common (but not constantly present) in the Ariflorae, are lacking in the Piperales.
Sieve tube plastids in the Piperales lack protein and contain starch grains only. This is an attribute of great phylogenetic significance in this part of the system. The Arales have typical cuneate *P*-type plastids (see p. 85).
The anthers are principally extrorse in the Ariflorae but not in the Piperales, although in both groups they may dehisce apically.

Tapetum is amoeboid in the Ariflorae but secretory in the Piperales; the tapetal cells are mostly uninucleate in the Ariflorae and mostly binucleate in the Piperales.

Microsporogenesis is successive in the Ariflorae, simultaneous in the Piperales.

The ovules are orthotropous in the Piperales, but very rarely so in the Ariflorae.

The endosperm usually develops a large unicellular chalazal haustorium in most Ariflorae but not in the Piperales.

The seeds are filled with a large perisperm in the Piperales, but perisperm is lacking altogether in the Ariflorae.

Embryogeny in the Ariflorae is the Caryophyllad or Onagrad types, but in the Piperales is of the Piperad or Asterad types. This character is however poorly known and may be of low significance.

The seeds are generally chlorophyllous in the Ariflorae, but not so in the Piperales.

It appears that most of these differences are of greater significance than are the similarities, and we are of the opinion that the two groups exhibit extraordinarily good examples of convergent evolution. However, certain similarities are not achieved by convergence but are certainly ancestral traits found in the common ancestors, and the Piperales no doubt evolved from a group of primitive angiosperms with close connection to the Magnolialean as well as the monocotyledonous (including very likely the Ariflorean) ancestors. This will be further commented upon on p. 336.

Superorder Liliiflorae

The delimitation of the Liliiflorae from the Areciflorae, Ariflorae and Alismatiflorae (excluding Triuridales, see p. 290) is straightforward. In contrast there is no clearcut boundary between the Liliiflorae and the Commeliniflorae, certain orders being assigned variously to one or the other superorder according to the source consulted. Furthermore, the delimitation of families within certain orders of the Liliiflorae raises problems which are discussed later.

The main impression is that there is a series of taxa whose members form a gradual transition between the extreme members of each superorder. At one end of this series are the Poales and Cyperales of the Commeliniflorae and at the other end the Liliales, Asparagales, Dioscoreales and Taccales of the Liliiflorae. Orders combining significant features of both these "ends" are, for example, the Haemodorales, Philydrales, Pontederiales, Velloziales and Bromeliales, all placed here in the Liliiflorae, and the Commelinales and Typhales, placed here in the Commeliniflorae.

One attempt to resolve the taxonomic problem on the high levels is due to Huber (1977), who erected a series of superorders for these intermediate groups: viz. the Haemodoriflorae (Haemodorales); the Pontederiiflorae (Pontederiales and Philydrales) and the Bromeliiflorae (Bromeliales and Velloziales). Alternatively the orders may be partitioned amongst the Liliiflorae and Commeliniflorae.

We will first consider two separate but interrelated problems. These are as follows:

1. Which of a number of character states are to be considered more representative of the Liliiflorae and which more representative of the Commeliniflorae?
2. On the bases of these character states, to which of the two superorders should the orders in dispute be assigned?

For each of the superorders a "core group" has been envisaged which neglects those orders whose affinities are in dispute. From a consideration of these "core groups" the following characters are more concentrated in the Liliiflorae than in the Commeliniflorae.

LILIIFLOREAN AND COMMELINIFLOREAN CHARACTERS

The following characters are more concentrated in the core groups of the Liliiflorae than in those of the Commeliniflorae.

(1) Secondary thickening growth is restricted to the Asparagales and Liliales, being absent in all the Commeliniflorae.
(2) Root epidermis is not differentiated into two kinds of cells, of which a short kind bears the root hairs in many Commeliniflorae.
(3) Presence of velamen is liliiflorean; absence, however, is of lesser taxonomic importance.
(4) Opposite, verticillate and compound leaves, reticulate venation and petioles of the Dioscorea type (p. 65) although relatively rare attributes are all indications of liliiflorean affinity, but their absence is useless for indicating superordinal affinities.
(5) Silica bodies absent.
(6) Oxalate raphides present.
(7) Vessels present or absent, when present with scalariform perforation plates in stems and leaves.
(8) Stomata anomocytic or more rarely paracytic.
(9) Perianth with two whorls of petaline tepals, and thus also with non-sepaline outer tepals.
(10) Nectaries in septa or at the tepal or stamen bases.
(11) Endothecial thickenings are of girdle type in many Commeliniflorae, but mostly of spiral type in the Liliiflorae.
(12) Pollen grains are more often ulcerate in the Commeliniflorae, but mostly sulcate in the Liliiflorae.
(13) Pollen grains binucleate rather than trinucleate.
(14) Placentation axile.
(15) Ovules several per placenta.
(16) Fruits capsular or baccate rather than nutlets or caryopses, although capsules with few seeds occur in some orders in the Commeliniflorae.
(17) Endosperm starchless.

(18) Embryos linear or "rudimentary" rather than capitate, broad or lateral.

(19) Tricin and flavone sulphates absent or rare.

(20) Steroid saponins common.

(21) Chelidonic acid common.

Moreover, the Liliiflorae, in the cases tested, more rarely contained isoenzymes of dehydroquinate hydrolyase (DHQ-ase) than did the Commeliniflorae. Many genera of rusts and smuts are virtually restricted to orders of the Commeliniflorae as host plants, an example of fairly widespread smuts being the generic complex around *Tolyposporium*. It is also shown that various insects, like certain groups of plant-lice and chinch bugs are restricted to hosts among the Commeliniflorae.

The opposite conditions in several, but not all, of the respects tabled are indicative of commeliniflorean affinity. Numerical judgement based on the above criteria therefore cannot be made indiscriminately but a careful assessment for each of the questionable orders would be relevant. Negative scores for attributes numbers 1, 3, 4, 16 and possibly 18 and 20 thus would not imply a commelinalean affinity even if the positive score is typical of liliiflorean groups, and absence of tricin, for example, is not a positive score for liliiflorean affinity.

LILIIFLOREAN VERSUS COMMELINIFLOREAN PROPERTIES OF SOME ORDERS

Haemodorales. This group shows positive liliiflorean scores for attributes no. 5 (no silica bodies), 6 (oxalate raphides), 7 (no vessels or vessels with scalariform peroration in stem), 9 (petaline tepals), 10 (presence of septal nectaries), 11 (sulcate or foraminate pollen grains), 13 (binucleate pollen grains), 14 (axile placentation), 15 (several ovules per placenta), 16 (capsular fruit), 18 ("rudimentary" embryo type) and no. 21 (chelidonic acid), but Haemodorales scores positively for a commeliniflorean affinity for attributes no. 1 (root epidermis), 8 (paracytic stomata), 11 (endothecial thickenings) and no. 17 (starchy endosperm) while attributes no. 2–4, 19 and 20 (uncertain occurrence of steroid saponins) are of no value in this comparison.

On the basis of this survey it is evident that Haemodorales can be referred with strong justification to the superorder Liliiflorae.

Philydrales has been carefully surveyed by Hamann in a series of papers (Hamann 1962c, 1966a and b) and is thus a well known order. Like the Haemodorales, to which it is probably rather closely allied it is intermediate between the core groups of Liliiflorae and those of Commeliniflorae. While neglecting attributes no. 1, 2, 3, 4, 19 and 21, the order shows liliiflorean agreement in attributes no. 5 (absence of silica bodies), 6 (presence of oxalate raphides), 7 (scalariform vessel perforation), 9 (petaline tepals), 11 (endothecial thickenings), 12 (pollen grains sulcate), 13 (pollen grains binucleate), 14 (axile or parietal placentation), 15 (several ovules per placenta), 16 (capsular fruit) and 18 (linear embryo type). The order shows commeliniflorean agreement in attributes no. 8 (paracytic stomata), 10 (nectaries probably lacking), 17 (starchy endosperm) and possibly no. 20 (? steroid saponins absent). However, the endosperm starch is not "mealy" because oil and protein are also present, indicating liliiflorean affinity. Also the helobial endosperm formation is indicative of this affinity.

Thus, as judged according to the above attributes, the Philydrales agree with the Commeliniflorae in some characters, but its superordinal affinity can hardly be questioned because of the excess of liliiflorean attributes.

Rübsamen and Hamann (personal communication) have recently found that the endosperm in unripe seeds of some Burmanniales contains ample amount of starch grains, which together with other similarities indicate a possible connection with the Philydrales.

Pontederiales (Fig. 97) shows the following tendency according to the same criteria. Liliiflorean attributes are numbers 5 (absence of silica bodies), 6 (presence of oxalate raphides), 7 (vessels when present in the stem having scalariform perforation plates), 9 (petaline tepals), 10 (septal nectaries present), 12 (pollen grains sulcate), 13 (pollen grains binucleate), 14 (axile placentation), 15 (several ovules per placenta), 16 (mostly capsular fruit) and 18 (linear embryo), while commeliniflorean attributes are numbers 8 (paracytic stomata), 11 (girdle type of endothecial thickenings), 17 (starchy, though not mealy, endosperm) and possibly 20 (absence of steroid saponins), the remaining characters being neutral or of no taxonomic significance. Thus it seems that this order to an equally high degree as the Philydrales merits a position in the Liliiflorae, and close to the latter order, with which it agrees in most respects.

Velloziales. Whereas Huber (1977) treated this together with the Bromeliales in a new superorder, the Bromeliiflorae, some taxonomists regard the Velloziaceae as closely related to the Hypoxidaceae, with which it shares tough,

fibrous leaf bases (mostly), conduplicative ptyxis, paracytic (to tetracytic) stomata, epigynous flowers, sulcate pollen grains and loculicidal capsules. There are important differences between the Velloziaceae and Hypoxidaceae, however, which can be shown from a study of the above diagrams, and we agree with Huber (1969) in regarding the Velloziaceae as an isolated family worthy of ordinal rank but closest to the Bromeliales.

According to Hamann (personal communication) there are, in addition to other conspicuous similarities between these two orders, the peculiar type of helobial endosperm formation where the chalazal chamber becomes cellular before the micropylar chamber. The two families also agree in possessing forms with a "trunk" and in having stems covered with a thick envelope of cauline roots, which proceed either in the stem cortex (in the Bromeliales) or outside this (in the Velloziales).

Liliiflorean attributes found in the Velloziales are the following: numbers 5 (absence of silica bodies), 6 (presence of oxalate raphides), 7 (no vessels or vessels with scalariform perforation plates in the stem), 9 (petaline tepals), 10 (presence of septal nectaries), 11 (spiral endothecial thickenings), 12 (sulcate pollen grains), 13 (binucleate pollen grains), 14 (axile placentation), 15 (several ovules per placenta) and 16 (capsular fruit).

In addition to this it may be mentioned that the epigynous flowers are indicative of liliiflorean rather than commeliniflorean affinity. Commeliniflorean attributes are numbers 8 (paracytic to tetracytic stomata) and 17 (starchy endosperm). However the starchy endosperm is not "mealy" as in Commeliniiflorae. It is not known whether steroid saponins may occur in the order.

From the above data it seems that the Velloziales is best retained in the Liliiflorae.

Bromeliales. This order is the one that would seem to be most intermediate between the strict Liliiflorae and the Commeliniflorae. It is definitely referred to the Commeliniflorae by Thorne (1968, 1976) and to (the Commelinanae of) subclass Commelinidae by Takhtajan (1969) and Cronquist (1968). One of us (Dahlgren, 1975) preferred to place it with some misgivings in the Liliiflorae, while Huber (1977) placed the Bromeliales and Velloziales in a separate superorder, the Bromeliiflorae, between the Liliiflorae and Commeliniflorae.

The following indications are given by our data. Liliiflorean attributes are numbers 2 (root epidermis), 6 (occurrence of oxalate raphides), 7 (vessels with scalariform perforation plates), 9 (the differentiation of tepals into petaline and sepaline is not really a liliiflorean attribute), 10 (presence of septal nectaries), 11 (spiral type of endothecial thickenings), 12 (prevalence of sulcate pollen grains), 13 (mostly binucleate pollen grains), 14 (axile placentation), 15 (several ovules per placenta), 16 (capsular or baccate fruits), 18 (linear embryo) and 20 (occurrence of steroid saponins), while commeliniflorean attributes are numbers 5 (occurrence of silica bodies), 8 (paracytic stomata) and 17 (starchy and "mealy" endosperm). In addition the flowers in a great part of the family are epigynous (a liliiflorean attribute). The occurrence of biforaminate pollen grains in some Bromeliales is hardly a commeliniflorean attribute (in which case they would be ulcerate). Besides, the endosperm formation is helobial, which may be regarded a liliiflorean attribute.

The above criteria clearly place the Bromeliales in the Liliiflorae, and with even more justification than is generally accepted.

Commelinales. Although this is the nominal group of the Commeliniflorae, it exhibits some liliiflorean attributes. An analysis of this order in terms of the criteria used above show the following results.

Liliiflorean attributes are numbers 6 (presence of oxalate raphides), 12 (sulcate pollen grains), 13 (binucleate, rarely trinucleate pollen grains), 14 (axile placentation), 15 (several ovules per placenta), 16 (capsular fruits), whereas commeliniflorean attributes include numbers 1 (root epidermis), 5 (occurrence, though not universal, of silica bodies), 7 (presence of vessels with simple perforation plates), 8 (tetracytic or rarely paracytic stomata), 9 (perianth; see below), 10 (lack of septal nectaries), 11 (girdle type of endothecial thickenings), 17 (starchy endosperm), 18 (broad or capitate embryo) and 20 (steroid saponins lacking). To this should be added the orthotropous or hemitropous ovules, a mainly commeliniflorean attribute. The petaline inner and sepaline outer tepals is not a liliiflorean attribute either, but can be regarded a commeliniflorean trait.

These characters indicate a certain preference for the commeliniflorean affinity which justifies us in retaining the Commelinales in that superorder. Admittedly the commeliniflorean attributes also seem to be more significant than the liliiflorean, some of the latter of which

Fig. 97. Representatives of the Pontederiaceae (Pontederiales). A–E: *Echhornia crassipes:* A, plant; B, stamen; C, pistil; D, capsule; E, seed. F–I; *Pontederia lanceolata*: F, plant; G, flower; H, stamen; I, pistil; J, nutlet. K, pollen grains of *Pontederia cordata*. (A–J from Cabrera, 1968; K from Erdtman, 1952.)

express a level rather than a phylogenetic affinity (sulcate pollen grains, axile placentation, several ovules, capsular fruits.)

Typhales. This is another order whose taxonomic position has been much disputed in the literature. While some see a similarity to the Ariflorae, others believe it to be related to the Pandanales (Areciflorae), neither of which positions is supported by our investigations. However, there seems to be some agreement with, for example, the Juncales, although certain attributes are liliiflorean. Hutchinson (1934) laid most emphasis on the latter, believing that the Typhales evolved from spike or spadix bearing Liliiflorae.

The following attributes, some of which are taxonomically highly significant, are liliiflorean viz. numbers 1

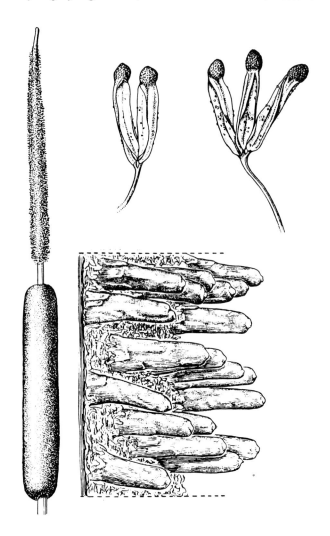

Fig. 98. *Typha angustifolia* (Typhaceae), showing inflorescences, the secondary axes of the female part of the inflorescence and two male flowers. (From Correa, 1969.)

(root epidermis), 5 (absence of silica bodies), 6 (presence of oxalate raphides), 7 (vessels with scalariform perforation plates in the stem), 11 (spiral endothecial thickenings), 13 (binucleate pollen grains), 18 (linear embryo) and, besides, the helobial endosperm formation. Maybe somewhat less significant are the commeliniflorean characters, some of which reflect an adaptation to wind pollination. These include attributes number 8 (paracytic stomata), 9 (non-petaline or lacking tepals), 10 (absence of nectaries), 12 (ulcerate pollen grains), 14 (non-axile placenta), 15 (one ovule per placenta), 16 (non-capsular and non-baccate fruit), 17 (starchy endosperm) and 20 (absence of steroid saponins).

Our conclusion is that the Typhales show distinct liliiflorean affinities and approach orders like the Philydrales, Pontederiales, Velloziales and Bromeliales in important features. This is, for example, the case in the type of helobial endosperm formation (see p. 206). It is likely from these premises that the similarities to the wind pollinated Commeliniflorae may have evolved independently, i.e. by convergent evolution, and therefore the Typhales is probably best treated as liliiflorean. Pollen tetrads occur in the Velloziales and Philydrales, as does this type of helobial endosperm formation, and the Philydrales and Pontederiales are also largely swamp plants with reductions in the floral parts as are the Typhales. However, one should not neglect the affinity to the Juncales, which share most of these properties and, besides, have anemogamous flowers. Considering all this, however, we prefer to transfer the Typhales to the Liliiflorae, but to a peripheral position as shown in Diagrams 106–107.

Hydatellales (Fig. 99). In this small monofamilial group, which includes minute, reduced plants, we can, however, hardly apply the above standards for a comparison. In the decidedly restionalean Centrolepidaceae (Fig. 107), for example, with a similar reduction, silica bodies are lacking and vessels have scalariform rather than simple perforation plates which may very well have to do with their diminutive size. The Hydatellaceae consist of submerged plants and we do not know to what extent the environmental conditions may have influenced the various morphological details. Liliiflorean attributes are the occurrence of anomocytic stomata which are wholly absent in all other Commeliniflorae; however, stomata are lacking in some Hydatellaceae and perhaps are always without function in submerged plants. Sulcate pollen grains, besides in this group are restricted in the Commeliniflorae to the Commelinales and Eriocaulales (Xyridaceae and Rapateaceae), but are common in the

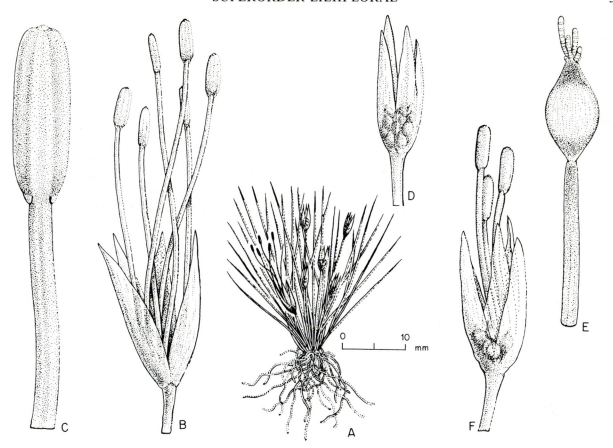

Fig. 99. *Hydatella inconspicua* (Hydatellaceae, Hydatellales). A, plant; B, male inflorescence; C, stamen (with tetrasporangiate anther); D, female inflorescence; E, female flower (note the uniseriate stigmatic hairs replacing a style); F, bisexual inflorescence. (All from Edgar, 1966.)

Liliiflorae. In terms of both these attributes the Hydatellaceae differ from the Centrolepidaceae (Hamann, 1976).

Also, the Hydatellaceae have anatropous ovules while the Centrolepidaceae have orthotropous (like most other Restionales), and the Hydatellaceae lack the starchy endosperm found in the Centrolepidaceae and all other Commeliniflorae. The endosperm is restricted to a few cells which do not serve as storage tissue. This reduced endosperm is compensated for by a starch-rich well developed perisperm having its counterpart only in some families of Zingiberales.

Further, Hamann (op. cit.) found the testa of the Hydatellaceae to be formed mainly from the outer layers of the outer integument, not as in the Centrolepidaceae from the inner integument. Moreover, the pollen grains seem to be binucleate (rather than trinucleate as in the Centrolepidaceae), although this needs verification. As is evident from the previous discussions these characters indicate that the Hydatellaceae are most uncertain in their affinity, but one cannot, of course, exclude that the

attributes considered here as liliiflorean have evolved independently in a separate (commeliniflorean) line of evolution under strong selective pressure in extreme environments (temporarily wet ground).

These interesting features may indeed suggest a liliiflorean affinity for the Hydatellaceae (sulcate pollen grains, anomocytic stomata, anatropous ovules, absence of silica bodies). The perisperm may be taken to indicate a zingiberalean connection, and other attributes could be used to argue for a place near the Ariflorae. However, considering the great possibility that the attributes of Hydatellaceae have evolved as a response to the immersed life we prefer to place it in a separate order *incertae sedis* near the Centrolepidaceae of Poales (Diagrams 107–108). The substitution of starchy endosperm by starchy perisperm according to Hamann (personal communication) leads to an acceleration of formation of seed nutrient developed in unstable habitats; cf. the Podostemaceae, which have a nucellar periplasmodium substituting the endosperm.

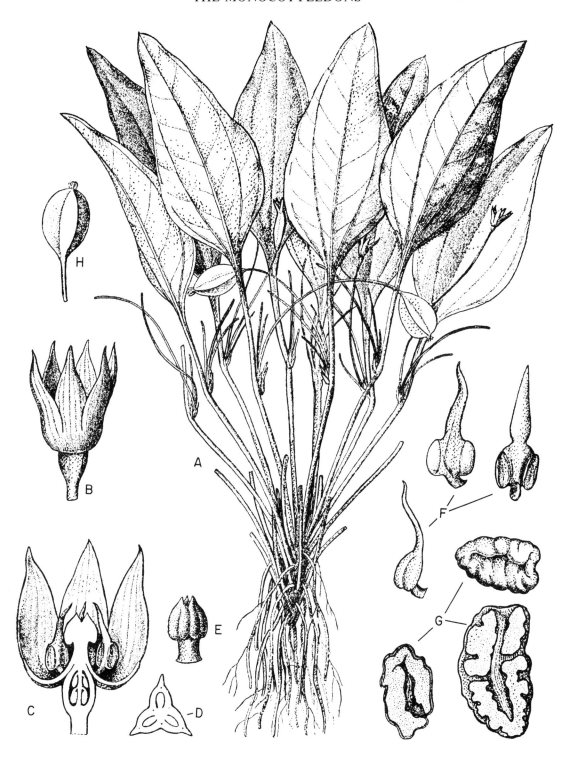

Fig. 100. *Trichopus zeylanicus*, an interesting member of the Dioscoreaceae combining features of *Dioscorea* with those of Stemonaceae, Taccaceae and Trilliaceae. Several features are probably ancient. A, plant; B, flower; C, flower, longitudinal section; D, ovary, transverse section; E, style and stigma; F, stamens with exserted connective tips; G, seed, below in transverse and longitudinal sections, showing the ruminate endosperm; H, fruit. (From Knuth, 1930.)

DELIMITATION OF DIOSCOREALES

The order Dioscoreales is poorly delimited from the Asparagales. Huber (1969) treated the families Stemonaceae (Roxburghiaceae) and Trilliaceae in a separate order from the Dioscoreales, viz. the Stemonales (Roxburghiales). However, from our data it does not seem that the Stemonaceae and Trilliaceae are more closely allied to each other than each of them, and especially the former, to Dioscoreaceae.

For example the Stemonaceae agree with the Dioscoreaceae in their petiolate, reticulately veined leaves, the presence of vessels in the stem, the sometimes epigynous (or half epigynous) flowers, the often connate filaments, the sulcate or sulculate pollen grains, the capsular fruit and occasional presence of seeds with an elaiosome. Dioscoreaceae subfam. Trichopodoideae which includes shade-growing herbs with bisexual flowers having anthers with protracted connective tips (Fig. 100), is in particular similar to Stemonaceae, e.g. *Croomia*, although this has hypogynous flowers.

An examination of the monogeneric Taccaceae (Fig. 101; see also Drenth, 1972) reveals numerous similarities especially with the Dioscoreaceae, for example in the petiolate, alternate leaves, the epigynous flowers, the simultaneous microsporogenesis, the sulcate pollen grains, the erect anatropous ovules, the capsular fruit type (though the Taccaceae more often have berries), the tropical distribution etc. Differences include the lack of vessels in the stems of *Tacca* (present in at least some Dioscoreaceae), the lack of septal nectaries, the parietal placentas and the stigma surface being wet (not dry as reported once for Dioscoreaceae). The outer integument is also two-layered in *Tacca* unlike any Dioscoreales. These differences are mostly of low significance, some of the former may even be negligible. Attributes, like the characteristic inflorescence and stout, bisexual flowers of the Taccaceae are hardly sufficient for ordinal distribution, nor are the bean-shaped, although very characteristic, longitudinally ridged seeds. Therefore it seems appropriate to include the Taccaceae as a family in the Dioscoreales.

Other families, which may seem to be better placed in the Dioscoreales than in the Asparagales are the Smilacaceae and Petermanniaceae (Fig. 102). They are however retained in the Asparagales because of integumentary (seed coat) characters in which they agree better with the berry-fruited asparagalean families (see Huber, 1969).

The Trilliaceae, from the detailed studies by Berg (1959, 1962a and b) turn out to approach the Liliaceae and Colchicaceae of the Liliales. *Medeola*, which is usually placed in Trilliaceae has so great embryological affinities to Liliaceae that it is best removed to this family (as done here) in spite of its berries (see Berg, 1962a). Thus, the Trilliaceae can no doubt be considered a link between the Dioscoreales (Stemonaceae, Dioscoreaceae and Trichopodaceae, if this is acknowledged), the Liliales (especially Liliaceae, Colchicaceae) and, although maybe less closely, to the Convallariaceae of the Asparagales. It seems that at least *Clintonia* and *Disporum* are best excluded from this last mentioned family, as proposed by Björnstad (1970).

Alstroemeriaceae and Philesiaceae (Fig. 103)
The Philesiaceae comprise a small disjunct South Hemisphere family divisible into different parts, viz. subfam. Philesioideae with the two South American genera *Philesia* and *Lapageria*, and the subfamily Luzuriagoideae with *Luzuriaga* and *Behnia* which have a disjunct South Hemisphere distribution, centred on Australasia. To them should, perhaps, be added *Drymophila* generally placed in Convallariaceae. Provisionally also *Geitonoplesium* and *Eustrephus* have been included in the Philesiaceae; their phytomelan-encrusted seed coats more "convallariaceous" leaves etc. give support to the views that the Philesiaceae are best placed in the Asparagales. However, various of these genera show agreement with the Alstroemeriaceae in the the following features:

Both families include scandent vines (*Lapageria*, *Bomarea*).

In both the leaves are subsessile to shortly petiolate and have an *inverted lamina*.

Both contain oxalate raphides (Alstroemeriaceae, and in addition some Melanthiaceae, being unusual in the Liliales in having raphides).

Both have vessels with scalariform perforation plates in the stems.

Both may have a large, campanulate to tubular-infundibular perigon sometimes with tepals more or less spotted inside.

Both have 3 + 3 stamens with basifixed or pseudobasifixed introrse anthers.

The pollen grains are generally sulcate in both families although in Philesiaceae subfam. Philesioideae they are inaperturate.

The ovary is tri- or unilocular, both conditions being found in each of the two families; in the latter case the placentation is parietal.

The style is simple and apically tribrachiate or trilobate.

The stigma surface is wet in both *Lapageria* and *Alstroemeria*.

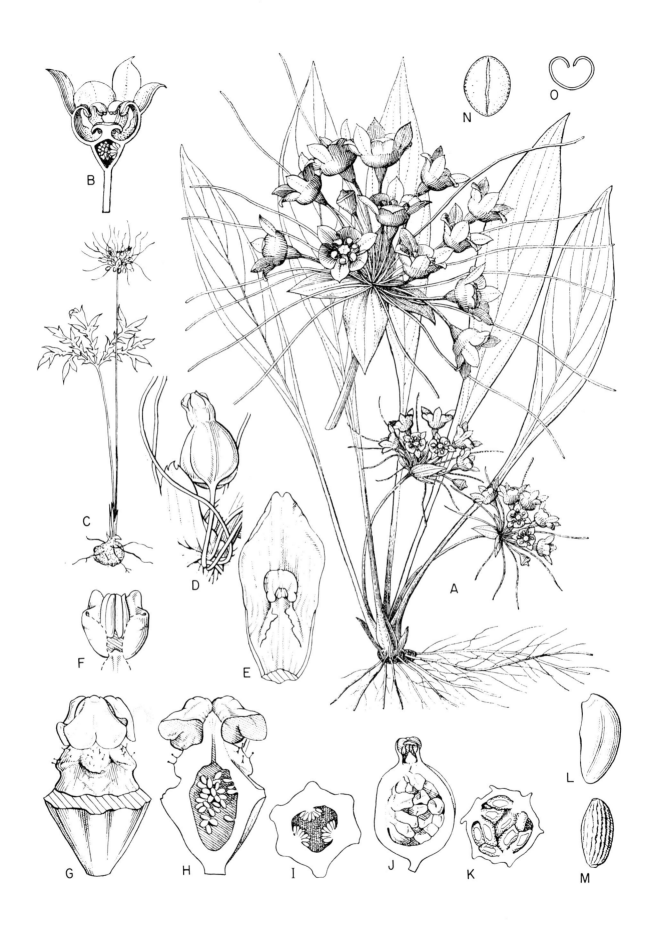

The ovules are anatropous (or campylotropous), with Polygonum type of embryo sac formation.

The seeds are globose or avoid and generally (except in *Geitonoplesium* and *Eustrephus*) lack phytomelan.

Endosperm is non-starchy.

Some of these similarities are significant but others are common attributes in the Liliiflorae.

Beside these similarities there are also some notable differences between the two groups. These include:

The members of the Philesiaceae are woody shrubs or vines, those of Alstroemeriaceae herbaceous.

The leaves in the Philesiaceae tend to have reticulate venation between the main longitudinal veins, while such is lacking in the Alstroemeriaceae.

The pedicels in most Philesiaceae carry small, bracteate prophylls.

The flowers in the Philesiaceae are hypogynous, those in Alstroemeriaceae epigynous.

There are septal nectaries in some Alstroemeriaceae, but probably not in Philesiaceae.

The different fruit type, capsules in most Alstroemeriaceae, berries in most of the Philesiaceae.

The ovules are crassinucellate in the Philesiaceae and tenuinucellate in the Alstroemeriaceae.

Several differences in the structure of the layers and their cells in the seed coat and endosperm (Huber, 1969).

It is interesting to note that the Philesiaceae (treated as one or two families) have an acceptable place among the woody-stemmed, berry-fruited taxa of Asparagales, but deviates from these in a few conspicuously alstroemeriaceous features. The Alstroemeriaceae likewise are tenable in the Liliales in spite of differing from the other families in some features. Thus, this family pair seems to bridge the gap between the two orders. The similarity between the two families was recognized by Hutchinson (1959), who erected the order Alstroemeriales for Alstroemeriaceae, Philesiaceae and Petermanniaceae, the latter being however more smilacaceous in habit and probably also in affinity.

In spite of acknowledging these affinities between the Philesiaceae *sensu lato* (especially *Lapageria* and *Luzuriaga*) and the Alstroemeriaceae we have decided to retain the families in separate orders, the Asparagales and the Liliales. The division between the Dioscoreales and these is indeed quite unclear, but it is convenient to maintain it; the other possibility is to make a very large order, Liliales *sensu lato*, that would be difficult to comprehend (Diagram 108).

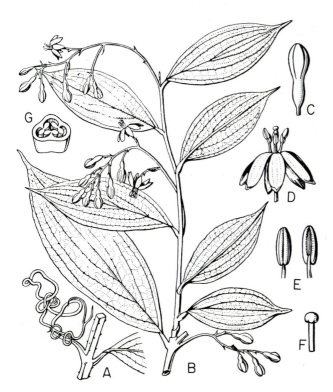

Fig. 102. *Petermannia cirrhosa* (Petermanniaceae, Asparagales). A, , tendril, doubtless homologous with a terminal inflorescence, the shoot system being sympodial; B, branchlet with inflorescences; C, floral bud; D, flower; E, stamens; F, style apex and stigma; G, ovary in transverse section, showing the parietal placentation. (After Hooker, from Emberger, 1960.)

LILIALES AND ASPARAGALES

These orders as defined by Huber (1969) make up groups which probably reflect best the differentiation in the "core groups" of "Liliales *sensu lato*". The differences are well illustrated by Huber (Huber, op. cit., tables on his pp. 510–513). However, the three orders, Dioscoreales, Asparagales and Liliales are by no means distinct as

Fig. 101. Taccaceae, habits and details. A–B: *Tacca plantaginea*, plant and flower, transverse section. C–M: *Tacca leontopetaloides*; C: plant; D: flower in centre of inflorescence; E: tepal and attached stamen sectioned out from the syntepalous perianth; F: stamen from inside the "hood"; G–H: ovary and style: H in longitudinal section; I: ovary, transverse section. J–K: mature fruit, longitudinal and transverse sections respectively; L: seed enclosed in aril; M: seed with aril removed. N–O: pollen grain of *T. laevis*. (A–B from Limpricht, 1928; C–M from Hutchinson and Dalziel, 1968; N–O from Erdtman, 1952.)

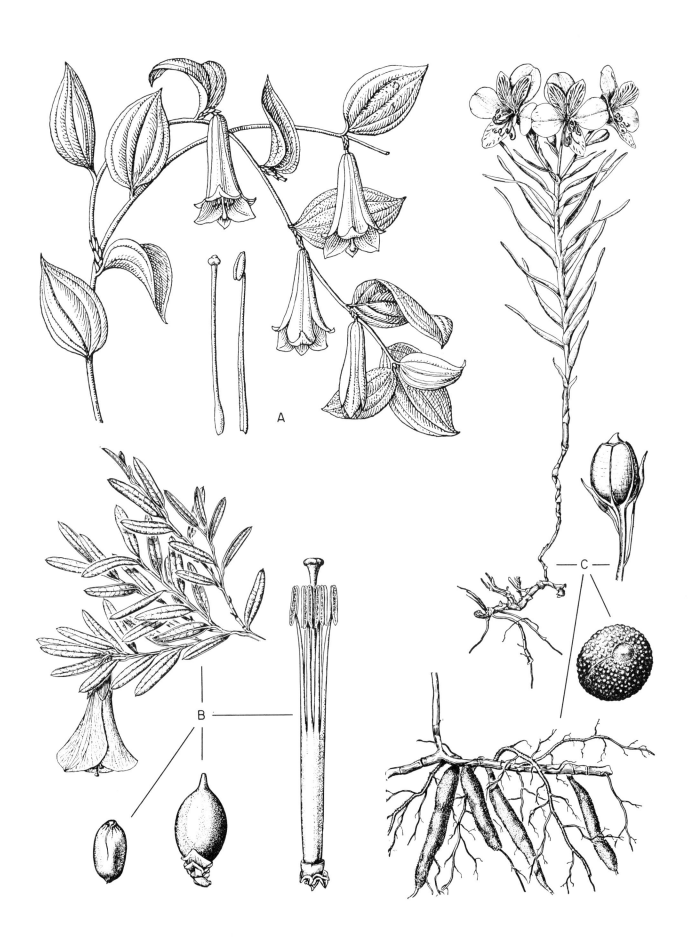

A

B

C

illustrated by the family pair Philesiaceae *sensu lato*—Alstroemeriaceae. Besides, the distinction between the orders becomes increasingly complex if embryological details are taken into consideration, which can be illustrated by the study of some convallariaceous genera by Björnstad (1970).

The following table, with her data shows that the genera *Disporum* and *Clintonia* mostly referred to the tribes Polygonateae or Convallarieae of a Liliaceae *sensu lato*, agree much more closely with genera of Liliaceae *sensu stricto* (*Medeola*, *Tulipa*), with which also the Colchicaceae have most features in common. *Disporum*, *Clintonia* and maybe more genera therefore should no doubt be transferred to Liliaceae or Colchicaceae.

Genus	*Septal nectaries*	*Oxalate raphides*	*Nucellus type*	*Parietal cell*
Convallaria	○	+	Muilla	+
Polygonatum	+	+	Muilla	+
Smilacina	+	+	Muilla	+
Maianthemum	+	+	Muilla	+
Disporum	○	○	Scoliopus	○
Clintonia	(○)*	○	Scoliopus	○
Medeola	○	○	Scoliopus	○
Tulipa	○	○	?	○

(○)* Septal nectaries present according to Daumann (1970).

An interesting similarity between the Liliaceae *sensu stricto* and Alstroemeriaceae is the occurrence in both families of tuliposides (Slob *et al.*, 1975) while these compounds were not detected in numerous other genera of Liliiflorae investigated.

The noteworthy similarity in karyotype between the Agavaceae (McKelvey and Sax, 1933) and *Hosta* (Funkiaceae) is not supported by other obvious similarities, and thus possibly has developed by convergence. The families are here placed separately and at some distance from each other in the Asparagales.

The genus *Walleria* is highly interesting in its peculiarities and needs to be reconsidered for its closest affinities, which may be more with Colchicaceae than with Tecophilaeaceae or Eriospermataceae.

ORCHIDALES AND BURMANNIALES

The Orchidales comprise a highly variable group of plants in regard to size, habit, vegetative structures and floral peculiarities. The adaptations related to pollination are manifold and still unexpected convergences are discovered in the structure and homology of "stipes", "caudicula" and similar structures (F. Rasmussen, personal communication). Presence of silica bodies, distribution of various stoma types, colour of seeds and cuticle structures of the seed epidermis contribute variable attributes useful in the classification of the orchids (H. Rasmussen, W. Barthlott, personal communication).

The following attributes of Orchidales are shared with taxa of the Liliales, such as in Iridaceae, Colchicaceae, Liliaceae or Melanthiaceae or in several of these families.

Corms (although different in shape and characters) are often present.

Ensiform leaves turn up repeatedly in Orchidaceae and occur in many Liliales.

Vessels are often absent; if present they normally have scalariform perforation plates.

Epigyny is found in the Orchidaceae and—along parallel lines—in the Iridaceae and Alstroemeriaceae.

Spotted tepals are found in many orchids and in a great proportion of the Liliales: Tricyrtidaceae, Liliaceae, some Colchicaceae, Calochortaceae and Iridaceae.

Septal nectaries are absent in the Orchidales and in most of the Liliales (except, for example, Iridaceae subfam. Ixioideae), but tepal nectaries occur in a number of taxa in both orders.

Tapetum is secretory in both orders.

Microsporogenesis is simultaneous in the Orchidales and in Iridaceae of the Liliales.

The pollen grains are generally binucleate in both orders.

The ovules are tenuinucellate in the Orchidales as in many Liliales, however not in the Iridaceae.

Endosperm formation is nuclear, when present, in the Orchidales as in most Liliales (endosperm formation is rare in the Orchidales, however).

The seeds as far as known are not provided with a layer of phytomelan in any taxa of the two orders.

Both orders often contain saponins although steroid saponins have only rarely been verified for the orchids.

These characters in combination indicate that the orchids evolved from ancestors of the lilialean type. There are many peculiarities in the orchids, however, such as the frequent occurrence of silica bodies in the

---

Fig. 103 Representatives of the order Alstroemeriales. A: *Lapageria rosea* (Philesiaceae) with pistil and stamen. B: *Philesia* (Philesiaceae) with staminal column, berry and seed. C: *Alstroemeria diazii* (plant only) and *A. aurantiaca* (details: roots, capsule and seed). (A from Krause, 1930; B–C from Correa, 1969.)

tropical forms, the greatly variable, often paracytic, stomata, the girdle type of endothecial thickenings in the anthers, the pollen tetrads, the reduced or more frequently lacking endosperm, and the wet stigma type (more often dry in the Liliales).

Garay (1960) argued in favour of a close relationship between the Orchidales and the Hypoxidaceae of the Asparagales. This is contradicted especially by the structure of the seed coat, although there are a few notable similarities between the families (epigyny, lack of septal nectaries and rarely nuclear endosperm formation). The androecial reductions found in the Tecophilaeaceae, the Alliaceae subfam. Gilliesioideae and the Philydraceae, judging from the syndrome of characters probably has nothing to do with that in the Orchidales.

Within the Orchidales, Apostasiaceae seems to form a distinct family with distinctly different pollen grains and seeds (Schill, 1978; Rauh *et al.*, 1975). The Cypripediaceae may be more closely allied to Orchidaceae *sensu stricto*, and are distinguished here largely by virtue of the staminal differences. The latter family shows an enormous variation; and within it, beside the Orchidoideae, Neottioideae and Epidendroideae, also the *Vanilla* group might deserve subfamily status.

The order Burmanniales has generally been regarded as most closely allied to the Orchidales, especially by virtue of the following attributes:

Both orders contain a fairly high number of chlorophyll-less mycotrophic genera ("saprophytes").
Both often have distichous leaves and a similar habit.
The flowers are epigynous in both orders.
They also tend to have unequal tepal whorls and (in Burmanniales rarely) a tendency for zygomorphy.
Both generally have parietal placentation and capsular fruit.
The ovules in both orders are tenuinucellate.
The seeds in both groups are very small and lack phytomelan in the seed coat.
Both are tropical groups.

In the concluding diagrams we have accepted these similarities as indicating affinity, but recent data make this conclusion somewhat ambiguous. It turns out that the Burmanniales agrees quite closely with some of the liliiflorean orders with a starchy endosperm, especially Philydrales.

The following attributes are shared between the Philydrales and Burmanniales (Hamann, personal communication).

Distichous leaves occur in both groups.
The stem in at least the green species are provided with vessels that have scalariform perforation plates.
Zygomorphy occurs in both groups.
Tepals are unequal in the two whorls and lack a spotted pattern.
Microsporogenesis is successive in both orders.
Endosperm formation is helobial in both orders.
Capsules are present in both orders.
The seed shape in some Burmanniales is somewhat reminiscent of that in Philydrales.
A well developed starchy endosperm is found in unripe seeds of Burmanniales, and in the ripe as well as unripe seeds of Philydrales.

Most of these attributes comprise differences between Burmanniales and Orchidales. Septal nectaries occur in Burmanniales and, although absent in Philydrales, are present in all the adjacent orders. On the other hand, stomata when present in Burmanniales so far are reported to be anomocytic while they are paracytic in Philydrales and related orders. The flowers are also epigynous in the Burmanniales, the placentation parietal (although basally often axile), the ovules numerous and tenuinucellate, the seeds very small and the distribution tropical. Some of these attributes are the ones mostly referred to for claiming relationships between Orchidales and Burmanniales.

The affinities of the latter order thus cannot yet be considered to be ultimately assessed.

Superorder Zingiberiflorae

There is no disagreement with regard to the composition of this superorder which consists only of the order Zingiberales. There are problems, however, with regard to its affinities with other groups of monocotyledons.

Among the groups which have been mentioned in this respect and call for a comparison are the orders Commelinales (Commeliniflorae), Bromeliales, Philydrales and Pontederiales (Liliiflorae) and Arecales and Cyclanthales (Areciflorae). The superficial resemblance to the Orchidales is not confirmed by a detailed consideration of similarities.

AFFINITIES WITH THE COMMELINALES

The following attributes represent similarities between the Zingiberales and Commelinales.

Both are herbaceous plants often with brittle nodes.
Ptyxis is often supervolute in both groups.
Silica bodies (though rare in the Commelinales) are found in both superorders.
Oxalate raphides are common in the Commelinales and present in the 5- or 6-staminate families of the Zingiberales.
Vessels are often present at least in the stems of the Marantaceae (Zingiberales), where they have simple perforation plates; in the Commelinales they almost constantly occur in stems and leaves and have simply perforated end plates.
Stomata are of the tetracytic type in the Commelinales and in many Zingiberales (especially the 5- or 6-staminate families and the Costaceae).
Inflorescences are composed of determinate (cymose) partial inflorescences (double cincinni) in the Commelinaceae and at least certain zingiberalean groups, such as the Marantaceae (in other groups of Zingiberales they are spicate).
The flowers are often zygomorphic (to asymmetric), and have at least an inner whorl which is petaline.
The anthers are generally basifixed and introrse.

Amoeboid tapetum, found in all Commelinales studied, also occurs in some Zingiberales, e.g. in the Cannaceae.
The pollen grains are generally binucleate.
The pistil in both groups has a simple, rarely apically, branched style.
Placentation is usually axile and the ovules few to numerous per locule (solitary in most Marantaceae and Heliconiaceae).
The ovules are generally anatropous (but other types occur especially in the Commelinaceae) and are always crassinucellate and a parietal cell is cut off in the nucellus in both orders.
The fruits are generally loculicidal capsules in both orders.
Arillar structures are widely distributed in both orders (but are probably not homologous).
Endosperm is well-developed and starchy in the Commelinales and in some of the 5–6-staminate Zingiberales (at least the Heliconiaceae); this possibly being primitive in the latter order.
The seedlings are mostly of the same type, viz. with plumular leaves emerging through a collar-like structure.

At a first glance this list may appear impressive, but the following facts should be noted, implying, as they do, considerable and important differences between the two orders:

The Zingiberales have mostly petiolate leaves quite different from those in the Commelinales.
The silica bodies, though occurring in both groups, are rare and less varied in the Commelinales, where they are mostly spinulose or minute.
Hairs are common and varied in the Commelinales, but very rare in the Zingiberales.
Vessels are more advanced in most Commelinales (i.e. in the Commelinaceae) than in the Zingiberales and also found in leaves, where they are constantly lacking in the Zingiberales.
The outer tepals are mostly green and sepaline in the Commelinales, but hardly ever so in the Zingiberales.
The flowers are epigynous in all Zingiberales (though incompletely so in the Lowiaceae), but always hypogynous in the Commelinales.

Septal nectaries which occur in most families of the Zingiberales are lacking in the Commelinales.

Petaloid staminodia are common in the Zingiberales and lacking in the Commelinales, although staminodia with sterile anthers occur in several genera of Commelinaceae.

The pollen grains are constantly sulcate in the Commelinales, inaperturate in the Zingiberales.

Endosperm formation is more frequently helobial than nuclear in the Zingiberales, but constantly nuclear in the Commelinales.

Perisperm is found in rich quantities in all taxa of the Zingiberales having one anther, but is lacking in Commelinales.

Therefore it does not seem that the Zingiberales are particularly closely allied to the Commelinales, nor to any other order of the Commeliniflorae.

AFFINITIES WITH THE BROMELIALES, PONTEDERIALES AND PHILYDRALES

A comparison between the Zingiberales on the one side and the Bromeliales, Pontederiales or Philydrales on the other also reveals a number of similarities:

In all these groups there are small to very large herbs, all of which lack secondary growth.

The leaves are mostly distichous in the Zingiberales as in the Philydrales, but rarely so in the Bromeliales (except in the inflorescence) or the Pontederiales.

The leaves are petiolate, with a well developed lamina in the Zingiberales as well as the Pontederiales.

Ligules may occur rarely in the Pontederiales and the Zingiberales, but they are not necessarily homologous.

Silica bodies are found in the Zingiberales and Bromeliales (but not in the other two orders).

Oxalate raphides are found in some Zingiberales and also in all of the Pontederiales, Philydrales (tapetum cells, elsewhere styloids; Hammann, 1966) and Bromeliales.

Vessels are often absent but when present generally have scalariform perforation plates in all these orders.

Stomata are basically paracytic in most taxa of all the orders.

The tepals are more or less petaline in all the orders, and in most Zingiberales and all Bromeliales there is conspicuous difference in size (and often in shape) between the outer and inner tepals.

Epigyny occurs in many Bromeliales and all Zingiberales (though not completely in the Lowiaceae).

Septal nectaries occur in most families of the Zingiberales and most Pontederiales, Haemodorales and Bromeliales.

Staminal reductions (convergence) are common, in particular, in the Philydrales, as is the case in the Zingiberales.

Amoeboid tapetum occurs in the Pontederiales and Haemodorales as in some (relatively few) Zingiberales, though otherwise the secretory type prevails in most of the orders compared here.

Microsporogenesis is successive in all groups (but also in Commelinales).

The pollen grains are usually binucleate.

There is usually a single style in all four orders compared.

Placentation is also axile in representatives of all four orders.

The ovules are also chiefly anatropous (except in the Haemodorales), a parietal cell is cut off and endosperm formation tends to be helobial in all four orders; yet nuclear endosperm formation is most common in the Zingiberaceae.

The fruits are mostly capsular in all the orders, most often with loculicidal dehiscence.

The seeds are provided with starchy endosperm in all orders where the endosperm is richly developed. This is substituted by perisperm in many Zingiberales, this being, however, probably a secondary state.

The embryos are often linear in the Zingiberales as in the Bromeliales, Pontederiales and Philydrales.

Flavonols are varied in both Zingiberales and Bromeliales.

It seems to us that these similarities, although balanced by a number of differences, are more significant than those between the Zingiberales and the Commelinales.

Conspicuous differences would be the specializations of petaloid staminodia and the perisperm of the families in Zingiberales with one functional anther only. Also, the aril structures of Zingiberales have no certain counterpart in the Liliiflorean orders mentioned. The pollen grains of the Zingiberales also normally lack apertures.

COMPARISON WITH THE ARECIFLORAE

Finally, a comparison between the Zingiberiflorae and the Areciflorae may be relevant, in particular as a consequence of the superficial similarity of leaf (petiolate, with a well defined apparently pinnate lamina which is often torn to appear "compound").

The following similarities emerge:

There are large herbs in the Cyclanthales and trees (*Ravenala*) in the Zingiberiflorae which bridge the general difference in life form.

The petioles and leaf bases are of similar types in the two superorders.

Silica bodies are found in the Zingiberiflorae and within the Areciflorae at least in the Arecales.

Oxalate raphides are found in the Areciflorae and also in those families of the Zingiberiflorae having 5–6 anthers.

Hairs are rare in both groups.

Vessels in the stems of both superorders have simple and/or scalariform perforation plates, if present. However, while the leaves of all Areciflorae have vessels with scalariform perforation plates the leaves in the Zingiberiflorae never have vessels.

Septal nectaries occur in many members of both superorders.

The pollen grains in both superorders are binucleate.

The ovules are usually anatropous and a parietal cell is formed in most taxa of both groups; nuclear endosperm formation also occurs in both superorders, although the helobial type is common in the Zingiberiflorae and in the Cyclanthales.

The differences include the following:

The Zingiberiflorae more often has distichous leaves than does the Areciflorae.

Ptyxis is supervolute in the Zingiberiflorae, plicate or conduplicate in the Areciflorae.

Stomata are mostly paracytic in the Zingiberiflorae (rarely so in the Arecales and Pandanales); tetracytic stomata occur in both groups.

Zygomorphic and asymmetric flowers, which are universal in the Zingiberiflorae, are lacking in the Areciflorae, is rare in the Areciflorae (found in Cyclanthaceae).

Epigyny, which is likewise universal in the Zingiberiflorae, is rare in the Areciflorae (found in Cyclanthaceae).

The flowers are generally bisexual in the Zingiberiflorae, unisexual in the Areciflorae.

Anther dehiscence is more often latrorse than introrse in the Areciflorae, the reverse being the case in the Zingiberiflorae.

The pollen grains are generally inaperturate in the Zingiberiflorae, sulcate, trichotomosulcate or ulcerate in the Areciflorae.

The style is generally simple in the Zingiberiflorae, mostly tribrachiate from the base or middle in the Areciflorae (where there may also be apocarpy).

The stigmas tend to be wet in the Zingiberales, probably more often dry in at least the Arecales of the Areciflorae.

Placentation is generally axile in the Zingiberiflorae, rarely or never so in the Areciflorae.

The fruits are mainly capsules in the Zingiberiflorae but are drupes or berries in the Areciflorae.

Arils are universal in the Zingiberiflorae but lacking in the Areciflorae (which have a carnose pericarp).

The Zingiberiflorae have a starchy endosperm or perisperm, while the Areciflorae lack perisperm and normally have a starchless endosperm.

Ruminate endosperm is common in the Areciflorae but is lacking in the Zingiberiflorae.

The flavonoid chemistry, as far as is known, is different in the two superorders. That in the Arecales is reminiscent of the one in grasses and sedges. Tricin, for example, is common in palms but lacking in the Zingiberiflorae.

The conclusion from the above comparisons is that there are few reasons for supposing a close affinity between the Zingiberiflorae and the Areciflorae.

The most probable bridge between the Zingiberiflorae and other monocotyledons would be among Liliiflorae with starchy endosperm, and the ancestors of the Zingiberiflorae probably developed as an early offshoot from the common ancestors of the Pontederiales, Philydrales and, perhaps, Typhales (see below).

We believe that the primitive Zingiberiflorae had well differentiated leaves with paracytic stomata, flowers with petaline tepals and six functional stamens, sulcate (or already inaperturate?) pollen grains, and a capsular fruit with arillate seeds containing copious starchy endosperm. The petaloid staminodia and voluminous starchy perisperm represented later specializations, the latter being combined with reduction of endosperm.

ON A POSSIBLE DIVISION OF THE ZINGIBERALES

In the above discussion it has often been necessary to refer separately to the families of the Zingiberales with 5–6 functional stamens and to those with one or only a half functional stamen respectively. There is no doubt that within the otherwise very natural superorder Zingiberiflorae hitherto regarded as comprising a single order, the Zingiberales, one can discern these two groups of families, the former (Group I) comprising the Lowiaceae, Musaceae, Heliconiaceae and Strelitziaceae and the latter (Group II) the Zingiberaceae, Costaceae, Cannaceae and Marantaceae.

A comparison between these groups, here called Groups I and II, reveals the following differences:

Oxalate raphides are found in Group I but are lacking in Group II.

Laticifers are found rarely in Group I (at least Musaceae), but not in Group II.

Stomata are generally more typically tetracytic in Group I, paracytic with 2–4 additional modified subsidiary cells in Group II, the Costaceae being however rather tetracytic; a difference of little significance.

Group I has 5–6 functional stamens, in Group II there is but one, the other being transformed into petaloid staminodia or missing.

Perisperm is lacking or poorly developed in Group I, very copious and starchy in Group II. In Group I the endosperm is better developed instead, and starchy in several genera.

While oxalate raphides, stamen and perisperm/endosperm conditions are significant differences, the others are not so. Hence although the families of Zingiberales form two natural groups we do not regard the differences as sufficiently important to award them formal taxonomic status. Lowiaceae is still somewhat incompletely known and its position in the order uncertain. The Zingiberaceae perhaps deviate from the other Zingiberales in having androecial rather than septal nectaries.

transformed to hairs or bristles in the Cyperales.

The pistils are syncarpous, superior and, these plants being anemogamous, lack septal nectaries.

The anthers are basifixed and tetrasporangiate.

Tapetum is secretory.

The pollen grains are united in tetrads, being, however, much more specialized in the Cyperales than in the Juncales.

They are ulcerate (in the Cyperales sometimes with additional lateral foramina or tenuitates beside the distal ulcus), and they are usually trinuclear.

The style is normally bi- or tribrachiate.

The stigma has dry surface.

The ovules are anatropous, a parietal cell is cut off, and the embryo sac development is of the Polygonum type.

The endosperm is starchy, mealy, non-ruminate.

The embryo formation follows the Onagrad type (*Juncus* variation) in both orders.

The embryo is broad in both groups (or often capitate in Cyperales).

"Diffuse centromeres" and post-reductional meiosis (Battaglia, 1955; Battaglia and Boyes, 1955) are restricted to these two orders only in the angiosperms.

Salt tolerance is also a feature of taxa of both orders.

The overall flavonoid spectrum is conspicuously similar.

Certain parasitic fungi (e.g. *Enthorrhiza*) occur only on plants of these two orders.

Both families are attacked by plant lice of the psyllid subfamily Liviinae and the aphid tribe Saltusaphidinae.

The differences are relatively few but conspicuous.

Silica bodies are absent in the Juncales (*Thurnia* excepted) but occur in all or almost all taxa of the Cyperales where they are generally of a particular conical type unique in the monocotyledons.

The tepals are better developed in the Juncales (cf. above, however).

The stamens are often 6 (but sometimes 3) in the Juncales, normally not more than 3 in the Cyperales.

In the pollen tetrads of the Juncales all 4 pollen grains develop, but in the Cyperales 3 of them degenerate.

The ovary is generally trilocular with axile multiovulate placentae in the Juncales (*Luzula* having, however, unilocular ovary with 3 basal ovules); in the Cyperales the ovary is unilocular with one basal ovule.

Endosperm formation is of the helobial type in the Juncales, of the nuclear type in the Cyperales.

The fruit is capsular in the Juncales, generally a nutlet in the Cyperales. (Drupes occur in *Scirpodendron*.)

The embryo is slightly bent in the Cyperales, not so in the Juncales.

The seedlings differ in the orders, the Juncales having type A and the Cyperales type C (pp. 246–247).

The above comparison demonstrates the close agreement between the two orders while the dissimilarities may be only merely regarded as sufficient for ordinal distinction.

The position of the two orders within the Commeliniflorae, in which they undoubtedly belong, is probably not far from Poales (or Eriocaulales), judging from the considerable number of shared features, including the flavonoid patterns. The relationships with the Typhales have already been dealt with, and are probably not as close as may have been suspected.

It is generally assumed in botanical literature where the Juncales and Cyperales are compared, that the latter order is derived from the former. In the light of logical phylogenetic analysis the question may be found irrelevant, as the two groups are no doubt derived from common ancestors and represent parallel evolutionary lines, each having evolved along its own courses and under its own selection pressure.

From the above differences it seems likely, for example, that pollen tetrads with all four pollen grains being functional, as in the Juncales, must represent a more primitive state than the condition in Cyperales, and that the capsular fruit represents a more primitive state than the indehiscent nutlets. Moreover, the 3-staminate condition of the Cyperales seems to be derived from the 6-staminate, as in the Juncales. Thus it is logical to assume that the common ancestors of the Juncales and the Cyperales would rather resemble (and would be classified as) juncalean.

Savile (1979) found the rusts of *Puccinia* and *Uromyces* (which were retained by him as two genera) parasitizing juncalean hosts to be more derived on the average in aeciospore surface sculpturing than those of the species parasitizing the Cyperales.

Looking however at the table XV in Savile (1979, p. 485), the evidence for this is slender: the "Group 2" fungal species forming the evidence that the parasite species of the Juncales are more advanced than those of the Cyperales consisting only of 3 species. The correlation between the hostal taxa and the parasitizing fungi with a particular sculpturing pattern on the spores thus forms a dubious basis for Savile's conclusion that the Juncales originated out of the Cyperales.

THE RELATIONS
POALES-RESTIONALES

There may be some divergences of opinion as to how to distribute the families between the orders Restionales and Poales. The Restionales in this work has provisionally included the families Restionaceae, Centrolepidaceae and some monogeneric families, Anarthriaceae, Ecdeiocoleaceae, Flagellariaceae, Joinvilleaceae and Hanguanaceae, the latter three being the most problematic. This circumscription of Restionales is somewhat problematical: the borderline between, and the distinctness of. the orders Restionales and Poales should be reconsidered, and the position of the Hanguanaceae may be questioned. A serological investigation (Lee *et al.*, 1975) indicates that the three families Hanguanaceae, Flagellariaceae and Joinvilleaceae are not close to each other nor to the grasses.

The core groups of the Restionales are the families Restionaceae (into which Anarthriaceae and Ecdeiocoleaceae could be included) and Centrolepidaceae (Fig. 107). A comparison between the Restionaceae and Centrolepidaceae on the one side and the Poaceae on the other (these being the best known families) may be of interest in estimating the positions of Flagellariaceae and Joinvilleaceae.

The following differences some of which are of very low significance, have occasionally been referred to (those being of the least significance being in brackets).

[1. The leaves are regularly distichous in the Poaceae (except for *Micraira*), sometimes (or often?) so in the Restionaceae and rarely in the Centrolepidaceae.]

[2. Characteristically shaped silica bodies are lacking in Centrolepidaceae and rare in the leaves of the Restionaceae (*Lepyrodia*) although granular amorphous silica bodies occur in most genera. In the Poaceae typical silica bodies are of regular occurrence and are mostly localized in the short cells of the epidermis.]

3. Vessels are present in the leaves of grasses where they have simple or mixed simple and scalariform perforation plates, but are mostly absent from the leaves of the Restionaceae and present but provided exclusively with scalariform perforation plates in the Centrolepidaceae.

[4. Stamens are generally six or three in the Poaceae,

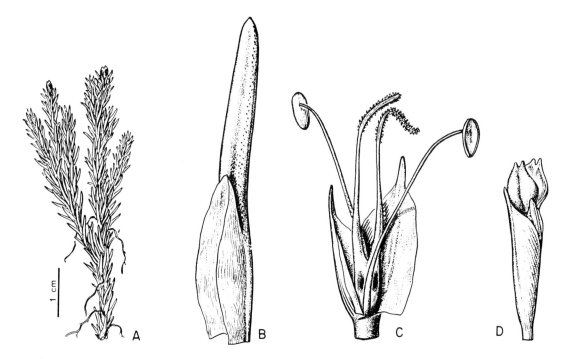

Fig. 107. *Gaimardia australis* (Centrolepidaceae). A, whole plant; B, leaf; C, inflorescence; D, fruits. (From Correa, 1969.)

Fig. 108. *Flagellaria guineensis* (Flagellariaceae). A, branch with inflorescence; B, flower; C, stamen with supporting tepal; D, pistil; E, fruit; F, seed. (From Napper, 1971.)

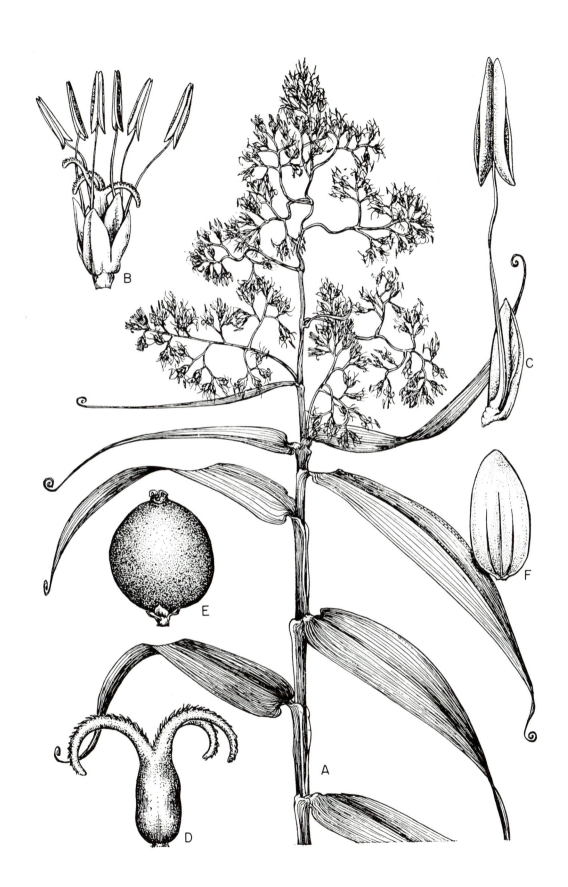

three or one in the Restionaceae and Centrolepidaceae respectively.]

[5. Anther dehiscence is latrorse in the Poaceae, mostly
introrse in the Restionaceae and undefinable in the
unistaminate male flowers in the Centrolepidaceae.]

6. The anthers are tetrasporangiate in the Poaceae, but
bisporangiate in most Restionaceae and in all Centrolepidaceae.

7. The embryo is lateral in relation to the longitudinal
axis of the seed and localized outside this in the
Poaceae, but situated at one end of the seed and partly
enclosed by the endosperm in the Restionaceae and
Centrolepidaceae.

8. In the grasses the nucellus epidermis cells divide
periclinally to form a more or less extensive
"nucellus cap". In the Centrolepidaceae and at least
some Restionaceae the epidermis cells at the top of the
nucellus elongate radially, but do not divide periclinally.

The frequent presence of ligules in Restionaceae, as in
practically all grasses, the similarity of the stomata (in
particular in the more or less dumb-bell-shaped stoma
cells), the similar pollen morphology, the mostly orthotropous ovules, the antipodal multiplication etc., all
indicate great affinity between the grasses and the
Restionales, and most of the above attributes comprise
no absolute differences.

It is now of great interest to investigate the relation-
ships of the monogeneric families Flagellariaceae and
Joinvilleaceae to the Poaceae and the Restionaceae-
Centrolepidaceae.

The Flagellariaceae (Fig. 108) agree with the grasses in
practically all the above characters except attribute
number 7 (and possibly 8). Although the placentation is
sometimes recorded to be axile, it is probably better described as apical-pendulous, as in the Restionaceae. The
fruit is either a drupe or (more likely) a berry.

The silica bodies found in *Flagellaria* are not deposited
in epidermis cells of the leaves as in the Poaceae, which
decreases the value of this attribute for indicating
affinities with the grasses. It should also be added that
the leaf sheath margins are fused in Flagellariaceae,
whereas they are unfused in the majority of the grasses.
In spite of this, the Flagellariaceae agree as closely with
the Poaceae as with the Restionaceae-Centrolepidaceae
and so bridge even more the slight gap between these
complexes.

The Joinvilleaceae are even more grass-like than the
Flagellariaceae and agree with the grasses in the same
characters mentioned as this family. *Joinvillea* has lateral
ligular lobes, and the leaf sheaths are unfused as in most

grasses. Also, silica bodies are present in epidermal short
cells, as in the grasses. Moreover, the internodes are long
and hollow as in the majority of the grasses. As in
Flagellaria, the fruit is somewhat baccate (a drupe or
berry ?) and trilocular, while the fruit in grasses is
unilocular. A difference from the grasses is that the style
is obsolete, the stigmatic lobes being situated on the
slightly elevated tip of the ovary. The inner three tepals
(lodicule homologues?) in *Joinvillea* are somewhat
smaller than the outer. Similar reductional steps are
found in Restionaceae. Thus, though being very grass-
like, *Joinvillea* shows differences from the Poaceae only
with respect to a few characters which do not militate
against its inclusion in the Poales.

The Hanguanaceae agree with the Poaceae in a few
significant characters such as presence of silica bodies
in the leaf epidermis. A comparison between Hanguanaceae and either the Poales or the Restionaceae-Centrolepidaceae shows few similarities. The male flowers agree
with the flowers of the Flagellariaceae in having 3 + 3
tepals and 3 + 3 stamens. The female flowers have a
trilocular ovary with uniovulate locules with axillary
placentation. As in the Joinvilleaceae the pistil has sessile
stigmatic lobes. Like both of these families it has a
carnose (but most likely drupaceous) fruit. The pollen
grains of *Hanguana* are inaperturate and the ovules
hemianatropous. In the overall spectrum of characters,
it agrees equally well in these characters and nearly as well
in the vegetative ones with, for example, the Doryanthaceae or Asteliaceae where it seems to be best placed until
conclusive investigations have been made, although it
may still prove to be related with the previous groups.

On the basis of the above we have preferred to treat
the provisional orders Restionales and Poales as one
single order, the Poales *sensu lato*, in which we acknow-
ledge the families Restionaceae (including Ecdeiocoleaceae and Anarthriaceae), Centrolepidaceae (excluding
Hydatellaceae), Flagellariaceae, Joinvilleaceae and Poaceae. Hanguanaceae is provisionally proposed as a
member of the Asparagales awaiting a comprehensive,
all-round study.

RELATIONSHIPS BETWEEN THE POALES (AND OTHER COMMEL-INIFLORAE) AND THE ARECALES

In a treatise of the monocotyledon classification, one
of us (Clifford, 1970) found that his computerized data
indicated a distinct agreement between the grasses (Poa-

ceae), sedges (Cyperaceae), *Flagellaria* (= Flagellariaceae) and palms (Arecaceae).

In our present study, agreement between the palms and the grasses (and also in most cases, the sedges) was found in the following attributes:

The frequently tree-like growth in both groups, not combined with secondary thickening growth.

Silica bodies are of wide occurrence in both groups and may show similar shapes.

Vessels in the stems have scalariform and simple perforation plates (the latter more frequently in the grasses).

Flowers are mainly hypogynous with non-showy perianth (however the palms are often whereas the grasses are only rarely insect pollinated).

Anthers are chiefly latrorse.

Tapetum is secretory.

Stigmas are dry.

There is only one ovule per placenta.

Endosperm formation is nuclear.

The seeds have copious endosperm, although this is starchy in grasses, non-starchy in palms.

The flavonoid pattern, especially with regard to the occurrence of luteolin/apigenin, tricin and flavone sulphates is similar.

There are similarities of plant-lice (Aphideae) between palms and (especially bambusoid) grasses.

In particular silica bodies and flavonoid chemistry comprise significant similarities. However, the above shared features are counterbalanced by a fairly high number of much more significant differences.

These include the following:

The general palm habit, which is not matched in the grasses.

The lateral roots in grasses are often developed opposite the phloem, which is not known in palms (few investigations, however).

The leaves are mostly distichous in grasses, not so in palms.

The ptyxis is supervolute or conduplicate in grasses, plicate in palms.

The leaves of palms are differently constructed, with long petiole and mostly dissolved lamina.

Ligules are present in grasses but lacking in palms; the hastulas of the palms are probably not homologous with ligules.

Oxalate raphides are lacking in grasses, present in palms.

The stomata are of paracytic "grass type" in the grasses, mainly tetracytic in palms or when paracytic of different construction from that in grasses.

Tepals are different in grasses and palms, and generally fewer in the former.

Septal nectaries are lacking in grasses, but mostly present in palms.

The anthers are latrorse in grasses, introrse in palms.

Microsporogenesis is successive in grasses, but usually simultaneous in palms.

Pollen morphology is very different in the two groups, ulcerate in grasses, otherwise in palms.

The pollen grains are trinucleate in grasses, mostly binucleate in palms.

The pistils are different, unilocular with two or three long stylodia in grasses and when syncarpous trilocular with short or no stylodia in palms.

The ovules are mostly orthotropous in grasses, anatropous in palms, and the former lack a parietal cell, while this is formed in palms.

The fruit is normally a nutlet or caryopsis in grasses, a berry or drupe in palms.

The endosperm is starchy in grasses, non-starchy in palms, and it is often ruminate in palms but not in grasses.

Embryogenesis in grasses is typical, with oblique walls, not so in palms.

The embryo is lateral in grasses, not in palms.

Seedling types are different in the two groups.

Rust and smut parasites are common on grasses, but do not or hardly ever occur on palms.

This list will no doubt remove any doubt about the fact that the two groups are distinct and probably remotely related. If one includes, however, in this comparison the Joinvilleaceae among the Poales then the degree of similarity will increase somewhat, as this family has more complete flowers than the grasses proper and lack many of the specializations of these. It represents, no doubt, a more primitive level of evolution than the grasses.

Superorder Areciflorae

This superorder has been recognized in most current systems and we find no reasons to question its "classical" circumscription so long as the Typhales and Arales are excluded, both being undoubtedly only distantly related to any of the three orders Arecales, Cyclanthales and Pandanales which comprise the Areciflorae (see also Stone, 1972). The lack of affinity between the Ariflorae and Areciflorae has been demonstrated under the former superorder. Also the degree of similarity between grasses and palms has been dealt with above.

The Pandanales have conspicuously different leaves from the Arecales and Cyclanthales and also differ in having constantly ulcerate pollen grains (rarely, ulcerate pollen grains occur also in the Arecales and Cyclan- thales), but otherwise agree either with the Arecales or with the Cyclanthales in practically every character surveyed here. The Cyclanthales show some singular features, like being partly herbaceous, by having rarely distichous leaves and by lacking vessels in the stem but not in the leaves and by having helobial endosperm formation. Also, there is a record (Harling, 1958) of starch in the endosperm of one genus, *Dicranopygium*. This, however, does not necessarily suggest that the Cyclanthales have had connections with other groups possessing starchy endosperm.

The Areciflorae seem to form a relatively isolated group in the monocotyledons.

A Revised Classification of the Monocotyledons

Based on the conclusions presented above we propose the following classification for the monocotyledons and associated groups of dicotyledons. (see also pp. 328–329.)

MONOCOTYLEDONS

Alismatiflorae

Hydrocharitales: Butomaceae, Aponogetonaceae, Hydrocharitaceae (incl. Thalassiaceae and Halophilaceae)
Alismatales: Alismataceae (incl. Limnocharitaceae)
Zosterales: Scheuchzeriaceae, Juncaginaceae (incl. Lilaeaceae), Njadaceae, Potamogetonaceae (incl. Ruppiaceae), Zosteraceae, Posidoniaceae, Cymodoceaceae, Zannichelliaceae

Triuridiflorae

Triuridales: Triuridaceae

Ariflorae

Arales: Araceae, Lemnaceae

Liliiflorae

Dioscoreales: Dioscoreaceae, Stenomeridaceae, Trichopodaceae, Taccaceae, Stemonaceae (incl. Croomiaceae), Trilliaceae
Asparagales: Smilacaceae (incl. Ripogonaceae and Petermanniaceae), Philesiaceae (incl. Luzuriagaceae), Geitonoplesiaceae, Convallariaceae, Asparagaceae, Ruscaceae, Herreriaceae, Dracaenaceae, Nolinaceae, Doryanthaceae, ? Hanguanaceae, Dasypogonaceae, Xanthorrhoeaceae, Agavaceae, Hypoxidaceae, Tecophilaeaceae, Cyanastraceae, Phormiaceae, Dianellaceae, Eriospermaceae, Asteliaceae, Aphyllanthaceae, Anthericaceae, Asphodelaceae, Hemerocallidaceae, Funkiaceae, Hyacinthaceae, Alliaceae (incl. Agapanthaceae and Gilliesiaceae), Amaryllidaceae
Liliales: Colchicaceae, Iridaceae, Geosiridaceae, Calochortaceae, Alstroemeriaceae, Tricyrtidaceae, Liliaceae, Melanthiaceae, ? Campynemataceae
Burmanniales: Burmanniaceae, Thismiaceae, Corsiaceae.
Orchidales: Apostasiaceae, Cypripediaceae, Orchidaceae.
Velloziales: Velloziaceae

Bromeliales: Bromeliaceae
Haemodorales: Haemodoraceae (incl. Conostylidaceae)
Pontederiales: Pontederiaceae
Philydrales: Philydraceae
Typhales: Sparganiaceae, Typhaceae

Zingiberiflorae

Zingiberales: Lowiaceae, Musaceae, Heliconiaceae, Strelitziaceae, Zingiberaceae, Costaceae, Cannaceae, Marantaceae

Commeliniflorae

Commelianales: Mayacaceae, Commelinaceae (incl. Cartonemataceae)
Eriocaulales: Rapateaceae, Xyridaceae, Eriocaulaceae
Juncales: Thurniaceae, Junaceae
Cyperales: Cyperaceae
Hydatellales: Hydatellaceae
Poales: Restionaceae (incl. Anarthriaceae and Ecdeiocoleaceae), Centrolepidaceae, Flagellariaceae, Joinvilleaceae, Poaceae

Areciflorae

Arecales: Arecaceae
Cyclanthales: Cyclanthaceae
Pandanales: Pandanaceae

ASSOCIATED DICOTYLEDONOUS GROUPS

Nymphaeiflorae

Nymphaeales: Cabombaceae, Nymphaeaceae (incl. Euryalaceae and Barclayaceae), Ceratophyllaceae
Piperales: Saururaceae, Piperaceae, Peperomiaceae

Magnoliiflorae

Magnoliales: Annonaceae, Myristicaceae, Eupomatiaceae, Canellaceae, Aristolochiaceae, Winteraceae, Degeneriaceae, and other families (should probably be subdivided into several orders)
Laurales: several families
Illiciales: Illiciaceae and Schisandraceae

CONCLUDING DIAGRAMS. ALTERNATIVE CLASSIFICATIONS OF THE MONOCOTYLEDONS

Alternative classifications expressing the interrelations between the monocotyledonous orders are given in Diagrams 106–108. These differ in various respects from the diagram used in the course of this survey. However, most of the main features of the latter diagram can be recognized. We have not found it necessary in the final diagrams to relocate the families of each order, as the main purpose for indicating their position was to achieve consistency in the mapping and evaluation of character states.

Some brief comments in regard to the diagrams and the novelties introduced in relation to the preliminary one used are given below.

Alternative 1 (Diagram 106)

The Alismatiflorae. The Hydrocharitales and the Alismatales are retained unchanged, although there is an indication that the Butomaceae in some features (such as the laminal placentation with many ovules approaches the Alismatales, especially Alismataceae subfam. Limnocharitoideae). The Aponogetonaceae in their spicate inflorescence approach the Juncaginaceae (and the Scheuchzeriaceae) of the Zosterales. The Najadales are included in Zosterales as there are insufficient reasons for retaining Najadaceae as a separate order. A position next to the Juncaginaceae is supported by the basal anatropous ovules. The Triuridales are excluded from the Alismatiflorae and established as a separate superorder. This shows similarities to the Alismatiflorae as well as to the Liliiflorae. The similarities between the Alismatiflorae and the dicotyledonous Nymphaeales are numerous although there are serious doubts about a close phylogenetic relationship.

The Ariflorae. The supposition on the outset of this study that the Ariflorae are closely connected with the Alismatiflorae has been supported and strengthened in the course of the study. The conspicuous similarities to the dicotyledonous Piperales remain, but to a great extent appear to be superficial or due to parallel adaptation to rainforest habitats.

The Triuridiflorae. This comprises the Triuridales. The order is undoubtedly isolated. Among the characters studied most agree with the Alismatiflorae, but a very close affinity to these is still in some doubt.

The Liliiflorae. This superorder contains the orders included in the preliminary classification with the addition of the Typhales and (very dubiously) the Hydatellales. The reasons for including the Typhales are explained above. In regard to the Hydatellales, the anomocytic stomata, lack of silica bodies, sulcate pollen grains and a few more details point to a liliiflorean position, but below (in the comments on Diagrams 107–108) will be given arguments for other views. The order Dioscoreales has been found to exhibit some dicotyledonous attributes and is retained in a position closest to the dicotyledons (the Magnoliales *sensu lato*, including the Aristolochiales). There seems no strong justification for upholding the Taccales as an order distinct from the Dioscoreales (nor Stemonales as in Huber, 1969, and Dahlgren, 1975).

The Asparagales are fairly diffusely delimited from Dioscoreales, with which especially Smilacaceae and Petermanniaceae (also to some extent Philesiaceae) show important similarities. The delimitation of the orders Asparagales, Dioscoreales and Liliales may need reconsideration in relation to these families in the future. Philesiaceae *sensu lato* is a heterogeneous family: with two genera (treated here as Geitonoplesiaceae) having phytomelan-coated seeds, it is well placed in the Asparagales; with others having perigonal nectaries (*Luzuriaga* and others) it approaches more the Liliales. The genera of Trilliaceae, Dioscoreales, show great affinity in embryological features and occasional presence of perigonal nectaries to the Liliales. Finally, the Alstroemeriaceae show affinity to the Philesiaceae (inverted leaves and other details) and in the septal nectaries approach the Asparagales, but otherwise seem to be closely allied to the Liliaceae. This may illustrate the close connections between the three orders concerned (which will support the treatment in Diagram 108).

The baccate-fruited families form a diffusely delimited but still conspicuous group in relation to the capsule-fruited taxa within the Asparagales which is indicated with a slight constriction of the ordinal bubble. In the course of the literature survey we have come to accept most of the segregate families of Huber (1969).

The Burmanniales and Orchidales as in the preliminary diagram approach the Liliales although both have been placed closer to the iridaceous end of this order.

The Velloziales, Bromeliales, Haemodorales, Philydrales and Pontederiales in this diagram have been retained as orders, although this high level may be questioned for all.

Thus, the Velloziales and Bromeliales approach each other, as pointed out by Huber (1969), and likewise the Pontederiales and Philydrales show obvious similarities,

(Hamann, 1966), both approaching the Haemodorales. With regard to the position of the Typhales more characters agree with the just mentioned orders than with orders of the Commeliniflorae, especially when laying less stress on characters in the wind pollination syndrome. The theory that wind pollination has evolved independently in the typhalean ancestors must not be disregarded. Therefore the Typhales have been removed from the Commeliniflorae to within the Liliiflorae, where a position close to the Pontederiales or Philydrales is feasible. On the other hand, there remain several similarities to the commeliniflorean Juncales, which has been taken into consideration in the diagram.

The Zingiberiflorae. Some families of this order are less specialized with regard to androecium development and storage tissue of seeds, having 5–6 functional stamens and starchy endosperm (but no perisperm), viz. the

Lowiaceae, Musaceae, Heliconiaceae, Strelitziaceae. In comparing these with other orders, we found the greatest concordance with the starch-endosperm orders of Liliiflorae (e.g. the Pontederiales, Philydrales, Bromeliales), and thus the diagram has been adapted accordingly.

The Commeliniflorae. With the (dubious) exclusion of the Hydatellales and Typhales, this superorder becomes somewhat more homogeneous. A connection with the orders of Liliiflorae having a starchy endosperm is obvious.

The Juncales-Cyperales form a closely allied pair of orders approaching, perhaps, the Typhales of the Liliiflorae as well as (with the juncalean Thurniaceae) the Eriocaulales (e.g. Rapateaceae). The Eriocaulales also show great similarity to the Commelinales, which are probably not so central and primary in evolution as indicated in the diagram used in the course of our study,

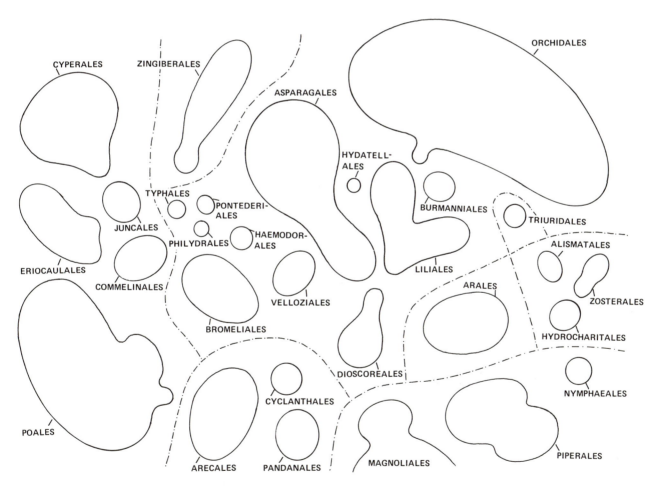

Diagram 106. An alternative phylogenetic diagram of the orders of monocotyledons based on the results of the present survey. Here, the Poales are placed close to the Arecales, and the Hydatellales are broken out from the Commeliniflorae to a position among the Liliiflorae. See further details in the text.

but could be secondarily adapted to entomogamy. The Restionales and Poales have been united to one order, the Joinvilleaceae and (to a lesser extent) Flagellariaceae being very similar to the grasses and both being similar to the Restionaceae.

The position of the Poales within the Commeliniflorae, nearest to the Arecales of Areciflorae in this diagram has been based on the evaluation of some floral similarities between palms and Flagellariaceae, the shared types of flavonoids etc., but is yet dubious (see Diagrams 107–108).

The Areciflorae. This superorder has been retained in its main features from the preliminary diagram, the Cyclanthales being placed closest to the Liliiflorae with starchy endosperm (this order having helobial endosperm, *Dicranopygium* having starchy endosperm), but

the similarities may not depend on a close affinity at all. However, the three orders show great similarity to each other.

Alternative 2 (Diagram 107)

This diagram is based on a somewhat different evaluation of the same data. The Alismatiflorae, Ariflorae and Triuridiflorae are wholly in agreement with Diagram 106, but some changes have been undertaken with regard to the other three superorders.

The Hydatellales have been retained in the Commeliniflorae, until more data become available; it seems possible that they are a specialized derivative with centrolepidaceous or other commeliniflorean affinity (in accordance with Hamann, personal communication). Lack of subsidiary cells, perisperm substituting endosperm, lack of

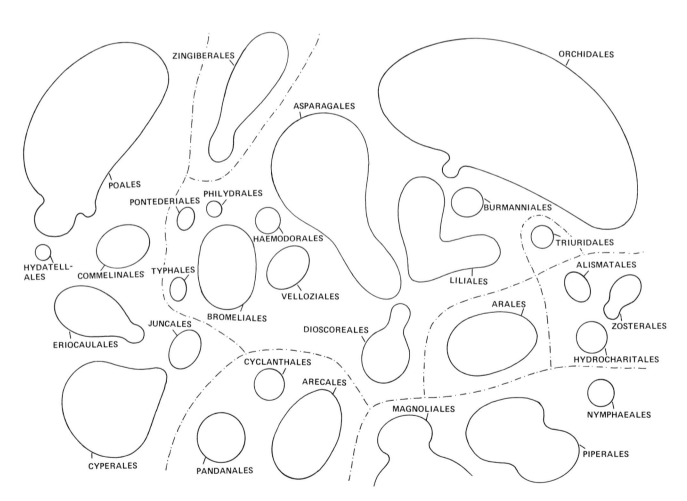

Diagram 107. An alternative phylogenetic diagram of the orders of monocotyledons. This diagram is that which corresponds best to the concepts of the present authors. As with Diagram 106, it takes a splitter's attitude to the families and orders. The Cyperales are here placed closer to the Areciflorae (especially the Pandanales) than are the grasses. The Hydatellales are considered a derivative of the Poales, especially the Centrolepidaceae. See further details in the text.

silica bodies etc. may very well be consequences of adaptations to a partly inundated life with short period for flowering and seed setting.

The Juncales and Cyperales have been placed in the lower part of the diagram instead of the upper, taking into consideration a possible affinity between the Cyperales and the Pandanales (expressed by, for example, Schultze-Motel, 1964). This does not violate the connections considerably, although the Poales are here distant from the Arecales (the similarities being also weak between these groups). The order Typhales, among the Liliiflorae, has been slightly relocated so as to admit connections with Juncales as well as with the Pontederiales and Bromeliales.

Alternative 3 (Diagram 108)

This diagram differs from the others in taking into consideration: the criticism of several contemporary taxonomists against splitting the orders too extensively; the different degree of affinity between different complexes of orders.

In its basic construction, this diagram agrees with Diagram 107. The fairly strong and easily defined difference between mono- and dicotyledons is illustrated with a broken line with dots.

The orders of the monocotyledons are divisible into three complexes, here at a level higher than the superorder. These are:

1. the Alismatiflorae and Ariflorae.
2. the Triuridiflorae, Liliiflorae, Zingiberiflorae and Commeliniflorae.
3. the Areciflorae.

The demarcation between these is indicated by broken lines in Diagram 108.

The superorders are separated with dotted lines, and are increased with one comprising the starch endosperm orders of the Liliiflorae, which are fairly well circumscribed.

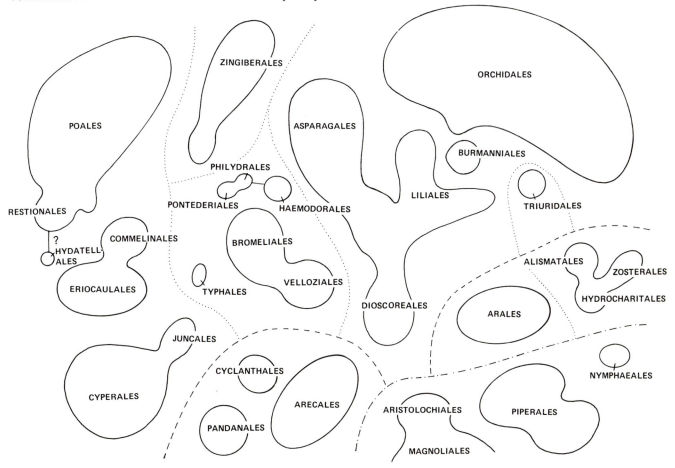

Diagram 108. An alternative phylogenetic diagram for the orders of the moncotyledons. In this, some of the orders upheld in Diagrams 106–107 are assembled to larger orders (not named in the diagram, but in the text). The phylogenetic interpretations are largely as in Diagram 107. See further details in the text.

Diagram 108 is constructed to guide those who prefer a wider circumscription of the orders than used in Diagrams 106 and 107, in which case the previously acknowledged orders could be united according to the following model:

1. The Alismatales, Hydrocharitales and Zosterales comprise one order: the Alismatales *sensu lato*.
2. The Dioscoreales, Asparagales and Liliales comprise one main order, Liliales *sensu lato*. Burmanniales in the future might prove to be best treated in or next to the Philydrales—Pontederiales complex.
3. The Velloziales and Bromeliales comprise the order Bromeliales *sensu lato*.
4. The Pontederiales and Philydrales comprise the Pontederiales *sensu lato*, and with a further amalgamation these orders could be united with the Haemodorales into one order, the Haemodorales *sensu lato*.
5. The Juncales and Cyperales comprise the Cyperales *sensu lato*, which would be based on substantial evidence from this study.
6. Further, there is evidence that the Commelinales and Eriocaulales (including Xyridaceae and Rapateaceae) could be united into one order, the Commelinales *sensu lato*.
7. As has already been referred to in Diagrams 107 and 108 the order Restionales is included in the Poales. A further inclusion in this order of the Hydatellales would be connected with great uncertainty.
8. Within the Areciflorae the three orders are distinct, although the Cyclanthales might be possible, with amalgamation in mind, to include in the Arecales.

This classification would result in only 14–18 orders of monocotyledons. The reason for proposing this is not that it is in accord with our own considerations—which favour equally Diagrams 106 and 107 or a combination of them—but because we are conscious of the fact that many taxonomists would be reluctant to accept the ordinal level and consequent degree of splitting that our alternative proposals imply. For such taxonomists our data can be used to support the classification implied in Diagram 108, combined with the family concept as presented below (pp. 328–329).

PROBLEMS OF FAMILY RANK IN THE LILIIFLORAE AND ELSEWHERE IN THE MONOCOTYLEDONS

With regard to the Liliiflorae in particular, a narrow approach has been chosen at the family level, largely in agreement with the concepts of Huber (1969, 1977). This has facilitated the work, as the smaller entities can thus be presented separately and their combinations of attributes more easily estimated in relation to each other. However, it may seem inflatory for a traditionalist used to the "Liliaceae *sensu lato*". An intermediary level between this wide family level and that practised by Huber may be feasible (and even preferred by most taxonomists). What is of crucial importance is, however, that the families comprise reasonably homogenous groups in an evolutionary sense and that they are arranged according to relationships as far as these can be ascertained from comparative studies.

In the classification given on p. 323 we proposed that the families should have a reasonable chance of representing natural groups of genera, even though these groups would become numerous and may quite often contain one or a few genera.

Should one wish to lay less stress on these criteria, yet have families more comparable in size to the average dicotyledon family (also in these, however, the splitting has often gone as far as in our classification, as witnessed by the large number of dicotyledon families recognized in Airy Shaw, 1973) then the following somewhat larger families would be a feasible alternative.
Ariflorae and *Triuridiflorae*: largely as above (p. 323).

Alismatiflorae
Hydrocharitales: Butomaceae, Aponogetonaceae, Hydrocharitaceae (incl. Thalassiaceae and Halophilaceae)
Alismataceae: Alismataceae (incl. Limnocharitaceae)
Zosterales: Juncaginaceae (incl. Scheuchzeriaceae and Lilaeaceae), Najadaceae, Potamogetonaceae (incl. Posidoniaceae, Ruppiaceae and Zosteraceae), Zannichelliaceae (incl. Cymodoceaeceae)

Liliiflorae
Dioscoreales: Dioscoreaceae (incl. Stenomeridaceae and Trichopodaceae), Taccaceae, Stemonaceae (incl. Croomiaceae), Trilliaceae
Asparagales: Philesiaceae (incl. Luzuriagaceae), Geitonoplesiaceae), Smilacaceae (incl. Ripogonaceae and Peter-

manniaceae), Convallariaceae (incl. Asparagaceae, Ruscaceae and Herreriaceae), Dracaenaceae (incl. Nolinaceae), Doryanthaceae, Hanguanaceae, Dasypogonaceae, Xanthorrhoeaceae, Agavaceae, Hypoxidaceae, Tecophilaeaceae (incl. Cyanastraceae), Asphodelaceae (priority dubious; incl. Phormiaceae, Dianellaceae, Eriospermaceae, Asteliaceae, Aphyllanthaceae, Anthericaceae, Aloëaceae and maybe Hemerocallidaceae), Funkiaceae, Hyacinthaceae, Alliaceae (incl. Agapanthaceae and Gilliesiaceae), Amaryllidaceae

Liliales: Iridaceae, Geosiridaceae (?), Colchicaceae (incl. Campynemataceae?), Alstroemeriaceae, Liliaceae (incl. Tricyrtidaceae and Calochortaceae), Melanthiaceae (incl. Petrosaviaceae)

Burmanniales: Burmanniaceae (incl. Thismiaceae), Corsiaceae

Orchidales: Apostasiaceae, Orchidaceae (incl. Cypripediaceae)

Typhales: Typhaceae (incl. Sparganiaceae)

Velloziales, Bromeliales, Haemodorales, Pontederiales and *Philydrales* are monofamilial containing but one (the nominal) family.

Zingiberiflorae

Zingiberales: Lowiaceae, Musaceae (incl. Heliconiaceae), Strelitziaceae, Zingiberaceae (incl. Costaceae), Cannaceae, Marantaceae

Commeliniflorae and **Areciflorae** as above.

The above classification, however tempting, camouflages the great distinctness of many generic assemblages worthy of being considered separately (e.g. the groups placed in the Asphodelaceae in the table above) and therefore even this mild "semi-lumping" approach is favoured neither by ourselves nor by Huber (personal communication) who has expended a great deal of effort in his study of the Liliiflorae. Alternatively, for those who wish to do so, the segregate families of the Asparagales and Liliales in particular could be regarded as subfamilies in which circumstances our treatment would slightly approach that of Thorne (1976) who makes considerable use of the sub-familial rank.

Relations between Monocotyledons and Dicotyledons: Evidence, Conclusions and Hypotheses

SOME RECENT CONTRIBUTIONS

During the last ten years several thought-provoking contributions dealing with the origin of monocotyledons and on the relations between these and the dicotyledons have been published. These include major papers by Huber (1969), Stebbins (1974), Meeuse (1975) and some other articles which will be briefly discussed below.

Huber (1969), who made an extensive survey of the Liliiflorae, has shown that the taxa of Dioscoreales exhibit a number of dicotyledonous features, which indicate that they are closely allied to the supposedly dicotyledonous ancestors of the monocotyledons. This is in concordance with Suessenguth (1921), who provided a pioneer comparison between monocotyledons and certain dicotyledons in numerous characters. Huber enumerated a number of features, including an embryo with a non-terminal cotyledon, young shoots which may have a single whorl of vascular strands, stems which contain vessels, and frequently opposite leaves with reticulate venation and with stipule-like appendages, all of which are suggestive of dicotyledons. Furthermore, their nectar secretion may rarely occur from a disc, and in his Dioscoreales and Stemonales the stamens sometimes (as in *Paris* and *Stenomeris*) have microsporangia located below the apices of slightly leaf-like stamens. Also, the microsporogenesis is often simultaneous as in most dicotyledons, the endosperm formation nuclear, and (in *Tamus*) the inner integument may be several-layered as in some rather primitive dicotyledons. The seeds also have a very small embryo with copious endosperm, a condition found more often in dicotyledons than in monocotyledons.

There is no doubt, however, that some of these features may in fact be secondary and advanced and therefore similar to those of dicotyledons by convergence. In spite of this, these views have much evidence to recommend them, and form an important approach to the question of monocotyledon-dicotyledon relationships.

Stebbins (1974) dealt more with the environmental conditions prevailing in the earliest evolutionary stages of the monocotyledons, leading to the selective advantage of a single tubular cotyledon (in accordance with Sargant, 1903). This he considered to be most likely the result of an "intercalary concrescence of the petioles of two original cotyledons". He regarded the monocotyledons to be derived from some now extinct, woody (probably shrubby) primitive dicotyledonous plant group with vessel-less xylem and "primitive" floral structure, having, for example, an apocarpous gynoecium. These plants probably evolved in such habitats as temporary swamps or pools and in a climate having a marked dry season. They were not presumed to be aquatic or marsh plants (in contrast to the hypothesis presented by Kudryashov (1964) and Takhtajan (1969)). Stebbins considered the aquatic as well as xeric forms, and also those with arboreal growth as secondary.

Meeuse (1975), like Huber, stressed the "repeatedly signalized connection between certain ranalean forms and some liliate taxa". This, he found, "is not unequivocal in that the magnolialean and nymphaealean dicots differ from all monocotyledonous orders in some essential, exclusively 'dicotyledonous' features whilst, on the other hand, exhibiting some rather convincing 'monocotyledonoid' traits". Similar conclusions were reached by ourselves in the course of the present work. Meeuse, however, declared, that "(unless the features shared by monocotyledons and dicotyledons are explained as the result of convergent evolution) the phylogenetic history of the monocotyledons goes back to a pre-angiospermous group of ancestors or to several such groups, and that this group, or part of it, was also progenitorial to the ranalean dicotyledons".

Other workers have contributed to the debate. Thus **Burger** (1978) found concentrated in the Piperales a suite of characters which are common among monocotyledons but rare or uncommon among dicotyledons. The concordance of the characters in the Piperales, he concluded, argues against convergence (Fig. 109). The Piperales, like the Nymphaeales, according to Burger, are more closely related to the monocotyledons than to other living dicotyledons, their nearest relatives being the Arales and some

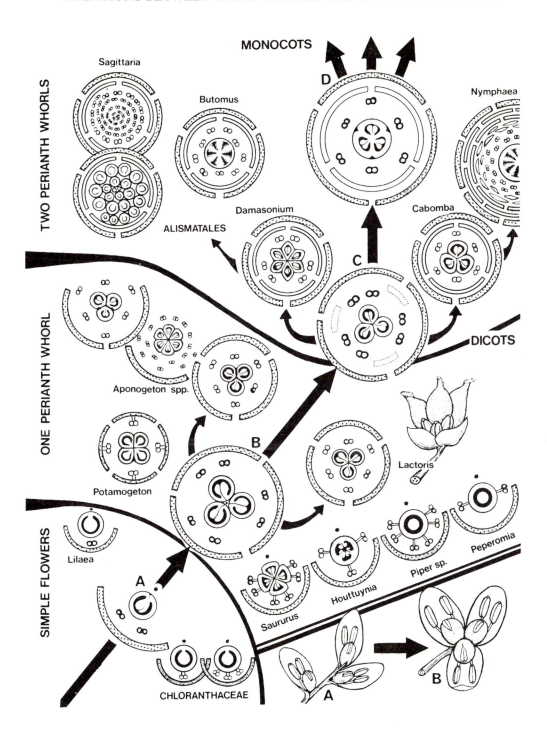

Fig. 109. Burger's (1977) interpretation of the phylogeny of the monocotyledons and most closely allied dicotyledons among the magnoliiflorean stock of dicotyledons. The Piperales are considered to be evolved along a separate line of evolution from that of the Nymphaeales, and *Aponogeton*, *Potamogeton* and other monocotyledons were derived from dicotyledonous ancestors independently from the Alismatales. These features are not supported by this survey.

Najadales, both of which Burger considered to represent very ancient lineages. In a manner reminiscent of Meeuse (1971), he derived the flowers of these groups from more simply constructed entities similar to the flowers of present Chloranthaceae which have come together by the contraction of internodes to form the three-parted flowers of the Piperales and many monocotyledons. The original flowers he suggested consisted of a single small bract, two stamens and a single adaxial, monocarpellate pistil.

Haines and Lye (1975) have suggested that the cotyledons of *Nuphar luteum* (Nymphaeaceae) may be interpreted either as a single bilobate cotyledon or as two cotyledons. The two cotyledonary lobes, they found, are joined at the radicular end, the union being rather more extensive on the face near the "coleoptile" than on the opposite side (Fig. 110). They concluded that the presence

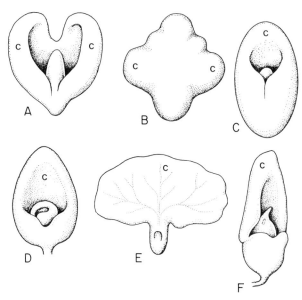

Fig. 110. Embryos from mature seeds in some monocotyledonous genera and some genera of dicotyledons related to the monocotyledons. A: *Nymphaea advena*, Nymphaeaoeae. B: *Ceratophyllum submersum*, Ceratophyllaceae. C: *Cttelia alismoides* (Alismataceae). D: *Tamus communis*, Dioscoreaceae. E: *Doryanthes palmeri* (Doryanthaceae). F: *Stipa juncea* (Poaceae). (Redrawn from Haines and Lye, 1975.)

of an operculum and of a "coleoptile" as well as other features of the seed and seedling are directly comparable to those in monocotyledons rather than dicotyledons. Therefore they recommended placing the Nymphaeales with the Helobiae, a view which might also gain support from various other evidence (see below).

In a similar way **El Gazzar and Hamza** (1975) suggested that the Alismataceae, Limnocharitaceae and Buto-

maceae should be placed in the dicotyledons on the basis of their close similarity to the Nymphaeales. Furthermore, they proposed that the Dioscoreaceae should be transferred to the dicotyledons and placed in the vicinity of Aristolochiaceae by virtue of some of the similarities between these groups.

A sceptical view with regard to the opinion that the Nymphaeales are closely allied to the monocotyledons is expressed by **Tomlinson** (e.g. 1976): The Helobiae (the Alismatiflorae as conceived here), he says,

are traditionally regarded as a primitive group [largely because of the evident apocarpy of some of their members]. However, this concept seems more strongly established as a matter of faith than reason. The putative common ancestry with the Nymphaeales in the dicotyledons should certainly be challenged, especially as research shows that the Nymphaeaceae are a very specialized group, any similarity of vascular system between them and monocotyledons is quite spurious. More accurate knowledge of the Helobiae is important in understanding the phylogeny of the monocotyledons, since it is difficult to derive other monocotyledons from a Helobial stock and their evident specialization in many biological features should be emphasized. It seems more reasonable to regard many of them as specialized end products.

Tomlinson has got strong support for these views from **Singh and Sattler** in a series of papers on the floral organogenesis of various Alismatiflorae (e.g. Singh, 1966; Singh and Sattler, 1977; Sattler and Singh, 1978). These have shown that in the Alismataceae (including Limnocharitaceae), but also in the Butomaceae, the secondary carpel and stamen initials are superimposed on the primary ones (which are in trimerous whorls), suggesting that the polymery is secondary in these groups. They also found similarity in the floral initials between *Aponogeton* (and *Scheuchzeria* and *Triglochin*, all in the Alismatiflorae), which have simply constructed trimerous flowers, and the Araceae (Ariflorae).

The background for a discussion of the monocotyledon-dicotyledon relationships would not be complete without mentioning the works by **Behnke** (1968, 1971) on sieve element plastid inclusions. The typical triangular monocotyledonous shapes of protein bodies have not been found in any other genus than *Asarum* (Aristolochiaceae); while several other shapes (quadratic, irregular, spherical) are represented in families of the Magnoliiflorae. The protein bodies in the sieve tube plastids together with the single cotyledon, have, therefore, unexpectedly reinforced the traditional delimitation of the monocotyledons in recent years.

Chemical characters of importance in this discussion include the distribution of benzylisoquinoline alkaloids, the isolated occurrence of ellagi-tannins in at least some taxa of Nymphaeales, and the fact that cyanogenic compounds are synthesized along the same (tyrosine) pathway in monocotyledons and, for example, some Magnoliiflorae, while other pathways are dominant in the remaining groups of dicotyledons (Hegnauer, 1977).

The conspicuous similarities between pairs of orders on either side of the monocotyledon-dicotyledon border have been stressed by Dahlgren (1974; 1979), where the Nymphaeales are compared with the Hydrocharitales-Alismatales, the Piperales with the Arales and the Aristolochiales (-Magnoliales) with the Dioscoreales.

In the course of this project several conditions have become apparent, which may be relevant in estimating the connections between monocotyledons and dicotyledons.

A NOTE ON RELATIVE PRIMITIVENESS

In giving a note on the concept of primitiveness we feel attempted to quote Sporne (1959), who transmits the following statement by Gilbert White: "Ingenious men will readily advance plausible arguments to support whatever theory they shall choose to maintain; but then the misfortune is, every one's hypothesis is each as good as another's since they are all founded on conjecture".

This indeed is applicable to much of what taxonomists consider as "primitive", "primary", "ancestral", or "original". We shall see, in connection with the monocotyledons and their possible ancestors, that what has generally been considered as primitive before, according to recent findings, may not necessarily be so. As an example may be mentioned: the numerous stamens and carpels in most Alismataceae, which would certainly qualify as a primitive attribute with Hutchinson (1943, 1959) and his contemporaries, but according to recent studies are indeed likely to be a derived specialization (Singh and Sattler, 1972, 1977 etc.). Such re-evaluations are of decisive importance when estimating the evolutionary direction and thus the most likely background of certain groups of monocotyledons.

Sporne (1956, 1974 etc.) has summarized some doctrines for establishing primitivity of a character state, but these doctrines in many cases give little guidance. Sporne's calculation of an advancement index for each family of monocotyledons is presented on p. 19.

In discussing differences and similarities between taxa it is particularly important to give an estimate as to whether the character states considered have a wide distribution or not, and whether they are likely to represent an ancient (primitive) or recent (derived) state in the group concerned.

Thus, in a comparison, for example between the Piperales and Arales, a feature like the occasional occurrence of sulcate pollen grains no doubt represents an ancient property, and thus should not necessarily be conceived as an indication of a particularly close relationship.

In this connection we will briefly introduce the concepts of Hennig (1950, 1966 etc.), partly because his methods are much debated and used among zootaxonomists and have been introduced by some plant taxonomists in the last years, partly because they provoke some interesting discussions on relations between, and the evolutionary background of, various groups of monocotyledons.

His methods call for a strict sequence in the course of evolution in the achievement of attributes which must be appreciated in relation to their distributions among plants. In fact, little is added by Hennig's concepts to the classical cladistic methods, but we will apply his concepts on the distribution patterns of character states given in the previous chapters in order to comment on some unexpected conclusions.

A feature shared by two or more groups and also found far back in a common ancestry is termed "plesiomorphous", and the joint possession of this feature "symplesiomorphy". The shared sulcate pollen grains in the Piperales and Arales is a supposed example of such a symplesiomorphy. Sulcate pollen grains are also known in many monocotyledons, Magnoliiflorae and Nymphaeiflorae.

Similarly a derived feature, evolved within an evolutionary line at a relatively late stage, is called "apomorphous". The joint possession of laticifers in Alismataceae *sensu stricto* and the occasionally recognized family Limnocharitaceae is a supposed synapomorphy; these families are also often (as here) regarded as subfamilies of the same family, Alismataceae *sensu lato*. Other probable synapomorphies of Alismataceae *sensu lato*, indicating a close relationship between its members, are the common presence of the Allium type of embryo sac formation, the curved embryos and the presence of pollen grains with two to numerous foramina.

Finally, by convergent or parallel evolution, a feature may evolve independently in two different lines of evolution. Such shared character states are "false synapomorphies". Basing the classification on such similarities leads to phylogenetically speaking heterogeneous groups.

Hennig's principles have recently been applied to angiosperm taxonomy by Bremer and Wanntorp (1978). Using them on the phylogenetic diagram of Takhtajan (1969), Bremer and Wanntorp found the Nymphaeales and the whole group of monocotyledons to be "sister groups", which gives a distorted view of their inter-relationships in the light of the evidence that will be presented on the following pages. A correct view of course can hardly be achieved in the absence of more fossil material, but meanwhile the nature of the attributes used for classification should be reconsidered with a view to determining their apparent plesiomorphy or apomorphy.

COMPARATIVE STUDIES OF MONOCOTYLEDONS AND CERTAIN DICOTYLEDONS

Comparison between the Nymphaeales and the Alismatiflorae

This comparison is presented in some detail earlier (pp. 288–289). Similarities are found in habit, atactostely, con-struction of root caps, occasional occurrence of stipules, common lack of vessels (especially in stem), occurrence of laticifers, occurrence of trimery in floral parts, occurrence of successive microsporogenesis and of sulcate pollen grains, apocarpy and occurrence of laminal placentation and helobial endosperm formation (the last-mentioned attribute rare in Nymphaeales).

Among the differences are to be mentioned the number of cotyledons (Haines and Lye's observations of 1975 do not diminish this difference substantially). Besides, the Nymphaeales have reticulate leaf venation, they lack intravaginal squamules, they lack protein accumulations in the sieve tube plastids, they have anomocytic stomata, their perianth parts and stamens are mostly spirally in-serted and they have seeds with endosperm as well as peri-sperm, all attributes of which deviate from the Alismati-florae. Also, they may accumulate ellagi-tannins which are absent in all monocotyledons.

Some of these differences, though few, are highly sig-nificant, and contradict the impression of a close relation-ship between the two groups. Whether the coincidence of so many similarities can be explained as parallel adapta-tions to an aquatic habitat is as yet unresolved.

Some such adaptations to an aquatic life in the Nymphaeales may very well be the atactostely and in connection with this the short-lived radicle and the tendency for reduction of vessels, the root cap similarities, the petiolate floating leaves and the lack of oxalate raphides.

Symplesiomorphous traits in this connection may be the trimerous and apocarpous flowers, the basifixed anthers, the successive microsporogenesis, the sulcate

pollen grains and the crassinucellate ovules. Similarities more difficult to explain are the occasional presence of stipules, the laminal placentation, the follicular fruits and (for Cabombaceae) the helobial endosperm formation.

It will be shown below that the Nymphaeales in some conspicuous characters approach other dicotyledons, notably the Piperales, and are not likely to be secondarily dicotyledonous derivatives from monocotyledonous an-cestors. Similarly, the Alismatiflorae approach other monocotyledons and do not share their direct dicotyle-donous ancestors with the Nymphaeales.

Comparison between the Alismatiflorae and the Ariflorae

The Ariflorae as conceived here of only one order, Arales, with the two families Araceae and Lemnaceae. Their superficial characteristics are mostly different from those of the Alismatiflorae, but some of the attributes reveal conspicuous similarities to this group.

The similarities, which are fully treated on p. 288, include the herbaceous habit, atactostely, ptyxis, usual absence of hairs, intravaginal squamules (found in Ari-florae only within *Philodendron*, however), absence of vessels in stems, occurrence of laticifers, frequently ex-trorse anther dehiscence, amoeboid tapetum, successive microsporogenesis, variable aperture conditions of pollen grains but with a sulcate basic type, crassi- or "pseudo-crassinucellate" ovules, lack of perisperm, frequent lack of endosperm and total lack of perisperm in the ripe seeds, the frequent Caryophyllad type of embryogenesis, the linear, often macropodous, embryo, which may be curved in both groups, and the general chemistry, e.g. the flavonol spectrum.

The differences are probably to a great degree con-nected with the fact that the Ariflorae mainly are rain-forest herbs, while the Alismatiflorae are marsh plants or aquatics. Also, however, the Ariflorae mostly accumu-late an abundance of oxalate raphides, while these are constantly lacking in the Alismatiflorae; the stomata are also generally tetracytic (rarely paracytic) in the Ariflorae while they are paracytic or lacking in the Alismatiflorae. The syncarpy and the cellular endosperm formation also contrast with the apocarpy and helobial or nuclear endo-sperm formation of the Alismatiflorae, and, finally, baccate fruits, as found in most Ariflorae, are practically absent in the Alismatiflorae.

Both groups give the impression of being derived (endosperm frequently lacking, pollen grains frequently inaperturate etc.) although there is in both a considerable "evolutionary depth".

Judging the similarities enumerated here that represent common ancient features inherited from past ancestors, i.e. symplesiomorphies, is difficult but these seem to be few. Rather, quite a number of similarities represent derived states, being either synapomorphies or apomorphies developed independently in the two lineages. Possible synapomorphies are the shared possession of intravaginal squamules, the lack of vessels in the stems, the amoeboid tapetum and the Caryophyllad type of embryogenesis, while the laticifers, the spadix like inflorescences, the lack of endosperm in ripe seeds and the curved embryos—most features of which are not consistent in either group—are more likely to have evolved by convergence. Lack of perisperm (mentioned as a contrast to the Piperales-Nymphaeales) and successive microsporogenesis no doubt represent symplesiomorphies.

Among the differences, the baccate fruit of the Ariflorae is doubtless an adaptation to rain forest habitat, while the cellular endosperm formation and tetracytic stomata are independent characters which may be apomorphies in the ariflorean line, as is the lack of stomata in many Alismatiflorae. Cellular endosperm formation in the Magnoliiflorae is probably an ancient feature and thus in the Ariflorae could represent a plesiomorphous attribute, but the cellular type is usually quite specialized in most Araceae being, for example, generally provided with a large unicellular chalazal chamber, and most likely could be derived from the helobial type.

For the Alismatiflorae the frequent occurrence of numerous stamens and carpels are probable apomorphies secondary in origin (see Sattler and Singh, 1978), and it is also possible that the laminal (dispersed) placentation in certain Alismatiflorae represent another apomorphy. For the Alismatales only, there are several apomorphies (foraminate pollen grains, Allium type of embryo sac etc.)

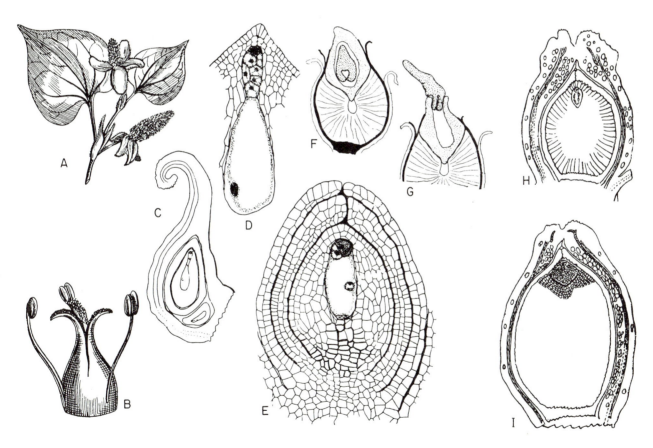

Fig. 111. Piperales: Saururaceae (A–G) and Piperaceae (H–I). A–B: part of plant and flower of *Houttuynia*. C–G: details of *Saururus cernuus*; C: carpel with an ovule in the stage of D, showing a few-celled endosperm with a large haustorial cell, all imbedded in the perisperm; E: ovule at the stage of fertilization; F–G: germinating seed with large perisperm (with radial streaks), a relatively copious endosperm and (dark-shaded) an embryo with two cotyledons. H–I: *Piper medium*, H showing a pistil in longitudinal section showing that the nucellar tissue (perisperm) is well developed already at the stage of the mature embryo sac. I shows a maturing seed (cf. the nymphaeaceous seed; Fig. 112) (A–B from Engler, 1889; C–G from Johnson, 1900; H–I from Johnson, 1902.)

Comparison between the Ariflorae (*Arales*) and the Piperales

Those familiar with plants in rainforest habitats have probably been struck by the similarity between certain araceous and piperaceous plants. Some members of Araceae are climbers, like species of *Pothos* and *Philodendron*, and thus resemble species of *Piper*, while most are low forest-floor plants, reminiscent of species of *Peperomia*. The similarities and differences between the Arales and the Piperales are more fully enumerated on p. 294, but will be considered here in their phylogenetic connection.

Besides the habit similarities between the two orders, the following attributes were mentioned: the atactostely, the petiolate leaves with a well developed, often cordate, lamina, the reticulate venation (rare in Arales, however), the tetracytic stomata, the inflorescence: a spadix with spathe (the spathe leaves of Piperales being rare and then several in number), the trimerous flowers, the similarities of reductions in floral parts, the occurrence of sulcate pollen grains, the common absence of a style, the dry stigma surface, the basal placentation, the cellular endosperm formation (although different in the details of endosperm), the baccate fruit and the presence of essential oils (rare in the Arales).

This list of similarities is counterbalanced by numerous differences, such as in the single cotyledon of the Arales, in the presence of vessels (even with simple perforation plates) in the Piperales where there is also secondary thickening growth, in ptyxis, in the orientation of stomata, in the occurrence of oxalate raphides (lacking in the Piperales), in the accumulations in the sieve tube plastids, in tapetum type, in microsporogenesis, in the ovules (ana- and orthotropous respectively), in the endosperm appearance (mostly with a large unicellular chalazal "chamber" in the Arales, in the occurrence of perisperm, in the embryogeny etc.)

An estimation of the above similarities does not appear to express so convincingly a close phylogenetic affinity in the light of the numerous significant differences.

A list of similarities between the Piperales and the Magnoliales (Annonales) would be if not equally long at least equally significant. The trimery of the flower, the general ovular morphology, the presence of essential oil cells and the occasional presence in *Piper* of benzylisoquinoline alkaloids, the sulcate pollen grains, the cellular endosperm formation and the copious endosperm in the ripe seed are some notable similarities.

It is also widely believed by taxonomists today that the Piperales represent a side-branch of the Magnolii-florae, a viewpoint which will be more fully discussed below.

Among the similarities between the Piperales and the Arales some were probably present in an ancient stock of plants representing the common ancestors of magnoliiflorous plants and monocotyledons. Such symplesiomorphies may be the petiolate reticulately veined leaves (found also in Arales), the trimerous flowers and the sulcate pollen grains, and perhaps also (but more doubtfully) the atactostely and the cellular endosperm formation (see above, however). Most other similarities enumerated above are likely to have evolved independently, some (like the reduced flowers, spicate inflorescence, spathes and baccate fruits) probably evolved as a response to the biological conditions of the rainforest habitat.

Among the differences noted, the vessel character, the different accumulations in the sieve tube plastids, the different number of cotyledons, the tapetum, the microsporogenesis, and the difference in development of perisperm, are so significant that one can hardly consider the Arales and Piperales as closely allied at all, the Arales being probably a fairly derived (rainforest adapted) order of an established monocotyledon stock, the Piperales a fairly distantly related, likewise derived (rainforest adapted) line with its origins in ancient, probably magnoliiflorean, stock of dicotyledons, which is indicated by the presence of oil cells and the (rare) occurrence of benzylisoquinoline alkaloids.

Comparison between the Piperales and the Nymphaeales

The members of Piperales and of Nymphaeales at a first impression may appear too dissimilar to be even considered as closely allied. This however is misleading, because of the number of striking shared features which include the following.

Both are herbs with scattered vascular strands.
Vessels with simply perforated plates occur in the Piperales and at least in *Cabomba* of the Nymphaeales (although vessels are normally known to be absent in the Nymphaeales).
Stomata are scattered (but of different types: tetra- and anomocytic respectively).
Hairs are largely lacking, and intravaginal squamules constantly lacking in both groups.
The sieve tube plastids are of the same type, lacking protein bodies but accumulating starch.
Oxalate raphides are lacking.
The flowers may be trimerous in both groups, although

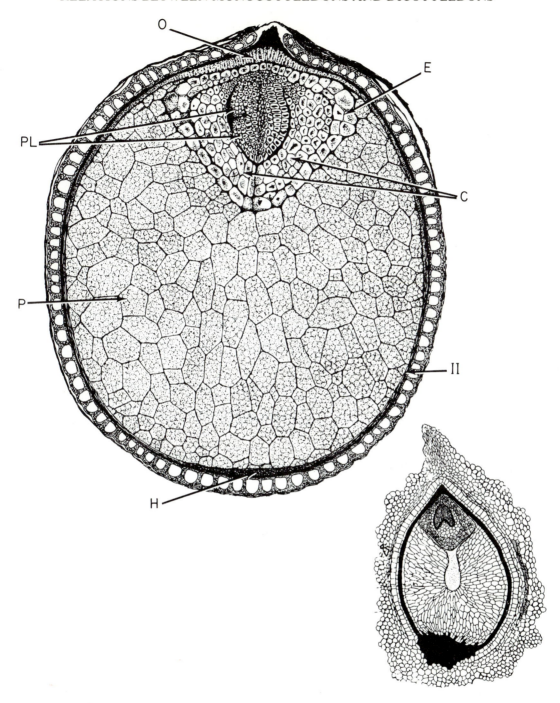

Fig. 112. Longitudinal section of a seed of *Ondinea purpurea* (Nymphaeaceae) with the aril removed. P = perisperm with starch granules; E = endosperm; C = cotyledons; O = operculum; H = hypostase; PL = plumular leaf; II = inner integument. (After Schneider and Ford, 1978.) For comparison: mericarp and seed of *Saururus cernuus* (Saururaceae; see also Fig. 111). (After Johnson, 1900.)

this is often not very obvious (cf. however, the trimerous stamens and carpels in Saururaceae, the trimerous organization in Cabombaceae).

The pollen grains are often sulcate in both groups (with a tendency to be inaperturate).

The tapetum is secretory.

Apocarpy is found in both orders: Saururaceae and Cabombaceae being the most obvious.

The stigmas are dry.

The ovules are crassinucellate.

Endosperm formation may be cellular in both groups.

Embryogeny conforms to the Asterad type in both groups (mentioned as a contrast to the Caryophyllad type in the Alismatiflorae and the Ariflorae).

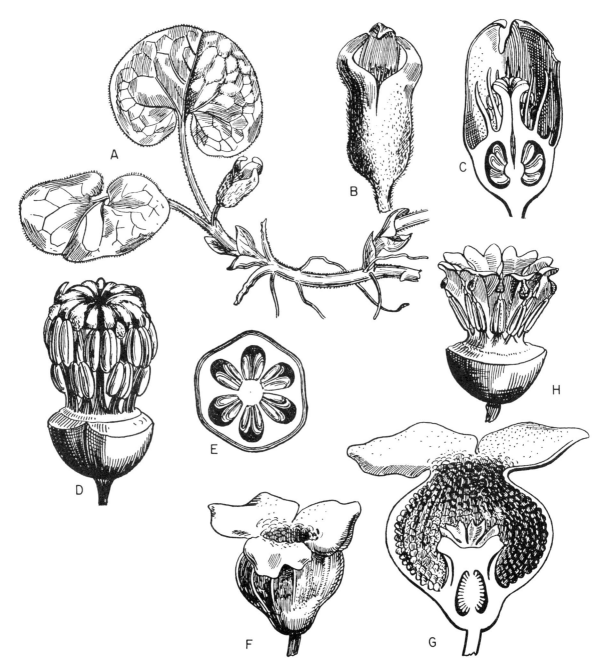

Fig. 113. *Asarum europaeum* (A–E) and *A. thunbergii* (F–H) (Aristolochiaceae) with floral details. B and F: flowers; C and G: flowers in longitudinal section; D and H: flowers with sepals removed; E: ovary in transverse section. This family shows surprisingly many monocotyledonous features. (From Lotsy, 1911.)

Perisperm is exceptionally well developed (Fig. 112) and contains copious amounts of starch and/or oil.

Endosperm is fairly well developed and has the same position subtended by the perisperm.

The embryo is small, dicotyledonous, and has a similar position.

The distributions agree in being not concentrated in the Southern Hemisphere.

Flavonoid chemistry is similar.

Contrasting with these similarities are, for example, the following differences.

The groups have very different appearances: the Piperales having small, reduced flowers in dense spikes, the Nymphaeales few, long-petiolate flowers, which are often large and provided with numerous floral parts in the perianth, androecium and carpels.

The vascular system is more reduced in the Nymphaeales than in the Piperales (adaptation to the aquatic life).

The stomata are mostly tetracytic in the Piperales, anomocytic in the Nymphaeales.

The ovules are orthotropous in the Piperales but anatropous in the Nymphaeales.

The Piperales have cells with essential oils, these are lacking in the Nymphaeales, which in contrast often synthesize ellagi-tannins.

Evaluating the above evidence, it seems obvious that the similarities in this connection cannot be dismissed. Many of these, however, represent probable symplesiomorphies, i.e. features likely to have occurred also in the past common ancestors of these orders and of the taxa in the Magnoliiflorae. Such probable symplesiomorphies include the dicotyledonary embryo, the scattered (non-parallel) stomata, the lack of intravaginal squamules and probably the lack of oxalate raphides, the trimerous flowers, the sulcate pollen grains, the secretory tapetum, the apocarpy, the dry stigmas, the crassinucellate ovules, the cellular endosperm formation, and the well developed endosperm and the small embryo. Thus the two groups are basically primitive in a number of features.

Synapomorphies, i.e. properties which are more derived and possibly evolved in the evolutionary branch common to the two orders may include the herbaceous habit and scattered vascular strands, the simply perforated vessels where these occur (although a parallel development of these is feasible), the lack of protein in the sieve tube plastids, and—maybe the most conspicuous similarity—the striking development of the perisperm and general plan of the seed (Fig. 112). The latter features are rare or lacking in the Magnoliiflorae.

Of the differences noted, much can be ascribed to

adaptations to different habitats and pollination strategies. The occurrence of essential oils in the Piperales is shared by many families of the Magnoliiflorae (Annonaceae, Aristolochiaceae, Magnoliaceae, Myristicaceae, Canellaceae etc.) and the rare occurrences of benzylisoquinoline alkaloids (in *Piper*) is no doubt also a "relic" attribute. Often, the large, polymerous flowers in the Nymphaeaceae have been regarded as primitive, while the oligomerous flowers of the Piperales have been considered to be derived—as was outlined by Hutchinson in his "Families of Flowering Plants" (1934, 1959, 1973). However, it seems more likely that the large flowers with polymerous parts (partly spirally set) are indeed derived, representing a compensatory achievement when the flowers became few, while the simply constructed flowers of the Piperales are no doubt also derived and represent a line of reduction. It is perhaps also this way that the laminar placentations in Nymphaeaceae should be looked upon; as a fairly advanced type evolved when ovules became more numerous within the carpels, thus representing an apomorphy rather than a plesiomorphy! It is the inability to decide whether features are orginal or derived and whether similarities are by common origin or evolved by convergence or parallelism that limit the cladistic methods of Hennig in creating radically better classifications than has been done before.

Thus, the postulated "bridges" between the Piperales and the Ariflorae and between the Nymphaeales and the Alismatiflorae seem to be rather unlikely in phylogenetic terms. Rather, the Alismatiflorae and the Ariflorae appear to be allied to each other on the monocotyledon side, both having some derived common features, and in the same way the Nymphaeales and Piperales show a number of similarities indicating a relatively close affinity, though some of their similarities also represent derived attributes. The latter affinity has appeared so convincingly to us, that it is suggested that the Piperales and Nymphaeales should be treated together as one superorder, the Nymphaeiflorae.

Status of the Piperales, Primitive or Advanced?

The order Piperales, which here consists only of the families Saururaceae and Piperaceae (incl. Peperomiaceae), has a clear connection with the Magnoliiflorae, as has been acknowledged before.

They comprise mostly woody climbers or shrublets which may have secondary thickening growth as in the Magnoliiflorae, although the vascular bundles are often scattered (as in the Nymphaeiflorae). In the secondary thickening growth the Piperales approach the Magnoliiflorae to a higher degree than do the Nymphaeiflorae.

Also, they more often have vessels in the stem. The vessel elements tend to have scalariform perforation plates in the Saururaceae but simple in the Piperaceae.

Cells with essential oils are in common between the Piperales and the Magnoliales, and benzylisoquinoline alkaloids have been detected in some species of *Piper* (Hänsel *et al.*, 1975), giving another connection with the Magnoliiflorae.

Finally, there should be mentioned the numerous, no doubt "plesiomorphic", attributes shared by the Piperales and Magnoliiflorae. These include the occurrence of sulcate binucleate pollen grains in both groups, the bitegmic, crassinucellate ovules with thick integuments (unitegmic ovules occur in *Peperomia*), the cellular endosperm formation, the copious endosperm and the small dicotyledonous embryo.

In a thought-provoking paper Meeuse (1972) has given a number of interesting comments on the Piperales (which he circumscribed more widely than here including also the Chloranthaceae and Lactoridaceae). He challenged a number of conventional ideas on the relation between Piperales and taxa of Magnoliales (e.g. *Magnolia*). These points are essential to the understanding of both groups, and therefore a recapitulation of his points compared with the above conclusions may be relevant.

The growth forms of the Piperales, viz. the herbaceous stems with peristent, subterranean tubers and the suffruticose habit, were considered by Meeuse to be most likely as primitive as the large arborescent growth forms. However, it appears to us that the combination of tubers and herbaceous stems is a more specialized state than the fruticose woody one although the most primitive state could be shrubby or arborescent. Meeuse also considered scattered vascular bundles to be as primitive as bundles in one whorl and nodes with several leaf traces to be as primitive as nodes with two or three leaf traces.

The *Peperomia* type of embryo sac formation in Piperaceae according to Meeuse may be interpreted as a primitive attribute, because it is somewhat reminiscent of the multinucleate condition of the embryo sac in *Gnetum*. However, the Saururaceae of the same order show the normal type of embryo sac formation, and *Piper* a type somewhat transitional between this and the Peperomia type (cell wall plates being formed during meiosis but later dissolved, so that a tetrasporic embryo sac is formed).

The primitivity of the perisperm was also stressed by Meeuse, and is presumed to be a retained gymnospermous character. Nonetheless the perisperm in the light of its limited distribution seems more likely to be a derived type of nutrient tissue, which is not developed in the Chloranthaceae, nor in other Magnoliiflorae (except Piperales or Nymphaeales).

Meeuse concluded that specialized families with reduced morphological features, such as Callitrichaceae and Hippuridaceae, usually retain "advanced" bioconstituents (in the case of these two families, iridoids), whereas the Piperales lack many compounds such as benzylisoquinoline alkaloids, polyhydroxylated acids and flavonoids, complex steroids and saponins. However, benzylisoquinoline alkaloids recently have been detected in a species of *Piper* (Hänsel *et al.*, 1975). Also, the flavonoid pattern of the Piperales is comparable to that in the Magnoliales, both groups being conspicuously poor in flavonoids. The Piperales is proportionally richer than many magnolialean families in O-methylated flavones, C-glycoflavones and chalcones (Gornall *et al.*, 1979).

As regards the organization of the flowers in the Piperales, this is generally accepted as a derived ("reduced") type, which is based on a comparison with, for example, *Saururus*, which has tri- or tetramerous flowers with 6 or 8 stamens and 3 or 4 carpels. Perianth members are lacking. As stated by Meeuse, there is no support for the theory that the Piperalean flower has evolved from forms so elaborate and polymerous as the flower of *Magnolia*. The *Magnolia* flower no doubt represents a derived and secondarily elaborate type. However the Piperalean flowers may have evolved from a trimerous, simply constructed but basically magnoliiflorean type with inconspicuous tepal members. This in no way militates against the basic concepts of Meeuse's "Anthocorm Theory". In most angiosperms, the flowers tend to be composed of lateral entities ("monandronal androclads and gynoclads", or maybe "monandronal and monogynonal androgynoclads") at a number per flower which is fairly fixed (to multiples of 3) in monocotyledons and (mostly to multiples of 4 or 5) in the main dicotyledonous groups. In the Magnoliiflorae these conditions are less stabilized and the "gonoclad" conditions maybe more complex than in the monocotyledons.

In conclusion, the above evidence supports a more "classical" derivation of the Piperales from an ancient, largely magnoliiflorean stock.

Comparison between certain Magnoliiflorae and certain Liliiflorae

This comparison concerns taxa of the Magnoliiflorae (especially the Annonaceae, Aristolochiaceae, Lactoridaceae and Chloranthaceae) on the dicotyledonous side, and some Liliiflorae, notably the Dioscoreales (Dioscoreaceae, Trichopodaceae, Taccaceae, Trilliaceae) and some Asparagales (Smilacaceae and a few more families), on the monocotyledonous side. As has been outlined above, Huber (1969, 1977), in particular, has stressed the links between these groups of families.

Some of the similarities to be stressed are the following.

Some members, in particular in the berry-fruited Asparagales, such as the Smilacaceae, Philesiaceae and Asparagaceae, are woody as are most primitive dicotyledons concerned.

The stems of certain Dioscoreales and of some of the families of the Asparagales (but also of the Velloziales, Haemodorales etc.) contain vessels with scalariform perforation plates as are found in many Magnoliiflorae.

In the Chloranthaceae (Magnoliiflorae) the vascular strands are disposed in two whorls and furthermore, as in *Sarcandra*, may lack vessels, which is otherwise only rarely the case in dicotyledons but a common feature in the Liliiflorae (such as in *Croomia*, Stemonaceae, all Trilliaceae, Taccaceae etc.).

Sieve tube plastids in a great many Magnoliiflorae investigated by Behnke (1971, 1972) have protein crystalloid bodies which may be square, rectangular or polygonal and in one genus, *Asarum* (Aristolochiaceae), even triangular as in the monocotyledons.

The leaves are opposite in many species of *Dioscorea* (Dioscoreaceae), in *Stemona* (Stemonaceae), a few species of *Smilax* (Smilacaceae) and in *Ripogonum* (Smilacaceae). Opposite leaves are otherwise rare in monocotyledons but fairly common in dicotyledons, although not particularly so in the Magnoliiflorae.

The leaves are petiolate in the Dioscoreales and certain Asparagales and have a rather short, almost non-sheathing base, as is normal in the Magnoliiflorae.

Stipule-like appendages (but maybe not true stipules?) occur in species of *Dioscorea* and *Tamus* (Dioscoreaceae) as well as in *Smilax* (Smilacaceae). Stipules are otherwise rare in monocotyledons (but occur in Zosterales). In dicotyledons stipules are widely distributed, although in the Magnoliiflorae they are rare, so this is a dubious similarity.

The leaf lamina is compound in some species of *Dioscorea* (Dioscoreaceae), in *Tacca* (Taccaceae) and, besides, in certain members of Arales. Otherwise compound leaves are much more common in dicotyledons, although not precisely in the Magnoliiflorae, this being therefore another dubious link between mono- and dicotyledons.

Venation of the leaves is reticulate in taxa of Dioscoreaceae, Stemonaceae, Trilliaceae (rarely), Taccaceae, Smilacaceae and Philesiaceae (rarely), the reticulate venation being otherwise typical to dicotyledons.

Stomata may also be somewhat variably directed in at least Smilacaceae, whereas normally they are parallel to the main veins in monocotyledons. In dictotyledons they are scattered in various directions.

The flowers in most Dioscoreales, in the Taccales and in many baccate-fruited Asparagales, as in some magnoliiflorean dicotyledons, are not particularly showy; their tepals are subequal, often greenish, and of small to moderate size.

The perianth is frequently trimerous in the Magnoliiflorae, in the Annonaceae there are mostly three such whorls; the Degeneriaceae and Canellaceae as a rule have 3 sepals but more numerous petals; in Aristolochiaceae (Fig. 114), *Saruma* and some species of *Asarum* have 3 sepals and 3 petals, while in most Aristolochiaceae there are only 3 more or less fused sepals; Lactoridaceae and most Myristicaceae have but 3 tepals. Trimery of perianth is more or less universal in monocotyledons where there are more regularly two tepal whorls than in the Magnoliiflorae.

The stamens may be trimerous in the Magnoliiflorae as in the monocotyledons, although only occasionally they have 3 + 3 stamens, as in *Lactoris* (Lactoridaceae) and rarely in Annonaceae (*Orophea* species). In *Chloranthus* (Chloranthaceae) they are 3 and in *Saruma* (Aristolochiaceae) the stamens tend to be 12, in two whorls. The trimery is well established in monocotyledons, where for example, most Liliiflorae have 3 + 3 or (no doubt secondarily) 3 stamens.

Certain Dioscoreales (such as *Trichopus*, *Stenomeris* and *Paris*) have stamens with a prolonged connective and microsporangia located far below the apex, such as is common in the Magnoliiflorae and especially the Magnoliales.

Anther wall formation in the dioscorealean *Tacca* is reported to be of the dicotyledonous type, which is the only case known of this type in the monocotyledons. Interestingly, the monocotyledonous type is known in *Piper* (Piperales).

Tapetum is secretory (glandular) in all Magnoliiflorae as in all Dioscoreales and most Asparagales.

Microsporogenesis is of the simultaneous type in some species of *Dioscorea* and in *Tacca*, as it is in most Magnoliiflorae and other dicotyledons. On the other hand the successive type or transitions between the two types are known in *Aristolochia*, *Annona*, *Cananga* etc. among the Magnoliiflorae, the successive type being otherwise the most widely distributed one in monocotyledons.

Monoaperturate (often sulcate) pollen grains, found in most monocotyledons, are also known in many Magnoliiflorae, although other derived types (inaperturate, 2-aperturate) occur as well; the occurrence in the Dioscoreaceae of bisulculate pollen grains is not a dicotyledonous attribute. However, sulcate pollen grains are known in other taxa of Dioscoreales.

Binucleate pollen grains are found in all or practically all Magnoliiflorae and in nearly all Liliiflorae.

The ovules are chiefly anatropous, bitegmic and provided with primary parietal cells in the Magnoliiflorae; this is generally also the case in the Dioscoreales and the most closely related families of Asparagales, although exceptions occur (orthotropous ovules in *Stemona*, lack of parietal cell in *Trichopus*). In *Dioscorea* the inner integument may be several-layered as in some Magnoliiflorae. In addition the normal (Polygonum) type of embryo sac formation is common in all these groups (although the Scilla type is known especially in the Trilliaceae).

Crystals are found in the inner epidermis of the outer integument in the Dioscoreaceae (and Trichopodaceae) as in many Magnoliiflorae.

The nuclear endosperm, which is widespread in the Dioscoreales and also in many berry-fruited Asparagales, may be regarded a dicotyledonous feature (Huber, 1969), but the Magnoliiflorae very often have cellular type endosperm formation; hence this character is a somewhat dubious indicator of relationship.

The seeds have copious endosperm, which may contain starch; starch is only occasionally found in the Dioscoreales (*Croomia*, sometimes *Paris* and *Trillium*) and related Asparagales (*Ripogonum*, *Streptopus*), but is more common in some remote orders, viz. the Velloziales, Bromeliales, Haemodorales and Pontederiales. In the Magnoliiflorae starchy endosperm is scattered and is found in some taxa of Annonaceae, Aristolochiaceae, Myristicaceae etc.

Ruminate endosperm, which is found in several members of the Magnoliiflorae, such as within the Annonaceae, Aristolochiaceae, Canellaceae, Degeneriaceae, Magnoliaceae, Eupomatiaceae and Myristicaceae is also found in *Avetra* and *Trichopus* (Dioscoreaceae).

The embryos are very small (rudimentary *sensu* Martin, 1946) in relation to the seed and endosperm in most Magnoliiflorae and in some of the Liliiflorae concerned with here (*Paris*, *Trillium*, *Medeola*) and also fairly small in most Dioscoreaceae.

Between the dicotyledonous embryo with a terminal plumule and the typically monocotyledonous one with a terminal cotyledon, there is a transitional, primitive monocotyledonous embryo type, where the plumule is not entirely pushed aside from its terminal position by the cotyledon; such embryos are found, according to Huber (1969) in, for example, *Dioscorea*, *Stenomeris*, *Trichopus* and *Tacca* (but also in certain other groups, e.g. certain Alismatiflorae and *Cocos*, according to Haccius and Philip, 1979).

Cyanogenic compounds seem to be of the same type (the tyrosine pathway; Hegnauer, 1977) in most Magnoliiflorae and in the monocotyledons; they are known in, for example, Magnoliaceae and Calycanthaceae as well as in Dioscoreaceae.

The flavonoid pattern is fairly similar and at least not conspicuously different between the Magnoliiflorae and those Liliiflorae concerned with here. Both have mainly the commonest types of flavonoids; in common between the Dioscoreales and the Aristolochiales are cyanidin and/or pelargonidin, and acylated anthocyanidin glucosides.

This evidence, although when listed as above, may seem more impressive than it is, and no doubt gives strong support for Huber's conclusion that there is a concentration of similarities across the borderline, Magnoliales *sensu lato*—Dioscoreales-Asparagales. Among the dicotyledons the Magnoliiflorae thus exhibit a multitude of monocotyledonous features, and among the monocotyledons the Dioscoreales and berry-fruited Asparagales of the Liliiflorae exhibit a number of dicotyledonous features.

One must not be misled, however, to conclude that the extant forms of Dioscoreales or Magnoliales represent missing links or relics of the common ancestors of the monocotyledons.

CONCLUSIONS

From the above comparisons, the following conclusions can be drawn.

1. The monocotyledons have a close connection with particular groups of dicotyledons. Thus it is very likely that their ancestors far enough back in time were in common with the ancestors of dicotyledons, and did not evolve from other (gymnospermous or proangiospermous) ancestors.

2. As the greatest concentration of similarities are found between the magnoliiflorous and certain liliiflorous groups of di- and monocotyledons respectively, it may be expected that a number of these similarities (namely those which are more widespread) were also found in

the ancestors of the monocotyledons. Other features shared between certain taxa in each of the groups are likely to have evolved by convergence.

3. The evidence raises doubts about the primitiveness of certain characters which are normally regarded as ancestral. These include, for example, the polymery of the flowers in many Hydrocharitaceae, Alismataceae and Nymphaeaceae, and the laminal placentation in some of these groups (in certain cases, maybe, even their apocarpous condition). It is likely that large polymerous flowers often can have evolved secondarily in different groups: in the Nymphaeaceae, Magnoliaceae, Illiciaceae etc., while the primitive members may have had fairly small and oligomerous flowers in which there was a tendency for the whorls to be trimerous.

4. The consideration that among the dicotyledons, the Lactoridaceae, Chloranthaceae, Annonaceae and Aristolochiaceae resemble (and have a certain connection with) the ancestors of the monocotyledons is corroborated by this study. This viewpoint has previously been stressed by Meeuse, Huber, Burger and others.

5. The monocotyledons in spite of their large number of species, exhibit a more stabilized organization of the flower (floral diagram) than do the Magnoliiflorae; also in other features (a typical character being the sieve tube plastid inclusions) the Magnoliiflorae show a greater variation than the monocotyledons.

6. The ancestors of the monocotyledons are not likely to have achieved all their monocotyledonous attributes simultaneously and along one line, but their differentiation has occurred successively, and with parallel diversifications, resulting in diverse patterns of variation.

7. The Ariflorae and the Alismatiflorae are probably not as distantly related as is usually believed, but may have a fairly recent common origin. They seem together to represent a bibrachiate side branch among monocotyledons, with possible roots in ancestors likely to agree most closely with primitive Liliiflorae (see below). The great difference between the Ariflorae and Alismatiflorae is no doubt connected with adaptations to rainforest habitat and swamps or semi- to totally aquatic habitat respectively. Some features of various Alismatiflorae, such as the polymery of the androecia and gynoecia (and perhaps even the laminar placentation and the apocarpy), may be secondary adaptations, although they are frequently considered as primitive attributes.

8. The Piperales and Nymphaeales on the dicotyledonous side are probably derived from a primitive Magnoliiflorean stock, features of which are more obvious in Piperales than in Nymphaeales, especially the chemical attributes. The similarity between the two orders in several significant characters, such as in the perisperm tissue and the starch type sieve element plastids, indicate that together they form a bibrachiate evolutionary branch and evolved in response to rainforest (Piperales) and aquatic (Nymphaeales) habitats respectively.

9. The groups with large, brightly coloured flowers in the Liliiflorae do not seem to be those with the most primitive attributes, and the nectar secretion has probably evolved along different evolutionary lines: some with septal nectaries, in the Dioscoreales, Asparagales and other liliiflorous orders but also in the Zingiberiflorae and Areciflorae, and others with nectar secretion at the base of the tepals or stamens in the Liliales and Orchidales. If the tepal-base nectaries did not succeed the septal ones (which is possible from their relative range of distribution and not altogether unlikely) then nectar secretion probably evolved simultaneously in different ways along different evolutionary lines, and was not yet established in the ancestors of the monocotyledons. These may have been insect pollinated, but in random and unspecialized ways, perhaps offering mainly pollen to the insects, or they may not have been insect pollinated at all (which would be more in line with the conclusions of Meeuse).

A Possible Model for the Monocotyledonous Ancestors as Based on the Evidence of this Study

Considering the evidence from the distribution of various characters in angiosperms in the light of general concepts as to what is likely to be primitive or not, the following very sweeping reconstruction of the monocotyledon ancestor may be attempted.

The ancestors of the monocotyledons were probably shrublets or subshrubs which by environmental conditions (a pronounced alternation between wet and dry periods) evolved compact underground stems, mainly short or long rhizomes from which herbaceous aerial stems were developed. In their adaptation to this life form they lost the cambial activity in the vascular strands, i.e. the ability to form secondary vascular tissue, as in most dicotyledons. This also influenced the growth of the radicle, which (when lacking ability for secondary thickening growth) became ephemeral, which was compensated for by the rich development of adventitious roots.

The cotyledons of the monocotyledon ancestors were probably two, which by intercalary concrescence (Stebbins, 1974) became adapted to a terminal tubular structure well suited to penetrate the ground after germination.

The stems of the monocotyledonous ancestors may already have had an atactostele with two or more whorls of vascular strands, as in present day Chloranthaceae and Piperaceae. It is also likely that they had vessels by ancestry, in the roots and possibly also in the stem, the vessels being probably of a primitive type with oblique walls and scalariform perforation plates. Further evolution in the monocotyledons went in two directions: against inhibited development of vessels (as in many Liliiflorae, the Ariflorae and the Alismatiflorae) and against more specialized vessels with shorter elements having simple perforation plates (as in many Commeliniflorae).

The leaves, as in many present day monocotyledons, were probably petiolate with a short, not or hardly sheathing base, and a well developed lamina with reticulate venation. They could have been alternate or sometimes opposite, and may have varied between being entire or palmately lobate or digitately compound. Stipules were probably lacking.

The sieve element plastids no doubt accumulated protein crystalloid bodies, which tended to have a triangular shape. Silica bodies were probably absent in the tissues, but oxalate raphides may have been present in the earliest forms. Intravaginal squamules were lacking, but other multicellular hairs may have occurred. The stomata were anomocytic or paracytic.

The flowers were hypogynous, actinomorphic and possibly fairly inconspicuous with tepals in two (to three) whorls; the tepals, stamens and carpels already in the past ancestors were more or less stabilized in trimerous whorls. The tepals were probably modest in size. Insect pollination may have occurred, although less specialized, the flowers being open. If nectar secretion occurred, this may have been from primitive types of nectaries along the septa of the carpels.

The stamens no doubt were more or less stabilized in a number of six, disposed in two trimerous whorls. The anthers were tetrasporangiate, basifixed and introrse or extrorse, maybe with a tendency to have a prolonged connective, as in present day *Trichopus* or *Asarum*. They dehisced by longitudinal slits and their walls tended to be formed according to the monocotyledonous rather than the dicotyledonous type, although this may not have been quite stabilized. Tapetum was secretory and consisted chiefly of binucleate cells. Microsporogenesis was of the successive type, and the pollen grains were binucleate and provided with a distal sulcus.

The gynoecium may have been apocarpous, but syncarpous types, fused in the ovary region soon evolved. Placentation then became central, axile. The stigma surface was dry. The ovules no doubt were anatropous, bitegmic with fairly thick inner integument, and crassinucellate, and a primary parietal cell was cut off from the archesporial cell before the meiotic division. Embryo sac formation proceeded according to the normal (Polygonum) type, while there is some doubt about whether the endosperm formation had already proceeded from a cellular (Wunderlich, 1959) to a helobial or nuclear type, the latter of which may have occurred along parallel lines or sometimes in two steps, the helobial type then preceding the nuclear.

The fruits may have been follicles or achenes when the pistil was apocarpous, or a capsule when it was syncarpous. It is not entirely impossible that syncarpous gynoecia later could have reverted back to apocarpous forms. The seeds had a well developed integumentary coat which was not covered with phytomelan. They had a copiously developed endosperm, maybe with starch (but not a substantial perisperm tissue) and the embryo was small and subterminal in position. The endosperm may have been ruminate.

The embryo, which shifted from a di- to a monocotyledonous state in the monocotyledon ancestors, went through a stage with a somewhat laterally placed cotyledon and with the plumule not wholly pushed aside. The embryo was chlorophyll-less and "rudimentary" (*sensu* Martin, 1946) as regards size class.

Chemically one may only guess that benzylisoquinoline alkaloids *could* have been developed as they are in most present day Magnoliiflorae with monocotyledonous traits and then perhaps lost their protective function taken over largely by steroid saponins, which evolved strongly in the liliiflorean line. Cyanogenic compounds (tyrosine pathway) are also likely to have been present in the monocotyledon ancestors.

This reconstruction is based on the most likely state for each attribute, but as evolution often takes unexpected courses the description is not likely to reflect the factual conditions in the ancestors in *all* respects. It is possible that several of the above characters soon achieved a wide variation by interaction with each other.

Probable monocotyledonous fossil pollen grains (*Liliacidites*) are known already from the Aptian (Cretaceous) and a possibly monocotyledonous leaf impression, *Alismaphyllum*, is known from the Albian (Hickey and Doyle, 1977). As primitively monocotyledon leaves with reticulate venation would probably be identified as dicotyledonous (and could have resembled *Menispermites* or *Celastrophyllum*), and as we do not know whether *Alismaphyllum* is at all monocotyledonous, we cannot base any far-reaching conclusions on them.

A POSTSCRIPT

The material presented in this book can be used for phylogenetic evaluations using principles other than those used here. Thus a cladistic analysis of this material will be presented in due course (Dahlgren and Rasmussen, in preparation), and a numerical analysis is planned (Clifford). These may each lead to slightly different conclusions from those presented above.

The data that we have used in the present study are meant to be displayed for the reader so that he will be in a position to form his own conclusions. Also these data should be complemented with more, as these become available, and then may be able to give a more conclusive picture of certain relationships.

A subject and name index was originally planned for the book, but at the last stage were found to be unnecessary, because the characters dealt with are presented in a rather logical sequence, with terminology and comments given under the appropriate titles. Families and genera are likewise mentioned in connection with the respective characters and are given in strict order in the base data lists. Thus it is our firm belief that the reader after making himself familiar with the contents will easily find any information in the book that he may need.

References

Ackerman, A. and Williams, N. H. (1980). Pollen morphology of the tribe Neottieae and its impact on the classification of the Orchidaceae. *Grana* **19**, 7–18.

Afzelius, K. (1918). Zur Entwicklungsgeschichte der Gattung Glorisa. *Acta Hort. Bergiani* **6** (3).

Airy Shaw, H. K. (1966). Diagnoses of new families, new names etc. for the seventh edition of Willis's "Dictionary". *Kew Bull.* **18**, 249–273.

Airy Shaw, H. K. (1973). "A dictionary of the Flowering Plants and Ferns". 8th ed. (J. C. Willis, ed.) Cambridge University Press, Cambridge.

Ancibor, E. (1979). Systematic anatomy of vegetative organs of the Hydrocharitaceae. *Bot. J. Linn. Soc.* **78**, 237–266.

Ankermann, F. (1927). Die Phylogenie der Monocotyledonen. *Bot. Archiv* **19**, 51–60.

Arber, A. (1923). On the "squamulae intravaginales" of the Helobieae. *Ann Bot.* **37**, 31–41.

Arber, A. (1925). Monocotyledons. A morphological study. Cambridge Univ Press, Cambridge.

Arekal, G. D. and Ramaswamy, S. N. (1973). Embryology of Burmannia pusilla (Wall. ex Miers) THW. and its taxonomic status. *Beitr. Biol. Pflanz.* **49**, 35–46.

Arekal, G. D. and Ramaswamy, S. N. (1980). Embryology of Eriocaulon hookerianum Stapf and the systematic position of Eriocaulaceae. *Bot. Notiser* **133**, 295–309.

Arnott, H. J. (1962). The seed, germination and seedling of Yucca. *Univ. Calif. Publ. Bot.* **35**, 7–144.

Ascherson, P. (1889). Potamogetonaceae. In "Die natürliche Pflanzenfamilien" (A. Engler and K. Prantl, eds), II (1).

Asplund, I. (1968). Embryological studies in the genera Sparganium and Typha. A preliminary report. *Svensk Bot. Tidskr.* **62**, 410–412.

Asplund, I. (1972). Embryological studies in the genus Typha. *Svensk Bot. Tidskr.* **66**, 1–17.

Asplund, I. (1973). Embryological studies in the genus Sparganium. *Svensk Bot. Tidskr.* **67**, 177–200.

Ayensu, E. S. (1972). Dioscoreales. In "Anatomy of the Monocotyledons", Vol. VI (C. R. Metcalfe, ed.). Oxford Univ Press, Oxford.

Ayensu, E. S. (1973). Biological and morphological aspects of the Velloziaceae. *Biotropica* **5**, 135–149.

Ayensu, E. S. (1974). Leaf anatomy and systematics of the New World Velloziaceae. *Smithsonian Contr. Bot.* **15**.

Ayensu, E. S. and Skvarla, J. J. (1974). Fine structure of Velloziaceae pollen. *Bull. Torrey Bot. Club* **101**, 250–266.

Baillon, J. (1894). "Familles des Plantes", Vol. 12. Paris.

Baillon, H. (1895). "Familles des Plantes", Vol. 13. Paris.

Bambacioni, V. (1928). Ricerche sulla ecologia e sulla embriologia di Fritillaria persica L. *Ann. di Bot.* **18**, 7–37.

Barthlott, W. (1976). Struktur und Funktion des Velamen Radicum der Orchideen. *Proc. 8th World Orchid Conf.*, 438–443.

Bate-Smith, E. C. (1962). The phenolic constituents of plants and their taxonomic significance. I. Dicotyledons. *J. Linn. Soc. (Bot.)* **58**, 34–54.

Bate-Smith, E. C. (1968). The phenolic constituents of plants and their taxonomic significance. II. Monocotyledons. *J. Linn. Soc. (Bot.)* **60**, 325–356.

Bate-Smith, E. C. (1973). Systematic distribution of ellagitannins in relation to the phylogeny and classification of the angiosperms. Nobel (25), Chemistry in Botanical Classification, 93–102.

Battaglia, E. (1955). A consideration of a new type of meiosis (mis-meiosis) in Juncaceae (Luzula) and Hemiptera. *Bull. Torrey Bot. Club* **82**, 383–396.

Battaglia, E. and Boyes, E. J. (1955). Postreductional meiosis: its mechanism and causes. *Caryologia* **8**, 87–134.

Batygina, T. B. (1969). On the possibility of separation of a new type of embryogenesis in angiosperms. *Revue Cyt. Biol. Veget.* **32**, 335–341.

Begum, M. (1968). Embryological studies in Eriocaulon quinquangulare. *Linn. Proc. Indian Acad. Sci., Sect. B*, **67**, 148–156.

Behnke, H.-D. (1968). Zum Feinbau der Siebröhren-Plastiden bei Monocotylen. *Naturwissenschaften* **55**, 140–141.

Behnke, H.-D. (1969). Die Siebröhren-Plastiden der Monocotyledonen. Vergleichende Untersuchungen über Feinbau und Verbreitung eines characteristischen Plastidentyps. *Planta (Berl.)* **84**, 174–184.

Behnke, H.-D. (1971). Zum Feinbau der Siebröhren-Plastiden von Aristolochia und Asarum (Aristolochiaceae). *Planta* **97**, 62–69.

Behnke, H.-D. (1972). Sieve-tube plastids in relation to angiosperm systematics. An attempt towards a classification by ultrastructural analysis. *Bot. Review* **38**, 155–197.

Behnke, H.-D. (1975). P-type sieve-element plastids: a correlative ultrastructural and ultrahistochemical study on the diversity and uniformity of a new reliable character in seed plant systematics. *Protoplasma* **83**, 91–101.

Behnke, H.-D. and Dahlgren, R. (1976). The distribution of characters within an angiosperm system, 2. Sieve element plastids. *Bot. Notiser* **129**, 287–295.

Bendz, G. and Santesson, J. (eds) (1974). "Chemistry in Botanical Classification." Nobel Symposium No. 25. Academic Press, New York and London.

Bentham, G. and Hooker, J. D. (1883). "Genera Plantarum", Vol. 3, Part 2. London.

Berg, R. Y. (1959). Seed dispersal, morphology and taxonomic position of Scoliopus, Liliaceae. *Skr. Norske Vidensk.-Akad. Oslo. I. Mat.-Naturvidensk. Kl. 1959*, No. 4.

Berg, R. Y. (1962a). Morphology and taxonomic position of Medeola, Liliaceae. *Skr. Norske Vidensk.-Akad. Oslo. I. Mat.-Naturvidensk. Kl. Ny ser.* **3**.

Berg, R. Y. (1962b). Contribution to the comparative embryology of the Liliaceae: Scoliopus, Trillium, Paris and Medeola. *Skr. Norske Vidensk.-Akad. Oslo. I. Mat.-Naturvidensk. Kl. Ny Ser.* **4**.

Berg, R. Y. (1978). Development of ovule, embryo sac and endosperm in Brodiaea (Liliales). *Norwegian J. Bot.* **25**, 1–7.

Berg, R. Y. and Maze, J. R. (1966). Contribution to the embryology of Muilla, with a remark on the taxonomic position of the genus. *Madroño* **18**, 143–151.

Bernard, Ch. and Ernst, A. (1909). Embryologie von Thismia javanica. *J.J.S. Ann. Jard. Bot. Buitenzorg* **23**, (*2 Ser.* **8**), 48–61.

Bernard, N. (1909). L'évolution dans la symbiose. Les orchidées et leur chamignons commensaux. *Ann. Sci. Nat., 9 Sér., Bot.* **9**, 1–196.

Bessey, C. E. (1915). The phylogenetic taxonomy of flowering plants. *Ann. Missouri Bot. Gard.* **2**, 109–164.

Bhandari, N. N. (1971). Embryology of the Magnoliales and comments on their relationships. *J. Arnold Arbor.* **53**, 1–39.

Bhojwani, S. S. and Bhatnagar, S. P. (1974). "The Embryology of Angiosperms". Vikas Publishing House, Delhi.

Billings, F. H. (1904). A study of Tillandsia usneoides. *Bot. Gazette* **38**, 99–121.

Björnstad, I. (1970). Comparative embryology of Asparagoideae-Polygonatae, Liliaceae. *Nytt Mag. Bot.* **17**, 169–207.

Boehm, K. (1931). Embryologische Untersuchungen an Zingiberacéen. Inaugural Dissertation, Berlin.

Boudet, A. M. and Lecusson, R. (1974). Généralité de l'association (5-déshydroquinate hydro-lyase, shikimate: NADP* oxydoréductase) chez les végétaux supérieurs. *Planta (Berl.)* **119**, 71–79.

Boudet, A. M., Lecusson, R. and Boudet, A. (1975). Mise en évidence et propriétés de deux formes de la 5-déshydroquinate hydro-lyase chez les végétaux supérieurs. *Planta (Berl.)* **124**, 67–75.

Boudet, A. M., Boudet, A. and Bouyssou, H. (1977). Taxonomic distribution of isoenzymes of dehydroquinate hydrolyase in the angiosperms. *Phytochemistry* **16**, 912–922.

Boyd, L. (1932). Monocotyledonous seedlings. *Trans. Proc. Bot. Soc. Edinburgh* **31**, 1–224.

Bremer, K. and Wanntorp, H.-E. (1978). Phylogenetic systematics in botany. *Taxon* **27**, 317–329.

Bresinsky, A. (1963). Bau, Entwicklungsgeschichte und Inhaltsstoffe der Elaiosomen. *Bibliotheca Bot.* **126**, 1–54.

Brewbaker, J. L. (1967). The distribution and phylogenetic significance of binucleate and trinucleate pollen grains in the angiosperms. *Am. J. Bot.* **54**, 1069–1083.

Brongniart, A. and Gris, A. (1861). Note sur le genre Joinvillea de Gaudichaud et sur la famille des Flagellariées. *Bull. Soc. Bot. France* **8**, 264–269.

Buchner, L. (1948). Vergleichende embryologische Studien an Scilloideae. *Österreich. Bot. Zeitschr.* **95**, 428–450.

Burger, W. C. (1977). The Piperales and the monocots. Alternative hypotheses for the origin of monocotyledonous flowers. *Bot. Review* **43**, 346–393.

Burkill, I. H. (1960). The organography and the evolution of Dioscoreaceae, the family of the Yams. *J. Linn. Soc. (Bot.)* **56**, 319–412.

Burtt, B. L. and Smith, R. M. (1972). Key species in the taxonomic history of Zingiberaceae. *Notes Roy. Bot. Gard. Edinburgh* **31**, 177–227.

Buxbaum, F. (1954). Morphologie der Blüte und Frucht von Alstroemeria und der Anschluss der Alstroemerioidéen bei

den echten Liliacéen. *Österreich. Bot. Zeitschr.* **101**, 337–352.

Cabrera, A. L. (1968). "Flora de la Provincia de Buenos Aires". Parte 1. Buenos Aires.

Calderon, C. E. and Soderstrom, T. R. (1973). Morphological and anatomical considerations of the grass subfamily Bambusoideae based on the new genus Maclurolyra. *Smithsonian Contrib. Bot.* **11**.

Calestani, C. (1933). Le origini e la classificazione delle Angiosperme. *Archo. Bot. Sist. Fito-Geogr. Genet.* **9**, 274–311.

Campbell, D. H. (1897). A morphological study of Naias and Zannichellia. *Proc. Calif. Acad. Sci. 3, Bot.* **1**, 1–71.

Campbell, D. H. (1898). The development of the flower and embryo on Lilaea subulata, H. B. K. *Ann. Bot.* **12** (45), 1–28.

Candolle, A. and C. de (1881). "Monographiae phanerogamarum", Vol. 3. Paris.

Capoor, S. P. (1937). Contribution to the morphology of some Indian Liliaceae. II. The gametopytes of Urginea indica Kunth. *Beih. Bot. Centralbl.* **56**, 156–170.

Carlquist, S. (1960). Anatomy of the Guayana Xyridaceae: Abolboda, Orectanthe and Achlyphila. *Mem. New York Bot. Gard.* **10**, 65–117.

Carlquist, S. (1961). Pollen morphology of Rapateaceae. *Aliso* **1**, 39–66.

Carlquist, S. (1976). Alexgeorgea, a bizarre new genus of Restionaceae from Western Australia. *Austr. J. Bot.* **24**, 281–295.

Carniel, K. (1952). Das Verhalten der Kerne im Tapetum der Angiospermen mit besonderer Berücksichtigung von Endomitosen und sogenannten Endomitosen. *Österreich. Bot. Zeitschr.* **99**, 318–362.

Carniel, K. (1963). Das Antherentapetum. *Österreich. Bot. Zeitschr.* **110**, 145–176.

Carter, S. (1960a). Butomaceae. "Flora of Tropical East Africa" (E. Milne-Redhead and R. M. Polhill, eds). London.

Carter, S. (1960b). Alismataceae. "Flora of Tropical East Africa" (E. Milne-Redhead and R. M. Polhill, eds). London.

Carter, S. (1966). Tecophilaeaceae. "Flora of Tropical East Africa" (E. Milne-Redhead and R. M. Polhill, eds). London.

Caspary, R. (1858). Die Hydrillen (Anacharidéen Endl.). *Pringsheim Jahrb. f. Wissensch. Bot.* **1**, 377–513.

Castellanos, A. (1958). Las Pontederiaceae de Brasil. *Arqu. Jard. Bot. (Rio de Janeiro)* **16**, 147–236.

Cave, M. S. (1941). Megasporogenesis and embryo sac development in Calochortus. *Am. J. Bot.* **28**, 390–394.

Cave, M. S. (1943). Cytology and embryology in the delimitation of genera. *Chronica Bot. (Plant Genera)* **14**, 140–153.

Cave, M. S. (1948). Sporogenesis and embryo sac development of Hesperocallis and Leucocrinum in relation to their systematic position. *Am. J. Bot.* **35**, 343–349.

Cave, M. S. (1955). Sporogenesis and the female gametophyte of Phormium tenax. *Phytomorphology* **5**, 247–253.

Cave, M. S. (1975). Embryological studies in Stypandra (Liliaceae). *Phytomorphology* **25**, 95–99.

Chanda, S. (1966). On the pollen morphology of the Centrolepidaceae, Restionaceae and Flagellariaceae, with special reference to taxonomy. *Grana Palyn.* **6**, 355–415.

Chanda, S. and Ghosh, K. (1976). Pollen morphology and its evolutionary significance in Xanthorrhoeaceae. *In* "The Evolutionary Significance of the Exine" (I. K. Ferguson and J. Muller, eds). *Linn. Soc. Sympos. Ser.*, No. 1, pp. 527–559.

Chanda, S., Ghosh, K. and Nilsson, S. (1979). On the polarity and tetrad arrangements in some mono- and diaperturate angiosperm pollen grains. *Grana* **18**, 21–31.

Chant, S. R. (1978). Cyclanthaceae. "Flowering Plants of the World" (V. Heywood, ed.). Oxford University Press, Oxford.

Charlton, W. A. (1976). Flower and inflorescence in Lilaea, Triglochin and Zostera. *Helobiae Newsletter* **1**, 10.

Chaudefaud, M. and Emberger, L. (1960). See Emberger (1960).

Cheadle, V. I. (1937). Secondary growth by means of a thickening ring in certain monocotyledons. *Bot. Gazette* **98**, 535–555.

Cheadle, V. I. (1943a). The origin and certain trends of specialization of the vessel in the Monocotyledoneae. *Am. J. Bot.* **30**, 11–17.

Cheadle, V. I. (1943b). Vessel specialization in the late metaxylem of the various organs in the Monocotyledoneae. *Am. J. Bot.* **30**, 484–490.

Cheadle, V. I. (1969). Vessels in Amaryllidaceae and Tecophilaeaceae. *Phytomorphology* **19**, 8–16.

Cheadle, V. I. and Kosakai, H. (1971). Vessels in Liliaceae. *Phytomorphology* **21**, 320–333.

Cheah, C. H. and Stone, B. C. (1975). Embryo sac and microsporangium development in Pandanus (Pandanaceae). *Phytomorphology* **25**, 228–238.

Chenery, E. M. (1950). Aluminium in the plant world. Part II. Monocotyledons and gymnosperms. *Kew Bull.* 1949, 463–473.

Clarke, C. B. (1881). Commelinaceae. *In* "Monographiae Phanerogamarum", Vol. 3 (A. and C. de Candolle, eds), pp. 113–324.

Clifford, H. T. (1970). Monocotyledon classification with special reference to the origin of grasses (Poaceae). *In* "New Research in Plant Anatomy" (N. K. B. Robson, D. F. Cutler and M. Gregory, eds). Suppl. 1 to *Bot. J. Linn. Soc.*, pp. 25–34.

Clifford, H. T. (1977). Quantitative studies of inter-relationships amongst the Liliatae. *Plant Syst. Evol.*, Suppl. 1, 77–96.

Clifford, H. T. and Smith, W. K. (1969). Seed morphology and classification of Orchidaceae. *Phytomorphology* **19**, 133–139.

Clifford, H. T. and Stephenson, W. (1975). "An Introduction to Numerical Taxonomy". Academic Press, London and New York.

Clifford, H. T. and Watson, L. (1978). "Identifying Grasses. Data, Methods and Illustrations". University of Queensland Press, St. Lucia.

Cook, C. D. K., Gut, B. J., Rix, E. M., Schneller, J. and Seitz, M. (1978). "Water Plants of the World". W. Junk, The Hague.

Cook, M. T. (1906). The embryology of some Cuban Nymphaeaceae. *Bot. Gazette* **42**, 376–392.

Cooper, D. C. (1933). Nuclear divisions in the tapetal cells of certain angiosperms. *Am. J. Bot.* **20**, 358–364.

Cooper, D. C. (1936). Development of the male gametes of Lilium. *Bot. Gazette* **98**, 169–177.

Corner, E. J. H. (1976). "The Seeds of Dicotyledons", Vols 1 and 2. Cambridge University Press.

Correa, M. N. (1969). "Flora Patagonica", Part II. Buenos Aires.

Cronquist, A. (1968). "The Evolution and Classification of Flowering Plants". Nelson, London and Edinburgh.

Cronquist, A. (1978). The Zingiberidae, a new subclass of Liliopsida (Monocotyledons). *Brittonia* **30**, 505.

Cronquist, A. (1979). "How to Know the Seed Plants". The pictured Key Nature Series. Wm. C. Brown Co., Dubuque.

Cullen, J. (1978). A preliminary survey of ptyxis (vernation) in the angiosperms. *Notes Roy. Bot. Gard. Edinburgh* **37**, 161–214.

Cutler, D. F. (1969). "Anatomy of the Monocotyledons. IV. Juncales" (C. R. Metcalfe, ed.). Clarendon Press, Oxford.

Cutler, D. F. and Airy Shaw, H. K. (1965). Anarthriaceae and Ecdeiocoleaceae: two new monocotyledonous families, separated from the Restionaceae. *Kew Bull.* **19**, 489–499.

Czaja, A. T. (1978). Structure of starch grains and the classification of vascular plant families. *Taxon* **27**, 463–470.

Dahlgren, K. V. O. (1927). Die Morphologie des Nuzellus mit besonderer Berücksichtigung der deckzellosen Typen. *Jahrb. Wissensch. Bot.* **67**, 347–426.

Dahlgren, R. (in cooperation with B. Hansen, K. Jakobsen and K. Larsen) (1974). "Angiospermernes Taxonomi". Bind 1. Akademisk Forlag, København.

Dahlgren, R. (1975a). A system of classification of the angiosperms to be used to demonstrate the distribution of characters. *Bot. Notiser* **128**, 119–147.

Dahlgren, R. (1975b). Current topics. The distribution of characters within an angiosperm system. 1. Some embryological characters. *Bot. Notiser* **128**, 181–197.

Dahlgren, R. (in cooperation with B. Hansen, K. Jakobsen and K. Larsen) (1976). "Angiospermernes Taxonomi". Bind 4. Akademisk Forlag, København.

Dahlgren, R. (1977). A commentary on a diagrammatic presentation of the angiosperms in relation to the distribution of character states. *Plant Syst. Evol.*, Suppl. 1, 253–283.

Dahlgren, R. (in cooperation with B. Hansen, K. Jakobsen, S. R. Jensen, K. Larsen and B. J. Nielsen) (1980a). "Angiospermernes Taxonomi". Bind 1 (Ed. 2). Akademisk Forlag, København.

Dahlgren, R. (1980b). The taxonomic significance of chlorophyllous embryos in angiosperm seeds. *Bot. Notiser* **133**, 337–341.

Daumann, E. (1970). Das Blütennektarium der Monocotyledonen unter besonderer Berücksichtigung seiner systematischen und phylogenetischen Bedeutung. *Feddes Repert.* **80**, 463–590.

Davis, G. L. (1966). "Systematic Embryology of the Angiosperms". J. Wiley, New York.

Davis, J. S. and Tomlinson, P. B. (1974). A new species of Ruppia in high salinity in Western Australia. *J. Arnold Arbor.* **55**, 59–66.

Decrock, P. (1911). Recherches morphologiques et anatomiques sur la graine des Ravenala. *Ann. Mus. Colon. Marseille* (1911), 23–50.

Degener, O. (1936). Liliaceae: Sansevieria guineensis. "New Illustrated Flora of the Hawaiian Islands".

Dellert, R. (1933). Zur systematischen Stellung von Wachendorfia. *Österreich. Bot. Zeitschr.* **82**, 335–345.

Den Hartog, C. (1970). The sea grasses of the world. *Verh. Konikl. Nederl. Akad. Wetensch., Afd. Naturkunde, Tweede Reeks* **59**, (1).

De Vos, M. (1956). Studies in the embryology and relationships of South African genera of the Haemodoraceae: Dilatris Berg. and Wachendorfia Burm. *S. Afr. J. Bot.* **22**, 41–63.

De Vos, M. (1963). Studies on the embryology and relationships of South African genera of the Haemodoraceae: Lanaria Ait. *S. Afr. J. Bot.* **29**, 79–90.

Deyl, M. (1955). The evolution of plants and the taxonomy of monocotyledons. *Acta Mus. Nat. Pragae*, XI B, No. 6, Bot. No. 3, 1–143.

Diels, L. (1942). Die Stämme des Pflanzenreichs. *Handb. Biol.* **4**, 268–426.

Di Fulvio, T. E. and Cave, M. S. (1964). Embryology of Blandfordia nobilis Smith (Liliaceae) with special reference to its taxonomic position. *Phytomorphology* **14**, 487–499.

Domin, K. (1911). Morphologische und phylogenetische Studien über die Stipularbildungen. *Ann. Jard. Bot. Buitenzorg Ser. 2* **9**, 117–326.

Drenth, E. (1972). A revision of the family Taccaceae. *Blumea* **20**, 367–406.

Dressler, R. L. (1974). Classification of the orchid family. *In* "Proceedings of the 7th World Orchid Conference" (M. Ospina, ed.), pp. 257–278. Medellin, Colombia.

Dressler, R. L. and Dodson, C. H. (1960). Classification and phylogeny in the Orchidaceae. *Ann. Missouri Bot. Gard.* **47**, 25–68.

Dutt, B. S. M. (1970). Comparative embryology of Angiosperms: Haemodoraceae, Cyanastraceae, Amaryllidaceae, Hypoxidaceae, Velloziaceae. *Bull. Ind. Nat. Sci. Acad.* **41**, 358–374.

Dyer, R. A. (1976). Gymnosperms and monocotyledons. *In* "The Genera of South African Plants", Vol. 2. Pretoria Dept. Agric. Techn. Services.

Eastop, V. (1979). Sternorrhyncha as angiosperm taxonomists. *Symbolae Bot. Upsal.* **22**, 120–134.

Eckardt, T. (1964). Die natürliche Verwandtschaft bei den Blütenpflanzen. *Umschau* **16**.

Edgar, E. (1966). The male flowers of Hydatella inconspicua (Cheesem.) Cheesem. (Centrolepidaceae). *N. Zeal. J. Bot.* **4**, 153–158.

Ehrendorfer, F. (1978). "Strasburger's Lehrbuch der Botanik", 31 Aufl. (D. von Denffer, F. Ehrendorfer, K. Mägdefrau and H. Ziegler, eds). G. Fischer, Jena.

El-Gazzar, A. and Hamza, M. K. (1975). On the monocots-dicots distinction. *Publ. Cairo Univ. Herb.* **6**, 15–28.

Emberger, L. (1960). *In* "Traité de Botanique", Vol. II, Les végétaux vasculaires (M. Chaudefaud and L. Emberger, eds). Masson et Cie, Paris.

Engler, A. (1884). Beiträge zur Kenntniss der Araceae, V. *Engler's Bot. Jahrb.* **5**, 141–336.

Engler, A. (1892). Syllabus der Vorlesungen über spezielle und medizinishpharmaceutische Botanik. Bornträger, Berlin.

Engler, A. (1919). Araceae. *In* "Das Pflanzenreich" (A. Engler, ed.), Vol. 4, 23 A.

Engler, A. (1930). Cyanastraceae. *In* "Die Naturlichen Pflanzenfamilien" (A. Engler and K. Prantl, eds), 2 Aufl., Vol. 15 a, pp. 188–190.

Erdtman, G. (1952). "Pollen Morphology and Plant Taxonomy. Angiosperms". Almqvist and Wiksell, Stockholm.

Esau, K. (1953). "Plant Anatomy". J. Wiley, New York.

Esau, K. (1965). "Plant Anatomy". Ed. 2. J. Wiley, New York.

Esau, K. (1969). "Pflanzenanatomie". G. Fischer, Stuttgart.

Fagerlind, F. (1939). Kritische und revidierende Untersuchungen über das Vorkommen des Adoxa ("Lilium")—Typs. *Acta Hort. Berg.* **13** (1), 1–49.

Fagerlind, F. (1940). Stempelbau und Embryosackentwicklung bei einigen Pandanazeen. *Ann. Jard. Bot. Buitenzorg* **49**, 55–83.

Fahn, A. (1967). "Plant Anatomy". Pergamon Press, Oxford.

Faulks, P. J. (1964). "The Systematization of the Angiosperms". Moredale, London.

Friedrich, W. L. and Koch, B. E. (1972). Der Arillus der tertiären Zingiberacee Spirematospermum wetzleri. *Lethaia* **5**, 47–60.

Fries, Th. C. E. (1919). Der Samenbau bei Cyanastrum Oliv. *Svensk Bot. Tidskr.* **13**, 295–304.

Frohne, D. and Jensen, U. (1973). "Systematik des Pflanzenreichs unter besonderer Berücksichtigung chemischer Merkmale und pflanzlicher Drogen". G. Fischer, Stuttgart.

Garay, L. A. (1960). On the origin of the Orchidaceae. *Bot. Mus. Leafl. Harv.* **19**, 57–96.

Gardner, R. O. (1976). Binucleate pollen in Triglochin L. *N. Zeal J. Bot.* **14**, 115–116.

Garrigues, R. (1951). Sur les anomalies mitotiques du tapis des étamines. *Rev. Gén. Bot. Paris* **58**, 305.

Gibbs, R. D. (1974). "Chemotaxonomy of Flowering Plants". Vol. 1–4. Mc Gill-Queen's University Press, Montreal and London.

Gibson, R. J. H. (1905). The axillary scales of aquatic monocotyledons. *J. Linn. Soc. (Bot.)* **37**, 228–237.

Glück, H. (1901). Die Stipulargebilde der Monocotyledonen. *Verh. Naturhist.-Med. Vereins zu Heidelberg, N. F.*, **7**, 1–96.

Glück, H. (1919). "Blatt- und Blütenmorphologische Studien". Jena.

Goebel, K. (1923). Organographie der Pflanzen. 2 Aufl. 3. Spezielle Organographie der Samenflanzen. G. Fischer, Jena.

Goebel, K. and Süssenguth, K. (1924). Beiträge zur Kenntnis der südamerikanischen Burmanniacéen. *Flora* **117**, 55–90.

Goldblatt, P. (1971). Cytological and morphological studies in the Southern African Iridaceae. *S. Afr. J. Bot.* **37**, 317–460.

Goldblatt, P. (1977). The genus Moraea in the winter rainfall region of Southern Africa. *Ann. Missouri Bot. Gard.* **63**, 657–786.

Good, R. (1947). "The Geography of the Flowering Plants". Longmans, London.

Gornall, R. J., Bohm, B. A. and Dahlgren, R. (1979). The distribution of flavonoids in the angiosperms. *Bot. Notiser* **132**, 1–30.

Grassmann, P. (1884). Die Septaldrüsen. Ihre Verbreitung, Entstehung und Verrichtung. *Flora* **67**, 113–136.

Guignard, J.-L. (1963). Embryogénie des Aracées. Developpement de l'embryon chez le Pistia stratiotes L. *C. r. Acad. Sci.* **257**, 1139–1142.

Guillarmod, A. J. and Marais, W. (1972). A new species of Aponogeton (Aponogetonaceae). *Kew Bull.* **27**, 563–565.

Gulliver, G. (1864). On Onagraceae and Hydrocharitaceae as elucidating the value of raphides as natural characters. *J. Bot.* **2**, 67–70.

Gupta, S. C., Rajeswari, V. M. and Agarwal, S. (1975). Stomatal complex in Butomus umbellatus. *Phytomorphology* **25**, 305–309.

Guttenberg, H. von (1960). Grundzüge der Histogenese höherer Pflanzen. I. Die Angiospermen. *Handb. Pflanzenanatomie* **8**, (3).

Guttenberg, H. von (1968). Der Primäre Bau der Angiospermen-Wurzel. *Handb. Pflanzenanatomie* **8** (5), 134–137 and 318–333.

Haccius, B. and Philip, V. J. (1979). Embryo development in Cocos nucifera L.: a critical contribution to a general understanding of palm embryogenesis. *Plant Evol. Syst.* **132**, 91–106.

Hagerup, O. (1944). On fertilization, polyploidy and haploidy in Orchis maculatus L. sensu lato. *Dansk Bot. Arkiv* **11** (5), 1–26.

Hagerup, O. and Petersson, V. (1956). "Botanisk Atlas. Danmarks Daekfrøede planter". Munksgaard, København.

Haines, R. W. and Lye, K. A. (1975). Seedlings of Nymphaeaceae. *Bot. J. Linn. Soc.* **70**, 255–265.

Håkansson, A. (1954). Meiosis and pollen mitosis in X-rayed and untreated spikelets of Eleocharis palustris. *Hereditas* **40**, 325–345.

Hallier, H. (1903). Vorläufiger Entwurf des natürlichen (phylogenetischen) Systems der Blütenpflanzen. *Bull. l'Herb. Boissier*, 2 Ser., No. 4, 306–317.

Hallier, H. (1905a). Provisional scheme of the natural (phylogenetic) system of flowering plants. *New Phytol.* **4**, 151–162.

Hallier, H. (1905b). Ein zweiter Entwurf des natürlichen (phylogenetischen) Systems der Blütenpflanzen. *Ber. Deutsch. Bot. Ges.* **23**, 85–91.

Hallier, H. (1912). L'origine et le systeme phyletique des angiospermes. *Arch. Neerl. Sci. Exact. Nat., Ser. III B*, **1**, 146–234.

Hamann, U. (1961). Merkmalsbestand und Verwandtschaftsbeziehungen der "Farinosae". *Willdenowia* **2**, 639–768.

Hamann, U. (1962a). Beitrag zur Embryologie der Centrolepidaceae mit Bemerkungen über den Bau der Blüten und Blütenstände und die systematische Stellung der Familie. *Ber. Deutsch. Bot. Ges.* **75**, 153–171.

Hamann, U. (1962b). Weiteres über Merkmalsbestand und Verwandtschaftsbeziehungen der "Farinosae" *Willdenowia* **3**, 169–207.

Hamann, U. (1962c). Über Bau und Entwicklung des Endosperms der Philydraceae und über die Begriffe "mehliges Nährgewebe" und "Farinosae". *Bot. Jahrb.* **81**, 397–407.

Hamann, U. (1963). Neue Metoden der Dokumentation in der systematischen Botanik. *Ber. Deutsch. Bot. Ges.* **76**, Sondernr. 1, 80–91.

Hamann, U. (1966a). Embryologische, morphologisch-anatomische und systematische Untersuchungen an Philydracéen. *Willdenowia*, Beiheft 4.

Hamann, U. (1966b). Nochmals zur Embryologie von Philydrum lanuginosum. *Beitr. Biol. Pflanzen* **42**, 151–159.

Hamann, U. (1974). Embryologie und Systematik am Beispiel der Farinosae. *Ber. Deutsch. Bot. Ges.* **77**, 45–54.

Hamann, U. (1975). Neue Untersuchungen zur Embryologie und Systematik der Centrolepidaceae. *Bot. Jahrb.* **96**, 154–191.

Hamann, U. (1976). Hydatellaceae—a new family of Monocotyledoneae. *N. Zeal. J. Bot.* **14**, 193–196.

Hänsel, R., Leuschke, A. and Gomez-Pompa, A. (1975). Aporphine-type alkaloids from Piper auritum. *Lloydia* **38**, 529–530.

Harada, I. (1948). Karyotyp und Entwicklungsgeschichte des fadenförmigen Pollens von Gattung Zostera. *Jap. J. Genet.* **23**, 13–14.

Harborne, J. B. (1973). Flavonoids as systematic markers in the angiosperms. Nobel (25), Chemistry in Botanical Classification, 103–115.

Harborne, J. B. (1977). Flavonoid sulphates: a new class of natural products of ecological significance in plants. *Prog. Phytochem.* **4**, 189–208.

Harling, G. (1958). Monograph of the Cyclanthaceae. *Acta Hort. Berg.* **18**.

Harms, H. (1930). Bromeliaceae. *In* "Die natürliche Pflanzenfamilien" (A. Engler and K. Prantl, eds), 2 Aufl., 15 a, pp. 65–159.

Hegi, G. (1935). "Illustrierte Flora von Mitteleuropa." Band I (2 Aufl.). C. Hanser, München.

Hegnauer, R. (1963). "Chemotaxonomie der Pflanzen". Band 2. Birkhäuser, Basel.

Hegnauer, R. (1973). Die cyanogenen Verbindungen der Liliatae und Magnoliatae-Magnoliidae: zur systematischen Bedeutung der Cyanogenese. *Biochem. Syst.* **1**, 191–197.

Hegnauer, R. (1977). Cyanogenic compounds as systematic markers in Tracheophyta. *Plant Syst. Evol.*, Suppl. 1, 191–209.

Hennig, W. (1950). "Grundzüge einer Theorie der phylogenetischen Systematik". Deutscher Zentralverlag, Berlin.

Hennig, W. (1965). Phylogenetic systematics. *Ann. Rev. Entomology* **10**, 97–116.

Hennig, W. (1966). "Phylogenetic Systematics". University of Illinois Press, Urbana.

Heslop-Harrison, Y. and Shivanna, K. R. (1977). The receptive surface of the angiosperm stigma. *Ann. Bot.* **41**, 1233–1258.

Hickey, L. J. and Doyle, J. A. (1977). Early Cretaceous fossil evidence for angiosperm evolution. *Bot. Review* **43**, 3–104.

Holm, L. (1966). Études urédinologiques, 4. Sur les Puccinia caricicoles et leurs alliés. *Svensk Bot. Tidskr.* **60**, 23–32.

Holm, T. (1891). Contributions to the knowledge of the germination of some North American plants. *Mem. Torrey Bot. Club* **11** (3), 57–108.

Holttum, R. E. (1948). The spikelet in Cyperaceae. *Bot. Review* **14**, 525–541.

Hong, D-Y. (1974). Revisio commelinacearum sinicarum. *Acta Phytotax. Siniaca* **12**, 459–483.

Horn af Rantzien, H. (1946). Notes on Mayacaceae of the Regnellian Herbarium in the Riksmuseum, Stockholm. *Svensk Bot. Tidskr.* **40**, 405–424.

Hubbard, C. E. (1954). "Grasses". Penguin, Harmondsworth.

Huber, H. (1969). Die Samenmerkmale und Verwandtschaftsverhältnisse der Liliiflorae. *Mitteil. Bot. Staatssamml. München* **8**, 219–538.

Huber, H. (1977). The treatment of the monocotyledons in an evolutionary system of classification. *Plant Syst. Evol.*, Suppl. 1, 285–298.

Humphrey, J. E. (1896). The development of the seed in the Scitamineae. *Ann. Bot.* **10**, 1–39.

Hutchinson, J. (1934). "The Families of Flowering Plants", Vol. II, Monocotyledons. Macmillan, London.

Hutchinson, J. (1959). "The Families of Flowering Plants", Vol. II (Ed. 2), Monocotyledons. Clarendon Press, Oxford.

Hutchinson, J. (1973). "The Families of Flowering Plants" (Ed. 3). Clarendon Press, Oxford.

Hutchinson, J. and Dalziel, J. M. (1968). "Flora of West Tropical Africa" Vol. III (1), (Ed. 2). London.

Huynh, K.-L. (1974). La morphologie microscopique de la feuille et la taxonomie du genre Pandanus. *Bot. Jahrb. Syst.* **94**, 190–256.

Inamdar, J. A. and Aleykutty, K. M. (1979). Studies on Cabomba aquatica (Cabombaceae). *Plant Syst. Evol.* **132**, 161–166.

Irmisch, T. (1858). Über das Vorkommen von schuppen- oder haarförmigen Gebilden innerhalb der Blattscheiden bei monocotylischen Gewächsen. *Bot. Zeit.* **16**, 177–179.

Johansen, D. A. (1950). "Plant Embryology. Embryology of the

Spermatophyta". Chronica Bot. Co, Waltham, Massachusetts.

Johnson, D. S. (1900). On the development of Saururus cernuus L. *Bull Torrey Bot. Club* **27**, 365–372.

Johnson, D. S. (1902). On the development of certain Piperaceae. *Bot. Gazette* **34**, 321–340.

Johnson, A. M. (1932). "Taxonomy of the Flowering Plants". Century Press, New York.

Johri, B. M. (1935a). Studies in the family Alismaceae. I. Limnophyton obtusifolium Miq. *J. Ind. Bot. Soc.* **14**, 49–66.

Johri, B. M. (1935b). Studies in the family Alismaceae. II. Sagittaria sagittifolia L. *Proc. Indian Acad. Sci* **1**, (7) 340–348.

Johri, B. M. (1935c). Studies in the family Alismaceae. III. Sagittaria guayanensis H. B. K. and S. latifolia Willd. *Proc. Indian Acad. Sci.* **2**, (1), Sect. B, 33–48.

Johri, B. M. (1938). The embryo sac of Hydrocleys nymphoides Buchen. *Beih. Bot. Centralbl.* **58**, 165–172.

Jones, S. B. and Luchsinger, A. E. (1979). "Plant Systematics". McGraw-Hill.

Kandeler, R. (1979). *In* "Illustrated Flora von Mitteleuropa" (G. Hegi, ed.), Band II (1) (3 Aufl.), p. 337.

Kaul, R. B. (1968), Floral morphology and phylogeny in the Hydrocharitaceae. *Phytomorphology* **18**, 13–35.

Kimura, Y. (1956). Système et phylogénie des monocotyledones. *Notulae Syst.* **15**, 137–159.

Knox, R. B. and Ducker, S. C. (1976). Submarine pollination in seagrasses. *Helobiae Newsletter* **1**, 25–26.

Knuth, R. (1930). Dioscoreaceae. *In* "Die natürliche Pflanzenfamilien" (A. Engler and K. Prantl, eds) 2 Aufl., 15 a, pp. 438–462.

Koch, B. E. and Friedrich, W. L. (1971). Früchte und Samen von Spirematospermum aus der miozänen Fasterholt-Flora in Dänemark. *Paleontographia* **136**, Abt. B, 1–46.

Koechlin, J. (1964). Scitaminales: Musacées, Strelitziacées, Zingiberacées, Cannacées, Marantacées. *In* "Flore du Gabon" (A. Aubreville, ed.). Paris.

Kosakai, H., Moseley, M. F. and Cheadle, V. I. (1970). Morphological studies of the Nymphaeaceae, V. Does Nelumbo have vessels? *Am. J. Bot.* **57**, 487–494.

Koyama, T. (1963). Studies in the Flora of Thailand: Smilacaceae. *Dansk Bot. Arkiv* **23**, (1), 9–18.

Kracauer, P. (1930). Die Haploidgeneration von Canna indica L. Inagural-Dissert., Friedrich Wilhelm University, Berlin.

Krajnčič, B. and Devidé, Z. (1979). Flower development in Spirodela polyrrhiza (Lemnaceae). *Plant Syst. Evol.* **132**, 305–312.

Kränzlin, F. (1912). Cannaceae. *In* "Das Pflanzenreich" (A. Engler, ed.), Vol. 4.

Krause, K. (1930). Liliaceae. *In* "Die natürliche Pflanzenfamilien" (A. Engler and K. Prantl, eds), 2. Aufl., 15 a, pp. 227–386.

Kress, W. J., Stone, D. E. and Sellers, S. C. (1978). Ultrastructure of exine-less pollen: Heliconia (Heliconiaceae). *Am. J. Bot.* **65**, 1064–1076.

Kudryashov, L. V. (1964). The origin of monocotyledony (as illustrated by the example of Helobiae). *Bot. Zhurnal SSSR* **49**, 473–486.

Kuhn, E. (1908). Über den Wechsel der Zelltypen im Endothecium der Angiospermen. Inagural-Dissert., Universität Zürich.

Lakshmanan, K. K. (1970). Comparative embryology of angiosperms: Hydrocharitaceae, Scheuchzeriaceae, Juncaginaceae, Potamogetonaceae, Zannichelliaceae, Najadaceae. *Bull. Indian Nat. Sci. Acad.* **41**, 336–357.

Larsen, K. (1961). Studies in the Flora of Thailand: Liliaceae, Triuridaceae, Trilliaceae, Iridaceae, Polygonaceae. *Dansk Bot. Arkiv* **20** (1), 37–54.

Larsen, K. (1977). "Kormofyternes Taxonomi" (2 udg.). Akademisk Forlag, København.

Lavarack, P. S. (1971). The taxonomic affinities of the Neottioideae, Vols 1 and 2. Doctoral thesis, University of Queensland.

Lawrence, G. H. M. (1951). "Taxonomy of Vascular Plants". Macmillan, New York.

Lazarte, J. E. and Palser, B. F. (1979). Morphology, vascular anatomy and embryology of pistillate and staminate flowers of Asparagus officinalis. *Am. J. Bot.* **66**, 753–764.

Leavitt, R. G. (1904). Trichomes of the root in vascular cryptogams and angiosperms. *Proc. Boston Nat. Hist. Soc.* **31**, 273–313.

Lee, D. W., Yap Kim Pin and Liew Foo Yew (1975). Serological evidence on the distinctness of the monocotyledonous families Flagellariaceae, Hanguanaceae and Joinvilleaceae. *Bot. J. Linn. Soc.* **70**, 77–81.

Limpricht, W. (1928). Taccaceae. *In* "Das Pflanzenreich" (A. Engler, ed.), Vol. 4 (42).

Lindley, J. (1853). "The Vegetable Kingdom" (Ed. 3.) Bradbury and Evans, London.

Link, H. F. (1843). "Anatomia Plantarum Iconibus Illustrata". Berlin.

Loesener, T. (1930). Zingiberaceae. *In* "Die natürliche Pflanzenfamilien" (A. Engler and K. Prantl, eds), 2. Aufl., 15 a.

Lotsy, J. P. (1911). "Vorträge über botanische Stammesgeschichte. Cormophyta Siphonogamia", Band 3. G. Fischer, Jena.

Lourteig, A. (1965). Maiacáceas. *In* "Flora Illustrata Catarinense". Itajai, Santa Catarina, Brasil.

Lowe, J. (1961). The phylogeny of monocotyledons. *New Phytologist* **60**, 355–387.

Maguire, B. and Wurdack, J. J. (1958). The botany of the Guayana Highland. Part III. *Mem. N. York Bot. Gard.* **10** (1), 1–156.

Maguire, B. *et al.* (1965). The botany of the Guayana Highland. Part IV. *Mem. N. York Bot. Gard.* **12** (3), 1–285.

Mahabale, T. S. and Biradar, N. V. (1968). Studies on palms: embryology of Phoenix sylvestris Roxb. *Proc. Indian Acad. Sci.* **67** B, 77–96.

Mahalingappa, M. S. (1977). Gametophytes of Eleusine compressa. *Phytomorphology* **27**, 231–239.

Maheshwari, P. (1946). The Adoxa Type of embryo sac: a critical review. *Lloydia* **9**, 73–113.

Maheshwari, P. (1950). "An Introduction to the Embryology of Angiosperms". McGraw-Hill, New York.

Maheshwari, S. C. (1955). The occurrence of bisporic embryo sacs in angiosperms—a critical review. *Phytomorphology* **5**, 67–99.

Maheshwari, S. C. and Kapil, R. N. (1963). Morphological and embryological studies on the Lemnaceae. I. The floral structure and gametophytes of Lemna paucicostata. *Am. J. Bot.* **50**, 577–686.

Maheshwari, S. C. and Khanna, P. P. (1956). The embryology of Arisaema wallichianum Hook. fil. and the systematic position of the Araceae. *Phytomorphology* **6**, 379–388.

Maheshwari, S. C. and Maheshwari, N. (1963). The female

gametophyte, endosperm and embryo of Spirodela polyrrhiza. *Beitr. Biol. Pflanz.* **39**, 179–188.

Malme, G. O. A. (1930). Xyridaceae. *In* "Die natürliche Pflanzenfamilien" (A. Engler and K. Prantl, eds), 2. Aufl., 15 a.

Mansfeld, R. (1937). Über das System der Orchidaceae-
Monandrae. *Notizbl. Bot. Gart. Berlin-Dahlem* **13**, 666–676.

Martin, A. C. (1946). The comparative internal morphology
of seeds. *Am. Midland Nat.* **36**, 513–660.

McClure, F. A. (1973). Genera of bamboos native to the New
World (Gramineae, Bambusoideae). *Smithsonian Contr. Bot.*
9, 1–148.

McKelvey, S. D. and Sax, K. (1933). Taxonomic and cytological
relationships of Yucca and Agave. *J. Arnold Arbor.* **14**, 76–81.

Meeuse, A. D. J. (1971). Interpretative gynoecial morphology
of the Lactoridaceae and Winteraceae—a re-assessment. *Acta
Bot. Neerl.* **20**, 221–238.

Meeuse, A. D. J. (1972). Taxonomic affinities between Piperales
and Polycarpicae and their implications in interpretative
floral morphology. *Adv. Plant Morph.* (1972), 3–27.

Meeuse, A. D. J. (1975). Aspects on the evolution of the monocotyledons. *Acta Bot. Neerl.* **24**, 421–436.

Melchior, H. (ed.) (1964). "A. Engler's Syllabus der Pflanzenfamilien", 12 Aufl. Band II. Angiospermen. Bornträger,
Berlin.

Menezes, N. L. de (1973). Natureza dos apendices petaloides
em Barbacenioideae (Velloziaceae). *Bol. Zool. Biol. Mar.*
(*San Paulo*), *N. S.* **30**, 713–755.

Mepham, R. H. and Lane, G. R. (1969). Formation and
development of the tapetal periplasmodium in Tradescantia
bracteata. *Protoplasma* **68**, 175–192.

Metcalfe, C. R. (1967). Distribution of latex in the Plant
Kingdom. *Econ. Bot.* **21**, 115–127.

Metcalfe, C. R. (1971). "Anatomy of the Monocotyledons".
Vol. V. Cyperaceae. Clarendon Press, Oxford.

Metcalfe, C. R. and Chalk, L. (1950). "Anatomy of the Dicotyledons", Vol. 1–2. Clarendon Press, Oxford.

Meyer, K. (1909). Untersuchungen über Thismia clandestina.
Bull. Nat. Moscou (1909), 1–18.

Milne-Redhead, E. (1975). Dioscoreaceae. *In* "Flora of
Tropical East Africa" (E. Milne-Redhead and R. M. Polhill,
eds). London.

Monteiro-Scanavacca, W. R. and Mazzoni, C. (1978). Embryological studies in Leiothrix fluitans (Mart.) Ruhl. (Eriocaulaceae). *Revista Brasil. Bot.* **1**, 59–64.

Moore, H. E. (1953). The genus Milla (Amaryllidaceae-
Allieae) and its allies. *Gentes Herb.* **8**, 262–294.

Moore, H. E. (1955). The cultivated Alliums, III. *Baileya* **3**,
137–167.

Moore, H. E. (1960). Tripogandra grandiflora (Commelinaceae). *Baileya* **8**, 77–83.

Moore, H. E. (1973). The major groups of palms and their
distributions. *Gentes Herb.* **11**, 27–141.

Morton, J. K. (1966). A revision of the genus Aneilema
R. Brown (Commelinaceae). *J. Linn. Soc.* (*Bot.*) **59**, 431–
478.

Müller-Doblies, D. (1968). Über die Verwandtschaft von Typha
und Sparganium im Inflorescenz- und Blütenbau. *Bot. Jahrb.
Syst.* **89**, 451–562.

Müller-Doblies, U. (1969). Über die Blütenstände und Blüten
sowie zur Embryologie von Sparganium. *Bot. Jahrb.* Syst.
89, 359–450.

Nagaraja Rao, A. (1953). Embryology of Dioscorea oppositi-

folia. *Phytomorphology* **3**, 121–126.

Nanda, K. and Gupta, S. C. (1977). Development of tapetal
periplasmodium in Rhoeo spathacea. *Phytomorphology* **27**,
308–314.

Napper, D. M. (1971). Flagellariaceae. *In* "Flora of Tropical
East Africa" (E. Milne-Redhead and R. M. Polhill, eds).
London.

Netolitzky, F. (1926). Anatomie der Angiospermen-Samen. *In*
"Handbuch der Pflanzenanatomie". Band 10. Bornträger,
Berlin.

Newell, T. K. (1969). A study of the genus Joinvillea (Flagellariaceae). *J. Arnold Arbor.* **50**, 527–555.

Nietsch, H. (1941). Zur systematischen Stellung von Cyanastrum. *Österreich. Bot. Zeitschr.* **90**, 31–52.

Nijalingappa, B. H. M. and Devaki, N. (1978). Embryological
studies in Diplachrum caricinum. R. Br, *Beitr. Biol. Pflanz.*
54, 215–225.

Nordborg, G. (ed.), Björkqvist, I., Dahlgren, R., Nilsson, Ö.,
Runemark, H., Snogerup, S. and Weimarck, G. (1971).
"Systematisk Botanik". Gleerups, Lund.

Nordenskiöld, H. (1951). Cytotaxonomical studies in the genus
Luzula, I. *Hereditas* **37**, 325–355.

Nordenskiöld, H. (1956). Cytotaxonomical studies in the genus
Luzula, II. *Hereditas* **42**, 7–73.

Nordenstam, B. (1964a). Studies in South African Liliaceae, I.
New species of Wurmbea. *Bot. Notiser* **117**, 173–182.

Nordenstam, B. (1964b). Studies in South African Liliaceae, II.
Two small species of Bulbine. *Bot. Notiser* **117**, 183–187.

Novák, F. A. (1954). Systém angiosperm. *Preslia* **26**, 337–364.

Pace, L. (1909). The gametophytes of Calopogon. *Bot. Gazette*
48, 126–137.

Panchaksharappa, M. G. (1962). Embryological studies in some
members of Zingiberaceae, I. Costus speciosus Smith. *Phytomorphology* **12**, 418–430.

Panchaksharappa, M. G. (1966). Embryological studies in some
members of Zingiberaceae, II. Elettaria cardamomum,
Hitchenia caulina and Zingiber macrostachyum. *Phytomorphology* **16**, 412–417.

Pankow, H. and Guttenberg, H. von (1957). Vergleichende
Studien über die Entwicklung monocotyler Embryonen und
Keimpflanzen. *Bot. Stud.* (*Jena*) **7**, 1–39.

Pfeiffer, A. (1891). Die Arillargebilde der Pflanzensamen. *Bot.
Jahrb. Syst.* **13**, 492–540.

Pfitzer, E. (1887). "Entwurf einer natürlichen Anordnung der
Orchideen". Heidelberg.

Pfitzer, E. (1888). Orchidaceae. *In* "Die natürlichen Pflanzenfamilien". (A. Engler and K. Prantl, eds), Vol. II (6), pp. 52–
224.

Philipson, W. R., Ward, J. M. and Butterfield, B. G. (1971).
"The Vascular Cambium. Its Development and Activity".
Chapman and Hall, London.

Philip, V. J. and Haccius, B. (1976). Embryogenesis in Bambusa
arundinacea Willd. and structure of the mature embryo.
Beitr. Biol. Pflanz. **52**, 83–100.

Pijl, L. van der (1972). "Principles of Dispersal in Higher
Plants" (Ed. 2). Springer, Berlin.

Pilger, R. (1930). Mayacaceae, Thurniaceae, Rapateaceae und
Philydraceae. *In* "Die natürlichen Pflanzenfamilien" (A.
Engler and K. Prantl, eds), 2 Aufl., 15 a, pp. 33–35,
58–59, 59–65, 190–191.

Pizarro, C. M. (1959). "Sinopsis de la Flora Chilena". Santiago.

Posluszny, U. and Tomlinson, P. B. (1977). Morphology and

development of floral shoots and organs in certain Zannichelliaceae. *Bot. J. Linn. Soc.* **75**, 21–46.

Prakash, N. (1969). The floral development and embryology of Centrolepis fascicularis. *Phytomorphology* **19**, 285–291.

Prakash, N. and Lee-Lee, A. (1973). Life history of a common Malaysian orchid, Spathoglottis plicata. *Phytomorphology* **23**, 9–17.

Punt, W. (1968). Pollen morphology of the American species of the subfamily Costoideae (Zingiberaceae). *Revue Paleobot. Palyn.* **7**, 31–43.

Punt, W. and Wessels Boer, J. G. (1966a). A palynological study in cocosoid palms. *Acta Bot. Neerl.* **15**, 255–265.

Punt, W. and Wessels Boer, J. G. (1966b). A palynological study in geonomoid palms. *Acta Bot. Neel.* **15**, 266–275.

Rădulescu, D. (1973). Recherches morpho-palynologiques sur la famille Liliaceae. *Acta Bot. Hort. Bucurest.* (1972–1973), 133–283.

Raju, M. V. S. (1957). Some aspects on the embryology of Dianella nemorosa Lamk. *J. Indian Bot. Soc.* **36**, 223–226.

Ramaswamy, S. N. (1975). Embryology and systematic position of Eriocaulaceae. Ph. D. thesis, University of Mysore.

Ramstad, E. (1953). Über das Vorkommen und die Verbreitung von Chelidonsäure in einigen Pflanzenfamilien. *Pharm. Act. Helv.* **28**, 45–57.

Rao, A. N. (1967). Flower and seed development in Arundina graminifolia. *Phytomorphology* **17**, 291–300.

Rao, P. P. M. and Kaur, A. (1979). Embryology and systematic position of Ophiopogon intermedius. *Proc. Indian Nat. Sci. Acad.* **45** B, 175–187.

Rao, V. (1959). Contributions to the embryology of Palmae. II. (Ceroxylineae. *J. Ind. Bot. Soc.* **38**, 48–75.

Rauh, W., Barthlott, W. and Ehler, N. (1975). Morphologie und Funktion der Testa staubförmiger Flugsamen. *Bot. Jahrb. Syst.* **96**, 353–374.

Raven, P. H. and Axelrod, D. I. (1974). Angiosperm biogeography and past continental movements. *Ann. Missouri Bot. Gard.* **61**, 539–673.

Rendle, A. B. (1930). "The Classification of Flowering Plants", Vol. I. Cambridge University Press, Cambridge.

Riley, C. V. (1892). The Yucca moth and Yucca pollination. *Missouri Bot. Gard.*, St. Louis.

Robbins, W. W. and Borthwick, H. A. (1952). Development of the seed of Asparagus officinalis. *Bot. Gazette* **80**, 426–438.

Romanov, I. D. (1959). The embryo sac and pollen morphology in Tulipa. *9th Congr. Internat. Bot.* **2**, 331–332.

Rosenberg, O. (1901). Über die Pollenbildung von Zostera. *Meddel. Stockholms Högsk. Bot. Inst.*

Ross Craig, S. (1971). "Drawings of British Plants", Part 18. G. Bell, London.

Ross Craig, S. (1972). "Drawings of British Plants", Part 19. G. Bell, London.

Rosso, S. W. (1966). The vegetative anatomy of the Cypripedioideae (Orchidaceae). *J. Linn Soc. (Bot.)* **59**, 309–341.

Ruhland, W. (1930). Eriocaulaceae. *In* "Die natürliche Pflanzenfamilien" (A. Engler and K. Prantl, eds), 2 Aufl. 15 a, pp. 39–57.

Sachs, T. (1978). Phyletic diversity in higher plants. *Plant Syst. Evol.* **130**, 1–11.

Sanio, C. (1865). Einige Bemerkungen in Betreff meiner über Gefässbündelbildung geäusserten Ansichten. *Bot. Zeitung* **23**, 165–172, 174–180, 191–193, 197–200.

Sargant, E. (1903). A theory of the origin of monocotyledons founded on the structure of their seedlings. *Ann. Bot.* **17**, 1–92.

Sattler, R. and Singh, V. (1978). Floral organogenesis of Echinodorus amazonicus Rataj and floral construction of the Alismatales. *Bot. J. Linn Soc.* **77**, 141–156.

Savile, D. B. O. (1962). Taxonomic disposition of Allium. *Nature* **196**, 792.

Savile, D. B. O. (1971). Coevolution of the rust fungi and their hosts. *Quart. Rev. Biol.* **46**, 211–218.

Savile, D. B. O. (1979). Fungi as aids in higher plant classification. *Bot. Review* **45**, 377–503.

Schachner, J. (1924). Beiträge zur Kenntniss der Blüten und Samenentwicklung der Scitamineen. *Flora N. F.* **17**.

Schaeppi, H. (1939). Vergleichend-morphologische Untersuchungen an den Staubblättern der Monocotyledonen. *Nova Acta Leopold., N. F.* **6**, 389–447.

Schill, R. (1978). Palynologische Untersuchungen zur systematischen Stellung der Apostasiaceae. *Bot. Jahrb. Syst.* **99**, 353–362.

Schill, R. and Pfeiffer, W. (1977). Untersuchungen an Orchideenpollinien unter besonderer Berücksichtigung ihrer Feinskulpturen. *Pollen et Spores* **19**, 5–118.

Schlee, D. (1971). "Die Rekonstruktion der Phylogenese mit Hennigs Prinzip". W. Kramer, Frankfurt am Main.

Schlittler, J. (1949). Die systematische Stellung der Gattung Petermannia F. v. Muell. und ihre phylogenetische Beziehungen zur Luzuriagoideae Engl. und den Dioscoreaceae Lindl. Vierteljahrschr. *Naturforsch. Ges. Zürich* **94**, Beih. Nr. 1, 1–28.

Schlittler, J. (1951). Die Gattungen Eustrephus R. Br. ex Sims und Geitonoplesium (R. Br.) A. Cunn. *Mitteil. Bot. Mus. Univ. Zürich* **189**, 175–239.

Schnarf, K. (1931). "Vergleichende Embryologie der Angiospermen". Bornträger, Berlin.

Schnarf, K. (1948). Der Umfang der Lilioideae in natürlichen System. *Österreich. Bot. Zeitschr.* **95**, 257–269.

Schnarf, K. and Wunderlich, R. (1939). Zur vergleichenden Embryologie der Liliaceae-Asphodeloideae. *Flora, N. F.* **33**, 297–327.

Schneider, E. L. and Ford, E. G. (1978). Morphological studies of the Nymphaeaceae. X. The seed of Ondinea purpurea Den Hartog. *Bull. Torrey Bot. Club* **105**, 192–200.

Schneider, E. L. and Moore, L. A. (1977). Morphological studies of the Nymphaeaceae. VII. The floral biology of Nuphar lutea subsp. macrophylla. *Brittonia* **29**, 88–99.

Schnepf, E. (1964). Zur Cytologie und Physiologie pflanzlicher Drüsen. IV. Licht- und elektronenmikroskopische Untersuchungen an Septalnektarien. *Protoplasma* **58**, 137–171.

Schrader, F. (1935). Notes on the mitotic behaviour of long chromosomes. *Cytologia* **6**.

Schultze-Motel, W. (1959). Entwicklungsgeschichtliche und vergleichend-morphologische Untersuchungen im Blütenbereich der Cyperaceae. *Bot. Jahrb. Syst.* **78**, 129–170.

Schultze-Motel, W. (1964). Reihe Cyperales. *In* "A. Engler's Syllabus der Pflanzenfamilien" (H. Melchior, ed.), 12 Aufl., Band II. Bornträger, Berlin.

Schultze-Motel, W. (1966). Ordnung Cyperales. *In* "Illustrierte Flora von Mitteleuropa" (G. Hegi, ed.), 2. Aufl., Band II (1). München.

Setchell, W. A. (1924). Ruppia and its environmental factors. *Proc. Nat. Acad. Sci.* **10**, 286–288.

Sharp, L. W. (1912). The orchid embryo. *Bot. Gazette* **54**, 373–385.

Singh, V. (1966). Morphological and anatomical studies in Helobiae. VI. Vascular anatomy of the flower of Butomus umbellatus Linn. *Proc. Indian Acad. Sci., Sect. B*, **63**, 313–320.

Singh, V. and Sattler, R. (1972). Floral development of Alisma triviale. *Canadian J. Bot.* **50**, 619–627.

Singh, V. and Sattler, R. (1974). Floral development of Butomus umbellatus. *Canadian J. Bot.* **52**, 223–230.

Singh, V. and Sattler, R. (1977). Development of the inflorescence and flower of Sagittaria cuneata. *Canadian J. Bot.* **55**, 1087–1105.

Skottsberg, C. (1940). "Växternas Liv. Populärvetenskaplig Handbok", Band 5. Nord. Familjeboks Förl, Stockholm.

Skvarla, J. J. and Rowley, J. R. (1970). The pollen wall of Canna and its similarity to the germinal apertures of other pollen. *Am. J. Bot.* **57**, 519–529.

Slater, J. A. (1976). Monocots and chinch bugs: a study of host plant relationships in the Lygaeid subfamily Blissinae (Hemiptera: Lygidae). *Biotropica* **8**, 143–165.

Slob, A., Jekel, B. and de Jong, B. (1975). On the occurrence of tuliposides in the Liliiflorae. *Phytochemistry* **14**, 1997–2005.

Smith, L. B. and Ayensu, E. S. (1976). A revision of American Velloziaceae. *Smithsonian Contr. Bot.* **30**.

Smithson, E. (1957). The comparative anatomy of Flagellariaceae. *Kew Bull.* **18**, 491–501.

Sneath, P. H. A. and Sokal, R. R. (1973). "Numerical Taxonomy". W. H. Freeman, San Francisco.

Snogerup, S. (1971). Växtgeografi. In "Systematisk Botanik" (G. Nordborg, ed.), pp. 237–255. Gleerups, Lund.

Solereder, H. and Meyer. F. J. (1928). "Systematische Anatomie der Monocotyledonen", Heft III, Principes, Synanthae, Spathiflorae. Bornträger, Berlin.

Solereder, H. and Meyer, F. J. (1930). "Systematische Anatomie der Monocotyledonen", Heft VI. Scitamineae, Microspermae. Bornträger, Berlin.

Solereder, H. and Meyer, F. J. (1933). "Systematische Anatomie der Monocotyledonen", Heft I, Pandanales, Helobiae, Triuridales. Bornträger, Berlin.

Soó, C. R. De (1953). Die modernen Grundsätze der Phylogenie im neuen System der Blütenpflanzen. *Act. Biol. Acad. Sci. Hungar.* **4**, 257–306.

Soó, C. R. de (1961). Present aspect of evolutionary history of Telomophyta. *Ann. Univ. Sci. Budapest, Biol* **4**, 167–178.

Soó, C. R. de (1965). "Phylogenetic Plant Taxonomy". (Ed. 3).

Soó, C. R. de (1975). A review of the new classification systems of flowering plants (Angiospermatophyta, Magnoliophyta). *Taxon* **24**, 585–592.

Sporne, K. R. (1956). The phylogenetic classification of the angiosperms. *Bot. Review* **31**, 1–29.

Sporne, K. R. (1974). "The Morphology of Angiosperms". Hutchinson, London.

Stant, M. Y. (1964). Anatomy of the Alismataceae. *J. Linn. Soc. (Bot.)* **59**, 1–42.

Stant, M. Y. (1970). Anatomy of Petrosavia stellaris Becc., a saprophytic monocotyledon. *In* "New Research in Plant Anatomy" (N. K. B. Robson, D. F. Cutler and M. Gregory, eds), Suppl. 1 to *Bot. J. Linn. Soc.*, pp. 147–161.

Staudermann, W. Von (1924). Die Haare der Monocotyledonen. *Bot. Archiv* **8**, 105–184.

Stebbins, G. L. (1974). "Flowering Plants. Evolution above the Species Level". E. Arnold, London.

Stenar, H. (1925). "Embryologische Studien". Akad. Diss., Uppsala.

Stenar, H. (1927). Uber die Entwicklung des siebenkernigen Embryosackes bei Gagea lutea Ker. *Svensk Bot. Tidskr.* **21**, 344–360.

Stenar, H. (1928a). Zur Embryologie der Asphodeline-Gruppe *Svensk Bot. Tidskr.* **22**, 145–159.

Stenar, H. (1928b). Zur Embryologie der Veratrum- und Anthericum-Gruppen. *Bot. Notiser* **81**, 357–378.

Stenar, H. (1949). Zur Kenntnis der Embryologie und der Raphiden-Zellen bei Bowiea volubilis Harvey und anderen Liliazéen. *Acta Hort. Berg.* **15** (3), 45–63.

Stenar, H. (1950). Studien über das Endosperm bei Galtonia candicans (Bak.) Decne und andere Scilloidéen. *Acta Hort. Berg.* **15** (8), 169–184.

Sterling, C. (1978). Comparative morphology of the carpel in the Liliaceae: Hewardieae, Petrosavieae and Tricyrtieae. *Bot. J. Linn. Soc.* **77**, 95–106.

Stern, F. C. (1956). "Snowdrops and Snowflakes. A study of the Genera Galanthus and Leucojum". London.

Stevens, P. F. (1978). Generic limits in the Xeroteae (Liliaceae sensu lato). *J. Arnold Arbor* **59**, 129–155.

Stirton, J. Z. and Harborne, J. B. (1980). Two distinctive anthocyanin patterns in the Commelinaceae. *Biochem. System. Ecol.* **8**, 285–288.

Stone, B. C. (1972). A reconsideration of the evolutionary status of the family Pandanaceae and its significance in monocotyledon phylogeny. *Quart. Rev. Biol.* **47**, 34–45.

Stone, B. C. (1974). Towards an improved infrageneric classification of Pandanus (Pandanaceae). *Bot. Jahrb. Syst.* **94**, 459–540.

Strandhede, S.-O. (1958). Eleocharis subser. Palustres i Skandinavien och Finland. *Bot. Notiser* **111**, 228–236.

Strandhede, S.-O. (1965). Chromosome studies in Eleocharis, subser. Palustres. III. Observations on Western European taxa. *Opera Bot.* **9** (2).

Subramanyan, K. and Narayana, H. S. (1972). Some aspects of the floral morphology and embryology of Flagellaria indica Linn. *In* "Advances in Plant Morphology" (Y. S. Murthy *et al.*, eds), pp. 211–217. Sarita Prakashan, Meerut (India).

Suessenguth, K. (1920). Beiträge zur Frage des systematischen Anschlusses der Monocotylen. *Beih. Bot. Zentralbl.* **38** (II), 1–79.

Swamy, B. G. L. (1942). Female gametophyte and embryogeny in Cymbidium bicolor Lindl. *Proc. Indian Acad. Sci., Sect. B*, **15**, 194–201.

Swamy, B. G. L. (1947). On the life history of Vanilla planifolia. *Bot. Gazette* **108**, 449–456.

Swamy, B. G. L. (1949a). Embryological studies in the Orchidaceae, I. Gametophytes. *Am. Midland Nat.* **41**, 185–201.

Swamy, B. G. L. (1949b). Embryological studies in the Orchidaceae. II. Embryogeny. *Am. Midland Nat.* **41**, 202–232.

Swamy, B. G. L. (1964). Observations on the floral morphology and embryology of Stemona tuberosa Lour. *Phytomorphology* **14**, 458–468.

Swamy, B. G. L. and Krishnamurthy, K. V. (1974). The

Helobial endosperm: a decennial review. *Phytomorphology* **23**, 74–79.

Swamy, B. G. L. and Parameswaran, N. (1963). The Helobial endosperm. *Biol. Review* **38**, 1–50.

Takhtajan, A. (1959). "Die Evolution der Angiospermen". G. Fischer, Jena.

Takhtajan, A. (1969). "Flowering Plants. Origin and Dispersal". Oliver and Boyd, Edinburgh.

Tanaka, N. (1939). Chromosome studies in Cyperaceae. III. The maturation divisions in Scirpus lacustris L., with special reference to heteromorphic pairing. *Cytologia* **9**, 533–556.

Thorne, R. F. (1968). Synopsis of a putatively phylogenetic classification of the flowering plants. *Aliso* **6**, 57–66.

Thorne, R. F. (1976). A phylogenetic classification of the Angiospermae. *Evol. Biol.* **9**, 35–106.

Thorne, R. F. (1977). Some realignments in the Angiospermae. *Plant Syst. Evol., Suppl. 1*, 299–320.

Tieghem, Ph. van (1887). Structure de la racine et disposition des radicelles dans Centrolepidées, Eriocaulées, Joncées, Mayacées et Xyridées. *J. Bot.* **1**, 305–315.

Tieghem, Ph. van (1891). "Traité de Botanique", Ed. 2. Paris.

Tiegham, Ph. van and Duliot, H. (1888). Recherches comparatives sur l'origine des membres endogénés dans les plantes vasculaires. *Ann. Sci. Nat., Ser. 7*, **8**, 1–666.

Tillich, H. J. (1977). Vergleichend-morphologische Untersuchungen zur Identität der Gramineen-Primärwurzel. *Flora* **166**, 415–421.

Tischler, G. (1915). Die Periplasmodien-Bildung in den Antheren der Commelinaceen und Ausblicke auf das Verhalten der Tapetumzellen bei den übrigen Monocotylen. *Jahrb. Wiss. Bot.* **55**, 53–90.

Tomita, K. (1931). Über die Entwicklung des nackten Embryos von Crinum latifolium L. *Sci. Rep. Tôhoku Imp. Univ., Ser. 4, Biol.* **4** (2), 163–169.

Tomlinson, P. B. (1961). "Anatomy of the Monocotyledons. II. Palmae" (C. R. Metcalfe, ed.). Clarendon Press, Oxford.

Tomlinson, P. B. (1969). "Anatomy of the Monocotyledons. III. Commelinales-Zingiberales" (C. R. Metcalfe, ed.). Clarendon Press, Oxford.

Tomlinson, P. B. (1970). Monocotyledons—towards an understanding of their morphology and anatomy. *In* "Advances of Botanical Research" (R. D. Preston, ed.). Academic Press, London and New York.

Tomlinson, P. B. (1974). Development of the stomatal complex as a taxonomic character in the monocotyledons. *Taxon* **23**, 109–128.

Tomlinson, P. B. (1976). Systematics and biology of the Helobiae. *Helobiae Newsletter* **1**, 6–7.

Tomlinson, P. B. and Zimmerman, M. H. (1969). Vascular anatomy of monocotyledons with secondary growth. An introduction. *J. Arnold Arbor* **50**, 159–179.

Traub, H. P. (1963). The genera of Amaryllidaceae. *Am. Pl. Life Soc.*, La Jolla, California.

Troll, W. (1954). Praktische Einführung in die Pflanzenmorphologie. 1. Der Vegetative Aufban. G. Fischer, Jena.

Uhl, N. W. and Moore, H. E. (1980). Androecial development in six polyandreus genera representing five major groups of palms. *Ann. Bot.* **45**, 57–7?.

Untawale, A. G. and Bhasin, R. K. (1973). On endothecial thickenings in some monocotyledonous familes. *Current Sci.* **42**, 398–400.

Valen, F. van (1978a). Contribution to the knowledge of cyanogenesis in Angiosperms. 3. Communication. Cyanogenesis in Liliaceae. *Proc. Koninkl. Nederl. Akad. Wetensch.* **81**, 132–140.

Valen, F. van (1978b). Contribution to the knowledge of cyanogenesis in angiosperms. 6. Communication. Cyanogenesis in some Magnoliidae. *Proc. Koninkl. Nederl. Akad. Wetensch.* **81**, 355–361.

Vermeulen, P. (1966). The system of the Orchidales. *Acta Bot. Neerl.* **15**, 224–253.

Veyret, Y. (1965). Embryogénie comparée et blastogénie chez les Orchidaceae-Monandrae. *Mem. Orstrom* **12**.

Veyret, Y. (1974). Development of the embryo and the young seedling stages of orchids. *In* "The Orchids—Scientific Studies" (C. L. Withner, ed.), pp. 223–263.

Vijayaraghavan, M. R. and Kumari, A. V. (1974). Embryology and systematic position of Zannichellia palustris L. *J. Ind. Bot. Soc.* **53**, 292–302.

Vogel, S. (1963). Duftdrüsen im Dienste der Bestäubung. Über Bau und Funktion der Osmophoren. *Abh. Akad. Wiss. Lit. Mainz, Mat.-Nat. Kl. H* **10**, Jahrg. 1962, 600–763.

Wagner, P. (1977). Vessel types of monocotyledons: a survey. *Bot. Notiser* **130**, 383–402.

Walker, J. W. (1974). Aperture evolution in the pollen of primitive angiosperms. *Am. J. Bot.* **61**, 1112–1137.

Walker, J. W. (1975). Comparative pollen morphology and phylogeny of the Ranalean complex. *In* "The Origin and Early Evolution of Angiosperms" (C. B. Beck, ed.). Columbia University Press, New York.

Walker, J. W. (1976). Evolutionary significance of the exine in the pollen of primitive angiosperms. *In* "The Evolutionary Significance of the Exine" (I. K. Ferguson and J. Muller, eds). *Linn. Soc. Sympos. Ser.*, No. 1, 251–308.

Warming, E. (1912). "Frøplanterne". Kjøbenhavn.

Weberling, F. (1970). Weitere Untersuchungen zur Morphologie des Unterblattes bei den Dicotylen. V. Piperales. *Beitr. Biol. Pflanz.* **46**, 403–434.

Weberling, F. and Schwantes, H. O. (1972). "Pflanzensystematik". E. Ulmer, Stuttgart.

Wettstein, R. von (1901). "Handbuch der systematischen Botanik". Wien.

Wettstein, R. von (1924). "Handbuch der systematischen Botanik". 3. Aufl. Wien.

Wettstein, R. von (1935). "Handbuch der systematischen Botanik". 4. Aufl. Franz Deuticke, Leipzig.

Williams, C. A. and Harborne, J. B. (1977). The leaf flavonoids of the Zingiberales. *Biochem. Syst. Ecol.* **5**, 221–229.

Williams, C. A., Harborne, J. B. and Mayo, S. J. (1981). Anthocyanin pigments and leaf flavonoids in the family Araceae. *Phytochemistry* **20** (in press).

Williams, N. H. (1979). Subsidiary cells in the Orchidaceae: their general distribution with special reference to development in the Oncidieae. *Bot. J. Linn. Soc.* **78**, 41–66.

Wirth, M. and Withner, C. L. (1959). Embryology and development in the Orchidaceae. *In* "The Orchids, a Scientific Survey" (C. L. Withner, ed.), pp. 155–188.

Wirtz, H. (1910). Beiträge zur Entwicklungsgeschichte von Sciaphila spec. und von Epirrhizanthes elongata Bl. *Flora* **101**, 395–446.

Withner, C. L., Nelson, P. K. and Wejksnora, P. (1974). The anatomy of orchids. *In* "The Orchids, Scientific Studies" (C. L. Withner, ed.), pp. 267–347.

Wulff, H. D. (1939). Die Pollenentwicklung der Juncacéen nebst einer Auswertung der embryologischen Befunde hinsichtlich einer Verwandtschaft zwischen den Juncacéen und Cyperacéen. *Jahrb. Wissensch. Bot.* **87**, 533–556.

Wunderlich, R. (1937). Zur vergleichenden Embryologie der Liliaceae-Scilloideae. *Flora* **32**, 48–90.

Wunderlich, R. (1954). Über das Antheren-Tapetum mit besonderer Berücksichtigung seiner Kernzahl. *Österreich. Bot. Zeitschr.* **101**, 1–63.

Wunderlich, R. (1959). Zur Frage der Phylogenie der Endospermtypen bei den Angiospermen. *Österreich. Bot. Zeitschr.* **106**, 203–293.

Yakovlev, M. S. and Zhukova, G. Y. (1973). Angiosperms with green and colourless embryo. *Acad. Sci. USSR, V. L. Komarov Bot. Inst.*, Leningrad.

Yakovlev, M. S. and Zhukova, G. Y. (1980). Chlorophyll in embryos of angiosperm seeds, a review. *Bot. Notiser* **133**, 323–336.

Yamashita, T. (1972). Eigenartige Wurzelanlage des Embryos bei Ruppia maritima L. *Beitr. Biol. Pflanzen* **48**, 157–170.

Yamashita, T. (1973). Über die Embryo- und Wurzelentwicklung bei Zostera japonica Aschers. et Graebn. *J. Fac. Sci. Univ. Tokyo. Sect. III*, **11**, 175–193.

Yamashita, T. (1976). Über die Pollenbildung bei Halodule pinifolia und H. uninervis. *Beitr. Biol. Pflanzen* **52**, 217–226.

Yates, I. E. and Duncan, W. H. (1970). Comparative studies of Smilax, section Smilax, of the Southeastern United States. *Rhodora* **72**, 289–312.

Young, D. J. and Watson, L. (1970). The classification of dicotyledons: a study of the upper levels of the hierarchy. *Austr. J. Bot.* **18**, 387–433.

Taxonomic Index

References to illustrations printed in bold face type.